AF355990

Nanomaterials for Micro/Nano Devices

Nanomaterials for Micro/Nano Devices

Guest Editor

Amir Hussain Idrisi

Basel • Beijing • Wuhan • Barcelona • Belgrade • Novi Sad • Cluj • Manchester

Guest Editor
Amir Hussain Idrisi
Mechanical and Aerospace
Engineering Department
Michigan Technological University
Houghton
United States

Editorial Office
MDPI AG
Grosspeteranlage 5
4052 Basel, Switzerland

This is a reprint of the Special Issue, published open access by the journal *Micromachines* (ISSN 2072-666X), freely accessible at: www.mdpi.com/journal/micromachines/special_issues/O9O4033H54.

For citation purposes, cite each article independently as indicated on the article page online and using the guide below:

Lastname, A.A.; Lastname, B.B. Article Title. *Journal Name* **Year**, *Volume Number*, Page Range.

ISBN 978-3-7258-2766-4 (Hbk)
ISBN 978-3-7258-2765-7 (PDF)
https://doi.org/10.3390/books978-3-7258-2765-7

Contents

micromachines

Editorial

Editorial for the Special Issue on Nanomaterials for Micro/Nanodevices

Amir Hussain Idrisi 

Mechanical and Aerospace Engineering, Michigan Technological University, Houghton, MI 49931, USA; amirhus@mtu.edu

The Special Issue of *Micromachines*, titled "Nanomaterials for Micro/Nanodevices", comprehensively examines the intersection of nanotechnology and micro/nanodevices. This issue is a collection of ten papers that explores the cutting-edge developments and applications of nanomaterials in the fabrication and performance enhancement of micro- and nano-scale devices. As the demand for smaller, faster, and more efficient electronic and mechanical systems grows, nanomaterials play a pivotal role in revolutionizing device engineering [1]. Nanomaterials offer unique electrical [2], mechanical [3], and chemical [4] properties that make them ideal for pushing the boundaries of micro/nanodevice performance. By advancing the interface between materials science and nanotechnology, this issue showcases innovative research driving the future of micro-/nano-scale technologies. Additionally, it addresses challenges such as scalability, cost-effective manufacturing, and material–device compatibility. These featured studies provide insights into current limitations that nanomaterials can address through interdisciplinary collaboration, setting the stage for next-generation innovations in electronics, energy harvesting, and biomedical technologies.

The following contributions in this Special Issue cover a broad range of applications and technologies and address specific gaps in current knowledge:

The review article by Zhang et al. [5] explores the research on organic–inorganic hybrid thin-film packaging for flexible organic electroluminescent devices using plasma-enhanced atomic layer deposition (PEALD) and molecular layer deposition (MLD) techniques. The authors examine the methods and optimization strategies for preparing hybrid encapsulation layers, the importance of these devices in flexible electronics, and applications. The paper also evaluates the encapsulation effect, stability, and reliability of these layers, summarizing current progress and outlining future research directions and trends.

The exploration of bio-inspired nanomaterials (BINMs) is a significant advancement in the biomedical applications of micro/nanodevices discussed by Harun-Ur-Rashid et al. [6]. These nanomaterials offer exceptional biocompatibility and multifunctionality, with potential in biosensors, drug delivery, and tissue engineering. This review examines various BINMs from proteins, DNA, and biomimetic polymers, their integration into devices, and their impacts on the biomedical field. It also addresses challenges and proposes strategies to enhance efficiency and reliability, aiming towards maximizing the applications of BINMs in advanced biomedical devices.

Complementing the focus on BINMs, the issue also delves into surface engineering techniques that can further enhance device functionality. One such technique is the development of superhydrophobic surfaces, which have garnered significant attention for their versatile applications. A review article on superhydrophobic surfaces discussed by Barthwal et al. [7] comprises theoretical foundations, fabrication methods, applications, and challenges. Key principles such as Young's equation, Wenzel and Cassie–Baxter states, and wetting dynamics are examined. This article also highlights various fabrication techniques and applications in fields like healthcare and environmental protection. Furthermore, it addresses challenges such as durability, scalability, and environmental concerns, aiming to guide future research and innovation in this field.

Transitioning from theoretical reviews to experimental investigations, the Special Issue includes cutting-edge research on novel materials. A valuable insight into the synthesis, characterization, and possibility for the use of nanostructured spinel chromites in advanced capacitor technologies was discussed by Mykhailovych et al. [8]. This study reports the successful synthesis of dielectric $ZnCr_2O_4$ nanoparticles using a sol–gel auto-combustion method followed by heat treatments in air from 5 to 11 h, in air at 500 to 900 °C. The morphology of the nanoparticles was analyzed using various techniques, revealing that higher temperatures increased nanoparticle size from 10 to 350 nm and changed their shape to pseudo-octahedra. These larger nanoparticles exhibited high dielectric constants and low dielectric losses. The annealing temperatures, which resulted in the formation of single-phase $ZnCr_2O_4$ nanopowders, range between 700 °C and 900 °C and the annealing time is 7 h, whereas lower annealing temperatures resulted in the formation of a secondary phase, identified as Cr_2O_3. However, a well-defined octahedral morphology of the nanoparticles is obtained at an annealing temperature of 900 °C for at least 11 h.

Expanding on the theme of material enhancement, Halford et al. [9] investigated the use of (3-Aminopropyl) triethoxysilane (APTES) as a primer for aluminum alloy AA2024-T3 and focused on its stability and interaction with a polystyrene (PS) topcoat. The aluminum alloy samples were primed with APTES under various durations of concentrated vapor deposition (20, 40, or 60 min). The samples were optional post-heat treatment and/or PS topcoat and were comparatively characterized with the help of electrochemical impedance spectroscopy (EIS) and surface energy. Optimal corrosion resistance was achieved with a 40 min APTES primer and post heat treatment, attributed to increased surface energy and enhanced PS wettability. The results also suggest that a thinner APTES primer (deposited for 20 min) enhances protection against corrosion in the early stages of exposure to the corrosion solution. The findings highlight APTES's potential in microdevice applications.

In research by Stine et al. [10], the authors aimed to adapt existing tools to identify nanoplastics and differentiate by polymer type using deformation data from scanning electron microscopy (SEM). Polyvinyl chloride (PVC), polyethylene terephthalate (PET), and high-density polyethylene (HDPE) were analyzed and compared against common environmental materials like algae and clay. Distinct deformation patterns of plastic particles were observed in selected environmental media and successfully developed a computer vision algorithm to enhance identification efficiency.

Turcitu et al. [11] compared the compliance and deformation properties of three different characteristic-sized (Chip A, Chip B, and Chip C, with parallel channel widths of 100, 40, and 20 μm, respectively) microfluidic devices made of polydimethylsiloxane (PDMS) and Norland Optical Adhesive (NOA). Further comparison was made based on properties like Young's modulus, roughness, contact angle, deformation, flow resistance, and compliance. The results indicate that chip A was found to be 4 times longer for PDMS devices than NOA devices, 1.6 times for chip B, and 2.5 times for chip C. The modulus of elasticity was significantly higher for NOA (1743 MPa) compared to PDMS's (2 MPa). The lower compliance and deformation of NOA indicate it as a better material for microfluidic device fabrication.

Almodóvar et al. [12] introduced a cost-effective method to produce high-performance cathodes for aluminum-air batteries. Commercial fuel cell cathodes were modified using the electrodeposition of nickel and manganese species. Optimal electrodeposition conditions were identified using Raman, SEM, TEM, and electrochemical characterization techniques, confirming the successful incorporation of these species. The modified cathodes showed enhanced electrochemical activity and achieved capacities of 50 mA h cm^{-2} for the electrodeposited Ni:Mn (3:2) cathode in aluminum-air batteries. This method is cost-effective and scalable to enhance the electrochemical performance. The results highlight its potential for practical applications in emerging energy storage technologies.

Yu et al. [13] synthesized $CH_3NH_3PbI_3$/graphene (MG) materials using organic–inorganic hybrid perovskite (OIHP) and graphene. The phase composition, lattice morphology, and micro-morphology of MG materials (MG-1 to MG-5) with different proportions (24:1, 16:1, 12:1, 8:1, 6:1) were analyzed by XRD, Raman, PL, and SEM. It was found that for MG composites, MG materials at MG-1 and MG-2 ratios exhibited better electromagnetic wave (EMW) absorption performance than other ratios. The optimal component ratio of 16:1 resulted in an absorber thickness of 1.87 mm and an effective EMW absorption width of 6.04 GHz covering the Ku frequency band. The $CH_3NH_3PbI_3$ component enhanced polarization loss, while graphene improved electrical conductivity loss. This research broadens the EMW absorption frequency band of OIHP and advances new EMW-absorbing materials.

In the last contribution, Wang et al. [14] employed sodium anthraquinone-2-sulfonate (AQS) with high redox reactivity to enhance the accessibility of ions and electrolyte and the capacitance performance of niobium carbide (Nb2C) MXene. The Nb2C–AQS composite shows significantly higher electrochemical capacitance (36.3 mF cm^2) compared to pure Nb2C (16.8 mF cm^2) at a scan rate of 20 mV s^{-1}. The supercapacitors showed excellent flexibility and stability, with 99.5% capacitance retention after 600 cycles.

Overall, this Special Issue of *Micromachines* presents a thorough and insightful exploration of the current state of research in nanomaterials for micro/nanodevices. The diverse topics and innovative approaches covered in these papers underscore the vast potential of nanomaterials to drive future technological advancements across various fields. Researchers and practitioners can draw valuable insights from these studies to further enhance the development and application of nanomaterials in both technology and medicine.

Conflicts of Interest: The author declare no conflict of interest.

References

1. Torkashvand, Z.; Shayeganfar, F.; Ramazani, A. Nanomaterials Based Micro/Nanoelectromechanical System (MEMS and NEMS) Devices. *Micromachines* **2024**, *15*, 175. [CrossRef] [PubMed]
2. Van Toan, N.; Sui, H.; Li, J.; Tuoi, T.T.K.; Ono, T. Nanoengineered Micro-Supercapacitors Based on Graphene Nanowalls for Self-Powered Wireless Sensing System. *J. Energy Storage* **2024**, *81*, 110446. [CrossRef]
3. Gulab, H.; Fatima, N.; Tariq, U.; Gohar, O.; Irshad, M.; Khan, M.Z.; Saleem, M.; Ghaffar, A.; Hussain, M.; Jan, A.K. Advancements in Zinc Oxide Nanomaterials: Synthesis, Properties, and Diverse Applications. *Nano-Struct. Nano-Objects* **2024**, *39*, 101271. [CrossRef]
4. Wang, J.; Wang, Z.; Shi, J.; Zhang, C.; Zhou, Y.; Da, Z.; Bhatti, A.S.; Wang, M. Arrays of Triangular Au Nanoparticles with Self-Cleaning Capacity for High-Sensitivity Surface-Enhanced Raman Scattering. *ACS Appl. Nano Mater.* **2024**, *7*, 5841–5852. [CrossRef]
5. Zhang, B.; Wang, Z.; Wang, J.; Chen, X. Recent Achievements for Flexible Encapsulation Films Based on Atomic/Molecular Layer Deposition. *Micromachines* **2024**, *15*, 478. [CrossRef]
6. Harun-Ur-Rashid, M.; Jahan, I.; Foyez, T.; Imran, A.B. Bio-Inspired Nanomaterials for Micro/Nanodevices: A New Era in Biomedical Applications. *Micromachines* **2023**, *14*, 1786. [CrossRef]
7. Barthwal, S.; Uniyal, S.; Barthwal, S. Nature-Inspired Superhydrophobic Coating Materials: Drawing Inspiration from Nature for Enhanced Functionality. *Micromachines* **2024**, *15*, 391. [CrossRef]
8. Mykhailovych, V.; Caruntu, G.; Graur, A.; Mykhailovych, M.; Fochuk, P.; Fodchuk, I.; Rotaru, G.-M.; Rotaru, A. Fabrication and Characterization of Dielectric $ZnCr_2O_4$ Nanopowders and Thin Films for Parallel-Plate Capacitor Applications. *Micromachines* **2023**, *14*, 1759. [CrossRef] [PubMed]
9. Halford, J., IV; Chen, C.-F. The Role of APTES as a Primer for Polystyrene Coated AA2024-T3. *Micromachines* **2024**, *15*, 93. [CrossRef] [PubMed]
10. Stine, J.S.; Aziere, N.; Harper, B.J.; Harper, S.L. A Novel Approach for Identifying Nanoplastics by Assessing Deformation Behavior with Scanning Electron Microscopy. *Micromachines* **2023**, *14*, 1903. [CrossRef] [PubMed]
11. Turcitu, T.; Armstrong, C.J.K.; Lee-Yow, N.; Salame, M.; Le, A.V.; Fenech, M. Comparison of PDMS and NOA Microfluidic Chips: Deformation, Roughness, Hydrophilicity and Flow Performance. *Micromachines* **2023**, *14*, 2033. [CrossRef] [PubMed]
12. Almodóvar, P.; Sotillo, B.; Giraldo, D.; Chacón, J.; Álvarez-Serrano, I.; López, M.L. Commercially Accessible High-Performance Aluminum-Air Battery Cathodes through Electrodeposition of Mn and Ni Species on Fuel Cell Cathodes. *Micromachines* **2023**, *14*, 1930. [CrossRef] [PubMed]

13. Yu, H.; Liu, H.; Yao, Y.; Xiong, Z.; Gao, L.; Yang, Z.; Zhou, W.; Zhang, Z. A Highly Efficient Electromagnetic Wave Absorption System with Graphene Embedded in Hybrid Perovskite. *Micromachines* **2023**, *14*, 1611. [CrossRef] [PubMed]
14. Wang, G.; Yang, Z.; Nie, X.; Wang, M.; Liu, X. A Flexible Supercapacitor Based on Niobium Carbide MXene and Sodium Anthraquinone-2-Sulfonate Composite Electrode. *Micromachines* **2023**, *14*, 1515. [CrossRef] [PubMed]

Review

Recent Achievements for Flexible Encapsulation Films Based on Atomic/Molecular Layer Deposition

Buyue Zhang [1], **Zhenyu Wang** [2], **Jintao Wang** [3,*] **and Xinyu Chen** [1,*]

[1] School of Physics, Changchun University of Science and Technology, Changchun 130012, China
[2] State Key Laboratory on Integrated Optoelectronics, College of Electronic Science & Engineering, Jilin University, Changchun 130012, China; zyw20@mails.jlu.edu.cn
[3] School of Information Engineering, Yantai Institute of Technology, Yantai 264005, China
* Correspondence: wangjintao@yitsd.edu.cn (J.W.); chenxinyucust@163.com (X.C.)

Abstract: The purpose of this paper is to review the research progress in the realization of the organic–inorganic hybrid thin-film packaging of flexible organic electroluminescent devices using the PEALD (plasma-enhanced atomic layer deposition) and MLD (molecular layer deposition) techniques. Firstly, the importance and application prospect of organic electroluminescent devices in the field of flexible electronics are introduced. Subsequently, the principles, characteristics and applications of PEALD and MLD technologies in device packaging are described in detail. Then, the methods and process optimization strategies for the preparation of organic–inorganic hybrid thin-film encapsulation layers using PEALD and MLD technologies are reviewed. Further, the research results on the encapsulation effect, stability and reliability of organic–inorganic hybrid thin-film encapsulation layers in flexible organic electroluminescent devices are discussed. Finally, the current research progress is summarized, and the future research directions and development trends are prospected.

Keywords: atomic layer deposition; functional integration; steric hindrance; thin-film encapsulation; flexible organic light-emitting diodes

Citation: Zhang, B.; Wang, Z.; Wang, J.; Chen, X. Recent Achievements for Flexible Encapsulation Films Based on Atomic/Molecular Layer Deposition. *Micromachines* **2024**, *15*, 478. https://doi.org/10.3390/mi15040478

Academic Editor: Sadia Ameen

Received: 27 February 2024
Revised: 22 March 2024
Accepted: 29 March 2024
Published: 30 March 2024

1. Background of the Study

1.1. Introduction to Flexible Organic Electroluminescent Devices and Their Stability Issues

Organic light-emitting diodes (OLEDs) have now become mainstream products in the display and lighting fields, and advanced electronics industry companies, including Huawei, Samsung, LG, and Sony, have focused on OLED technology in recent years in the product research and development of TVs, tablets, smartphones, and lighting panels [1].

As shown in Figure 1, typical OLEDs often consist of a substrate/anode/hole injection layer/hole transport layer/light emitting layer/electron transport layer/electron injection layer/cathode stacked structure [2]. Electroluminescence is the basic principle of the operation of OLEDs, and Figure 2 depicts the energy band structure of the different functional layers in OLEDs, as well as the carrier transport behavior during operation. Electrons and holes are injected into the organic electron transport layer and the organic hole transport layer through the cathode and anode of the device, respectively, and they move toward each other in their respective transport layers under the action of an applied electric field. They compound to form an exciton in the organic light-emitting layer, and the exciton radiatively jumps back to the ground state and generates light emission, which is emitted from the transparent side of the device [3–5].

Compared with liquid crystal display (LCD), OLEDs reduce the backlight module and simplify the design structure [6,7], eliminate part of the light leakage caused by the backlight emission, and realize true all-black displays [8,9]. They also show a thinner and lighter product form [10,11]. In addition, the self-luminous nature of OLEDs results in lower power consumption and excellent stability over a wider temperature range, highlighting the potential for applications in harsh environments [12,13]. Wider color coverage [14] and wider viewing angles shape the product value of OLEDs [15].

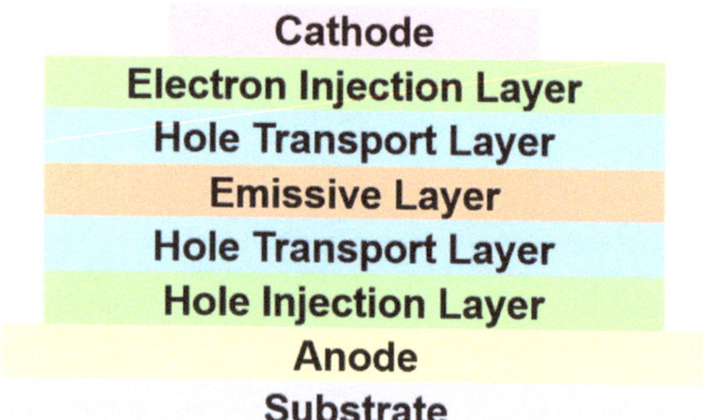

Figure 1. Basic structure of OLEDs.

LUMO = <u>L</u>owest <u>U</u>noccupied <u>M</u>olecular <u>O</u>rbital
HOMO = <u>H</u>ighest <u>O</u>ccupied <u>M</u>olecular <u>O</u>rbital

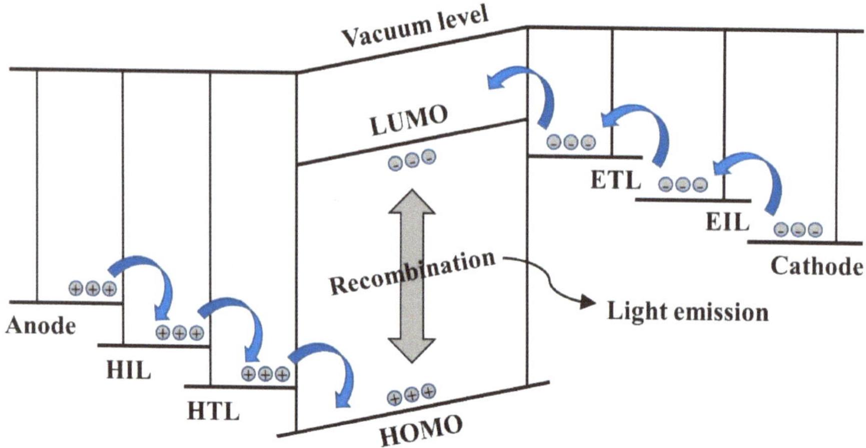

Figure 2. Schematic diagram of energy levels of OLEDs with carrier transport process.

In addition to applications in the display industry, OLEDs are also capable of being used in the production of efficient solid-state lighting devices. Compared to traditional inorganic light sources, such as incandescent or fluorescent lamps, OLEDs have higher energy efficiency and light quality while causing zero pollution to the environment [16]. The soft scattered light emitted by OLEDs possesses the characteristics closest to natural light; therefore, OLED light sources are considered to be the ideal light source that is the healthiest and safest [17]. In addition, the large-area feasibility of OLEDs brings great commercial value to them [18].

In recent years, flexible display and lighting technologies have gradually become available, and the wearable and foldable technologies of optoelectronic devices have become the current mainstream research hotspots [19]. In the preparation process, OLEDs can be effectively prepared on substrate materials other than glass substrates, including flexible substrate materials PET, PEN or PI, combined with the excellent mechanical properties of the organic materials themselves, which enables the flexible form of OLEDs to be realized.

The technological advantages of OLEDs in the field of display and lighting and the flexibility of the products have made flexible OLEDs become the final form of future display and lighting devices [20,21].

However, even though OLEDs have many advantages in the industry due to their technological features, their poor environmental stability has been troubling every research team [22,23]. Organic materials and metal electrodes are highly susceptible to degradation and failure upon contact with ambient water vapor. In addition, the ionization reaction between the electrodes and water vapor during operation will cause rapid degradation of the devices. All of these phenomena can lead to the degradation of the performance of OLEDs, or even failure. Figure 3 demonstrates the generation mechanism of black dots in OLEDs, where ambient water vapor fails the organic material while H_2 and O_2 generated by the ionization reaction causes a break between the metal electrodes and the organic layer, and carriers are unable to be transported and in the failed portion [24]. In 2018, the research team of H. Fukagawa at the Science and Technology Research Laboratory (STRL) of NHK, Japan, reported a stability study and molecular design scheme for red light OLEDs, and they monitored the change in their luminescence state with time. The device showed a circular failure region under the effect of ambient water vapor, and the area of the region gradually became larger with the increase in the time [25].

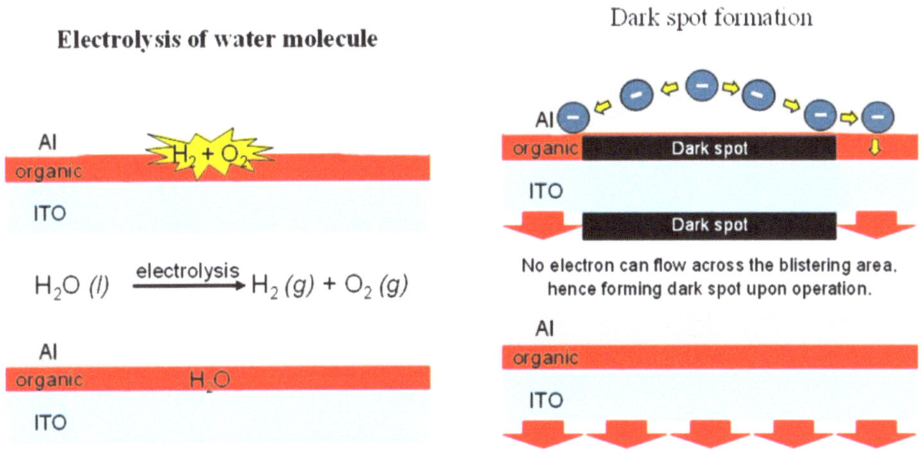

$$H_2O\ (l) \xrightarrow{\text{electrolysis}} H_2\ (g) + O_2\ (g)$$

Figure 3. Electrolysis of water molecules and formation of black dots in OLEDs (reproduced from [24] with permission form John Wiley and Sons, 2020).

Therefore, in order to enhance the operational stability of OLEDs, effective protection means are often used to isolate the device from environmental water vapor, avoid direct contact, and minimize device damage [24]. Among them, encapsulation technology is currently one of the most effective means of protection for organic optoelectronic devices, with a high-performance barrier layer covering the outer surface of the device, thus forming a permeable barrier to environmental water vapor [26,27]. The current academic and industrial consensus is that the water vapor transmittance rate (WVTR) of the barrier layer reaches 10^{-6} g/m^2/day, which is the basic condition to ensure the 10-year service life of organic optoelectronic devices. The standard applies to both rigid and flexible devices, and the water vapor barrier performance of different optoelectronic devices and the mechanical properties are shown in Figure 4 [28].

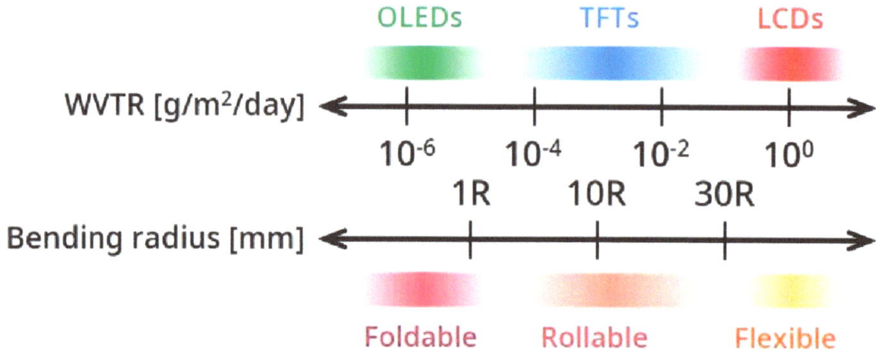

Figure 4. Water vapor barrier performance and mechanical property requirements of optoelectronic devices for different application scenarios (reproduced from [28] with permission from Taylor and Francis).

1.2. Thin-Film Packaging Technology and Its Flexible Applications

Encapsulation technology is an effective means to avoid OLEDs being damaged by environmental water vapor [29–31]. As shown in Figure 5a, the traditional OLED encapsulation process is realized by the cavity structure cover plate encapsulation by the glass cover plate. Between the substrate and the glass cover plate, the epoxy resin was used to seal the surrounding area. The entire encapsulation process was completed in the inert gas (such as argon, nitrogen, etc.) within the atmosphere, and the desiccant (such as calcium oxide, etc.) was placed in the cavity to minimize the water vapor [32–34]. The glass cover shows excellent protection performance in the early stage for organic optoelectronic devices and is widely used in various optoelectronic devices. However, the introduction of glass covers has significantly increased the overall weight of the devices, hindering the development of lightweight optoelectronic devices [35–37]. In addition, the physical properties of the glass cover make it difficult to realize the flexible form of OLEDs, and the three-dimensional mesh structure formed after the curing of the epoxy resin, which is prone to generating large internal stresses. Thus, the brittleness of the cover increases, and it can even crack or other damages can form in a short period of time. The cracks would provide a penetration path for the ambient moisture, resulting in the decline of the sealing performance. The desiccant in the sealing structure will swell after absorbing water vapor, affecting the encapsulation performance, and the protection efficiency of the cover encapsulation is reduced [26]. The iteration of display technology, from flat panel display, curved display, and then to the future of the curly, foldable display technology, means that the packaging technology should also be gradually flexible. Therefore, the disadvantages of the cover package technology are obvious, and it is a big obstacle to the process of lightweighting and flexibilization of optoelectronic devices, which makes its replacement by flexible packaging technology an inevitable trend for the future development of optoelectronic devices.

At present, thin-film encapsulation technology, as an advanced encapsulation process, has been gradually applied to various types of optoelectronic devices [38]. Inorganic thin films, such as SiNO, Al_2O_3, ZrO_2, etc., prepared using chemical vapor deposition (CVD) have undergone a series of encapsulation-related studies, and the obtained encapsulation models can meet the current application requirements.

As shown in Figure 5b, in addition to rigid devices, thin-film encapsulation technology can also be used to efficiently encapsulate and protect flexible OLEDs prepared based on flexible substrate materials (e.g., PET, PEN, PES, PI, etc.), which ensures that the devices are bendable and foldable, protecting the devices from damage brought about by the intrusion of ambient water vapor [39]. The encapsulated thin-film material is grown directly on the surface of the device, eliminating the edge penetration effect caused by the sealing process. The development and widespread use of thin-film encapsulation technology is a key step in the lightweighting process of flexible displays. Thin-film encapsulation technology can be divided into inorganic thin-film encapsulation technology, organic thin-film encapsulation technology, organic/inorganic hybrid thin-film encapsulation technology, and so on, according to the type of composition of the encapsulated thin-film material.

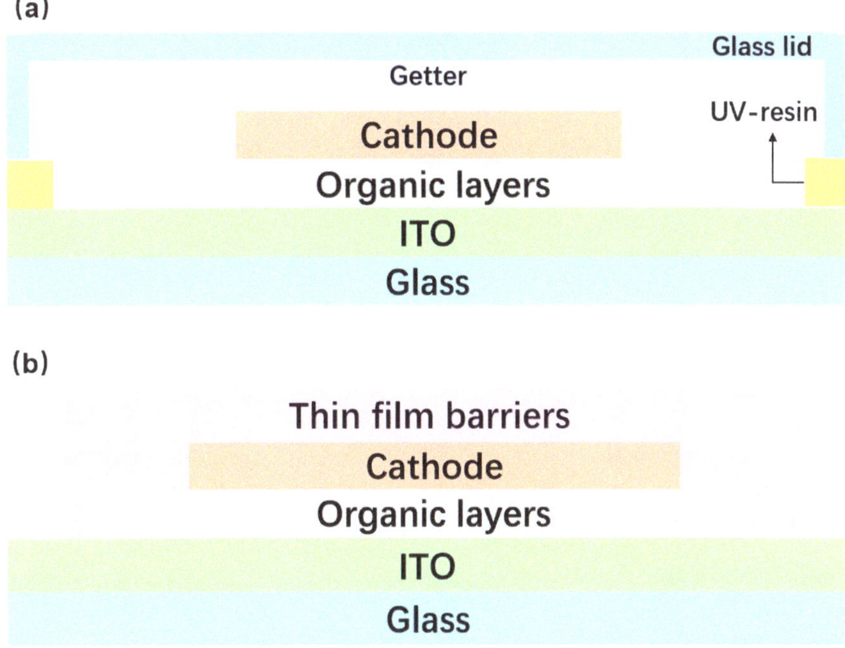

Figure 5. (**a**) Schematic diagram of glass cover package. (**b**) Schematic diagram of thin-film package.

In order to make thin-film encapsulation technology meet the water vapor barrier needs of organic optoelectronic devices, many advanced optoelectronics industry companies and research institutes have invested a great deal of money in the development of more advanced fabrication processes and encapsulation structures. In 2002, The company "Vitex Systems" developed a new flexible OLED encapsulation technology named "Barix" [40], shown in Figure 6. The technology employs a combination of UV-cured organics and sputtered inorganics to grow organic and inorganic films alternately on the surface of the protected optoelectronic device to form an organic/inorganic stacked-film structure. The water vapor barrier performance of single-layer inorganic films is limited by the water vapor permeation path generated by defects within the film, while in the Barix encapsulation structure, the introduction of the organic film sandwich structure can effectively couple the defects in the adjacent inorganic film to extend the water vapor permeation path, which will have a greater improvement in the water vapor barrier performance of the film. In 2011, South Korea's Samsung purchased a patent for the technology and collaborated with the Korea Institute of Technology KAIST to create the organic/inorganic stacked-film structure. In 2011, Samsung purchased the patent and cooperated with KAIST to apply this encapsulation system to the protection of OLEDs. By using multilayer thin-film wrapping encapsulation on substrate and the surface of OLED devices, forming alternating dense inorganic water vapor barrier layers with organic polymers in a vacuum environment, with the total thickness of the stacked structure of about 3 μm. The thin-film encapsulation process can directly grow thin films on the surface of OLEDs, avoiding the use of other encapsulation materials or mechanical encapsulation components, reducing the size and weight of the device, and minimizing the damage of the device caused by the environmental water vapor infiltration. Barix encapsulation structure embodies the excellent encapsulation performance, which can be effectively applied to flexible displays. However, the performance of OLEDs will be degraded by high-temperature and ultraviolet radiation [41]; in addition, the Barix encapsulation scheme is cumbersome in the growth process, the process flow is slow and the cost is high, which cannot fully meet the demand of the future device flexibilization process, and it has been gradually abandoned by the industry.

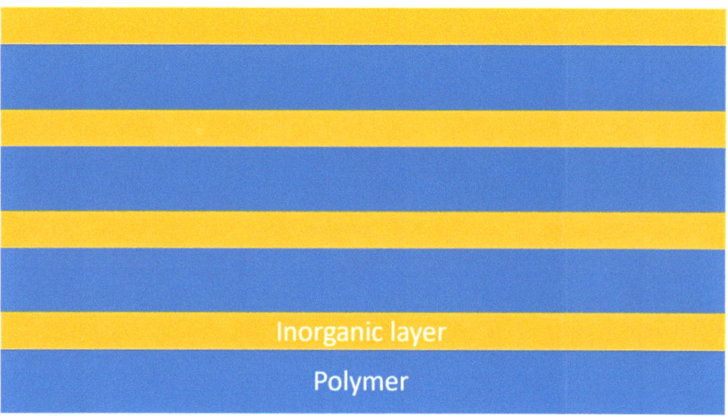

Figure 6. The image of Barix organic–inorganic stacked encapsulated film cross-section.

CVD technology is a process of thin-film growth by chemical reaction on the substrate surface in a vacuum environment, and the short process time as well as the high densities of the prepared films have led to the increasing application of CVD technology in the preparation of inorganic barrier layers in thin-film encapsulation processes [42]. However, to obtain a sufficiently dense thin-film material by CVD technology that can be effectively applied in the thin-film encapsulation of organic optoelectronic devices, the required process temperature often exceeds 150 °C [43]. OLEDs are a kind of electronic device that are sensitive to the process temperature, and higher process temperatures can degrade the organic functional materials [44], affecting the performance of the device, thereby hindering the development of CVD technology in the field of thin-film packaging. Plasma-enhanced chemical vapor deposition (PECVD) utilizes plasma to compensate for the problem of low reactivity caused by reactive precursors or process temperature [45,46], which makes the thin-film vapor growth process no longer limited by the process temperature and realizes the high quality of the thin-film growth process under low process temperature. The introduction of plasma enabled PECVD technology to meet the process requirements for the thin-film encapsulation of OLEDs, and its related encapsulation process link is practical to this day. In 2008, Kateeva, founded by C. Madigan et al. of MIT, proposed inkjet printing (IJP) technology [47]. The growth of thin-film materials is achieved by directly spraying an ink containing the corresponding nano-composition on a flexible or rigid substrate [48,49]. Since the ink-spraying path during the IJP process can be set by a software program, it is possible to prepare patterned thin-film materials without the help of a mask plate, which is a promising thin-film printing technology [48]. The IJP process has perfected the encapsulation structure of inorganic thin films, which makes the thin-film encapsulation technology truly applicable to the protection of optoelectronic devices, and the organic–inorganic stacking of PECVD/IJP/PECVD has been used for the protection of optoelectronics. PECVD organic–inorganic stacked encapsulation structure is also one of the most effective means of encapsulation for current devices [49], but the method requires a thicker organic layer coupled with the PECVD film, so that the position of the outer inorganic film is far from the neutral plane and generates a large surface deformation during the bending process, and the encapsulated film surface stresses can easily reach the modulus of rupture of its counterpart, at the expense of the mechanical properties of encapsulated films.

In summary, looking through the development path of the thin-film encapsulation process, with the iteration of display technology, the corresponding thin-film encapsulation process is also gradually moving towards high efficiency and advanced. Facing the future development plan of foldable and rollable display technology, the current thin-film encapsulation technology will not be able to continue to meet the future demand for the corresponding flexible applications. PECVD/IJP/PECVD encapsulation structure needs to prepare thicker film material to couple the internal defects of the film in order to achieve excellent encapsulation performance, at the expense of the mechanical properties of the encapsulated film. At present, many advanced optoelectronics companies combined with scientific research units have begun to look for new thin-film encapsulation processes, which will determine the arrival of the next generation of optoelectronic devices.

1.3. Introduction to Atomic Layer Deposition

Similar to CVD, atomic layer deposition (ALD) is a thin-film preparation technique based on chemical reactions on the substrate surface, and, in addition to similar film growth conditions, some precursor materials are common between the two processes [50]. The difference is that the CVD technique maintains the coexistence of the two precursor materials in a vacuum reaction chamber, which chemisorb on the substrate surface to form thin films [51]. In contrast, the surface chemical reaction established by ALD technology occurs independently and alternately for each precursor material, and each precursor material has a self-limiting reaction property, and the corresponding self-limiting surface half-reaction grows the material layer by layer on the substrate surface in the form of a single atomic layer [52], and the successive self-limiting surface reactions satisfy the need for single-atomic-layer control and co-conformal deposition in the process of thin-film growth.

The surface reaction process of ALD technology is characterized by continuous, self-limiting properties [53]. As shown in Figure 7, a typical ALD process tends to employ a binary reaction sequence for film growth, where two precursors complete their respective corresponding half-reactions in sequence on the substrate surface to achieve a monolayer deposition process of a binary compound film. The active sites on the substrate surface are the basis for ALD film growth [54]; therefore, before the film growth process starts, the substrate is often introduced with active sites, or it increases the active site density by means of some surface pretreatment [55]. For example, the number of hydroxyl groups (-OH) on the substrate surface can be greatly enhanced by oxygen plasma (O_2 plasma) or UV radiation, as shown in Figure 7a. The binary reaction sequence involved in the ALD process is divided into four steps, as shown in Figure 7b. First, precursor A is passed into the reaction chamber to undergo a self-limiting surface reaction with the active sites on the substrate surface, adsorbing a monoatomic layer and generating the corresponding by-products, followed by purging the entire chamber and piping using inert gas Ar to evacuate the residual precursor A and reaction by-products. Next, precursor B is passed into the reaction chamber and undergoes a self-limiting surface reaction with the active sites provided by precursor A, adsorbing another monatomic layer, and is accompanied by the production of by-products. Finally, Ar is again used as a purge gas to evacuate the residual precursor B with the corresponding by-products, and the re-exposed active sites are able to react with precursor A again. At this point, a cycle ends and the growth of a layer of product is completed. Repeat the above cycle N times and customize the ALD process parameters according to the usage requirements.

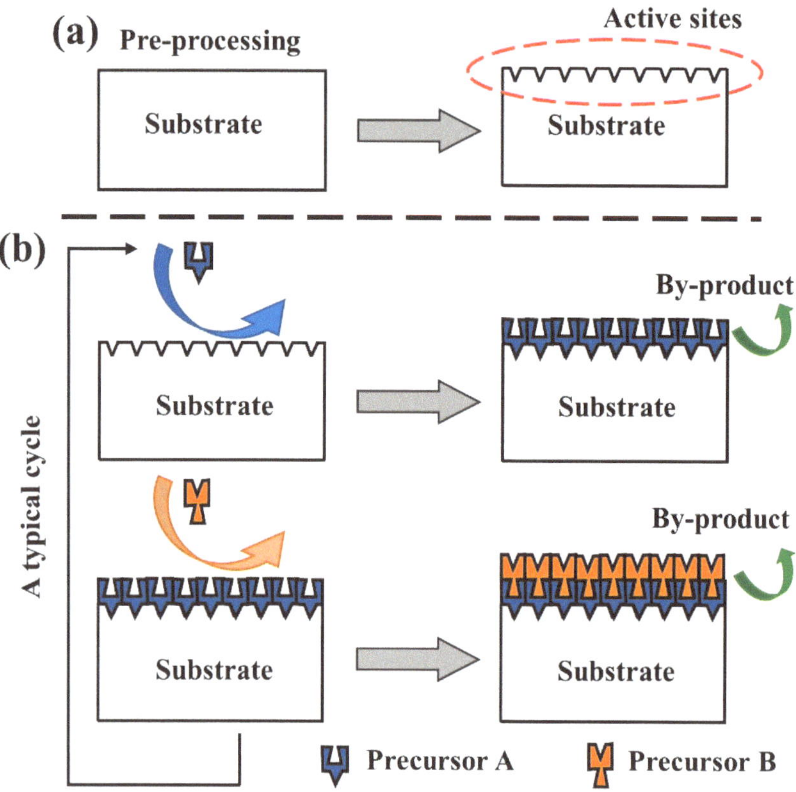

Figure 7. (**a**) Substrate surface pretreatment process. (**b**) Schematic of the ALD binary reaction sequence realized using precursors A and B for self-limiting surface reaction.

Since the number of active sites on the substrate surface is finite, the amount of surface material is deposited via the half-reactions, corresponding to each surface half-reaction having its own saturation state [56]. If both of the independent surface half-reactions are self-limiting, the two reactions can be carried out in a sequential, alternating manner to obtain a layer-by-layer deposition process of thin films and to satisfy atomic-level controllability [57]. The ALD process is governed by a surface chemistry reaction, and since its surface reactions are carried out in a sequential, alternating manner, the two precursors do not come into contact in the gas phase, and the separation would suppress the possible occurring CVD-like gas-phase reactions [58], avoiding the appearance of particulate products on the film surface.

Although precursor materials have self-limiting reaction properties, the reactions at the surface active sites are sequential due to different precursor gas fluxes. Precursors may physically adsorb in the form of van der Waals forces to the surface where the reaction has been completed and subsequently desorb from that region to continue to react with other unreacted surface regions and produce conformal deposition [59]. Because the ALD technique avoids the stochastic nature of the precursor flux, the self-limiting nature of the surface reactions also produces non-statistical deposition, which results in each surface half-reaction being driven to occur and reach near saturation. Therefore, ALD-grown films are very smooth and conformal to the original substrate [60]. The films tend to be continuous and pinhole-free because there are almost no surface active sites remaining during film growth [61]. This property is important for the preparation of excellent dielectric and water vapor barrier films.

Currently, ALD technology has great application prospects in the preparation of ultrathin and ultrafine films. Typical thin-film materials, such as Al_2O_3, SiO_2 and ZnO, have been used in various electronics industries [62–65]. In recent years, thin-film deposition and component modulation have been widely used in micro/nanofabrication technologies, such as mechanical structures, electrical isolation and connectivity [66–69]. The International Technology Roadmap for Semiconductors (ITRS) applies ALD technology to the preparation of high dielectric constant gate oxides in MOSFET structures and copper diffusion barrier layers in back-end interconnects [70]. The miniaturized layout of semiconductor processes and the resulting high depth-to-width ratio structure of the products make the precise control and conformal coating of thin-film deposition techniques a key technological need, and the ALD process provides an effective solution to this need.

In addition, due to the excellent densification of the film grown using ALD technology, it can form a good gas molecule barrier within a hundred nanometer thickness [71], and the ultrathin film morphology provides an important technical support for the application of flexible products. Therefore, the current ALD technology is widely recognized as one of the effective means of protecting optoelectronic devices in the future, and ALD-based thin-film encapsulation technology demonstrates a thinner and lighter encapsulation weight and superior flexibility compared to existing encapsulation means [72]. Prof. S. F. Bent of Stanford University believes that ALD will become an effective solution to the problem of thin-film packaging due to its precise and controllable atomic scale film growth [73]. At present, inorganic materials such as Al_2O_3, ZrO_2, SiO_2, HfO_2, etc., prepared using ALD technology have been subjected to a lot of research and achieved excellent encapsulation results [74–80]. However, thin-film encapsulation materials based on ALD technology are usually dominated by oxides, with stable binary bonds between metal and oxygen atoms in the molecular structure, leading to high Young's modulus for oxide films, which tend to be rigid as the density and thickness of the films increase [81]. In addition, plasma-assisted ALD technology (plasma-enhanced atomic layer deposition, PEALD) is often employed to compensate for the lack of low-temperature reactivity in order to meet the demand for low-temperature deposition [82–84]; however, the introduction of O_2 plasma introduces a large amount of residual stresses inside the films [85]. The inherent properties attributed to ALD-grown inorganic materials, such as low ductility, low fracture toughness, and high brittleness, limit the durability and reliability of inorganic encapsulation materials during mechanical motions [86–91], and the inorganic films are unable to maintain the encapsulation stability under rigorous mechanical motions, despite the excellent encapsulation properties.

Similar to the ALD technique, the molecular layer deposition (MLD) technique is capable of depositing single molecular layers layer-by-layer onto the substrate surface, and it is often used for the growth of organic or organic–inorganic hybrid materials [92,93]. It is worth noting that MLD techniques often have some organic components introduced, and the organic or organic–inorganic hybrid films prepared using them have excellent mechanical properties [94,95]. However, MLD often uses organic precursors as the surface growth unit of the monomolecular layer, which contains long-chain organic structures. This will lead to a large molecular size of the precursor material, which tends to form spatial site resistance on the substrate surface during the semi-reaction process and obscures some of the active sites, and thus limits the degree of saturation [96,97], and the residual active sites give rise to a higher number of defective states inside the film. The defective sites then have the opportunity to provide penetration paths for ambient water vapor, which greatly affects the water vapor barrier performance of the film [98,99]. In addition to this, during the mechanical movement of the film, additional stresses are concentrated at the film defect locations [100,101], and location-specific film stress release patterns are thus induced to occur. Whether it is inorganic or organic materials, using either one of them alone cannot meet the future packaging needs of flexible optoelectronic devices.

At present, domestic and foreign research teams and advanced optoelectronics enterprises hold the organic/inorganic stacked packaging structure as one of the key research directions, the excellent packaging characteristics of inorganic materials combined with the mechanical properties being brought about by organic materials, so that the organic/inorganic stacked packaging structure has become the current mainstream packaging program, supporting the most advanced thin-film packaging technology. In the future, in the face of foldable, wearable and other ultra-flexible optoelectronics, the innovation of thin-film packaging technology is bound to come soon.

CVD, ALD and MLD are effective film deposition technologies; they all own the advantages of uniform and conformal deposition, which allows for the deposition of thin films with excellent uniformity over complex and irregularly shaped surfaces, making them suitable for applications even on three-dimensional structures. That said, the ALD and MLD are limited by the slow deposition speed, which indicates the limited application scenarios. The CVD has been applicated in commercial encapsulation, while the fabrication may destroy the devices, and the particles would form during the deposition process.

2. Research History and Current Status of Development

ALD technology is an emerging film preparation technology. Due to its preparation of dense and conformal film, the lower film thickness can be achieved with the same level of water vapor barrier performance compared with other film preparation methods. Therefore, ALD technology has greater prospects for development in the field of flexible thin-film packaging, and it will be a favorable means of overall device thinning at the packaging level. At present, ALD technology in the field of thin-film encapsulation is mainly used to prepare highly dense inorganic encapsulation films, but the mechanical properties of ALD inorganic films are often poor. In order to realize the flexible encapsulation, it is highly desirable to introduce organic components to enhance the mechanical properties of the encapsulation structure. In this thesis, the research progress and status of the ALD process in the field of thin-film encapsulation will be introduced from the aspects of "inorganic encapsulation materials", "organic and inorganic composite encapsulation materials", and "flexible encapsulation structure design".

2.1. Analysis of Inorganic Thin-Film Encapsulation Materials and Their Properties

The inorganic encapsulation layers involved in ALD technology include Al_2O_3, SiO_2, ZrO_2, etc. Each of these inorganic materials possesses excellent densification, high conformality, and precise thickness control during the growth process. Among them, Al_2O_3, as the most typical and mature ALD inorganic material, is often used in the packaging of optoelectronic devices. In 2006, P.F. Carcia's research team at DuPont Research Institute's Development Experiment Station utilized ALD technology to grow a 25 nm layer on a PEN substrate by selecting H_2O and Trimethylaluminum (TMA) as the oxidizing agent and aluminum source, respectively. A 25 nm thick Al_2O_3 film was grown on a PEN substrate, and the barrier properties of the film were quantitatively analyzed by calcium testing. The calcium electrode was isolated by the encapsulation layers. With the penetration of oxygen and moisture, the color of calcium electrode was changed. By measuring the variation of transmission, the WVTR could be estimated. The optical transmittance curves of the calcium-tested components are shown in Figure 8.

It was found that the ambient temperature affects the water vapor barrier performance of encapsulated films. When the ambient humidity was maintained at 85% and the ambient temperature was set at 38 and 60 °C, the WVTR of the PEN/Al_2O_3 films exhibited 1.7×10^{-5} $g \cdot m^{-2} \cdot day^{-1}$ and 6.5×10^{-5} $g \cdot m^{-2} \cdot day^{-1}$, respectively. After the test conditions were maintained unchanged, and the ALD-Al_2O_3 film was replaced with a glass cover sheet and fixed by epoxy resin, the WVTR was increased. The research team was surprised by this result but did not give an explanation for it. A more reasonable explanation would be that the epoxy resin provides a penetration path for water vapor in this harsh test environment. The glass itself has excellent vapor barrier properties, but the epoxy resin joiner reduces the overall performance. On this basis, the research team estimated the WVTR of 25 nm thick ALD-Al_2O_3 film at 23 °C to be up to 6.5×10^{-5} $g \cdot m^{-2} \cdot day^{-1}$

based on the apparent activation energy theory, which is close to the water vapor barrier performance of the glass cover [102].

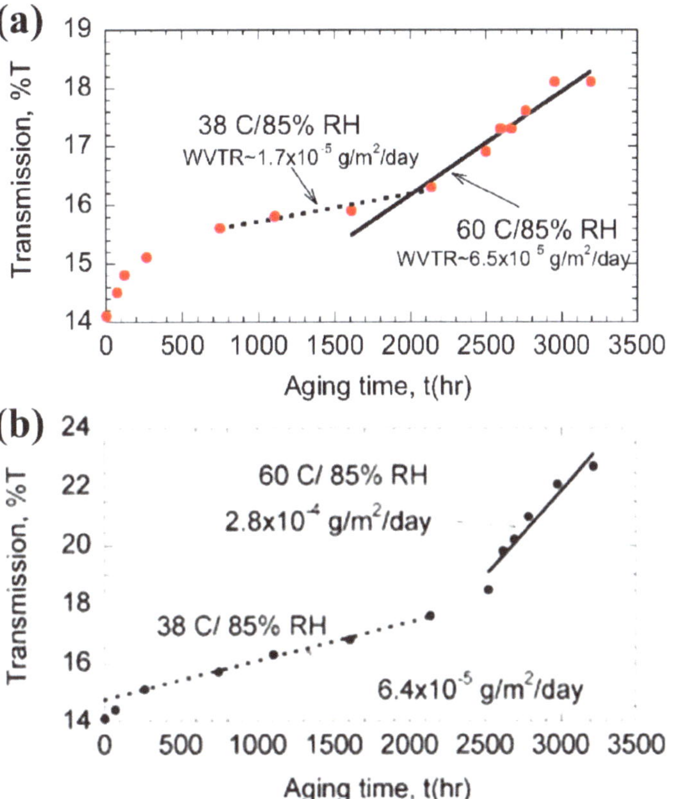

Figure 8. Optical transmittance curves with time for calcium test elements in different test environments with encapsulated films of (**a**) 25 nm Al_2O_3 and (**b**) glass cover (reproduced from [102] with permission from AIP Publishing).

In 2016, H. Jeon's research team at Hanyang University grew 50 nm thick Al_2O_3 on the surface of PEN substrates using ALD technology, and the film exhibited a WVTR of 3×10^{-3} g·m^{-2}·day^{-1} with suppressed the generation of black dots in flexible OLEDs [103]. In 2019, H. Kim's research team at Yonsei University in South Korea centered on the PEALD–SiO_2 film growth behavior on the surface of PEN substrate and found that the pretreatment of PEN substrate using O_2 plasma can enhance the adhesion between inorganic film and substrate, as well as reduce the difficulty of the film growth process of PEALD technology on the surface of flexible substrate [78].

At present, many research teams have grown inorganic films on the surface of flexible substrates using ALD technology and have made the evaluation and optimization of film quality. At the same time, the flexible devices protected by inorganic films also show improved stability. Although these works do not take into account the mechanical properties of the film, it is still a good practice and validation of the application of ALD technology in the field of flexible packaging.

CVD technology, as a traditional inorganic film growth process, is undoubtedly the biggest competitor of ALD technology in the field of thin-film encapsulation. In order to explore the differences in encapsulation performance between the two, in 2012, the research team of W.M.M. Kessels, Eindhoven University of Technology, carried out a comparative study on the encapsulation performance of two SiN_x:H and Al_2O_3 encapsulated thin films, which were prepared using PECVD and PEALD, respectively [104]. As shown in Figure 9a, both 20 nm and 40 nm thick PEALD-Al_2O_3 films have a WVTR of less than 2×10^{-6} g·m^{-2}·day^{-1} at 20 °C and 50% relative humidity (RH), while the 300 nm thick PECVD-SiN_x:H only exhibits 4×10^{-6} g·m^{-2}·day^{-1} WVTR in the same environment. Although the thickness of PECVD-SiN_x:H (300 nm) is about an order of magnitude higher than that of PEALD- Al_2O_3 (40 nm or 20 nm), the latter is still far ahead in terms of water vapor barrier performance. The team then encapsulated the OLEDs in four different structures: 300 nm PECVD-SiN_x:H, 20 nm PEALD-Al_2O_3, 40 nm PEALD-Al_2O_3, and 300 nm PECVD-SiN_x:H/40nm PEALD-Al_2O_3 stack. The encapsulated devices were placed in an environment of 20 °C and 50% RH for aging tests to observe the evolutionary behavior of the black spots on the organic materials produced by ambient water vapor with aging time. As shown in Figure 9b, after 114 days of aging treatment, the black dot density on the surface of the PECVD-SiN_x:H encapsulated OLEDs is higher, indicating that the encapsulated film has more internal defects, and multiple water vapor penetration paths are formed. This performance comparison work demonstrates the technical superiority of PEALD technology in the encapsulation field.

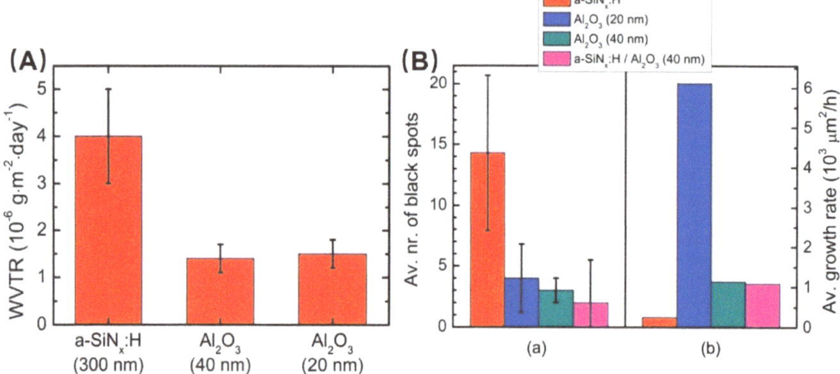

Figure 9. (**A**) Comparison of the water vapor barrier properties of a-SiN_x:H grown with a thickness of 300 nm by PECVD and Al_2O_3 films grown with a thickness of 20 nm versus 40 nm by PEALD in a test environment of 20 °C and 50% relative humidity for 70 days. (**B**) Average of OLEDs encapsulated with different barrier layers (a) and (b) in a 114-day aging experiment number of black dots generated and average black dot growth rate (reproduced from [104] with permission from American Vacuum Society).

Inorganic nanostacked structures exhibit superior water vapor barrier properties compared to single-layer inorganic films. In 2009, T. Riedl's team at the University of Technology Braunschweig suggested that alternating nanostacked structures can effectively inhibit the formation of microscopic pores and nanocrystals, which leads to a significant reduction in the probability of the occurrence of water vapor infiltration pathways [105]. In 2012, S.M. Cho's team at Sungkyunkwan University used ALD technology to prepare Al_2O_3/ZrO_2 nanostacked structures with a 350% improvement in water vapor barrier properties compared to the same thickness of inorganic films, and they pointed out that the $ZrAl_xO_y$ structure at the interface of the film was able to block the water vapor infiltration paths and that the water vapor barrier performance of the film improved with the increase in the number of stacked layers [106]. In 2014, the team of Pohang University of Science and Technology's C. E. park's team pointed out that the amorphous Al_2O_3 film is prone to degradation under the action of water vapor, and its WVTR exhibits 3.75×10^{-4} g·m^{-2}·day^{-1};

whereas the TiO_2 film possesses excellent corrosion-resistant passivation properties, and the Al_2O_3/TiO_2 nano-stacked layer structure effectively inhibits the degradation and reduces the WVTR to 1.81×10^{-4} g·m^{-2}·day^{-1} [107]. And, subsequently, the team grew dense and amorphous Al_2O_3/HfO_2 thin films using ALD technology in 2017 and showed excellent chemical stability, which exhibited a WVTR of 6.75×10^{-6} g·m^{-2}·day^{-1}, compared to Al_2O_3 monolayers (WVTR = 3.26×10^{-4} g·m^{-2}·day^{-1}), the WVTR decreased by two orders of magnitude, again demonstrating the enhanced water vapor barrier performance afforded by nanostacked films [74].

Although the inorganic films prepared using ALD technology can exhibit excellent water vapor barrier properties in monolayer or nanostacked structures, the ALD film growth process is susceptible to the process temperature, and process temperature can affect the film quality and degrade the water vapor barrier properties either too low or too high [108]. Each ALD film growth process has its corresponding temperature window, and when the process temperature is within the temperature window, the film quality is excellent and almost does not change with the process temperature, but once the process temperature jumps out of the temperature window, the film quality is temperature modulated [109,110]. When the process temperature is too high, the saturation chemisorption ratio of precursor molecules on the substrate surface decreases; when the process temperature is too low, the condensation of precursor molecules on the substrate surface increases, triggering surface CVD reactions. As a result, when the process temperature is not within the temperature window, the number of film defects increases, inducing a large number of water vapor penetration paths and a consequent decrease in water vapor barrier performance.

In 2017, S.G. Im's research team at the Korea Advanced Institute of Science and Technology (KAIST) grew 21.5 nm thick Al_2O_3 thin films by using the thermal ALD (Thermal ALD, T-ALD) technique, using H_2O and TMA as the oxidant and aluminum source, respectively. They investigated the effects of different process temperatures (60 to 120 °C) on the water vapor barrier properties of the films. As shown in Figure 10, the water vapor barrier performance of the films gradually increased with the gradual increase in the process temperature from 60 to 120 °C, corresponding to a gradual decrease in the WVTR from the order of 10^{-1} g·m^{-2}·day^{-1} to the order of 10^{-3} g·m^{-2}·day^{-1}. The research team concluded that the growth of inorganic thin films is achieved through the chemical reaction of reactive substances adsorbed on the substrate surface, and the decrease in substrate temperature inhibited the chemical combination between surface TMA and H_2O. The unreacted residue was inside the Al_2O_3 films, leading to the increase in the film defect density, and thereby the water vapor penetration path, and decreasing the water vapor barrier performance [42].

In addition to this, the authors had carried out similar experiments using process temperature as a research variable in their research in 2020, during which H_2O and TMA were selected as the oxidant and aluminum source, respectively, and 45 nm thick Al_2O_3 thin films were prepared using the T-ALD technique at process temperatures ranging from 50 to 110 °C. Subsequently, the electrical calcium test was utilized to measure the film's WVTR, and the results of the study are shown in Figure 11. The WVTR of the Al_2O_3 films exhibited 2.2×10^{-3} g·m^{-2}·day^{-1} when the process temperature was 50 °C, while the WVTR of the Al_2O_3 films exhibited 1.1×10^{-4} g·m^{-2}·day^{-1} when the process temperature was 110 °C. The increase in the process temperature from 50 to 110 °C decreased the WVTR of the films by one order of magnitude, and the water vapor barrier performance was significantly improved [111], and this trend is consistent with the results obtained by S. G. Im's research team.

In addition to the limitation of process temperature on film quality, the poor mechanical properties of inorganic films have been a key issue limiting their application in flexible packaging, such as low ductility, low fracture toughness, and high brittleness of the films, all of which limit the durability and reliability of the films when they are subjected to mechanical stresses or deformation [86–91].

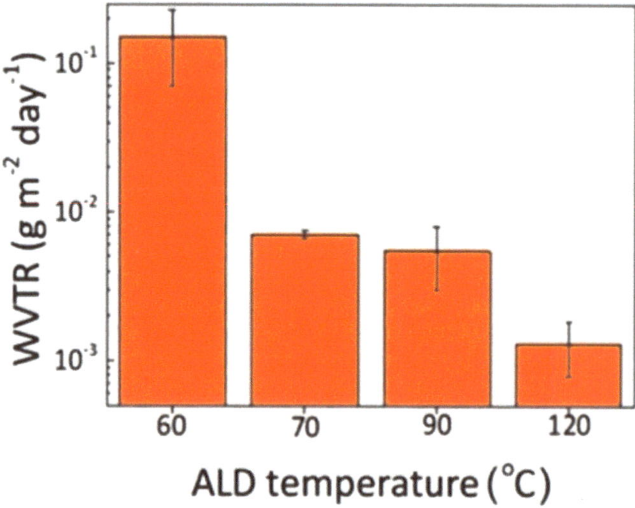

Figure 10. WVTR of ALD-Al$_2$O$_3$ thin films at different growth temperatures (reproduced from [42] with permission from John Wiley and Sons).

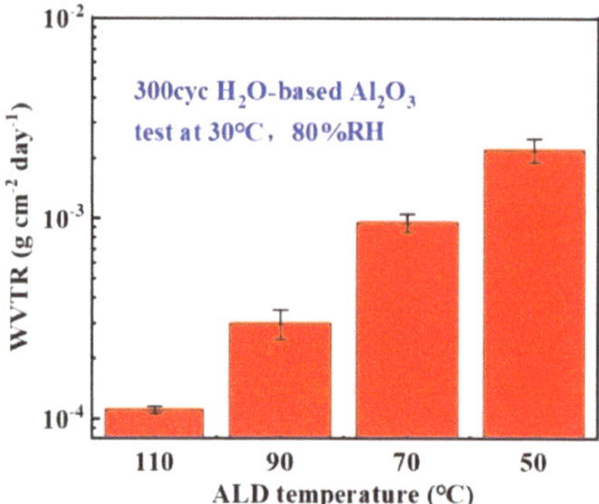

Figure 11. WVTR of Al$_2$O$_3$ thin films grown using T-ALD technique at different growth temperatures. The ALD temperature represents the reaction temperature (reproduced from [111] with permission from Elsevier).

The poor mechanical properties of inorganic thin films are attributed to the inherent properties of the material, the growth conditions, and the interaction between the film and the substrate. For example, some inorganic oxide or nitride materials lack the ability to move dislocations due to the strong ionic bonding structure contained within them, resulting in low fracture toughness and ductility. In addition, inorganic thin-film materials are often grown using physical vapor deposition (PVD) or CVD technology; these growth processes will introduce defects and residual stresses inside the film. In the process, the film is subjected to mechanical movement, the defects will lead to the concentration of the film stress, when the load stress and the accumulation of residual stresses in the film accumulate, and the defect location will preferentially produce fracture or delamination to release the stress, resulting in film damage, which leads to further degradation of the mechanical properties of the film. In 2011, S.M. George's research team analyzed the mechanical limits of Al$_2$O$_3$ thin films grown using ALD technology by means of strain analysis and pointed out that the critical tensile strain of the film decreases with the increase in the thickness, with the 40 nm alumina film having a lower tensile strain than the 40 nm alumina film, which is the highest tensile strain in the world, where the critical tensile strain of 40 nm Al_2O_3 films is $0.95 \pm 0.17\%$ and decreases to $0.52 \pm 0.22\%$ when the film thickness is increased to 80 nm [112].

In 2015, Y.-C. Chang's research team at Feng Chia University used the T-ALD technique to grow Al_2O_3 thin films with a thickness of 50 nm at a process temperature of 80 °C by selecting H_2O and TMA as the oxidant and aluminum source, respectively. Meanwhile, they applied them in the encapsulation of chalcogenide photovoltaic devices and investigated the mechanical stability of the films [113]. As shown in Figure 12, the films have good water vapor and oxygen barrier properties, with WVTR and Oxygen Transmission Rate (OTR) of 9.0×10^{-4} g·m^{-2}·day^{-1} and 1.9×10^{-3} cm^3·m^{-2}·day^{-1}, respectively. However, when the films were bent cyclically at a bending radius of 13 mm for 1000 times, its WVTR and OTR increased to 2.1×10^{-3} g·m^{-2}·day^{-1} and 2.6×10^{-3} cm^3·m^{-2}·day^{-1}, respectively. The films underwent a significant decline in water–oxygen barrier properties after the bending process, indicating that mechanical movement tends to damage the inorganic films and generate additional water–oxygen permeation paths.

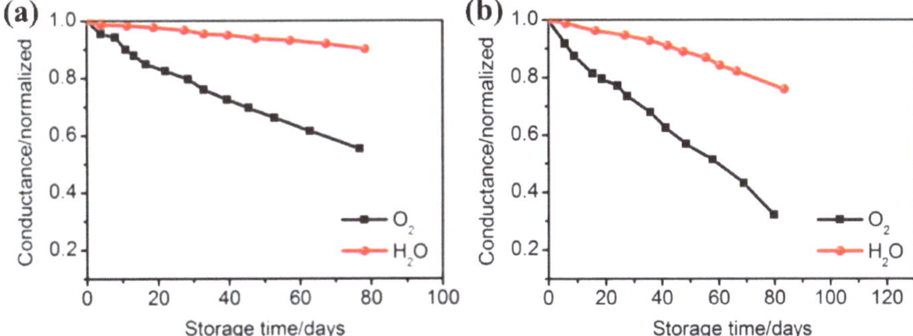

Figure 12. (**a**) Electrical calcium test curves of 50 nm Al_2O_3 film. (**b**) Electrical calcium test curves of 50 nm Al_2O_3 film after subjected to 1000 cyclic bends with a bending radius of 13 mm (reproduced from [113] with permission from American Chemical Society).

In addition, in 2018, a similar phenomenon was found by the research team of K.C. Choi at the Korea Institute of Technology. They found that a 60 nm thick Al_2O_3 film grown using the ALD technique yielded a deformation of 0.75% at a bending radius of 1.67 cm, which resulted in an increase in two orders of magnitude in the WVTR [114]. In 2021, the research team of Chen Rong at the Huazhong University of Science and Technology (HUST) investigated the PEN/40 nm Al_2O_3 film under cyclic bending for 100 times at bending radii of 7 and 5 mm, respectively. The obvious cracks on the Al_2O_3 surface in the SEM images could be found, the film completely failed and lost its water vapor barrier properties [115].

At present, in order to expand the scope of application of inorganic films in the field of flexible encapsulation, organic and inorganic composite structures as well as special flexible structure design are usually used to improve the mechanical properties of films. Organic and inorganic composite materials can combine the unique properties of the two materials: the encapsulation film has not only excellent water vapor barrier properties provided by inorganic materials, but also good mechanical properties provided by organic materials. In addition, the overall mechanical properties of the encapsulation structure can also be improved through some special structural design. For example, improvement could be made to the design of the stacked-film structure to achieve the superposition and neutralization of stress, to the introduction of the defect modification layer to reduce the number of defects within the film to reduce the probability of damage occurring in the defects of the film under mechanical movement, to the adjustment of the neutral surface position to the weak layer of the encapsulation structure to reduce the deformation of the layer in bending, maximizing the suppression of film damage, and to adjusting the position of the neutral plane to the weak layer of the encapsulation structure to reduce the deformation of the layer during bending to minimize film damage.

2.2. Organic–Inorganic Composite Package Structure

Organic–inorganic composite encapsulation structure is currently the most commonly used means of flexible thin-film encapsulation, due to the organic material containing a large number of carbon skeletons inside, so that the material itself has excellent mechanical properties [116]. The advantages of inorganic thin film grown using ALD technology lie in its excellent densification, but it also results in the existence of large residual stresses in the film, and, at the same time, the ALD process introduces a defective state inside the inorganic film. In the process of mechanical movement, the defective locations of the film will lead to a concentration of stress in the film, and when the load stress and the residual stress inside the film accumulate too much, the defective locations will preferentially produce a fracture or delamination to release the stress [117], leading to film damage and enhanced water vapor permeability. In contrast, the introduction of organic materials can release film stresses through coupling defects within the ALD inorganic films [32,118]. In addition, due to the excellent ductility of the organic film, it can bring the inorganic film closer to the actual neutral plane position when placed underneath the inorganic film. As a result, the organic–inorganic composite encapsulation structure is able to obtain better mechanical properties while ensuring excellent water vapor barrier properties.

Currently, many research teams have carried out related research around organic–inorganic composite encapsulation structures. In order to improve the mechanical properties of ALD-Al_2O_3 thin films, S. J. Kim's research team at Sungkyunkwan University, South Korea, 2020, introduced plasma polymer (PP) into Al_2O_3 thin films in the form of a sandwich structure. The performance of this organic/inorganic laminated structure was analyzed and studied, and the encapsulation structure was finally successfully applied to the encapsulation protection of flexible OLEDs [119]. The inorganic Al_2O_3 films were prepared using the ALD process, in which N_2O plasma and TMA were used as the oxidant and aluminum source, respectively. The organic layers were deposited by plasma polymerization in the same reaction chamber using CHF_3, benzene or cyclohexane precursors, and the stacked structure through the cyclohexane precursor exhibited better flexibility. A schematic of the Al_2O_3/PP stacked layer structure with the WVTR test results of the films before and after bending is shown in Figure 13.

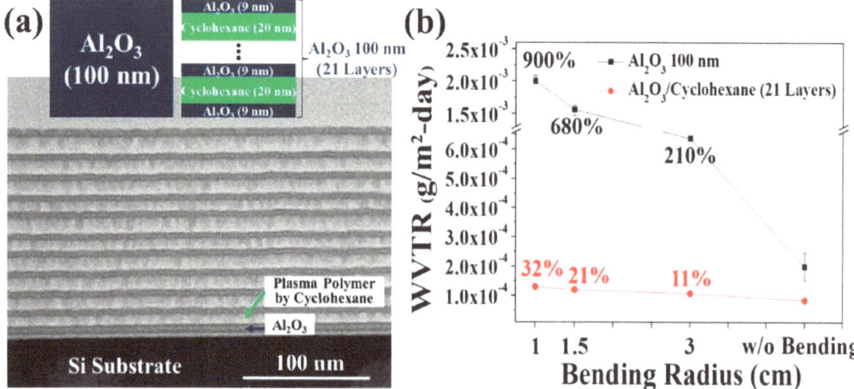

Figure 13. (a) Schematic structure of Al_2O_3/PP stack with cross-section SEM photographs. (b) WVTR before and after bending of films with different structures (reproduced from [119] with permission from American Vacuum Society).

The difference between the two film structures is that the stacked structure is made by dividing 100 nm Al_2O_3 into 11 monolayers and inserting 20 nm PP between every two monolayers. The stacked structure exhibits excellent water vapor barrier properties with a WVTR of 8.5×10^{-5} g·m^{-2}·day^{-1}, which is 58% lower than that of monolayer Al_2O_3, attributed to the ability of the PP layer to act as a water vapor barrier and its coupling effect on the defects within the Al_2O_3 layer. For the comparison of the mechanical properties of the two structures, the WVTR of monolayer Al_2O_3 increased by 900% after 1000 cycles of bending at a bending radius of 1 cm, whereas the WVTR increment of the organic/inorganic stacked-structure film was only 32% under the same conditions, implying that the organic component enhances the mechanical properties of the film.

In the same year, M. C. Gather's research team at the University of St. Andrews, UK, fabricated ultrathin flexible OLEDs, with a total thickness of only 12 μm, and was able to maintain its stability after immersion in water. The upper and surfaces of the devices were made of organic and inorganic stacked structural thin films as the high-resistance substrate and encapsulation layer. The inorganic films were 50 nm Al_2O_3/ZrO_2 inorganic nanostacked layers grown using ALD, and the organic films were 3 μm parylene C monolayers grown using CVD. After the encapsulated device was bent 5000 times with a bending radius of 1.5 mm, the device performance did not show any significant degradation [120].

IJP technology is a commonly used organic thin film fabrication technique, which is used in the field of thin-film encapsulation for the coupling of defects in inorganic encapsulated films, and the wrapping coverage of surface-attached particles. In addition, the IJP technology could provide an organic component for flexible encapsulation structures. In 2021, the research team of B.-H.K at ETRI Research Institute in South Korea prepared Al_2O_3 monolayers using the PEALD technique for the OLED encapsulation; meanwhile, acrylate polymer films were prepared using IJP technology to form an organic–inorganic stacked structure with the Al_2O_3 monolayer. Both the inorganic monolayer and the organic–inorganic stacked structure have excellent water vapor barrier properties, and the WVTRs of both of them are less than 5×10^{-5} g·m^{-2}·day^{-1}. However, the stacked structure exhibits better mechanical properties. The WVTR of the Al_2O_3/polymer film containing 1.5 cycles of stacking increased to only 2.31×10^{-4} g·m^{-2}·day^{-1} after 10 cycles of bending at a bending radius of 3.2 mm, whereas the WVTR of the Al_2O_3 monolayer increased to 4.26×10^{-1} g·m^{-2}·day^{-1} after the same mechanical motions, at which point the film had completely failed [121].

Graphene material is considered a valuable flexible encapsulation material with high permeability, excellent water vapor barrier properties and mechanical properties, and is also often used in organic–inorganic stacked structures. In 2017, H. Kim's research team at Yonsei University prepared ALD-Al_2O_3/CVD-Graphene stacked structures and found that the stacked structures significantly improved in terms of both water vapor barrier properties and mechanical stability compared to the Al_2O_3 monolayers [122].

The MLD technology is a monomolecular layer growth process on the substrate surface, and unlike ALD, the MLD process often employs long-chain organic precursors for the growth of organic or organo-inorganic hybridized films. Due to the similarity of the principles of the two processes and the self-limiting nature of the corresponding substrate surface chemistry, respectively, the ALD and MLD technologies are able to be used interchangeably in the same reaction chamber, while the film growth process of both processes is built on top of the substrate surface chemistry; thus, the two films can be combined at the interface in the form of chemisorption to achieve a large adhesion force. This process convenience and the advantage of the adhesion of the stacked structure have made the ALD/MLD technology an important research direction for organic–inorganic stacked encapsulation structures. In 2012, a research team from the University of Colorado, led by S.M. George, proposed that, compared to the ALD-Al_2O_3 or MLD-alucone monolayer thin-film structures, the ALD-Al_2O_3/MLD-alucone nanostructures can be used for the production of nanocrystalline nanostructures. The nanostacked structures are able to increase the critical tensile strain of the films by enhancing the degree of cross-linking within the films as well as reducing the film brittleness [123]. In addition, the MLD organic films can reduce the residual stresses in the ALD inorganic films by vectorial iteration of stresses or deformation generation to improve their mechanical properties [124,125], and the elastic modulus and hardness of the films in the nanostacked structure are smaller than those of the inorganic film monolayer, which further decrease with the increase in the thickness of the organic layer [126,127].

As the most widely used encapsulation structure, the organic–inorganic composite structure can combine the water vapor barrier properties of inorganic film and the mechanical properties of organic film, so that the advantages and disadvantages of the two films complement each other, showing the value of application in the field of flexible encapsulation. However, although this composite structure well reflects the important role of organic components in flexible encapsulation, the current effect is still difficult to meet the needs of future flexible applications.

In addition to the enhancement of mechanical properties, the organic components can also serve as desiccants for the encapsulation structure at the same time. In 2019, in order to prolong the lifetime of flexible OLEDs, the research team of Y.J. Choi at Sungkyunkwan University in South Korea chose N_2O plasma and TMA as the oxidizing agent and the aluminum source, respectively. They prepared 25 nm thick Al_2O_3 thin films as the encapsulation layer of the devices by the PEALD technique. Meanwhile, 100 nm TiMO and 100 nm Li(acac) organic films with hygroscopic effect were grown using vacuum thermal evaporation and placed in the middle of Al_2O_3 film to form a stacked encapsulation structure [128]. In addition, cyclohexane organic films without hygroscopic effect were added to the comparison experiments to verify the importance of the hygroscopic process, and the results of the film performance tests are shown in Figure 14.

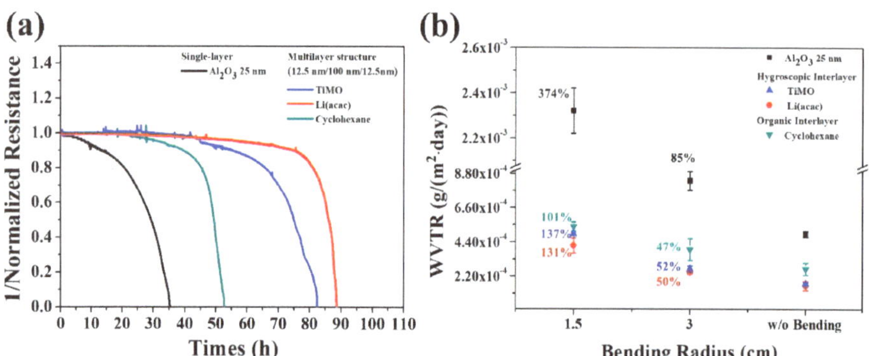

Figure 14. (**a**) Electrical calcium test curves of encapsulated films with inorganic monolayers and organic/inorganic stacked structures. (**b**) Changes in water vapor barrier properties of encapsulated films with inorganic monolayers and organic/inorganic stacked structures before and after bending (reproduced from [128] with permission from Elsevier).

The WVTR of the 25 nm Al_2O_3 film exhibits 4.5×10^{-4} g·m^{-2}·day^{-1}, which decreases to 2.7×10^{-4} g·m^{-2}·day^{-1} for the stacked encapsulated structure through the introduction of the cyclohexane organic film, which is attributed to the coupling effect of the organic film on the internal defects of the inorganic film to prolong the water vapor penetration path. By replacing the organic layer with TiMO and Li(acac), respectively, the WVTR of the films were further reduced to 1.7×10^{-4} g·m^{-2}·day^{-1} and 1.5×10^{-4} g·m^{-2}·day^{-1}, due to the film's water vapor absorption effect. In addition, the organic/inorganic stacked-film structures were both improved in terms of flexibility compared to the monolayer Al_2O_3. After performing 1000 cyclic bends with a bending radius of 1.5 cm, the WVTR increments of the three stacked-structure films were 101%, 137%, and 131%, respectively, which were much smaller than that of the pure inorganic film, which was 374%.

Although the introduction of organic films with hygroscopic effects can simultaneously improve the water vapor barrier properties and mechanical properties of inorganic encapsulation films, this hygroscopic process cannot be adopted as an effective way to optimize the film properties because organic polymer materials tend to expand or even break after absorbing a certain amount of water vapor, resulting in damage to the encapsulation film.

In organic–inorganic composite structures, designing inorganic films as nano-stacked structures can also enhance the mechanical properties by suppressing the defects' relay. In 2017, K.C. Choi's research team at the Korea Advanced Institute of Science and Technology (KAIST) compared the performances of Al_2O_3 and Al_2O_3/ZnO stacks in organic–inorganic composite encapsulant films, respectively, and investigated the defects' suppression mechanism of the inorganic stacked structures [129].

The inorganic thin films were grown using the T-ALD technique, with TMA, diethylzinc (DEZ) and H_2O as the aluminum source, zinc source and oxidant, respectively, at a process temperature of 70 °C. In the inorganic stacked structure, ZnO plays a role of defect coupling, effectively inhibiting the occurrence of cracks at the defect location of the Al_2O_3 film, and therefore the inorganic stacked structure exhibits excellent mechanical properties. When flexible OLEDs were encapsulated with the two structures and subjected to 1000 cycles of bending with a bending radius of 1 cm, the inorganic stacked-layer encapsulation structure effectively suppressed the generation of cracks, and the devices showed no streak-like corrosion. This work exemplifies the influence of defect states in inorganic thin films on mechanical properties and enhances the mechanical properties of the films by means of defect suppression. Although the inorganic stacked-layer structure was able to reduce the film failure caused by fracture at the defect location, the WVTR of the film increased from 7.87×10^{-6} $g \cdot m^{-2} \cdot day^{-1}$ to 7.78×10^{-5} $g \cdot m^{-2} \cdot day^{-2}$ in 1 cm bending experiments, and this order-of-magnitude degree of decay implies that the mechanical properties are still deficient.

Organic–inorganic composite packages can be introduced inside the inorganic film with organic components in addition to the stacked structure. In 2018, a research team of S.H. Yong from Sungkyunkwan University, South Korea, prepared a carbon-rich Al_2O_3 thin-film material by using the PEALD technique, selecting TMA and N_2O plasma as the aluminum source and oxidant, respectively, and combining it with an inorganic Al_2O_3 monolayer as stacked encapsulation. The mechanical properties of the 6.7 nm Al_2O_3/2.5 nm 15% C-Al_2O_3 film with 2.5 stacking cycles were investigated [130]. In this case, the carbon-rich structure of the Al_2O_3 film was able to play a similar role to that of the organic film in the stacked structure.

Carbon-rich growth of Al_2O_3 films is achieved by supplying an excess of TMA precursor during the growth of PEALD-Al_2O_3 films, during which the following surface reaction takes place: $Al(CH_3)_3 + N_2O \rightarrow \alpha Al_2O_3 + \beta C_2H_6 + \Upsilon N_2$. The WVTR of composite films with 2.5 laminating cycles exhibits a WVTR of 3.3×10^{-4} $g \cdot m^{-2} \cdot day^{-1}$, which is 36% lower compared to the 25 nm Al_2O_3 monolayer encapsulated film. In addition, the WVTR increment of the composite film after 1000 cycles of bending at a bending radius of 1.5 cm was only 86%, whereas the WVTR increment of the Al_2O_3 monolayer film was as high as 367% after subjecting it to the same mechanical motion.

Although this carbon-rich growth of Al_2O_3 can implant organic components into inorganic films and enhance the mechanical properties of the films, this oversaturated supply of precursors introduces a large site resistance during the film growth process, leading to a higher defect density in the film, which creates more water vapor permeation paths and reduces the upper limit of the film's water vapor barrier properties.

2.3. Flexible Package Structure Design

In addition to organic–inorganic composite packaging structures, there are many thin-film structure design options for flexible applications. For example, stress neutralization is achieved by symmetrically distributing the film on both sides of the neutral plane, mechanical degradation caused by defects in the inorganic film is suppressed by introducing a defect modification layer, and the deformation of the mechanically weak layer is reduced by the modulation of the position of the neutral plane. These structural design solutions are also of great significance in the field of flexible thin-film packaging.

2.3.1. Stress-Neutralizing Structural Design

Two films with the same stress are grown on both sides of the polymer substrate, and since the two sides have a symmetric stress distribution structure, the substrate does not need to provide additional elastic deformation to match the stress distribution. Thus, the equivalent total stress is zero, and this structural design is considered to be an effective means to be able to enhance the mechanical properties of thin films. In 2018, a research team led by J.S. Park from Sungkyunkwan University, South Korea, grew, in order to alleviate the films' residual stresses in them, SiN_x, prepared using PECVD, on one or both sides of the PEN substrate, respectively, and the total thickness of the SiN_x films was kept constant in the comparison experiments [131]. The important conclusions from the study are shown in Figure 15.

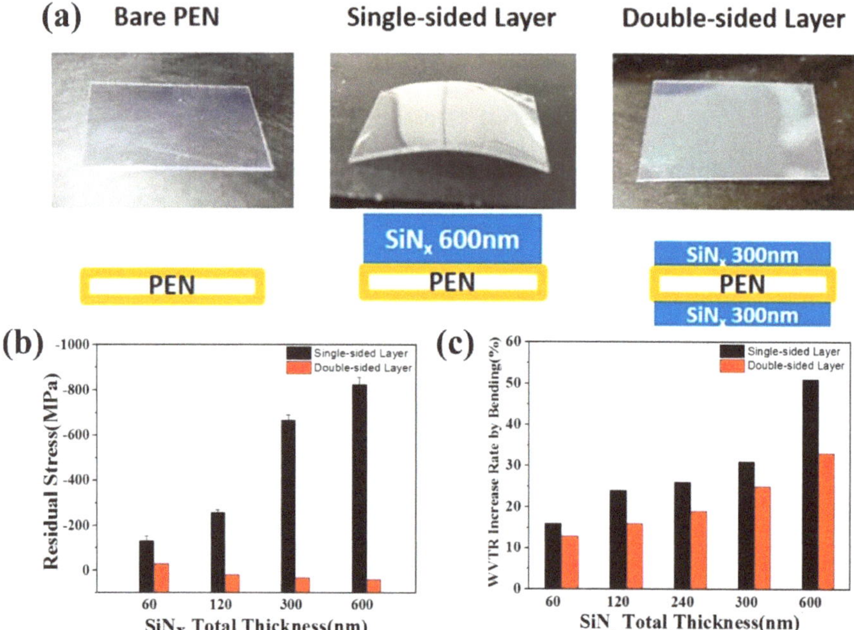

Figure 15. (**a**) Schematic of the unilateral or bilateral growth of the film. (**b**) Comparison of the stresses in the unilateral versus bilateral film. (**c**) Amount of change in the film's WVTR as a result of bending (reproduced from [131] with permission from AIP Publishing).

It was found that by growing SiN_x films on both sides of the PEN substrate bilaterally, the stresses on both sides can cancel each other out, so that the overall structure maintains a lower stress state, and this feature means that the total residual stress of the bilaterally thin-film structure does not increase with the enhancement of the film thickness. In addition, when the total thickness of the SiN_x film is 600 nm, the WVTR increment of the bilaterally thin-film structure is 33%, while that of the unilaterally thin-film structure is 51% after cyclic bending for 1000 times at a bending radius of 1.5 cm. Contrastingly, the effect of bending on the bilaterally thin-film structure is smaller than that of the unilaterally thin-film structure in other thicknesses, which proves that the bilaterally thin-film structure can effectively improve the mechanical properties of the film. However, in practical applications, the encapsulated film cannot be distributed symmetrically on both sides of the device. In addition, the biggest role of this structure is to keep the flexible substrate in the high stress film coverage, but it can still maintain the initial flat state to avoid curling. Although the symmetrical distribution of stress can play a total stress neutralization effect, the residual stress inside the single-layer film did not, in fact, weaken; too much residual stress will still limit the mechanical properties of the film.

2.3.2. Repair of Defects in Inorganic Films

Inorganic thin-film materials are often grown using PVD or CVD technology; these growth processes will introduce defective states and residual stresses within the film. In the mechanical movement process of the film, the defective location will lead to film stress concentration. When the load stress and the residual stresses within the film are too large, the defective location will be preferred to produce fracture or delamination to release the stresses, resulting in film damage. This phenomenon leads to further degradation of the mechanical properties of the film. Therefore, the elimination of defects within the film is particularly important.

In 2018, H.G. Kim's research team at Kyung Hee University in South Korea conducted research on the repair of defects in ALD-Al_2O_3 thin films using a self-assembled material, DDT [132], and confirmed that the film defects are also a key factor affecting the mechanical properties of the films, and the results of the thin-film structure and bending tests are shown in Figure 16.

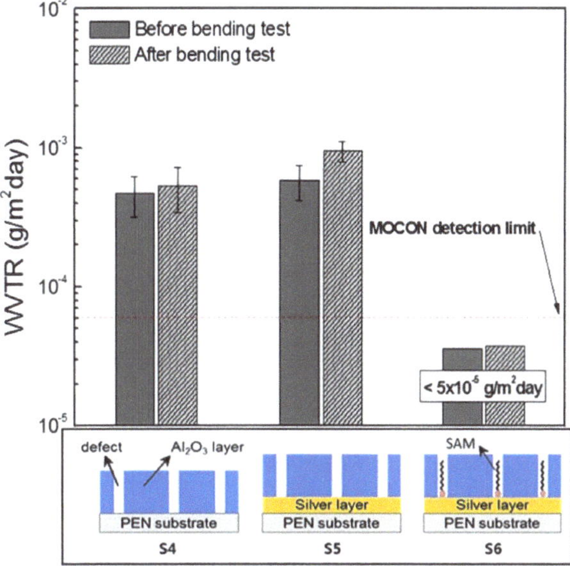

Figure 16. Schematic of different thin-film structures prepared with bending tests (reproduced from [132] with permission from Elsevier).

This work deposited an 8 nm thick Ag layer onto the surface of a PEN substrate by thermal evaporation, followed by the deposition of a 40 nm thick Al_2O_3 monolayer by LFALD using TMA and O_2 as the aluminum source and oxidizer, respectively. Self-assembled monolayers were formed using a DDT precursor, which consists of an organosulfur headgroup that interacts with the metal substrate, a methyl endgroup exposed to air, and a 9-carbon alkyl chain between the headgroup and the endgroup. The self-assembled monolayer based on the DDT precursor effectively covered the pinhole defects of the Al_2O_3 monolayer. The final multilayer encapsulated structure obtained had a WVTR of less than 5.0×10^{-5} g·m^{-2}·day^{-1} at 38 °C and 100% RH, and it remained unchanged after 25,000 cycles of bending at a 5 mm bending radius. The repair of defects did enhance the mechanical properties of the films, but this defect-filling process via self-assembled materials requires the introduction of additional characterizing materials, such as a metal layer, to allow the self-assembled molecules to undergo directional adsorption, which largely increases the process difficulty. In addition, the size of the self-assembled molecules is sometimes larger than the size of the film defects created by the ALD process, making the defect-repair process using self-assembled molecules even more difficult to carry out.

2.3.3. Neutral Surface Studies

The neutral plane is the plane where the invisible change occurs during the bending process of the film, and the film in the neutral plane position should be well protected. In 2013, the research team of S.-W. Seo at Sungkyunkwan University in South Korea studied the effect of the neutral plane position on the mechanical properties of the film. They prepared an organic/inorganic stacked composite structure, where the organic/inorganic stacked portion in the middle of the two layers of PENs is the traditional flexible encapsulation layer, and the bottom PEN serves as the substrate, and the top PEN serves as the neutral plane regulating layer. The film structure and properties were analyzed as shown in Figure 17 [133].

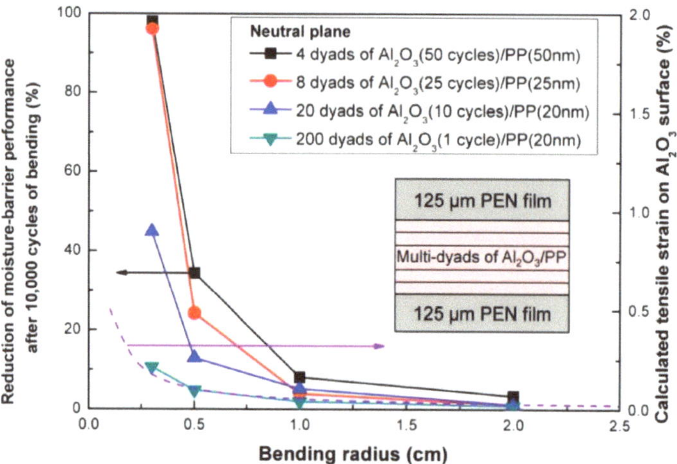

Figure 17. Mechanical properties analysis of the film before and after bending with the encapsulation layer in the neutral plane position (reproduced from [133] with permission from AIP Publishing).

In this work, an Al_2O_3 monolayer was grown using the ALD technique by selecting TMA and H_2O as the aluminum source and oxidizer, respectively. A PP layer was grown using the PECVD technique by selecting a C6H14 precursor, and the encapsulated film was placed in the neutral plane position through the modulation of the PEN layer. This encapsulation structure displayed a WVTR increment of only 10% after cyclic bending for 10,000 times at a bending radius of 3 mm.

In 2016, Y.C. Han's research team at the Korea Advanced Institute of Science and Technology (KAIST) prepared S-H nanocomposites/Al_2O_3 structures with 3.5 stacking cycles and tuned the neutral plane position by controlling the thickness of the Hybrimer layer [134]. In this work, 3.5 stacking cycles of encapsulated films, including organic–inorganic hybrid nanocomposite monolayers (S-H nanocomposites) and Al_2O_3 monolayers, were grown on PET substrates with a thickness of 125 µm. The nanocomposites and the

Al$_2$O$_3$ films were prepared using the solution method and ALD technique, respectively. The neutral plane position of the samples during bending can be changed by regulating the thickness of the uppermost Hybrimer layer, and the intervention of the Hybrimer layer drastically reduces the damage of the encapsulated films during bending, and the WVTR increment generated by 1 cm bending is reduced from 500 times to 2 times. In the final structure, the stacked encapsulant film is placed in the neutral plane position of the sample, and the performance of the encapsulated flexible OLEDs is basically maintained in the initial state after cyclic bending experiments with a bending radius of 1 cm and exposure to 30 °C and 90% RH for 30 days.

Neutral plane regulation is a means to improve the mechanical stability of thin films through structural optimization; however, the design of the neutral plane position needs to take into account all the functional layers of the device, which makes it difficult to design the neutral plane in the face of complex device structures. In addition, while protecting a structure from damage caused by bending by means of neutral plane regulation, the probability of damage to other structures far from the neutral plane position will increase.

In summary, the inorganic encapsulation film could render the superior WVTR, while the existence of strain limits the flexible applications. In contrast, the organic film is suitable for flexible and stretchable application, while the low WVTR could not render efficient protection for devices. Thus, the organic–inorganic hybrid thin-film encapsulation technology could render high WVTR and be suitable for flexible devices, while the multi-layer stacked films significantly enhance the complexity of the fabrication process. Contrastingly, the organic–inorganic hybrid thin-film encapsulation film is still the mainstream encapsulation technology for flexible application.

3. The Development of Flexible Encapsulation Films

The encapsulation technology needs further investigation for application in flexible optoelectronic devices. In 2023, Wang et al. demonstrated that by tailoring the fabrication process of ALD, a homogeneous organic/inorganic hybrid encapsulation could be achieved. By changing the EG/O plasma surface reaction ratio, the component unit of the film could achieve arbitrary ratios. Moreover, the organic/inorganic films were applied in flexible OLED encapsulation, the performance of OLED demonstrated little-to-no chance after bending 10,000 cycles under a 3 mm bending radius [135]. At the same year, the group claimed that by inserting the O plasma into the MLD process, the highly cross-linked, densified flexible encapsulation material AlOC could be fabricated with the WVTR of 1.44×10^{-5} g m^{-2} day^{-1}. The encapsulated perovskite solar cells could maintain 95% of its initial efficiency after 2400 h under 30°C and 80% RH [136]. In 2023, Chen et al. demonstrated flexible encapsulation technology based on PECVD. They found that by tuning the ration of N$_2$/H$_2$ during the fabrication process, the denser, more etch-resistant, higher compressive stress and lower hydrogen content film could be achieved. Moreover, they exhibited an inorganic/organic/inorganic sandwich structure for flexible OLED encapsulation. The encapsulated OLED demonstrated no dark spots under a 2 mm bending radius for 10,000 times [137]. Burger and co-workers thoroughly investigated the encapsulation performance between Si oxide/nitride fabricated by PECVD and metal oxides fabricated by PEALD. Moreover, they combined the inorganic encapsulation film with parylene to form the organic/inorganic hybrid encapsulation film, and they found that the parylene-AlO$_x$ demonstrated the most effective solutions. WVTR could reach up to 3.1×10^{-4} g m^{-2} day^{-1} under 38°C and 90% RH [138]. Choi et al. achieved the foldable and washable textile-based OLEDs based on the pV3D3 and Al$_2$O$_3$/TiO$_2$ bilayer encapsulation structure. The Al$_2$O$_3$ and TiO$_2$ were deposited by ALD at near-room-temperature, which was beneficial for the OLED devices. Combined with the CVD-based pV3D3 polymer, the flexibility and waterproof property was significantly enhanced, the wearable OLED could emit red light even folded under water during hand-washing [139], and foldable and washable textile-based OLEDs possessed a multi-functional near-room-temperature encapsulation layer for smart e-textiles. Chen et al. also demonstrated the advantages of the organic/inorganic hybrid encapsulation layer. They utilized the PDMS as the organic sublayers and utilized O_2 plasma to provide the nucleation sites for the deposition of the Al_2O_3 layer. Furthermore, the epoxy layer was used to improve the strain condition under the bending test. Finally, the hybrid encapsulation demonstrated superior isolation and mechanical reliability, and the lifetime of blue OLED could reach up to 370 h under 25 °C and 60% RH [115], Flexible PDMS/Al2O3 Nanolaminates for the Encapsulation of Blue OLEDs.

4. Summary and Outlook

Flexible OLEDs have many advantages, such as thinness, lightness, shock resistance, etc., which can provide more possibilities for the diversity of product forms, and they have potential application value in display, lighting, and medical industries. In recent years, domestic and foreign conferences on the theme of "wearable optoelectronics" have been favored. OLEDs, ranging from the rigid form to the curved form, are now foldable, in order to meet the increasing demand for flexibility and the minimum bending radius of the product to withstand the millimeter level, which is a great test for the thin-film encapsulation technology. This is a great test for thin-film packaging technology.

Existing thin-film encapsulation technologies are mainly based on traditional thin-film fabrication means, such as CVD, IJP, sputtering, spin-coating and thermal evaporation. Although inorganic thin-film materials have good water vapor barrier properties, their poor mechanical properties have also been a key issue limiting their application in flexible packaging, such as low film ductility, low fracture toughness, high brittleness, etc. Mechanical stress or deformation of the film will limit the durability and reliability of the film in the process. Organic materials are flexible, but due to the process characteristics of the preparation method, organic films contain more defects that provide environmental water vapor penetration paths, making their barrier properties unable to meet the demands of optoelectronic devices. The method of organic/inorganic stacking can combine the advantages of inorganic and organic materials, using the excellent barrier properties of inorganic materials to form a good barrier to water vapor permeation, and the organic layer can effectively relieve the film stress, separating the neutral surface, which brings good mechanical properties. However, the method requires a thicker organic film coupled with an inorganic film, which keeps the inorganic film position away from the neutral surface position. The deformation of the outer surface of the encapsulated film would increase during device bending, introducing higher additional stresses. The probability of film damage is thus elevated, lowering the upper limit of the mechanical properties. Moreover, the adhesion strength at the organic/inorganic interface is low and prone to delamination, which will be reflected in the strict mechanical motion process. Therefore, the organic/inorganic stacked layer structure can no longer meet the future needs of flexible device protection. Although placing the encapsulation layer in the neutral position can effectively protect it from mechanical movement damage, this method is only a compromise program. In the face of complex device structure, the neutral surface design difficulty is also higher.

Therefore, the authors believe that for the future development of flexible optoelectronic devices, including flexible OLEDs, the existing high level of thin-film fabrication and encapsulation technology can no longer satisfy, and research can no longer be limited by traditional process methods and empirical structures. The development of new processes, materials and structures is the only way to realize the future standard of flexible packaging.

Funding: This research was supported by the National Natural Science Foundation of China (Grant No. 11904031).

Conflicts of Interest: The authors declare no conflict of interest.

References

1. Hong, G.; Gan, X.; Leonhardt, C.; Zhang, Z.; Seibert, J.; Busch, J.M.; Braese, S. A Brief History of OLEDs-Emitter Development and Industry Milestones. *Adv. Mater.* **2021**, *33*, 2005630. [CrossRef] [PubMed]
2. Geffroy, B.; le Roy, P.; Prat, C. Organic light-emitting diode (OLED) technology: Materials, devices and display technologies. *Polym. Int.* **2006**, *55*, 572–582. [CrossRef]
3. Kulkarni, A.P.; Tonzola, C.J.; Babel, A.; Jenekhe, S.A. Electron transport materials for organic light-emitting diodes. *Chem. Mater.* **2004**, *16*, 4556–4573. [CrossRef]
4. Uoyama, H.; Goushi, K.; Shizu, K.; Nomura, H.; Adachi, C. Highly efficient organic light-emitting diodes from delayed fluorescence. *Nature* **2012**, *492*, 234–238. [CrossRef] [PubMed]
5. Sun, Y.R.; Giebink, N.C.; Kanno, H.; Ma, B.W.; Thompson, M.E.; Forrest, S.R. Management of singlet and triplet excitons for efficient white organic light-emitting devices. *Nature* **2006**, *440*, 908–912. [CrossRef] [PubMed]
6. Li, S.; Wang, L.; Tang, D.; Cho, Y.; Liu, X.; Zhou, X.; Lu, L.; Zhang, L.; Takeda, T.; Hirosaki, N.; et al. Achieving High Quantum Efficiency Narrow-Band beta-Sialon:Eu^{2+} Phosphors for High-Brightness LCD Backlights by Reducing the Eu^{3+} Luminescence Killer. *Chem. Mater.* **2018**, *30*, 494–505. [CrossRef]
7. Chiu, H.-J.; Cheng, S.-J. LED backlight driving system for large-scale LCD panels. *IEEE Trans. Ind. Electron.* **2007**, *54*, 2751–2760. [CrossRef]
8. Park, C.H.; Kim, J.G.; Jung, S.-G.; Lee, D.J.; Park, Y.W.; Ju, B.-K. Optical characteristics of refractive- index-matching diffusion layer in organic light-emitting diodes. *Sci. Rep.* **2019**, *9*, 8690. [CrossRef] [PubMed]
9. Choi, Y.; Oh, S.-W.; Choi, T.-H.; Yu, B.-H.; Yoon, T.-H. Formation of polymer structure by thermally-induced phase separation for a dye-doped liquid crystal light shutter. *Dye. Pigment.* **2019**, *163*, 749–753. [CrossRef]
10. Ma, H.; Yip, H.-L.; Huang, F.; Jen, A.K.Y. Interface Engineering for Organic Electronics. *Adv. Funct. Mater.* **2010**, *20*, 1371–1388. [CrossRef]
11. Gross, M.; Muller, D.C.; Nothofer, H.G.; Scherf, U.; Neher, D.; Brauchle, C.; Meerholz, K. Improving the performance of doped pi-conjugated polymers for use in organic light-emitting diodes. *Nature* **2000**, *405*, 661–665. [CrossRef] [PubMed]
12. Wang, S.; Zhang, H.; Zhang, B.; Xie, Z.; Wong, W.-Y. Towards high-power-efficiency solution-processed OLEDs: Material and device perspectives. *Mater. Sci. Eng. R-Rep.* **2020**, *140*, 100547. [CrossRef]
13. Li, C.-C.; Tseng, H.-Y.; Liao, H.-C.; Chen, H.-M.; Hsieh, T.; Lin, S.-A.; Jau, H.-C.; Wu, Y.-C.; Hsu, Y.-L.; Hsu, W.-H.; et al. Enhanced image quality of OLED transparent display by cholesteric liquid crystal back-panel. *Opt. Express* **2017**, *25*, 29199–29206. [CrossRef]
14. Vashishtha, P.; Ng, M.; Shivarudraiah, S.B.; Halpert, J.E. High Efficiency Blue and Green Light-Emitting Diodes Using Ruddlesden-Popper Inorganic Mixed Halide Perovskites with Butylammonium Interlayers. *Chem. Mater.* **2019**, *31*, 83–89. [CrossRef]
15. Zou, S.-J.; Shen, Y.; Xie, F.-M.; Chen, J.-D.; Li, Y.-Q.; Tang, J.-X. Recent advances in organic light-emitting diodes: Toward smart lighting and displays. *Mater. Chem. Front.* **2020**, *4*, 788–820. [CrossRef]
16. Bhagat, S.A.; Borghate, S.V.; Kalyani, N.T.; Dhoble, S.J. Novel Na+ doped Alq(3) hybrid materials for organic light-emitting diode (OLED) devices and flat panel displays. *Luminescence* **2015**, *30*, 251–256. [CrossRef] [PubMed]
17. Lian, C.; Piksa, M.; Yoshida, K.; Persheyev, S.; Pawlik, K.J.; Matczyszyn, K.; Samuel, I.D.W. Flexible organic light-emitting diodes for antimicrobial photodynamic therapy. *NPJ Flex. Electron.* **2019**, *3*, 18. [CrossRef]
18. Reineke, S.; Lindner, F.; Schwartz, G.; Seidler, N.; Walzer, K.; Luessem, B.; Leo, K. White organic light-emitting diodes with fluorescent tube efficiency. *Nature* **2009**, *459*, 234–238. [CrossRef] [PubMed]
19. White, M.S.; Kaltenbrunner, M.; Glowacki, E.D.; Gutnichenko, K.; Kettlgruber, G.; Graz, I.; Aazou, S.; Ulbricht, C.; Egbe, D.A.M.; Miron, M.C.; et al. Ultrathin, highly flexible and stretchable PLEDs. *Nat. Photonics* **2013**, *7*, 811–816. [CrossRef]
20. Xu, R.-P.; Li, Y.-Q.; Tang, J.-X. Recent advances in flexible organic light-emitting diodes. *J. Mater. Chem. C* **2016**, *4*, 9116–9142. [CrossRef]
21. Liu, Y.-F.; Feng, J.; Bi, Y.-G.; Yin, D.; Sun, H.-B. Recent Developments in Flexible Organic Light-Emitting Devices. *Adv. Mater. Technol.* **2019**, *4*, 1800371. [CrossRef]
22. Popovic, Z.D.; Aziz, H. Reliability and degradation of small molecule-based organic light-emitting devices (OLEDs). *IEEE J. Sel. Top. Quantum Electron.* **2002**, *8*, 362–371. [CrossRef]
23. Azrain, M.; Omar, G.; Mansor, M.; Fadzullah, S.; Lim, L. Failure mechanism of organic light emitting diodes (OLEDs) induced by hygrothermal effect. *Opt. Mater.* **2019**, *91*, 85–92. [CrossRef]
24. Swayamprabha, S.S.; Dubey, D.K.; Shahnawaz; Yadav, R.A.K.; Nagar, M.R.; Sharma, A.; Tung, F.-C.; Jou, J.-H. Approaches for Long Lifetime Organic Light Emitting Diodes. *Adv. Sci.* **2021**, *8*, 2002254. [CrossRef] [PubMed]
25. Fukagawa, H. Molecular Design and Device Design to Improve Stabilities of Organic Light-Emitting Diodes. *J. Photopolym. Sci. Technol.* **2018**, *31*, 315–321. [CrossRef]
26. Ghosh, A.P.; Gerenser, L.J.; Jarman, C.M.; Fornalik, J.E. Thin-film encapsulation of organic light-emitting devices. *Appl. Phys. Lett.* **2005**, *86*, 223503. [CrossRef]
27. Wang, R.; Mujahid, M.; Duan, Y.; Wang, Z.-K.; Xue, J.; Yang, Y. A Review of Perovskites Solar Cell Stability. *Adv. Funct. Mater.* **2019**, *29*, 1808843. [CrossRef]
28. Jeong, E.G.; Kwon, J.H.; Kang, K.S.; Jeong, S.Y.; Choi, K.C. A review of highly reliable flexible encapsulation technologies towards rollable and foldable OLEDs. *J. Inf. Disp.* **2020**, *21*, 19–32. [CrossRef]

29. Park, S.H.K.; Oh, J.; Hwang, C.S.; Lee, J.I.; Yang, Y.S.; Chu, H.Y. Ultrathin film encapsulation of an OLED by ALD. *Electrochem. Solid State Lett.* **2005**, *8*, H21–H23. [CrossRef]

30. Moro, L.; Krajewski, T.A.; Rutherford, N.M.; Philips, O.; Visser, R.J.; Gross, M.E.; Bennett, W.D.; Graff, G.L. Process and design of a multilayer thin film encapsulation of passive matrix OLED displays. In Proceedings of the Conference on Organic Light-Emitting Materials and Devices VII, San Diego, CA, USA, 4–6 August 2003; pp. 83–93.

31. Wu, J.; Fei, F.; Wei, C.; Chen, X.; Nie, S.; Zhang, D.; Su, W.; Cui, Z. Efficient multi-barrier thin film encapsulation of OLED using alternating Al2O3 and polymer layers. *RSC Adv.* **2018**, *8*, 5721–5727. [CrossRef]

32. Lee, S.; Han, J.H.; Lee, S.H.; Baek, G.H.; Park, J.S. Review of Organic/Inorganic Thin Film Encapsulation by Atomic Layer Deposition for a Flexible OLED Display. *JOM* **2019**, *71*, 197–211. [CrossRef]

33. Madogni, V.I.; Agbomahéna, M.; Kounouhéwa, B.B.; Douhéret, O.; Lazzaroni, R. Effects of Residual Oxygen in the Degradation of the Performance of Organic Bulk Heterojunction Solar Cells: Stability, Role of the Encapsulation. *Adv. Mater. Phys. Chem.* **2018**, *8*, 321. [CrossRef]

34. Park, M.-H.; Han, T.-H.; Kim, Y.-H.; Jeong, S.-H.; Lee, Y.; Seo, H.-K.; Cho, H.; Lee, T.-W. Flexible organic light-emitting diodes for solid-state lighting. *J. Photonics Energy* **2015**, *5*, 053599. [CrossRef]

35. Zhang, D.; Huang, T.; Duan, L. Emerging Self-Emissive Technologies for Flexible Displays. *Adv. Mater.* **2020**, *32*, 1902391. [CrossRef] [PubMed]

36. Gu, G.; Burrows, P.E.; Venkatesh, S.; Forrest, S.R.; Thompson, M.E. Vacuum-deposited, nonpolymeric flexible organic light-emitting devices. *Opt. Lett.* **1997**, *22*, 172–174. [CrossRef] [PubMed]

37. Ai, X.; Evans, E.W.; Dong, S.; Gillett, A.J.; Guo, H.; Chen, Y.; Hele, T.J.H.; Friend, R.H.; Li, F. Efficient radical-based light-emitting diodes with doublet emission. *Nature* **2018**, *563*, 536–540. [CrossRef] [PubMed]

38. Lu, Q.; Yang, Z.C.; Meng, X.; Yue, Y.F.; Ahmad, M.A.; Zhang, W.J.; Zhang, S.S.; Zhang, Y.Q.; Liu, Z.H.; Chen, W. A Review on Encapsulation Technology from Organic Light Emitting Diodes to Organic and Perovskite Solar Cells. *Adv. Funct. Mater.* **2021**, *31*, 2100151. [CrossRef]

39. Chwang, A.B.; Rothman, M.A.; Mao, S.Y.; Hewitt, R.H.; Weaver, M.S.; Silvernail, J.A.; Rajan, K.; Hack, M.; Brown, J.J.; Chu, X.; et al. Thin film encapsulated flexible organic electroluminescent displays. *Appl. Phys. Lett.* **2003**, *83*, 413–415. [CrossRef]

40. Park, J.S.; Chae, H.; Chung, H.K.; Lee, S.I. Thin film encapsulation for flexible AM-OLED: A review. *Semicond. Sci. Technol.* **2011**, *26*, 034001. [CrossRef]

41. Kwon, S.-K.; Baek, J.-H.; Choi, H.-C.; Kim, S.K.; Lampande, R.; Pode, R.; Kwon, J.H. Degradation of OLED performance by exposure to UV irradiation. *RSC Adv.* **2019**, *9*, 42561–42568. [CrossRef]

42. Lee, Y.I.; Jeon, N.J.; Kim, B.J.; Shim, H.; Yang, T.Y.; Seok, S.I.; Seo, J.; Im, S.G. A low-temperature thin-film encapsulation for enhanced stability of a highly efficient perovskite solar cell. *Adv. Energy Mater.* **2018**, *8*, 1701928. [CrossRef]

43. Choi, J.-H.; Ha, M.-J.; Park, J.C.; Park, T.J.; Kim, W.-H.; Lee, M.-J.; Ahn, J.-H. A Strategy for Wafer-Scale Crystalline MoS2 Thin Films with Controlled Morphology Using Pulsed Metal-Organic Chemical Vapor Deposition at Low Temperature. *Adv. Mater. Interfaces* **2022**, *9*, 2101785. [CrossRef]

44. Salameh, F.; Al Haddad, A.; Picot, A.; Canale, L.; Zissis, G.; Chabert, M.; Maussion, P. Modeling the Luminance Degradation of OLEDs Using Design of Experiments. *IEEE Trans. Ind. Appl.* **2019**, *55*, 6548–6558. [CrossRef]

45. Batey, J.; Tierney, E. Low-temperature deposition of high-quality silicon dioxide by plasma-enhanced chemical vapor deposition. *J. Appl. Phys.* **1986**, *60*, 3136–3145. [CrossRef]

46. Kim, J.; Hwang, J.H.; Kwon, Y.W.; Bae, H.W.; An, M.; Lee, W.; Lee, D. Hydrogen-assisted low-temperature plasma-enhanced chemical vapor deposition of thin film encapsulation layers for top-emission organic light-emitting diodes. *Org. Electron.* **2021**, *97*, 106261. [CrossRef]

47. Zikulnig, J.; Chang, S.; Bito, J.; Rauter, L.; Roshanghias, A.; Carrara, S.; Kosel, J. Printed Electronics Technologies for Additive Manufacturing of Hybrid Electronic Sensor Systems. *Adv. Sens. Res.* **2023**, *2*, 2200073. [CrossRef]

48. Derby, B. Inkjet Printing of Functional and Structural Materials: Fluid Property Requirements, Feature Stability, and Resolution. *Annu. Rev. Mater. Res.* **2010**, *40*, 395–414. [CrossRef]

49. Wang, T.; Sun, T.; Xie, C.; Wang, Y.; Qin, C.; Zhang, Z.; Zhou, W.; Zhang, S. *SID Symposium Digest of Technical Papers*; Wiley Online Library: London, UK, 2019; pp. 1881–1883.

50. Crowell, J.E. Chemical methods of thin film deposition: Chemical vapor deposition, atomic layer deposition, and related technologies. *J. Vac. Sci. Technol. A Vac. Surf. Film.* **2003**, *21*, S88–S95. [CrossRef]

51. Sabzi, M.; Anijdan, S.H.M.; Shamsodin, M.; Farzam, M.; Hojjati-Najafabadi, A.; Feng, P.; Park, N.; Lee, U. A Review on Sustainable Manufacturing of Ceramic-Based Thin Films by Chemical Vapor Deposition (CVD): Reactions Kinetics and the Deposition Mechanisms. *Coatings* **2023**, *13*, 188. [CrossRef]

52. Puurunen, R.L. Surface chemistry of atomic layer deposition: A case study for the trimethylaluminum/water process. *J. Appl. Phys.* **2005**, *97*, 121301. [CrossRef]

53. George, S.; Ott, A.; Klaus, J. Surface chemistry for atomic layer growth. *J. Phys. Chem.* **1996**, *100*, 13121–13131. [CrossRef]

54. Forte, M.A.; Silva, R.M.; Tavares, C.J.; Silva, R.F.E. Is Poly(methyl methacrylate) (PMMA) a Suitable Substrate for ALD? A Review. *Polymers* **2021**, *13*, 1346. [CrossRef]

55. Kovacs, R.L.; Csontos, M.; Gyongyosi, S.; Elek, J.; Parditka, B.; Deak, G.; Kuki, A.; Keki, S.; Erdelyi, Z. Surface characterization of plasma-modified low density polyethylene by attenuated total reflectance fourier-transform infrared (ATR-FTIR) spectroscopy combined with chemometrics. *Polym. Test.* **2021**, *96*, 107080. [CrossRef]
56. Clark, M.P.; Muneshwar, T.; Xiong, M.; Cadien, K.; Ivey, D.G. Saturation Behavior of Atomic Layer Deposition MnOx from Bis(Ethylcyclopentadienyl) Manganese and Water: Saturation Effect on Coverage of Porous Oxygen Reduction Electrodes for Metal-Air Batteries. *ACS Appl. Nano Mater.* **2019**, *2*, 267–277. [CrossRef]
57. George, S.M. Atomic Layer Deposition: An Overview. *Chem. Rev.* **2010**, *110*, 111–131. [CrossRef] [PubMed]
58. de la Huerta, C.M.; Huong, N.V.; Dedulle, J.-M.; Bellet, D.; Jimenez, C.; Munoz-Rojas, D. Influence of the Geometric Parameters on the Deposition Mode in Spatial Atomic Layer Deposition: A Novel Approach to Area-Selective Deposition. *Coatings* **2019**, *9*, 5. [CrossRef]
59. Shahmohammadi, M.; Mukherjee, R.; Takoudis, C.G.; Diwekar, U.M. Optimal design of novel precursor materials for the atomic layer deposition using computer-aided molecular design. *Chem. Eng. Sci.* **2021**, *234*, 116416. [CrossRef]
60. Fabreguette, F.H.; Wind, R.A.; George, S.M. Ultrahigh x-ray reflectivity from W/ Al^2O^3 multilayers fabricated using atomic layer deposition. *Appl. Phys. Lett.* **2006**, *88*, 013116. [CrossRef]
61. Wu, Y.; Yang, X.; Chen, H.; Zhang, K.; Qin, C.; Liu, J.; Peng, W.; Islam, A.; Bi, E.; Ye, F.; et al. Highly compact TiO2 layer for efficient hole-blocking in perovskite solar cells. *Appl. Phys. Express* **2014**, *7*, 052301. [CrossRef]
62. Graniel, O.; Weber, M.; Balme, S.; Miele, P.; Bechelany, M. Atomic layer deposition for biosensing applications. *Biosens. Bioelectron.* **2018**, *122*, 147–159. [CrossRef]
63. Yu, Y.; Zhang, Z.; Yin, X.; Kvit, A.; Liao, Q.; Kang, Z.; Yan, X.; Zhang, Y.; Wang, X. Enhanced photoelectrochemical effciency and stability using a conformal TiO2 film on a black silicon photoanode. *Nat. Energy* **2017**, *2*, 17045. [CrossRef]
64. Sheng, J.; Lee, J.-H.; Choi, W.-H.; Hong, T.; Kim, M.; Park, J.-S. Review Article: Atomic layer deposition for oxide semiconductor thin film transistors: Advances in research and development. *J. Vac. Sci. Technol. A* **2018**, *36*, 060801. [CrossRef]
65. Song, E.; Lee, Y.K.; Li, R.; Li, J.; Jin, X.; Yu, K.J.; Xie, Z.; Fang, H.; Zhong, Y.; Du, H.; et al. Transferred, Ultrathin Oxide Bilayers as Biofluid Barriers for Flexible Electronic Implants. *Adv. Funct. Mater.* **2018**, *28*, 1702284. [CrossRef]
66. Sheng, J.; Hong, T.; Kang, D.; Yi, Y.; Lim, J.H.; Park, J.-S. Design of InZnSnO Semiconductor Alloys Synthesized by Supercycle Atomic Layer Deposition and Their Rollable Applications. *ACS Appl. Mater. Interfaces* **2019**, *11*, 12683–12692. [CrossRef]
67. Multia, J.; Heiska, J.; Khayyami, A.; Karppinen, M. Electrochemically Active In Situ Crystalline Lithium-Organic Thin Films by ALD/MLD. *ACS Appl. Mater. Interfaces* **2020**, *12*, 41557–41566. [CrossRef] [PubMed]
68. Ponja, S.D.; Williamson, B.A.D.; Sathasivam, S.; Scanlon, D.O.; Parkin, I.P.; Carmalt, C.J. Enhanced electrical properties of antimony doped tin oxide thin films deposited via aerosol assisted chemical vapour deposition. *J. Mater. Chem. C* **2018**, *6*, 7257–7266. [CrossRef]
69. Islam, M.R.; Rahman, M.; Farhad, S.F.U.; Podder, J. Structural, optical and photocatalysis properties of sol-gel deposited Al-doped ZnO thin films. *Surf. Interfaces* **2019**, *16*, 120–126. [CrossRef]
70. International Technology Roadmap for Semiconductors. 2007 Edition. Available online: http://www.itrs.net/ (accessed on 3 December 2023).
71. Chou, C.-T.; Yu, P.-W.; Tseng, M.-H.; Hsu, C.-C.; Shyue, J.-J.; Wang, C.-C.; Tsai, F.-Y. Transparent Conductive Gas-Permeation Barriers on Plastics by Atomic Layer Deposition. *Adv. Mater.* **2013**, *25*, 1750–1754. [CrossRef]
72. Li, Y.; Xiong, Y.; Yang, H.; Cao, K.; Chen, R. Thin film encapsulation for the organic light-emitting diodes display via atomic layer deposition. *J. Mater. Res.* **2020**, *35*, 681–700. [CrossRef]
73. Johnson, R.W.; Hultqvist, A.; Bent, S.F. A brief review of atomic layer deposition: From fundamentals to applications. *Mater. Today* **2014**, *17*, 236–246. [CrossRef]
74. Kim, L.H.; Jang, J.H.; Jeong, Y.J.; Kim, K.; Baek, Y.; Kwon, H.-J.; An, T.K.; Nam, S.; Kim, S.H.; Jang, J. Highly-impermeable Al_2O_3/HfO_2 moisture barrier films grown by low-temperature plasma-enhanced atomic layer deposition. *Org. Electron.* **2017**, *50*, 296–303. [CrossRef]
75. Zhu, Z.; Merdes, S.; Ylivaara, O.M.E.; Mizohata, K.; Heikkila, M.J.; Savin, H. Al_2O_3 Thin Films Prepared by a Combined Thermal-Plasma Atomic Layer Deposition Process at Low Temperature for Encapsulation Applications. *Phys. Status Solidi A-Appl. Mater. Sci.* **2020**, *217*, 1900237. [CrossRef]
76. Oh, J.; Shin, S.; Park, J.; Ham, G.; Jeon, H. Characteristics of Al_2O_3/ZrO_2 laminated films deposited by ozone-based atomic layer deposition for organic device encapsulation. *Thin Solid Film.* **2016**, *599*, 119–124. [CrossRef]
77. Li, C.; Cauwe, M.; Yang, Y.; Schaubroeck, D.; Mader, L.; de Beeck, M.O. Ultra-Long-Term Reliable Encapsulation Using an Atomic Layer Deposited $HfO_2/Al_2O_3/HfO_2$ Triple-Interlayer for Biomedical Implants. *Coatings* **2019**, *9*, 579. [CrossRef]
78. Lee, Y.; Seo, S.; Oh, I.-K.; Lee, S.; Kim, H. Effects of O_2 plasma treatment on moisture barrier properties of SiO_2 grown by plasma-enhanced atomic layer deposition. *Ceram. Int.* **2019**, *45*, 17662–17668. [CrossRef]
79. Lee, U.S.; Choi, J.S.; Yang, B.S.; Oh, S.; Kim, Y.J.; Oh, M.S.; Heo, J.; Kim, H.J. Formation of a Bilayer of ALD-SiO_2 and Sputtered Al_2O_3/ZrO_2 Films on Polyethylene Terephthalate Substrates as a Moisture Barrier. *Ecs Solid State Lett.* **2013**, *2*, R13–R15.
80. Kukli, K.; Kemell, M.; Castan, H.; Duenas, S.; Seemen, H.; Rahn, M.; Link, J.; Stern, R.; Heikkila, M.J.; Ritala, M.; et al. Atomic Layer Deposition and Performance of ZrO_2-Al_2O_3 Thin Films. *ECS J. Solid State Sci. Technol.* **2018**, *7*, P287–P294. [CrossRef]
81. Yu, D.; Yang, Y.-Q.; Chen, Z.; Tao, Y.; Liu, Y.-F. Recent progress on thin-film encapsulation technologies for organic electronic devices. *Opt. Commun.* **2016**, *362*, 43–49. [CrossRef]

82. Yun, S.J.; Lim, J.W.; Lee, J.H. Low-temperature deposition of aluminum oxide on polyethersulfone substrate using plasma-enhanced atomic layer deposition. *Electrochem. Solid State Lett.* **2004**, *7*, C13–C15. [CrossRef]
83. Yang, Y.-Q.; Duan, Y.; Chen, P.; Sun, F.-B.; Duan, Y.-H.; Wang, X.; Yang, D. Realization of Thin Film Encapsulation by Atomic Layer Deposition of Al2O3 at Low Temperature. *J. Phys. Chem. C* **2013**, *117*, 20308–20312. [CrossRef]
84. Lim, J.W.; Yun, S.J. Electrical properties of alumina films by plasma-enhanced atomic layer deposition. *Electrochem. Solid State Lett.* **2004**, *7*, F45–F48. [CrossRef]
85. Iliescu, C.; Avram, M.; Chen, B.; Popescu, A.; Dumitrescu, V.; Poenar, D.P.; Sterian, A.; Vrtacnik, D.; Amon, S.; Sterian, P. Residual stress in thin films PECVD depositions: A review. *J. Optoelectron. Adv. Mater.* **2011**, *13*, 387–394.
86. Zhang, S.; Shi, W.; Siegler, T.D.; Gao, X.; Ge, F.; Korgel, B.A.; He, Y.; Li, S.; Wang, X. An All-Inorganic Colloidal Nanocrystal Flexible Polarizer. *Angew. Chem. Int. Ed.* **2019**, *58*, 8730–8735. [CrossRef]
87. Yuan, Y.; Xie, S.; Ding, C.; Shi, X.; Xu, J.; Li, K.; Zhao, W. Fabricating flexible wafer-size inorganic semiconductor devices. *J. Mater. Chem. C* **2020**, *8*, 1915–1922. [CrossRef]
88. Yeom, B.; Kim, S.; Cho, J.; Hahn, J.; Char, K. Effect of interfacial adhesion on the mechanical properties of organic/inorganic hybrid nanolaminates. *J. Adhes.* **2006**, *82*, 447–468. [CrossRef]
89. Yeom, B.; Jeong, A.; Lee, J.; Char, K. Enhancement of fracture toughness in organic/inorganic hybrid nanolaminates with ultrathin adhesive layers. *Polymer* **2016**, *91*, 187–193. [CrossRef]
90. Tu, N.; Jiang, J.; Chen, Q.; Liao, J.; Liu, W.; Yang, Q.; Jiang, L.; Zhou, Y. Flexible ferroelectric capacitors based on Bi3.15Nd0.85Ti3O12/muscovite structure. *Smart Mater. Struct.* **2019**, *28*, 054002. [CrossRef]
91. Niu, R.; Liu, G.; Ding, X.; Sun, J. Ductility of metal thin films in flexible electronics. *Sci. China Ser. E-Technol. Sci.* **2008**, *51*, 1971–1979. [CrossRef]
92. Zhao, Y.; Zhang, L.; Liu, J.; Adair, K.; Zhao, F.; Sun, Y.; Wu, T.; Bi, X.; Amine, K.; Lu, J.; et al. Atomic/molecular layer deposition for energy storage and conversion. *Chem. Soc. Rev.* **2021**, *50*, 3889–3956. [CrossRef] [PubMed]
93. Multia, J.; Karppinen, M. Atomic/Molecular Layer Deposition for Designer's Functional Metal-Organic Materials. *Adv. Mater. Interfaces* **2022**, *9*, 2200210. [CrossRef]
94. McIntee, O.M.; Welch, B.C.; Greenberg, A.R.; George, S.M.; Bright, V.M. Elastic modulus of polyamide thin films formed by molecular layer deposition. *Polymer* **2022**, *255*, 125167. [CrossRef]
95. Lee, B.H.; Lee, K.H.; Im, S.; Sung, M.M. Vapor-Phase Molecular Layer Deposition of Self-Assembled Multilayers for Organic Thin-Film Transistor. *J. Nanosci. Nanotechnol.* **2009**, *9*, 6962–6967. [CrossRef]
96. Muneshwar, T.; Cadien, K. Surface reaction kinetics in atomic layer deposition: An analytical model and experiments. *J. Appl. Phys.* **2018**, *124*, 095302. [CrossRef]
97. Wang, H.; Wang, Z.; Xu, X.; Liu, Y.; Chen, C.; Chen, P.; Hu, W.; Duan, Y. Multiple short pulse process for low-temperature atomic layer deposition and its transient steric hindrance. *Appl. Phys. Lett.* **2019**, *114*, 201902. [CrossRef]
98. Schaepkens, M.; Kim, T.W.; Erlat, A.G.; Yan, M.; Flanagan, K.W.; Heller, C.M.; McConnelee, P.A. Ultrahigh barrier coating deposition on polycarbonate substrates. *J. Vac. Sci. Technol. A* **2004**, *22*, 1716–1722. [CrossRef]
99. Perrotta, A.; Aresta, G.; van Beekum, E.R.J.; Palmans, J.; van de Weijer, P.; van de Sanden, M.C.M.R.; Kessels, W.M.M.E.; Creatore, M. The impact of the nano-pore filling on the performance of organosilicon-based moisture barriers. *Thin Solid Film.* **2015**, *595*, 251–257. [CrossRef]
100. Chen, D.H.; Ozaki, S. Stress concentration due to defects in a honeycomb structure. *Compos. Struct.* **2009**, *89*, 52–59. [CrossRef]
101. Gao, Y.; Chen, Y. Sawing stress of SiC single crystal with void defect in diamond wire saw slicing. *Int. J. Adv. Manuf. Technol.* **2019**, *103*, 1019–1031. [CrossRef]
102. Carcia, P.F.; McLean, R.; Reilly, M.; Groner, M.; George, S. Ca test of Al2O3 gas diffusion barriers grown by atomic layer deposition on polymers. *Appl. Phys. Lett.* **2006**, *89*, 031915. [CrossRef]
103. Choi, H.; Shin, S.; Jeon, H.; Choi, Y.; Kim, J.; Kim, S.; Chung, S.C.; Oh, K. Fast spatial atomic layer deposition of Al2O3 at low temperature (<100 degrees C) as a gas permeation barrier for flexible organic light-emitting diode displays. *J. Vac. Sci. Technol. A* **2016**, *34*, 01A121.
104. Keuning, W.; van de Weijer, P.; Lifka, H.; Kessels, W.M.M.; Creatore, M. Cathode encapsulation of organic light emitting diodes by atomic layer deposited Al2O3 films and Al2O3/a-SiNx:H stacks. *J. Vac. Sci. Technol. A* **2012**, *30*, 01A131. [CrossRef]
105. Meyer, J.; Görrn, P.; Bertram, F.; Hamwi, S.; Winkler, T.; Johannes, H.H.; Weimann, T.; Hinze, P.; Riedl, T.; Kowalsky, W. Al2O3/ZrO2 nanolaminates as ultrahigh gas-diffusion barriers—A strategy for reliable encapsulation of organic electronics. *Adv. Mater.* **2009**, *21*, 1845–1849. [CrossRef]
106. Seo, S.-W.; Jung, E.; Chae, H.; Cho, S.M. Optimization of Al2O3/ZrO2 nanolaminate structure for thin-film encapsulation of OLEDs. *Org. Electron.* **2012**, *13*, 2436–2441. [CrossRef]
107. Kim, L.H.; Kim, K.; Park, S.; Jeong, Y.J.; Kim, H.; Chung, D.S.; Kim, S.H.; Park, C.E. Al2O3/TiO2 nanolaminate thin film encapsulation for organic thin film transistors via plasma-enhanced atomic layer deposition. *Acs Appl. Mater. Interfaces* **2014**, *6*, 6731–6738. [CrossRef]
108. Choi, H.; Lee, S.; Jung, H.; Shin, S.; Ham, G.; Seo, H.; Jeon, H. Moisture Barrier Properties of Al2O3 Films deposited by Remote Plasma Atomic Layer Deposition at Low Temperatures. *Jpn. J. Appl. Phys.* **2013**, *52*, 035502. [CrossRef]

109. Pilz, J.; Perrotta, A.; Leising, G.; Coclite, A.M. ZnO Thin Films Grown by Plasma-Enhanced Atomic Layer Deposition: Material Properties Within and Outside the "Atomic Layer Deposition Window". *Phys. Status Solidi A-Appl. Mater. Sci.* **2020**, *217*, 1900256. [CrossRef]

110. Jang, W.; Jeon, H.; Kang, C.; Song, H.; Park, J.; Kim, H.; Seo, H.; Leskela, M.; Jeon, H. Temperature dependence of silicon nitride deposited by remote plasma atomic layer deposition. *Phys. Status Solidi A-Appl. Mater. Sci.* **2014**, *211*, 2166–2171. [CrossRef]

111. Wang, H.; Zhao, Y.; Wang, Z.; Liu, Y.; Zhao, Z.; Xu, G.; Han, T.-H.; Lee, J.-W.; Chen, C.; Bao, D.; et al. Hermetic seal for perovskite solar cells: An improved plasma enhanced atomic layer deposition encapsulation. *Nano Energy* **2020**, *69*, 104375. [CrossRef]

112. Jen, S.-H.; Bertrand, J.A.; George, S.M. Critical tensile and compressive strains for cracking of Al_2O_3 films grown by atomic layer deposition. *J. Appl. Phys.* **2011**, *109*, 084305. [CrossRef]

113. Chang, C.-Y.; Lee, K.-T.; Huang, W.-K.; Siao, H.-Y.; Chang, Y.-C. High-Performance, Air-Stable, Low-Temperature Processed Semitransparent Perovskite Solar Cells Enabled by Atomic Layer Deposition. *Chem. Mater.* **2015**, *27*, 5122–5130. [CrossRef]

114. Kwon, J.H.; Jeong, E.G.; Jeon, Y.; Kim, D.-G.; Lee, S.; Choi, K.C. Design of highly water resistant, impermeable, and flexible thin-film encapsulation based on inorganic/organic hybrid layers. *Acs Appl. Mater. Interfaces* **2018**, *11*, 3251–3261. [CrossRef]

115. Li, Y.; Xiong, Y.; Cao, W.; Zhu, Q.; Lin, Y.; Zhang, Y.; Liu, M.; Yang, F.; Cao, K.; Chen, R. Flexible $PDMS/Al_2O_3$ nanolaminates for the encapsulation of blue OLEDs. *Adv. Mater. Interfaces* **2021**, *8*, 2100872. [CrossRef]

116. Casillas, G.; Mayoral, A.; Liu, M.; Ponce, A.; Artyukhov, V.I.; Yakobson, B.I.; Jose-Yacaman, M. New insights into the properties and interactions of carbon chains as revealed by HRTEM and DFT analysis. *Carbon* **2014**, *66*, 436–441. [CrossRef]

117. Shugurov, A.R.; Panin, A.V. Mechanisms of stress generation and relaxation in thin films and coatings. In Proceedings of the International Conference on Physical Mesomechanics of Multilevel Systems 2014, Tomsk, Russia, 3–5 September 2014; pp. 575–578.

118. Li, Y.-S.; Tsai, C.-H.; Kao, S.-H.; Wu, I.W.; Chen, J.-Z.; Wu, C.-I.; Lin, C.-F.; Cheng, I.C. Single-layer organic-inorganic-hybrid thin-film encapsulation for organic solar cells. *J. Phys. D-Appl. Phys.* **2013**, *46*, 435502. [CrossRef]

119. Kim, S.J.; Yong, S.H.; Choi, Y.J.; Hwangbo, H.; Yang, W.-Y.; Chae, H. Flexible Al_2O_3/plasma polymer multilayer moisture barrier films deposited by a spatial atomic layer deposition process. *J. Vac. Sci. Technol. A* **2020**, *38*, 022418. [CrossRef]

120. Keum, C.; Murawski, C.; Archer, E.; Kwon, S.; Mischok, A.; Gather, M.C. A substrateless, flexible, and water-resistant organic light-emitting diode. *Nat. Commun.* **2020**, *11*, 6250. [CrossRef]

121. Kwon, B.-H.; Joo, C.W.; Cho, H.; Kang, C.-M.; Yang, J.-H.; Shin, J.-W.; Kim, G.H.; Choi, S.; Nam, S.; Kim, K. Organic/Inorganic Hybrid Thin-Film Encapsulation Using Inkjet Printing and PEALD for Industrial Large-Area Process Suitability and Flexible OLED Application. *Acs Appl. Mater. Interfaces* **2021**, *13*, 55391–55402. [CrossRef]

122. Nam, T.; Park, Y.J.; Lee, H.; Oh, I.-K.; Ahn, J.-H.; Cho, S.M.; Kim, H. A composite layer of atomic-layer-deposited Al_2O_3 and graphene for flexible moisture barrier. *Carbon* **2017**, *116*, 553–561. [CrossRef]

123. Jen, S.-H.; Lee, B.H.; George, S.M.; McLean, R.S.; Carcia, P.F. Critical tensile strain and water vapor transmission rate for nanolaminate films grown using Al_2O_3 atomic layer deposition and alucone molecular layer deposition. *Appl. Phys. Lett.* **2012**, *101*, 234103. [CrossRef]

124. Chen, G.; Weng, Y.; Sun, F.; Zhou, X.; Wu, C.; Yan, Q.; Guo, T.; Zhang, Y. Low-temperature atomic layer deposition of Al_2O_3/alucone nanolaminates for OLED encapsulation. *RSC Adv.* **2019**, *9*, 20884–20891. [CrossRef]

125. Chen, T.; Wuu, D.; Wu, C.; Chiang, C.; Chen, Y.; Horng, R.-H. High-performance transparent barrier films of SiO_x/SiN_x stacks on flexible polymer substrates. *J. Electrochem. Soc.* **2006**, *153*, F244. [CrossRef]

126. Leterrier, Y. Durability of nanosized oxygen-barrier coatings on polymers. *Prog. Mater. Sci.* **2003**, *48*, 1–55. [CrossRef]

127. Lewis, J.S.; Weaver, M.S. Thin-film permeation-barrier technology for flexible organic light-emitting devices. *IEEE J. Sel. Top. Quantum Electron.* **2004**, *10*, 45–57. [CrossRef]

128. Choi, Y.J.; Yong, S.H.; Kim, S.J.; Hwangbo, H.; Cho, S.M.; Pu, L.S.; Chae, H. Hygroscopic interlayers for multilayer Al_2O_3 barrier films. *Thin Solid Film.* **2019**, *690*, 137524. [CrossRef]

129. Jeong, E.G.; Kwon, S.; Han, J.H.; Im, H.-G.; Bae, B.-S.; Choi, K.C. A mechanically enhanced hybrid nano-stratified barrier with a defect suppression mechanism for highly reliable flexible OLEDs. *Nanoscale* **2017**, *9*, 6370–6379. [CrossRef]

130. Yong, S.H.; Kim, S.J.; Park, J.S.; Cho, S.M.; Ahn, H.J.; Chae, H. Flexible Carbon-rich Al_2O_3 Interlayers for Moisture Barrier Films by a Spatially-Resolved Atomic Layer Deposition Process. *J. Korean Phys. Soc.* **2018**, *73*, 40–44. [CrossRef]

131. Park, J.S.; Yong, S.H.; Choi, Y.J.; Chae, H. Residual stress analysis and control of multilayer flexible moisture barrier films with SiN_x and Al_2O_3 layers. *AIP Adv.* **2018**, *8*, 085101. [CrossRef]

132. Kim, H.G.; Lee, J.G.; Kim, S.S. Self-assembled monolayers as a defect sealant of Al_2O_3 barrier layers grown by atomic layer deposition. *Org. Electron.* **2018**, *52*, 98–102. [CrossRef]

133. Seo, S.-W.; Jung, E.; Seo, S.J.; Chae, H.; Chung, H.K.; Cho, S.M. Toward fully flexible multilayer moisture-barriers for organic light-emitting diodes. *J. Appl. Phys.* **2013**, *114*, 14. [CrossRef]

134. Han, Y.C.; Jeong, E.G.; Kim, H.; Kwon, S.; Im, H.-G.; Bae, B.-S.; Choi, K.C. Reliable thin-film encapsulation of flexible OLEDs and enhancing their bending characteristics through mechanical analysis. *RSC Adv.* **2016**, *6*, 40835–40843. [CrossRef]

135. Wang, Z.; Chen, Z.; Wang, J.; Shangguan, L.; Fan, S.; Duan, Y. Realization of an autonomously controllable process for atomic layer deposition and its encapsulation application in flexible organic light-emitting diodes. *Org. Express* **2023**, *13*, 21672–21688. [CrossRef] [PubMed]

136. Wang, Z.; Wang, J.; Li, Z.; Chen, Z.; Shangguan, L.; Fan, S.; Duan, Y. Crosslinking and densification by plasma-enhanced molecular layer deposition for hermetic seal of flexible perovskite solar cells. *Nano Energy* **2023**, *109*, 108232. [CrossRef]
137. Chen, Z.; Wang, J.; Lin, J.; Shen, Y.; Wang, M.; Duan, Y. Optimizing the gradient stress sandwich structure thin-film encapsulation for super flexible organic light-emitting devices. *Appl. Phys. Lett.* **2023**, *123*, 083506. [CrossRef]
138. Buchwalder, S.; Bourgeois, F.; Leon, J.J.; Hogg, A.; Burger, J. Parylene-AlO$_x$ Stacks for Improved 3D Encapsulation Solutions. *Coatings* **2023**, *13*, 1942. [CrossRef]
139. Jeong, S.Y.; Shim, H.R.; Na, Y.; Kang, K.S.; Jeon, Y.; Choi, S.; Jeong, E.G.; Park, Y.C.; Cho, H.E.; Lee, J.W.; et al. Foldable and washable textile-based OLEDs with a multi-functional near-room-temperature encapsulation layer for smart e-textiles. *NPJ Flex. Electron.* **2012**, *5*, 15. [CrossRef]

micromachines

Review

Nature-Inspired Superhydrophobic Coating Materials: Drawing Inspiration from Nature for Enhanced Functionality

Subodh Barthwal [1], Surbhi Uniyal [2] and Sumit Barthwal [3],*

[1] Department of Mechanical Engineering, Amity University, Greater Noida 201308, India; subodh_barthwal23@yahoo.co.in
[2] Department of Mechanical Engineering, Graphic Era University, Dehradun 248002, India; surbhiuniyal.87@gmail.com
[3] Nanomechatronics Lab, Kookmin University, Seoul 02707, Republic of Korea
* Correspondence: sumitb3@yahoo.co.in or sumitb3@kookmin.ac.kr

Abstract: Superhydrophobic surfaces, characterized by exceptional water repellency and self-cleaning properties, have gained significant attention for their diverse applications across industries. This review paper comprehensively explores the theoretical foundations, various fabrication methods, applications, and associated challenges of superhydrophobic surfaces. The theoretical section investigates the underlying principles, focusing on models such as Young's equation, Wenzel and Cassie–Baxter states, and the dynamics of wetting. Various fabrication methods are explored, ranging from microstructuring and nanostructuring techniques to advanced material coatings, shedding light on the evolution of surface engineering. The extensive applications of superhydrophobic surfaces, spanning from self-cleaning technologies to oil–water separation, are systematically discussed, emphasizing their potential contributions to diverse fields such as healthcare, energy, and environmental protection. Despite their promising attributes, superhydrophobic surfaces also face significant challenges, including durability and scalability issues, environmental concerns, and limitations in achieving multifunctionality, which are discussed in this paper. By providing a comprehensive overview of the current state of superhydrophobic research, this review aims to guide future investigations and inspire innovations in the development and utilization of these fascinating surfaces.

Keywords: superhydrophobic; fabrication methods; wetting models; self-cleaning; coatings

Citation: Barthwal, S.; Uniyal, S.; Barthwal, S. Nature-Inspired Superhydrophobic Coating Materials: Drawing Inspiration from Nature for Enhanced Functionality. *Micromachines* **2024**, *15*, 391. https://doi.org/10.3390/mi15030391

Academic Editor: Giampaolo Mistura

Received: 18 December 2023
Revised: 1 March 2024
Accepted: 6 March 2024
Published: 13 March 2024

1. Introduction

For many years, the interaction between surfaces and water has been a subject of keen interest. Nature provides numerous distinctive surface patterns designed to interact with water in ways that enhance functionality and adaptability to the environment. These natural designs serve as a foundational source for structural concepts extensively studied and implemented in the development of surfaces with specific water affinities. Various applications require surfaces with either high or low water affinity, underscoring the importance of investigating surface wettability and developing strategies to control it according to the desired outcome [1–3].

Wettability, a fundamental property describing the interaction between a solid surface and a liquid, is intricately linked to surface roughness and water contact angle. The water contact angle, a vital parameter quantifying wettability, reflects the degree of hydrophobicity, with higher angles denoting increased water resistance and lower surface energy. The terms "hydrophobic" and "hydrophilic" categorize materials based on their wetting behavior, describing materials as hydrophobic if they resist water and are challenging to wet and hydrophilic if they readily interact with water. These terms are derived from the Greek word "hydro", meaning "water", along with the suffixes "phobos" and "philia", representing "fear" and "love", respectively. Analyzing the static contact angle, which measures the angle formed by a static liquid droplet with the interphase of the solid surface and surrounding vapor, provides insight into the water–surface interaction.

The water contact angle (WCA) is determined using a contact angle goniometer, providing insights into the interaction between a water droplet and a solid surface.

A material is categorized as hydrophilic when its water contact angle (WCA) is below 90°, while materials with a WCA exceeding 90° are termed hydrophobic. The term "superhydrophobic" is used to describe materials with a WCA ranging between 150° and 180°. "Superhydrophilic" surfaces feature a minimal contact angle, causing droplets to spread and completely wet the surface, while superhydrophobic surfaces showcase a maximal contact angle, resulting in droplets forming nearly spherical shapes, as illustrated in Figure 1. One of the remarkable features associated with superhydrophobic surfaces is their inherent self-cleaning ability. The combination of high water repellency and appropriate surface roughness allows water droplets to easily roll off the surface, carrying away contaminants and dust particles. This self-cleaning property has significant implications for various applications, ranging from anti-fouling surfaces in marine environments to low-maintenance coatings in architectural settings. Understanding the intricate relationships among surface roughness and surface chemistry is pivotal for advancing the design and application of functional surfaces with tailored wetting characteristics.

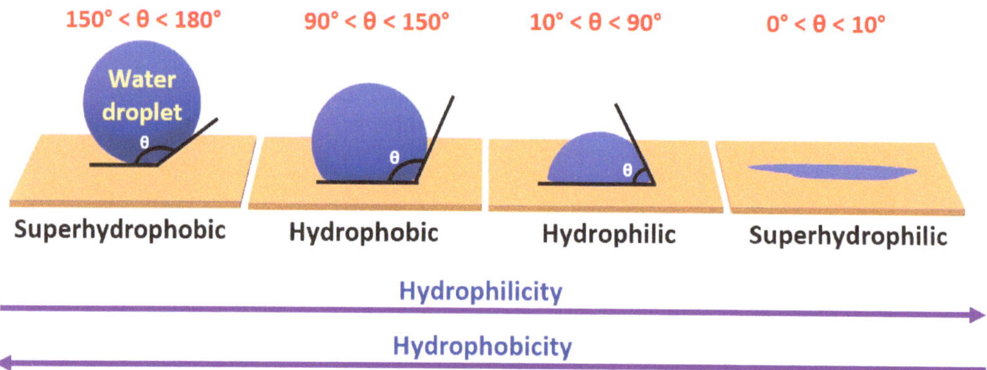

Figure 1. Diagram illustrating various surface wettability's through contact angle measurements.

The origins of superhydrophobic materials trace back to the early 20th century, where historical records document contact angles of 160° and above on coated surfaces treated with stearic acids and soot [4,5]. The fundamental principles of superhydrophobic behavior were subsequently established by Wenzel, Cassie, and Baxter in the ensuing decades. Although research progressed gradually from the 1940s to the 1990s, the crucial moment came in 1996 when T. Onda et al. validated artificial superhydrophobic surfaces [4–10]. The term "lotus effect" was officially patented by Neinhuis and Barthlott in 1997, referring to the naturally occurring superhydrophobic feature observed in lotus leaves [11]. This groundbreaking discovery ignited a surge in research, as evidenced by Chen et al.'s comprehensive review of superhydrophobicity in 1999 [12]. Figure 2 illustrates the steady increase in the development of "superhydrophobic coatings" over the past decade, based on the number of papers retrieved from the Web of Science. The notable rise in the number of publications in recent years highlights the growing interest and concern surrounding superhydrophobic coatings in our society.

Throughout the process of evolution, nature has ingeniously developed various biological systems with superhydrophobic surface properties, as shown in Figure 3. One prominent example of such surfaces is found in lotus leaves (*Nelumbo nucifera*), recognized for their exceptional superhydrophobicity. The discovery of the self-cleaning mechanism of lotus leaves can be attributed to botanists Barthlott and Ehler in 1977 [13]. Their investigation unveiled a sophisticated combination of hierarchical roughness, encompassing both microstructures and nanostructures, all covered beneath a layer of low surface energy wax. This unique combination of intricate roughness and a waxy coating contributes to the lotus leaves' superhydrophobic nature, characterized by a water contact angle (WCA) of 161°. The synergy of high roughness and the presence of a waxy layer facilitates water droplets to effortlessly roll off the surface, effectively cleansing it by dislodging dust particles. This phenomenon is widely recognized as the "lotus effect", as shown in Figure 4a,b.

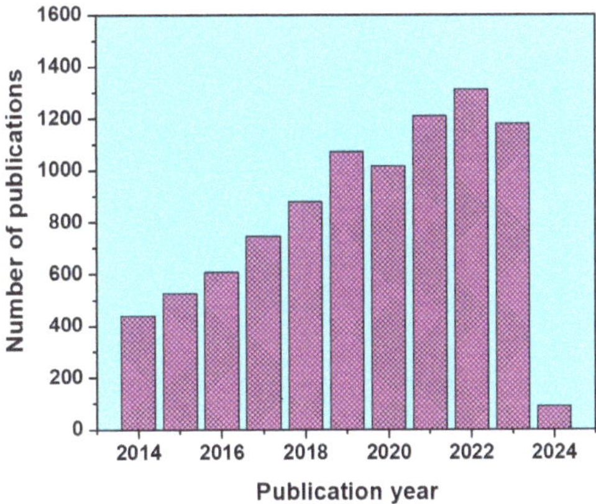

Figure 2. Statistics obtained from the Web of Science, revealing the total number of publications from 2014 to 2024 when using "superhydrophobic coating" as a keyword in the search.

Figure 3. Superhydrophobic surfaces available in nature.

Superhydrophobicity is not exclusive to lotus leaves but can be observed across various plants, insects, and birds, showcasing a diverse range of natural adaptations. Examples include rice (*Oryza sativa*) and taro (*Colocasia esculenta*) leaves, mosquito eyes, butterfly wings, desert beetles, gecko's feet, water strider legs, and shark skins [14–28], as depicted in Table 1. Guo et al. [23] discovered that rice leaves possess remarkable superhydrophobicity, preventing water droplets from wetting their surface, as shown in Figure 4c. Their surface exhibits a binary microstructure and nanostructure similar to lotus leaves, featuring papillae with an average diameter of 5–8 mm in a one-dimensional order. The sublayer beneath the surface contains uniformly distributed nanometer-scale pins, enhancing air-trapping capability. This surface achieves superhydrophobicity with a WCA of 157.28°, facilitating the effortless rolling off of water droplets.

Table 1. Superhydrophobic surfaces present in nature.

Natural Surface	WCA	Properties	References
Rice leaf	164°	Superhydrophobic, self-cleaning, antifouling, and low drag	Bixler et al. (2014) [14]
Mosquito compound eyes	155°	Superhydrophobic and anti-fog	Gao et al. (2007) [16]
Butterfly wings	152 ± 1.7°	Superhydrophobic and self-cleaning	Zheng et al. (2007) [17]
Gecko foot	150°	Superhydrophobic and anti-adhesion	Stark et al. (2016) [19]
Water strider legs	167.6 ± 4.4°	Superhydrophobic and anti-adhesion	Gao et al. (2004) [20]
Lotus leaf	>150°	Superhydrophobic and self-cleaning	Barthlott et al. (1997) [24]
Indian cress	180°	Superhydrophobic and self-cleaning	Otten et al. (2004) [25]
Shark skin	160°	Superhydrophobic, self-cleaning, and anti-fouling	Liu et al. (2012) [26]
Desert beetle	>150°	Superhydrophobic and fog-collection behavior	Kostal et al. (2018) [27]
Rose petal	154.6°	Superhydrophobic and high surface adhesion	Feng et al. (2008) [28]

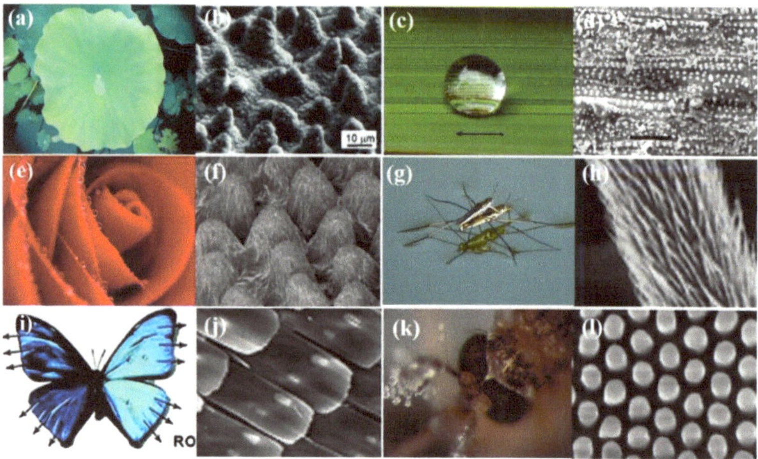

Figure 4. Diagram illustrating organisms with their SEM images: (**a**,**b**) lotus leaf; (**c**,**d**) rice leaf; (**e**,**f**) rose petal; (**g**,**h**) water strider; (**i**,**j**) butterfly wing; and a (**k**,**l**) mosquito compound eye [22].

Similarly, an insightful study by Gao and Jiang [20] highlighted the remarkable superhydrophobic properties exhibited by water striders, focusing on their legs. The water striders' ability to stand and glide swiftly on water is attributed to the complex combination of hierarchical microstructures and nanostructures. These structures include numerous oriented, tiny hairs known as microsetae, which possess fine nanogrooves. These needle-shaped setae exhibit diameters ranging from 3 microns down to several hundred nanometers, with an average length of about 50 μm and an inclined angle of approximately 20° from the leg's surface, as illustrated in Figure 4g,h. This unique adaptation allows water striders to stand and move swiftly on the water's surface.

Zheng et al. [17] observed that butterfly wings exhibit overlapping quadrate scales (150 mm in length and 70 mm in width), forming a periodic hierarchy in one direction (Figure 4i). Each scale's surface features separate ridging stripes, consisting of fine nano-stripes with multi-layers of cuticle lamellae of different lengths and tilted slightly upward with emerging nano-tips. These distinctive structures create air pockets that significantly reduce the surface contact area with water on the wings, showcasing superhydrophobicity with a water contact angle (WCA) of 152°.

In recent years, several high-quality review articles in the superhydrophobic coatings field have summarized recent advancements, challenges, and applications [29–39]. Wang et al. [29] specifically focused on the development and mechanism of superhydrophobic and antimicrobial coatings, addressing potential applications and associated challenges in commercial adoption. Nomeir et al. [31] offered new perspectives on self-cleaning superhydrophobic coatings for solar energy applications, exploring the impact of dust accumulation on solar panel efficiency. Table 2 provides an overview of current reviews on bio-inspired superhydrophobic coatings across various applications.

Table 2. Representative review articles on superhydrophobic coatings in recent years (2020–2024).

S. No.	Title	Journal	Publication Date	Features	Ref.
1	Recent advances in superhydrophobic and antibacterial coatings for biomedical Materials	*Coatings*	October 2022	Summarized development trends in medical device coatings, focusing on superhydrophobic and antimicrobial coatings. Addressed potential applications and challenges in the commercial adoption of antimicrobial coatings.	Wang et al. [29]
2	3D-printed biomimetic structures for energy and environmental applications	*DeCarbon*	March 2024	An overview of the current state of 3D-printed biomimetic structures and their applications in energy and environment.	Li et al. [30]
3	Recent progress on transparent and self-cleaning surfaces by superhydrophobic coatings deposition to optimize the cleaning process of solar panels	*Solar Energy Materials and Solar Cells*	August 2023	An overview of the latest studies on self-cleaning superhydrophobic coatings for solar energy applications.	Nomeir et al. [31]
4	Bioinspired marine antifouling coatings: antifouling mechanisms, design strategies and application feasibility studies	*European Polymer Journal*	May 2023	Summarization of the design strategy and development trend of bionic marine antifouling coatings, evaluates the antifouling performance, antifouling mechanism, and antifouling effect of the coatings.	Li et al. [32]
5	Recent advances in bioinspired sustainable sensing technologies	*Nano-Structures & Nano-Objects*	April 2023	A comprehensive overview concerning the innovative approach for identifying, recognizing and showcasing the current advancements and key milestones accomplished in biosensing technologies based on bioinspired/bioderived materials.	Mishra et al. [33]
6	Recent advances in bio-inspired multifunctional coatings for corrosion protection	*Progress in Organic Coatings*	July 2022	An overview of research findings on bioinspired organic/inorganic superhydrophobic and slippery coatings for the corrosion protection of metal substrates.	George et al. [34]
7	Icephobic/anti-icing properties of superhydrophobic surfaces	*Advances in Colloid and Interface Science*	June 2022	The research progress of superhydrophobic materials on icephobictiy in recent years is reviewed from the aspects of ice formation and propagation.	Huang et al. [35]
8	Bioinspired and green synthesis of nanoparticles from plant extracts with antiviral and antimicrobial properties: A critical review	*Journal of Saudi Chemical Society*	September 2021	The advancement in green synthesis of nanoparticles using natural compounds such as plant extracts, fruit juices, and other relevant sources have been highlighted. A deep insight into antiviral and antimicrobial activities of these nanoparticles provided.	Naikoo et al. [36]
9	Superhydrophobic and superoleophilic membranes for oil–water separation application: A comprehensive review	*Materials & Design*	June 2021	A comprehensive review of the SHSO membranes, fabrication and characterization methods, the advantages and disadvantages of the fabrication techniques, current status and prospects of SHSO surfaces, and potential future research directions.	Rasouli et al. [37]
10	Bioinspired materials for water-harvesting: focusing on microstructure designs and the improvement of sustainability	*Materials Advances*	November 2020	Comprehensive insights into the bioinspired water-harvesting materials, focusing on the microstructure designs and improvements of sustainability	Zhang et al. [38]
11	Bioinspired polymers for lubrication and wear resistance	*Progress in Polymer Science*	November 2020	Bioinspired lubrication using novel polymeric structures, which has led to producing a myriad of new systems with effective and sustainable antifriction and wear resistant properties.	Adibnia et al. [39]
12	Nature-inspired nano-coating materials: drawing inspiration from nature for enhanced functionality	*Micromachines*	2024	Comprehensive overview of the current state of superhydrophobic research. This review aims to guide future investigations and inspire innovations in the development and utilization of these fascinating surfaces.	Present work

This review delves into the physics of bio-inspired surfaces, relevant theories, and the fabrication of artificial superhydrophobic surfaces. Additionally, we discuss potential applications of superhydrophobic coatings along with current issues and challenges.

2. Theoretical Background of Wetting

Surface wettability is often assessed through the measurement of water contact angles (WCA), a widely accepted method for characterizing hydrophilic and hydrophobic surfaces. The contact angle is typically determined using contact angle goniometers, with a 2 µL or 5 µL water droplet positioned on the surface. Despite its reliability, variations in WCA readings can occur due to the heterogeneous nature of surfaces, both in terms of chemical composition and topography, leading to discrepancies of up to 20° on certain surfaces. Superhydrophobic surfaces, while often characterized by high static water contact angles (WCAs), may not always exhibit efficient water droplet roll-off. Some surfaces, like the petal surfaces of red roses, demonstrate a "pinning effect", where water droplets remain spherical but do not easily roll off due to high adhesion forces [28]. Conversely, certain surfaces showcase a "slippery" behavior, challenging droplets to maintain stability. Therefore, beyond static contact angle, dynamic contact angle is also employed to characterize surface wettability. Contact angle hysteresis (CAH) is assessed through dynamic experiments involving droplets, providing insights into the adhesive properties between water droplets and the surface [40]. CAH ($\Delta\theta$) is defined as the difference between the advancing contact angle (θ_A) and the receding contact angle (θ_R). θ_R, indicating solid−liquid adhesion, is always smaller than or equal to θ_A, representing solid–liquid cohesion. The advancing angle (θ_A) and receding angle (θ_R) are determined by tilting the substrate or changing the droplet volume. In the substrate tilting method, the solid surface is inclined at a specific angle. As the surface is tilted, the droplet on it begins to move, and the advancing contact angle (θ_A) is measured when the droplet is in motion. Subsequently, the receding contact angle (θ_R) is measured when the droplet either comes to rest or starts to retract, as shown in Figure 5a.

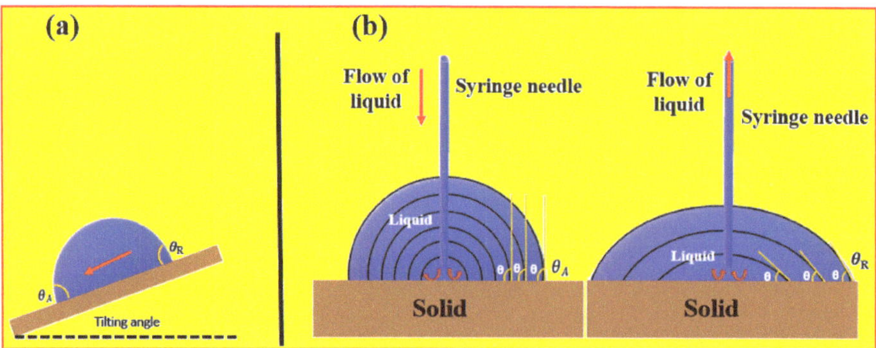

Figure 5. Measuring the advancing and receding angles by (**a**) tilting the substrate; and (**b**) changing the droplet volume by gradually introducing or extracting liquid from a sessile droplet.

Alternatively, the droplet volume change technique involves intentionally altering the volume of the droplet, either by adding or removing liquid. As the droplet undergoes volume change, the contact angle also changes. θ_A is measured as the droplet increases in volume, and θ_R is measured as the droplet decreases in volume, as shown in Figure 5b. Both techniques offer dynamic insights into the interaction between liquid droplets and solid surfaces, contributing to a comprehensive understanding of superhydrophobic and self-cleaning properties. Superhydrophobic surfaces with self-cleaning properties necessitate both high contact angles and low CAH. Advanced tools like contact angle goniometers, high-speed cameras, and image analysis software are often employed for precise measurement and analysis of the dynamic behavior of droplets on surfaces.

Various expressions of contact angle hysteresis are frequently discussed in the scientific literature, and it is crucial to make a clear distinction among them. A commonly employed experimental method for determining the contact angle of a droplet involves gradually increasing its volume until spreading initiates. The angle at which spreading commences, indicating that the droplet can overcome any energy barriers hindering its movement, is referred to as the advancing contact angle. Conversely, when the droplet volume is quasi-statically reduced, the contact line initially shifts at the receding contact angle. The disparity between the advancing and receding contact angles defines the contact angle hysteresis, which is zero for an ideal substrate but can be $10°$ or more for a real surface, posing a notorious challenge in measurement. This configuration is designated as unforced static hysteresis. Alternatively, it is feasible to exert force to propel a liquid drop across a surface and measure both the advancing angle at the front and the receding contact angle at the rear when the motion initiates. This scenario is denoted as forced, static hysteresis. It is crucial to highlight that the measured unforced and forced contact angle hysteresis for a specific drop on a particular surface may not necessarily align. This discrepancy arises because a forced drop undergoes deformation, and the free energy barriers are contingent on the shape of the drop.

Quantitatively understanding contact angle hysteresis poses challenges due to its dependence on the intricate characteristics of surface inhomogeneities, which are typically random in both position and size. Nevertheless, recent progress in surface engineering has enabled the creation of surfaces featuring well-defined chemical patterning with distinct areas exhibiting varying contact angles.

Another noteworthy advancement involves the development of superhydrophobic surfaces. When surfaces with an inherent hydrophobic contact angle are covered with micron-scale posts, the macroscopic contact angle can, in specific cases, approach close to $180°$. This innovation presents exciting possibilities for manipulating surface properties for various applications. Drops exhibit two distinct states on surfaces: a suspended state, known as the Cassie–Baxter state [41], where they rest on top of the posts, and a collapsed state, referred to as the Wenzel state [42], where they fill the interstices between the posts. Additionally, on these surfaces, drops can easily roll [43–45], highlighting the significance of contact angle hysteresis in comprehending this behavior.

Kusumaatmaja and Yeomans [46] conducted an investigation into contact angle hysteresis on chemically patterned and superhydrophobic surfaces by quasi-statically altering the drop volume. Their study encompassed both two and three dimensions, employing analytical and numerical methods to minimize the free energy of the drop. In two dimensions, on a surface striped with regions exhibiting different equilibrium contact angles, θ, they observed a slip, jump, and stick motion of the contact line. The advancing and receding contact angles corresponded to the maximum and minimum values of θ, respectively. In three dimensions, these values served as bounds, with contact angle hysteresis being mitigated by the free energy associated with surface distortion [47,48]. Although stick, slip, and jump behavior persisted, it is important to note that defining a single macroscopic contact angle for patterns around the drop size becomes problematic. The position and magnitude of the contact line jumps proved to be sensitive to the details of surface patterning and could vary in different directions relative to that patterning.

On superhydrophobic surfaces, the behavior of drops in two dimensions reveals specific contact angles. In the ideal scenario, the advancing contact angle is $180°$, as confirmed by previous studies [49]. Simultaneously, the receding angle corresponds to θ_R, which represents the intrinsic contact angle of the surface. When considering collapsed drops in two dimensions, the advancing contact angle remains $180°$. However, the receding angle is $\theta_R - 90°$ due to the necessity for the contact line to dewet the sides of the posts.

In the realm of three dimensions, both suspended and collapsed drops exhibit an advancing angle close to $180°$, as supported by various sources [45,50,51]. However, there is an increase in the receding contact angle. Despite this increase, the receding angle remains notably smaller in the collapsed state, indicating stronger contact line pinning. Consequently, the hysteresis observed in suspended drops is generally much smaller than that observed in collapsed drops on the same surface.

Wettability is a phenomenon observed when water comes into contact with surfaces, wherein some surfaces readily attract water while others repel water droplets. This behavior is intricately influenced by the interplay of molecular, chemical, and physical forces occurring at the interface of liquid, solid, and gas phases. The interactions at the solid–liquid–air interfaces play a pivotal role in determining the behavior of water droplets on a surface. When a liquid encounters a surface, cohesive and adhesive forces come into play, dictating the shape of the liquid. If adhesive forces between the liquid and the surface are dominant over cohesive forces within the liquid, the liquid will spread, forming a thin film across the surface. Conversely, if cohesive forces within the liquid outweigh adhesion to the surface, a droplet will form, maintaining minimal contact with the surface. The theoretical foundation of wetting involves various models and principles designed to comprehend and predict these intricate interactions. Thomas Young, a British scientist in 1805, introduced the first comprehensive understanding of droplet-wetting characteristics on an ideally smooth, solid surface [52]. Young's equation establishes a relationship between the contact angle (θ), surface tension (γ), and interfacial tensions among the three phases (solid, liquid, and gas) at the droplet's contact point on a solid surface, as shown in Figure 6a.

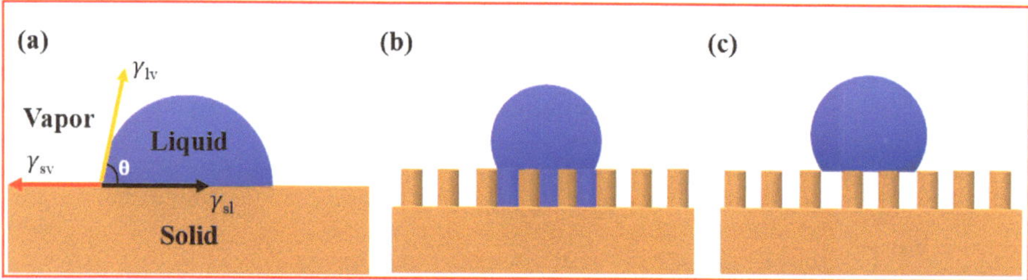

Figure 6. Schematics of wetting models: (**a**) Young model, (**b**) Wenzel model, and (**c**) Cassie–Baxter model.

According to Young's equation, the contact angle of a liquid on a flat surface is given by:

$$\cos\theta = \frac{\gamma_{sv} - \gamma_{sl}}{\gamma_{lv}} \tag{1}$$

where γ_{sv}, γ_{sl}, and γ_{lv} refer to the interfacial tensions of the solid–vapor, solid–liquid, and liquid–vapor phases, respectively.

Despite its significance, Young's equation has limitations, primarily being applicable only to surfaces with uniform chemical composition and smoothness. In reality, most surfaces exhibit an uneven chemical composition and varying degrees of roughness, rendering Young's equation unsuitable in such scenarios.

Therefore, the early work by Wenzel (1936) and later by Cassie and Baxter in 1944 has led to the development of two distinct models explaining wetting states on rough surfaces.

Wenzel modified Young's equation by introducing a roughness factor to elucidate wettability on textured surfaces [53]. In the Wenzel state, the liquid fully wets the textured surface, maximizing the liquid–solid contact area, as shown in Figure 6b. According to Wenzel's theory, the apparent contact angle (θ_a) of a liquid droplet placed on a rough, solid surface is given by the following equation:

$$\cos\theta_a = r\cos\theta \tag{2}$$

Here, r represents the roughness factor, defined as the ratio of the actual contact area of the solid–liquid to the apparent contact area of the solid–liquid, and θ is the equilibrium contact angle of the liquid on a smooth surface of the same material.

Since r is always greater than one for a rough surface, Wenzel's equation predicts that if ($\theta > 90°$), ($\theta_a > \theta$), indicating a hydrophobic surface, and if ($\theta < 90°$), ($\theta_a < \theta$), indicating a hydrophilic surface. Therefore, in the Wenzel state, the surface roughness will make intrinsically hydrophobic surfaces more hydrophobic and hydrophilic surfaces more hydrophilic. However, the equation has limitations and does not apply when the chemical composition of the solid surface differs.

The Cassie–Baxter equation describes the wetting behavior of liquids on rough surfaces and is an extension of the Wenzel model. Proposed by Cassie and Baxter in 1944, this model accounts for the presence of air pockets trapped within surface asperities [54]. In the Cassie state, liquid droplets are supported by the tops of the rough features, and air is trapped underneath, creating a composite interface, as displayed in Figure 6c. The Cassie–Baxter equation is given by:

$$\cos \theta^* = f_1 \cos \theta - f_2 \tag{3}$$

where θ^* is the apparent contact angle for a droplet on a rough surface; θ is the equilibrium contact angle obtained on an ideally flat surface of the same chemical composition; f_1 is the fraction of solid substrate in contact with the liquid; and f_2 is the fraction of air in contact with the liquid (i.e., $f_1 + f_2 = 1$). Thus, Equation (3) can be simplified into Equation (4):

$$\cos \theta^* = f_1 (\cos \theta + 1) - 1 \tag{4}$$

If f_1 is very small, it reveals there will be a large amount of air trapped under the water droplet, and $\cos \theta^*$ can approach -1 with the water apparent contact angle (θ^*) becoming close to 180°. Hence, to achieve a high apparent contact angle, the practical contact area between the solid and liquid droplets should be as small as possible.

The Cassie–Baxter state is characterized by enhanced liquid repellency due to the trapped air, preventing complete wetting of the surface. This model is particularly relevant for superhydrophobic surfaces, where the combination of surface roughness and air entrapment leads to extraordinary water-repellent properties.

A set of equations was solved to characterize the dynamics and thermodynamics of the considered drops by using the lattice Boltzmann algorithm [46].

2.1. The Model Drop

The authors opted to describe the equilibrium properties of the drop using a continuum free energy model [55]:

$$\Psi = \int_V \left(\psi_b(n) + \frac{K}{2}(\partial_\alpha n)^2 \right) dV + \int_S \psi_s(n_s) dS. \tag{5}$$

$\psi_b(n)$ is a bulk free energy [31]:

$$\psi_b(n) = p_c \, (v_n + 1)^2 \left(v_n^2 - 2v_n + 3 - 2\beta\tau_w \right) \tag{6}$$

where $v_n = (n - n_c)/n_c$, $\tau_w = (T_c - T)/T_c$, and n, n_c, T, T_c, and p_c are the local density, critical density, local temperature, critical temperature, and critical pressure of the fluid, respectively.

In Equation (5), the second term represents the free energy attributed to interfaces within the system [55]. The final term in Equation (5) characterizes the interactions between the fluid and the solid surface. Cahn [56] defined the surface energy density as $\psi_s(n) = -\phi n_s$, where n_s represents the fluid density value at the surface. The parameter ϕ governs the strength of interaction, influencing the local equilibrium contact angle. The minimization of the free energy results in the establishment of the boundary condition at the surface:

$$\partial_{\perp} n = -\phi/\kappa \tag{7}$$

and a relation between ϕ and the equilibrium contact angle θ [55]:

$$\phi = 2\beta\tau_{\mathrm{w}}\sqrt{2p_c\kappa}\,\mathrm{sign}\left(\frac{\pi}{2}-\theta_e\right)\sqrt{\cos\frac{\alpha}{3}\left(1-\cos\frac{\alpha}{3}\right)} \tag{8}$$

where $\alpha = \cos^{-1}(\sin^2\theta)$, and the function sign is employed to determine the sign of its argument. Equivalent boundary conditions can be applied to non-flat surfaces. A method for addressing corners and ridges essential for modeling superhydrophobic surfaces is outlined in [57]. The governing equations for the drop's motion are the continuity and Navier–Stokes equations:

$$\partial_t n + \partial_\alpha(n u_\alpha) = 0 \tag{9}$$

$$\partial_t(n u_\alpha) + \partial_\beta(n u_\alpha u_\beta) = -\partial_\beta P_{\alpha\beta} + \upsilon\partial_\beta\left[n(\partial_\beta u_\alpha + \partial_\alpha u_\beta + \delta_{\alpha\beta}\partial_\gamma u_\gamma)\right] + n a_\alpha, \tag{10}$$

Here, u, P, υ, and a represent the local velocity, pressure tensor, kinematic viscosity, and acceleration, respectively.

To simulate unforced static hysteresis, a gradual increase or decrease in drop volume is necessary. This is achieved by adjusting the drop liquid density by approximately $\pm 0.1\%$. Consequently, the drop volume is influenced as the system returns to its coexisting equilibrium densities. For forced hysteresis, a body force $n a_\alpha$ is introduced into the Navier–Stokes Equation (6) to address the phenomenon.

2.2. Three-Dimensional Drop on a Chemically Patterned Surface

Illustrated diagrammatically in Figure 7, the chemically patterned surface under consideration consists of squares with a side length $a = 12$, separated by a distance $b = 5$. In the first scenario (surface A), the squares' equilibrium contact angle is $\theta_1 = 110°$, while that of the channels between them is $\theta_2 = 60°$. In the second case (surface B), exchange of the equilibrium contact angles takes place, i.e., $\theta_1 = 60°$ for the squares and $\theta_2 = 110°$ for the channels [46]. Despite the nearly identical macroscopic contact angle for both surfaces, $\theta_{\mathrm{CB}} \approx 85.5°$, calculated using the Cassie–Baxter formula:

$$\cos\theta_{\mathrm{CB}} = f_1\cos\theta_1 + f_2\cos\theta_2 \tag{11}$$

which averages over the surface contact angles. The parameters f_1 and f_2 represent the fractions of the surface with intrinsic equilibrium contact angles θ_1 and θ_2, respectively. At first glance, one might anticipate the two surfaces to exhibit very similar behavior. However, this assumption proves incorrect unless the strength of the heterogeneities falls below a specific threshold [58,59]. This condition implies negligible contact angle hysteresis.

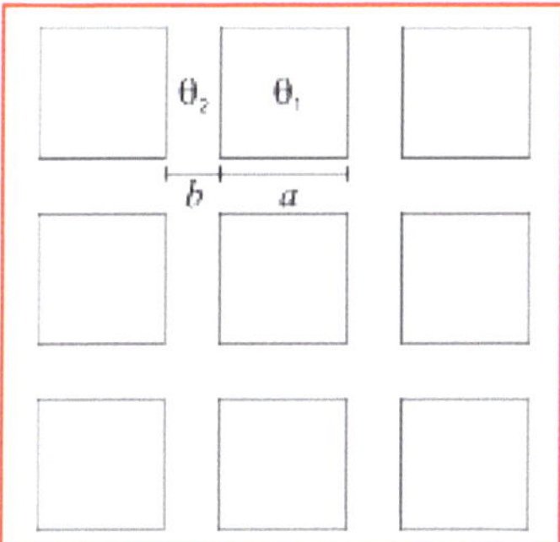

Figure 7. Figure of a chemically patterned surface [46].

2.3. Three-Dimensional Suspended Drop on a Topologically Patterned Surface

The results from three-dimensional lattice Boltzmann simulations are presented in this section with the aim of investigating contact angle hysteresis on three-dimensional topologically patterned surfaces.

For the suspended state, we selected parameters $a = 3$, $b = 7$, $l = 5$, and $\theta = 110°$. According to the Cassie–Baxter formula, this configuration provides an estimated macroscopic contact angle of $\theta = 160°$. The extremely high value of the drop contact angle poses a challenge for simulations, as small changes in the contact angle necessitate significant adjustments in the drop volume. Additionally, to ensure the drop is suspended on an adequate number of posts, a substantial simulation box is required. Lattice with dimensions $168 \times 168 \times 168$ is used, and the largest simulated drop had a volume of 16.8×10^5.

The susceptibility of the drop shape near the surface (deviating from a spherical cap) to small variations can lead to substantial uncertainties in contact angle measurements. The equation used for that is:

$$\theta\text{macro} = 2\,\tan^{-1}\left(\frac{H}{r_{max}}\right) \tag{12}$$

where H represents the height of the drop and r_{max} is the maximum base radius, offering an alternative definition for the macroscopic contact angle.

Figure 8a illustrates the drop contact angle as a function of volume. In the simulation, we increased the volume from 6.8×10^5 to 16.8×10^5, resulting in a macroscopic contact angle θ (as per Equation (12)) ranging from 161.8 to 166.6. In Figure 8b–e, contour plots, top views, and side views of the drop at the beginning and end of the simulation are presented.

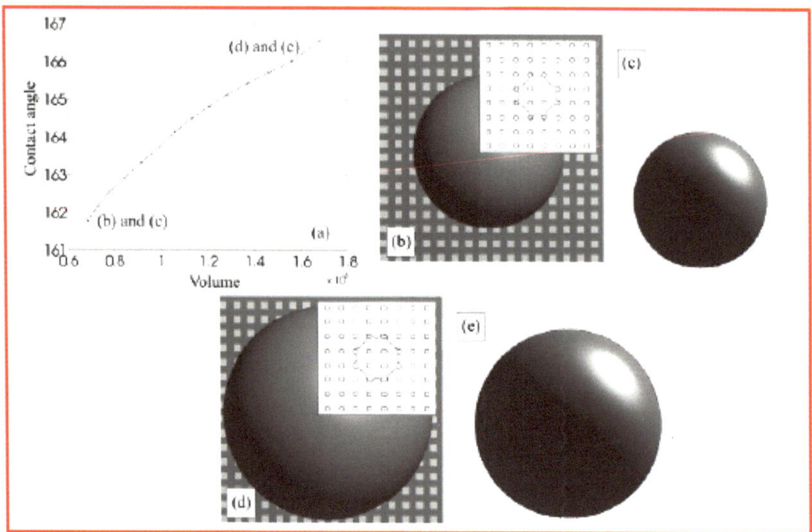

Figure 8. Contact line dynamics for a suspended drop on a topologically patterned surface. (**a**) Contact angle as a function of volume. (**b,c**) Top and side view of the drop at $V = 6.8 \times 10^5$. (**d,e**) Top and side view of the drop at 16.8×10^5 [46].

2.4. Three-Dimensional Collapsed Drop on a Topologically Patterned Surface

Now, we shift our focus to the collapsed state, where the space between the posts is filled with liquid. We chose the parameters $a = 4$, $b = 6$, $l = 5$, and $\theta = 120°$, and according to the Wenzel formula, this configuration yields a contact angle $\theta_W = 154°$.

Figure 9a depicts the macroscopic contact angle of the drop, as a function of increasing volume. Initially, the contact angle rises because, despite the contact line moving outward diagonally with respect to the posts, it remains pinned in the horizontal and vertical directions. This can be observed in the contour plots displayed in Figure 9b,c. Once the drop contacts the four neighboring posts along the diagonals, it wets the tops of these posts (Figure 9d,e), causing a slight decrease in the contact angle, approximately by 5°. At this stage, the contact line is pinned again until it becomes energetically favorable to jump and wet the neighboring posts in the vertical and horizontal directions. However, this contact line jump did not occur in the presented simulations due to the prohibitive computational cost of increasing the size of the simulation box.

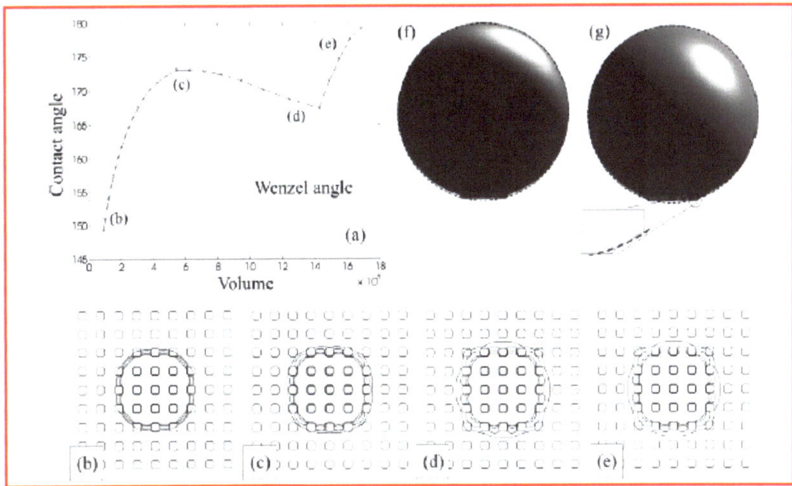

Figure 9. Contact line dynamics for a collapsed drop on a topologically patterned surface. (**a**) Contact angle as a function of volume. (**b–e**) Contour plots of the drop at positions indicated in panel (**a**). (**f**) The drop cross section in the horizontal direction. (**g**) The drop cross section in the diagonal direction [46].

In Figure 9f,g, typical side views of the drops are shown, with the dashed line representing the corresponding spherical fits. The fitted curves align well with drop profiles above the posts but not with profiles between the posts.

3. Fabrication of Superhydrophobic Surfaces

Artificial superhydrophobic surfaces, characterized by extremely high water repellency, can be engineered by precisely controlling surface roughness and surface energy—two critical parameters influencing solid surface wettability. Achieving this involves the creation of hierarchical micro/nanostructures on the surface and chemical modification using low-surface energy materials. Various methods are employed to achieve superhydrophobicity, often relying on two main approaches: top-down and bottom-up methods, as shown in Figure 10. Both approaches offer flexibility and can be combined to create intricate hierarchical structures. While top-down methods are often faster and more scalable, bottom-up methods provide enhanced control over material composition and can facilitate the incorporation of diverse functionalities. The selection between these approaches depends on the specific requirements of the application and the desired characteristics of the superhydrophobic surface.

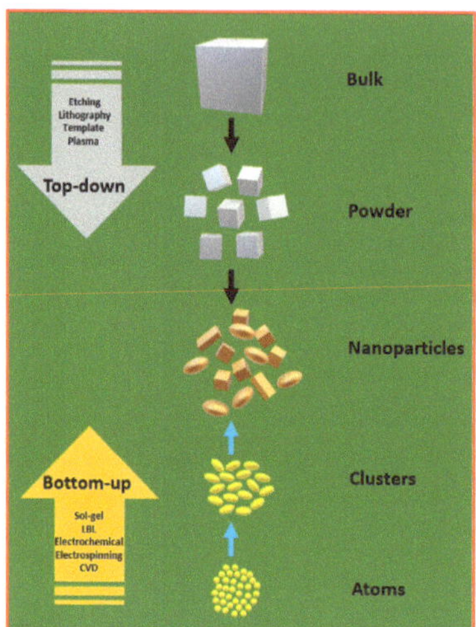

Figure 10. Diagram showing top-down and bottom-up methods for nanomaterial synthesis.

3.1. Top-Down Synthesis

Top-down synthesis for superhydrophobic surfaces involves the fabrication of such surfaces by modifying existing structures or materials at the macroscopic level to achieve the desired properties. This approach typically begins with a bulk material, and through various techniques such as etching, abrasion, or lithography, the surface is altered to create microstructures and nanostructures. The top-down synthesis approach provides precise control over the size, shape, and arrangement of nanostructures. Despite its advantages in precision and reproducibility, top-down synthesis may face challenges in terms of scalability and cost-effectiveness for large-scale production.

3.1.1. Chemical Etching

Chemical etching stands out as a highly effective method for crafting superhydrophobic surfaces, utilizing a controlled chemical reaction to selectively remove material and engineer intricate microstructures and nanostructures on a substrate. In the superhydrophobic context, this process begins with the application of a specific etchant solution that interacts with the material surface (almost any metal or alloy), inducing reactions that result in the creation of tailored surface textures. The choice of etchant (strong acids, such as HCl and H_2SO_4, and a strong base, such as NaOH), along with parameters like concentration and duration, plays a pivotal role in shaping the final surface morphology. Chemical etching offers several advantages, including cost-effectiveness, scalability, and applicability to a diverse range of materials, making it a versatile option for creating superhydrophobic surfaces.

The success of chemical etching in fabricating superhydrophobic surfaces lies in its ability to generate hierarchical structures that replicate natural water-repellent surfaces found in certain plant leaves and animal skins. The resulting surface roughness, coupled with low surface energy coatings, leads to enhanced water repellency and low adhesion. This approach finds diverse applications, ranging from self-cleaning materials to anti-icing coatings, demonstrating the versatility and potential of chemical etching as a top-down method for precisely tailoring surface properties at the microscale and nanoscale [60–65]. For example, Kumar et al. [60] developed a robust and self-cleaning superhydrophobic coating for aluminum surfaces via a chemical etching method with a mixture of hydrochloric and nitric acids, followed by treatment with hexadecyltrimethoxysilane (HDTMS). The generated surface with rough rectangular pit-like microstructures displayed a WCA of $162.0 \pm 4.2°$ and a sliding angle of $4 \pm 0.5°$. The coating displayed excellent thermal, chemical, and mechanical stability. Kim et al. [61] used a hydrogen fluoride (HF) solution to prepare superhydrophobic surfaces from austenitic stainless steel using a simple two-step chemical etching process. After fluorination, the surface showed a WCA of 166° and a SA of 5°. To further enhance the superhydrophobicity, the prepared stainless steel surface was dipped in a 0.1 wt.% NaCl solution at 100 °C, and the WCA was increased to 168° and the sliding angle was decreased to ~2°. The prepared surface displayed self-cleaning properties with excellent superhydrophobicity after 1 month, as shown in Figure 11.

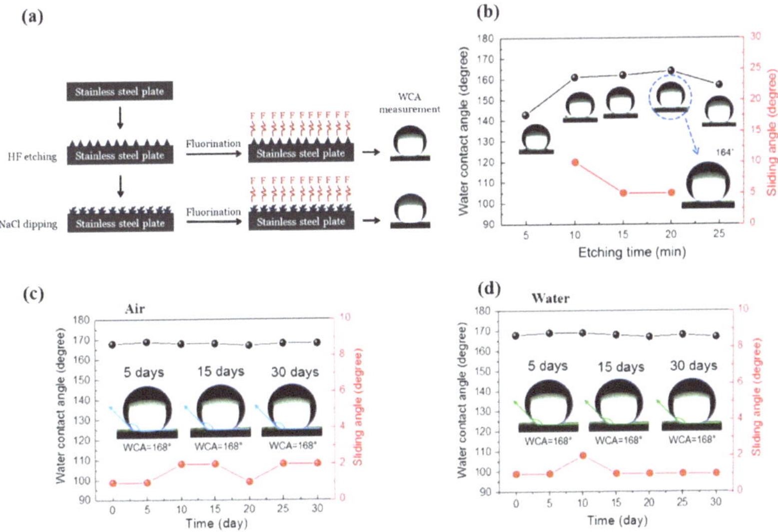

Figure 11. (**a**) Schematic depiction outlining the various steps involved in preparing superhydrophobic stainless steel surfaces; (**b**) changes in the WCA and SA of stainless steel following different etching durations in HF; durability assessment of the superhydrophobic stainless steel sample after exposure to (**c**) air and (**d**) water for a period of 1 month [61].

Similarly, Zhu et al. [62] fabricated superhydrophobic magnesium alloy substrates by employing a one-step etching process and fluoride modification. The prepared magnesium surface with multiple hierarchical micro-nano structures displayed a water contact angle (WCA) of 173.3° and a sliding angle (SA) of 1°. Furthermore, the surface exhibited excellent properties, including self-cleaning, chemical stability even when exposed to air, chemically aggressive solutions (including strong acids and bases), and cyclic icing/melting treatments. Rodič et al. [63] outlined a process for producing superhydrophobic films on aluminum surfaces through a two-step method. This involved an initial etching step in a FeCl$_3$ solution to create a hierarchical micro/nano-structured aluminum surface. Subsequently, grafting was carried out directly in an ethanol solution of 1H,1H,2H,2H-perfluorodecyltriethoxysilane (FAS-10) as a low-surface-energy material at ambient temperature, as displayed in Figure 12a,b. Optimizing the etching time to 20 min and the grafting duration to 30 min in FAS-10 solution yielded a superhydrophobic aluminum surface with a water contact angle exceeding 150° and a minimal sliding angle below 10°, as shown in Figure 12c. This superhydrophobic aluminum surface demonstrated effective self-cleaning against solid pollutants and enhanced anti-icing performance with delayed melting.

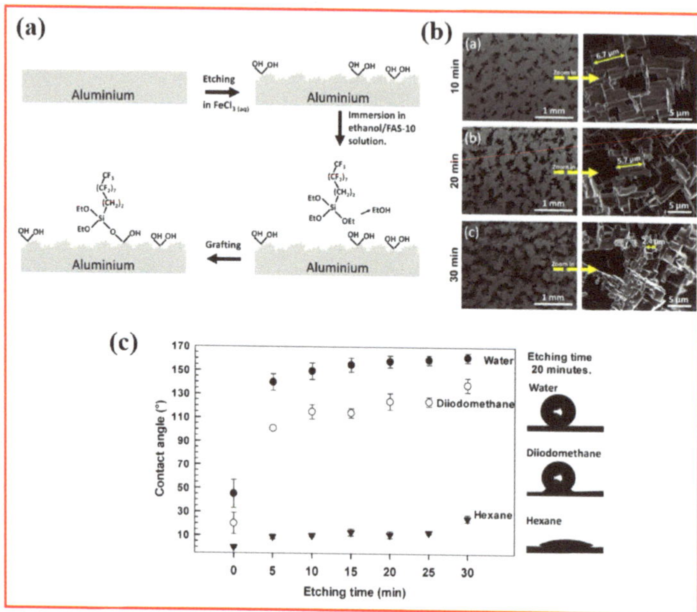

Figure 12. (**a**) Schematic diagram of the fabrication process of a superhydrophobic aluminum surface, including etching in a FeCl$_3$ solution and subsequent modification with FAS-10; (**b**) surface morphologies of the surface etched for varying time intervals; and (**c**) contact angles measured for water, diiodomethane, and hexane on aluminum surfaces etched for different durations and treated for 30 min with 1 wt.% ethanol FAS-10 solutions [63].

3.1.2. Lithography Technique

Lithography emerges as a sophisticated and precise technique for generating superhydrophobic surfaces, relying on the principles of light and pattern transfer to create microstructures and nanostructures on a substrate. In lithography, a photosensitive material is exposed to light through a mask or a photomask, which carries the desired pattern. The exposed material undergoes chemical or physical changes, creating a positive or negative replica of the pattern on the substrate. The resulting surfaces are characterized by intricate patterns that effectively trap air pockets, reducing the contact area between water droplets and the surface. This, in turn, leads to the manifestation of superhydrophobicity.

The lithography-based technique can be categorized into various forms depending on the type of light source utilized. These include photolithography with ultraviolet (UV) and X-ray sources [66,67], nanoimprinting [68], electron beam lithography (EBL) [69], nanosphere lithography (NSL) [70], and laser lithography [71]. Photolithography, one of the primary forms of lithography, provides exceptional control over surface structures, allowing for the meticulous design of features at the microscale and nanoscale. This technique offers high resolution, excellent reproducibility, and scalability, making it an invaluable tool for tailoring surfaces to exhibit superhydrophobic properties. Nevertheless, ensuring a clean and smooth initial surface is crucial and potentially requires the use of a cleanroom.

Nanoimprint lithography involves pressing a mold with nanoscale features onto a material surface, transferring the pattern, and creating microstructures and nanostructures. The versatility of lithography makes it applicable to various substrates, and when combined with post-processing steps or additional coatings, it enhances the durability and multifunctionality of superhydrophobic surfaces. Lithography's precision and versatility make it a crucial tool in the development of advanced materials for applications ranging from self-cleaning surfaces to microfluidic devices [72–74]. Ghasemlou et al. [72] prepared robust superhydrophobic surfaces featuring engineered lotus leaf-mimetic multiscale hierarchical structures through a hybrid approach that combines soft imprinting and spin-coating. The process involved direct soft-imprinting lithography onto starch/polyhydroxyurethane/cellulose nanocrystal (SPC) films, creating micro-scaled features resembling lotus leaves. Subsequently, low-surface-energy poly(dimethylsiloxane) (PDMS) was spin-coated over these microstructures. Further enhancement of the PDMS@SPC film involved modification with vinyltriethoxysilane (VTES) functional silica nanoparticles (V-SNPs), resulting in the fabrication of a superhydrophobic interface with a WCA of approximately 150° and a SA of less than 10°. The outcomes demonstrated that the prepared coating displayed outstanding moisture barrier properties, self-cleaning capabilities, and superior mechanical durability against various tests, including knife scratches, finger-rubbing, jet water impact, a sandpaper abrasion test for 20 cycles, and a tape-peeling test for approximately 10 repetitions.

Guo et al. [73] developed a flexible superhydrophobic microarray utilizing photolithography technology, achieving a maximum WCA of 151.1 ± 0.9° through the optimization of microcolumn spacing and height. To enhance its functionality, carbon black nanoparticles were incorporated into the superhydrophobic microarray, imparting outstanding photothermal conversion performance. This prepared microarray found applications in anti-icing and deicing, significantly delaying the complete freezing of a water droplet by 87%. Moreover, it demonstrated efficiency by requiring 43.1% less time for the removal of an ice pellet and 28.5% less time for an ice sheet. Wang et al. [74] presented a one-step approach for creating a durable superhydrophobic surface on stainless steel using direct ultrafast laser microprocessing. The laser texturing method generated dense hierarchical micro/nanostructures on the stainless steel surface, showcasing commendable thermal stability and excellent anti-icing performance.

3.1.3. Template Method

The template method has emerged as a simple and versatile approach for crafting superhydrophobic surfaces by utilizing predefined templates or molds to impart well-defined microstructures and nanostructures onto a substrate. This methodology involves the replication of surface features found in nature or artificially designed templates onto a chosen material, creating intricate patterns that contribute to superhydrophobic characteristics. The template method can be outlined in three distinct steps: template creation, molding, and demolding. Natural templates, such as lotus leaves, or synthetic structures, serve as blueprints for the desired surface morphology. The substrate, often composed of polymers or metals, is then shaped to mimic the surface features of the template [75–80].

This approach is cost-effective, scalable, and reproducible, making it a widely adopted method for producing polymeric surfaces. Additionally, the template method enables the incorporation of various materials and functionalities, providing flexibility in the design of surfaces with multifunctional properties. As a result, the template method emerges as a promising avenue for fabricating customized superhydrophobic surfaces tailored to specific industrial and technological applications.

Wang et al. [75] manufactured a superhydrophobic fluororubber surface featuring a systematically layered microprotrusion structure by directly replicating the surface microstructure of stainless steel mesh. In the process, the treated stainless steel mesh was positioned onto a fluororubber sheet, and following vulcanization, the mesh was carefully removed. The resulting superhydrophobic fluororubber surface showcased a well-defined concave–protrusion structure, demonstrating outstanding superhydrophobicity with a contact angle of 153.93 ± 0.39°. Additionally, the fabricated surface exhibited remarkable thermostability and mechanical durability, enhancing its potential for practical applications. Xu et al. [76] showcased an efficient method for producing superhydrophobic PDMS films with low adhesion, employing a chemically etched template and subsequent thermal curing. The chemically etched nickel template was replicated to attain a large-scale superhydrophobic PDMS film with micro/nanostructures through a roll-to-roll thermal curing process. The cured PDMS film not only exhibited impressive liquid-repelling and self-cleaning characteristics but also demonstrated high transparency, along with robust chemical and mechanical durability. Wang et al. [77] generated hydrophobic templates with varied wettability by adjusting the laser etching distance on the surface of 6061 aluminum (Al) alloy tubes. Afterward, superhydrophobic flexible tubes were prepared, utilizing the prepared Al alloy tubes as templates and polydimethylsiloxane (PDMS) for replication, resulting in a water contact angle (WCA) of 162.8°. Remarkably, the superhydrophobic tubes exhibited exceptional durability and resistance to abrasion, and the prepared surface also demonstrated blood repellency, as shown in Figure 13.

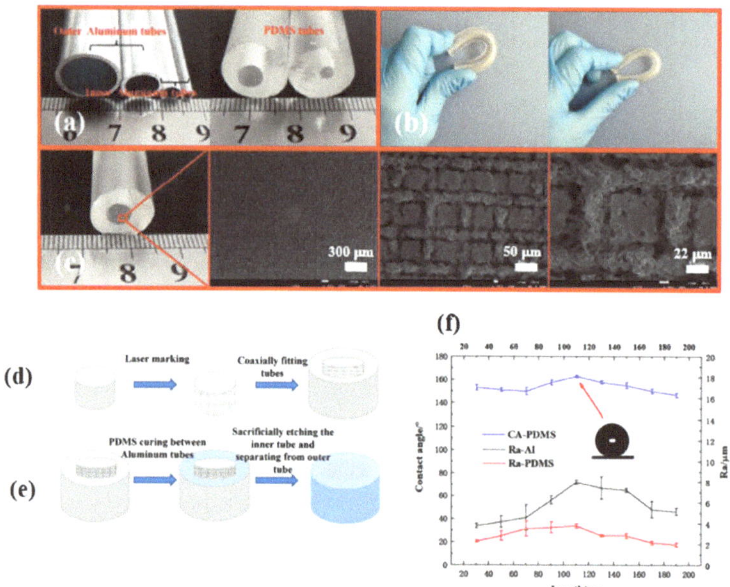

Figure 13. Prepared (**a**) various sizes of templates and PDMS tubes; (**b**) flexibility assessment of the PDMS tubes; (**c**) SEM images of the flexible tubes with corresponding scales of 300 μm, 50 μm, and 22 μm, respectively. The flowchart outlines the preparation of flexible tubes: (**d**) template preparation; (**e**) superhydrophobic flexible tube fabrication; (**f**) impact of processing spacing on roughness (Ra) and contact angle [77].

3.1.4. Plasma Treatment

The plasma method represents an effective approach for fabricating superhydrophobic surfaces, leveraging the controlled modification of surface properties through plasma treatment. In this technique, a substrate material is subjected to a plasma discharge, inducing various chemical and physical changes on its surface. The controlled exposure to plasma allows for the creation of microstructures and nanostructures, contributing to the development of superhydrophobic characteristics. Plasma-based surface modification majorly involves plasma etching, plasma sputtering, and plasma polymerization [81].

Plasma etching, also known as dry etching, is a simple method for modifying the surface properties. During plasma etching, a low-pressure gas is introduced into a chamber, and an electric field is applied to ionize the gas, forming a plasma. This plasma contains energetic ions and electrons that react with the material's surface. The specific chemistry of the gas, along with the parameters of the plasma (such as power, pressure, and temperature), determines the etching characteristics. This fabrication method can alter the wettability of the substrate by increasing roughness [82] or changing the functional group of the surface [83–86]. For example, Somrang et al. [86] reported the fabrication of nanostructured patterns through CF4 plasma etching on SiO_2-based substrates, followed by treatment with Teflon coatings to achieve superhydrophobicity and antireflection. Initially, nickel-thin films were deposited on the substrates using the PVD sputtering process with varying deposition times. To promote the agglomeration of nickel nanoparticulate patterns, serving as etch-resistant masks, the thin films underwent rapid thermal annealing at 500 °C. These patterned substrates were subsequently subjected to CF_4 plasma etching, followed by nickel mask removal through a nitric acid rinse. Finally, polytetrafluoroethylene films were applied to the nanostructured surface to achieve superhydrophobicity.

Plasma sputtering, or physical vapor deposition (PVD), is a highly precise method used to deposit thin films onto surfaces. In this process, a high-energy plasma is generated in a vacuum chamber, and a target material is bombarded with energetic ions from the plasma. These ions dislodge atoms from the target material, creating a vapor that condenses on the substrate, forming a thin, uniform film. Becker et al. [87] introduced a method utilizing laser-assisted magnetron sputtering as a smart approach to achieve superhydrophobic properties on a poly(ethylene terephthalate) (PET) substrate. This innovative technique enables the creation of hierarchical surfaces with precise control. In this process, plasma sputter-deposited fluoropolymers are applied to PET substrates using the Nd/YAG laser-assisted magnetron sputtering technique, employing a poly(tetrafluoroethylene) (PTFE) target.

Similarly, plasma polymerization is a versatile technique for depositing thin polymer films on various substrates using plasma-enhanced chemical vapor deposition (PECVD). During this process, a monomer gas is introduced into a vacuum chamber containing plasma. The plasma breaks down the monomer molecules, initiating polymerization on the substrate surface. Plasma polymerization is utilized in applications ranging from biomedical coatings to surface modifications due to its ability to create uniform and adherent polymer layers on diverse materials [88–90]. Siddig et al. [90] successfully created a robust flame-resistant and highly hydrophobic phosphorus–fluoride coating on aramid fabrics using plasma-induced graft polymerization. Initially, the aramid fabrics underwent activation and roughening through low-pressure plasma treatment. Subsequently, a phosphorus–fluoride emulsion copolymer mixture was sequentially applied as a coating. In flame tests, the fabric demonstrated remarkable resilience with a char length of only 0.68 cm, showing no observable after-flame or after-glow times.

Ussenkhan et al. [91] demonstrated the successful production of superhydrophobic films using the hexamethyldisiloxane (HMDSO) plasma polymerization method in an RF discharge plasma jet at atmospheric pressure. These films were applied using 3D printing technology, enabling large-scale application of the superhydrophobic coating, as shown in Figure 14. Their study revealed that the contact angle of the films was directly influenced by the RF discharge power, with HMDSO concentration and the number of cycles having no effect on the superhydrophobic properties but impacting optical properties, leading to reduced light transmission. While the morphology of the films remained consistent at one and ten deposition cycles on silicon substrates, there was a significant difference in film thickness. Films deposited in one cycle had a thickness of approximately 3 μm, whereas those deposited in ten cycles measured around 24.5 μm, as depicted in Figure 14b–e. This suggests that increasing the number of deposition cycles results in thicker films, affecting optical transparency. Under optimal conditions, the films achieved a contact angle of approximately 165°, indicating a highly hydrophobic surface. When applied as a single layer, the transmittance at 700 nm was about 88%, compared to 92% for pure glass. Surface morphology studies revealed that the plasma jet created microstructured and nanostructured coatings with high surface roughness, imparting superhydrophobic properties to the surface. Additionally, the resulting superhydrophobic films demonstrated stability and resistance to chemical solutions, high temperatures, UV irradiation, and weathering, as shown in Figure 14f.

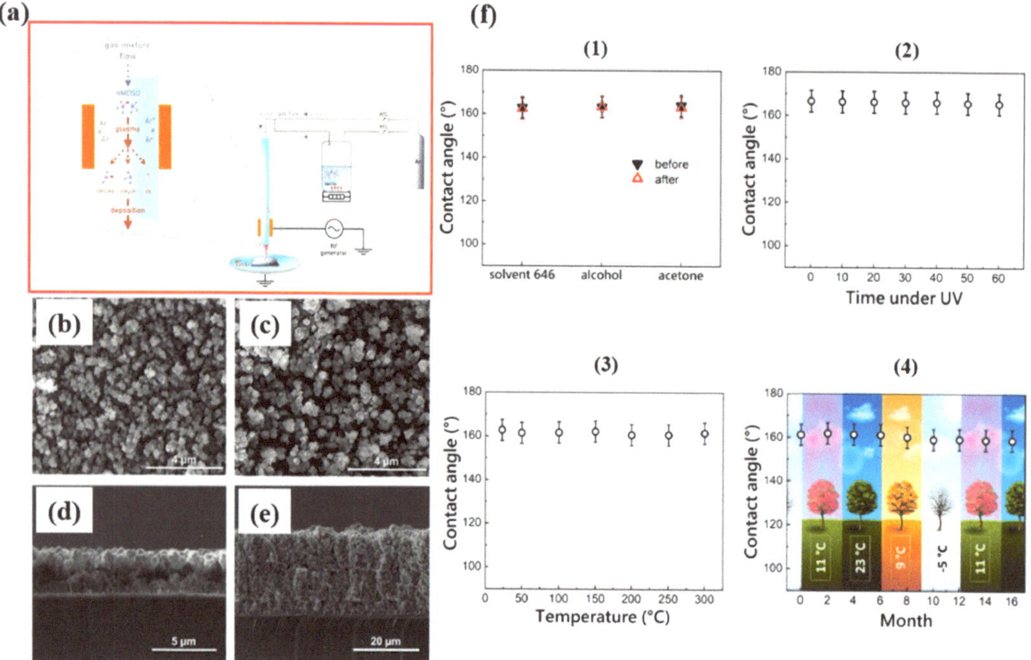

Figure 14. (**a**) Experimental setup schematic for superhydrophobic film deposition via atmospheric pressure plasma polymerization. SEM images depicting the film surface on a silicon substrate after one (**b**) and ten (**c**) cycles of plasma polymerization. Cross-sectional views of thin films after one (**d**) and ten (**e**) cycles. Image (**f**) displays contact angle graphs obtained from chemical tests (1), UV tests (2), thermal tests (3), and over time (4) [91].

3.2. Bottom-Up Synthesis

Bottom-up synthesis is a method employed for the preparation of superhydrophobic surfaces that involves building up nanostructures or materials from smaller components or molecules. In contrast to top-down approaches, which start with larger structures and reduce them, bottom-up synthesis starts at the molecular or nanoscale level and assembles these components to create the desired structures. This technique often relies on self-assembly processes driven by molecular interactions, such as van der Waals forces, hydrogen bonding, or chemical reactions. By carefully designing and manipulating these interactions, researchers can create nanostructures with specific properties that contribute to superhydrophobicity. Bottom-up approaches are advantageous for their ability to produce intricate and precisely controlled nanostructures, providing a high level of customization for superhydrophobic surface properties. Common bottom-up techniques for superhydrophobic surface preparation include hydrothermal techniques, chemical vapor deposition, sol–gel processes, electrochemical deposition, and layer-by-layer deposition.

3.2.1. Sol–Gel Method

The sol–gel method is a versatile and widely used technique for fabricating superhydrophobic surfaces, offering control over surface properties and morphology. In this process, a sol, which is a colloidal suspension of nanoparticles, is formed by the hydrolysis and condensation of precursor compounds such as metal alkoxides or organometallic inorganic salts. The sol is then applied to a substrate, and subsequent gelation leads to the formation of a three-dimensional network. The resulting gel undergoes a drying and curing process to produce a solid film on the substrate.

To achieve superhydrophobicity, various modifications can be introduced during or after the sol–gel process. Incorporating hydrophobic agents such as silanes or fluorinated compounds enhances the water-repellency of the surface. Additionally, creating micro/nanostructures within the sol–gel matrix further contributes to the superhydrophobic characteristics by reducing surface energy and promoting the Cassie–Baxter state, where water droplets rest on the surface with minimal contact [92–95].

The sol–gel method stands out for its simplicity, cost-effectiveness, and scalability. It has been applied to diverse substrates, including glass, metal, and polymers, making it a promising approach for developing superhydrophobic surfaces with tailored functionalities for applications in self-cleaning, anti-fouling, and corrosion resistance. However, this method exhibits several drawbacks, such as the fact that the process can be time-consuming, particularly for thick coatings, and achieving precise control over microstructure and purity is challenging, affecting reproducibility. Compatibility issues with certain substrates and the potential for shrinkage and cracking during drying are additional limitations. Liang et al. [96] synthesized superhydrophobic silica powder through the sol–gel process and incorporated it into PDMS to formulate a transparent and resilient SiO_2/PDMS composite coating. The use of the PDMS polymer aimed to improve the adhesion between the coating and the substrate, enhancing thermal, chemical, and mechanical stability. The spin-coated glass surface exhibited a transmittance exceeding 80% and a WCA of 158°. The resulting SiO_2/PDMS composite coating maintained its superhydrophobicity even under challenging environmental conditions, including varying temperatures and exposure to strong chemicals.

Similarly, Li et al. [97] achieved superhydrophobic methylated silica with a core–shell structure via a sol–gel process. They initially prepared a silica gel (ca. 110 nm) using tetraethylorthosilicate (TEOS) as a precursor. The superhydrophobic methylated silica sol was then prepared by grafting methyl groups onto the silica gel surface using methyltrimethoxysilane (MTMS), as shown in Figure 15a. The initial silica particles, measuring 100 nm in diameter, exhibited a uniform size distribution and minimal aggregation. Following treatment with MTMS, a thin layer of 5~10 nm formed on the methylated silica, resulting in increased surface roughness compared to the pristine silica, as illustrated in Figure 15b,c. The surface morphology of the methylated silica coating on a PET film, depicted in Figure 15d, revealed a rugged surface with domains composed of numerous small particles. This coating exhibited a highly porous structure with a hierarchical micro/nano design reminiscent of the lotus leaf. The hierarchical pinecone-like structures created numerous grooves, promoting air entrapment and contributing to a more hydrophobic surface. This process enables the direct construction of superhydrophobic and superoleophilic surfaces on various substrates through dip-coating or the doctor-blading method, with a maximum WCA and WSA of 161° and 3°, respectively. The treated substrates demonstrated outstanding water repellence, oil absorption, and oil–water separation efficiency exceeding 96%, as shown in Figure 15e,f.

Liu et al. [98] achieved the synthesis of superhydrophobic graphene aerogel beads (SGA beads) using a one-step, in situ sol–gel method within a coagulation bath containing octadecylamine (ODA). Modification with ODA introduced long alkyl chains with low surface energy onto graphene oxide (GO) sheets, resulting in numerous nanoscale wrinkles on the graphene surface and a remarkable WCA of 153°. The SGA beads exhibited a low density (14.4 mg/cm^3), high porosity (96.9%), a specific surface area (18.49 m^2/g), rapid adsorption rates, high adsorption capacity for oils, and recyclability. These characteristics indicate promising applications in the treatment of oils and organic pollutants.

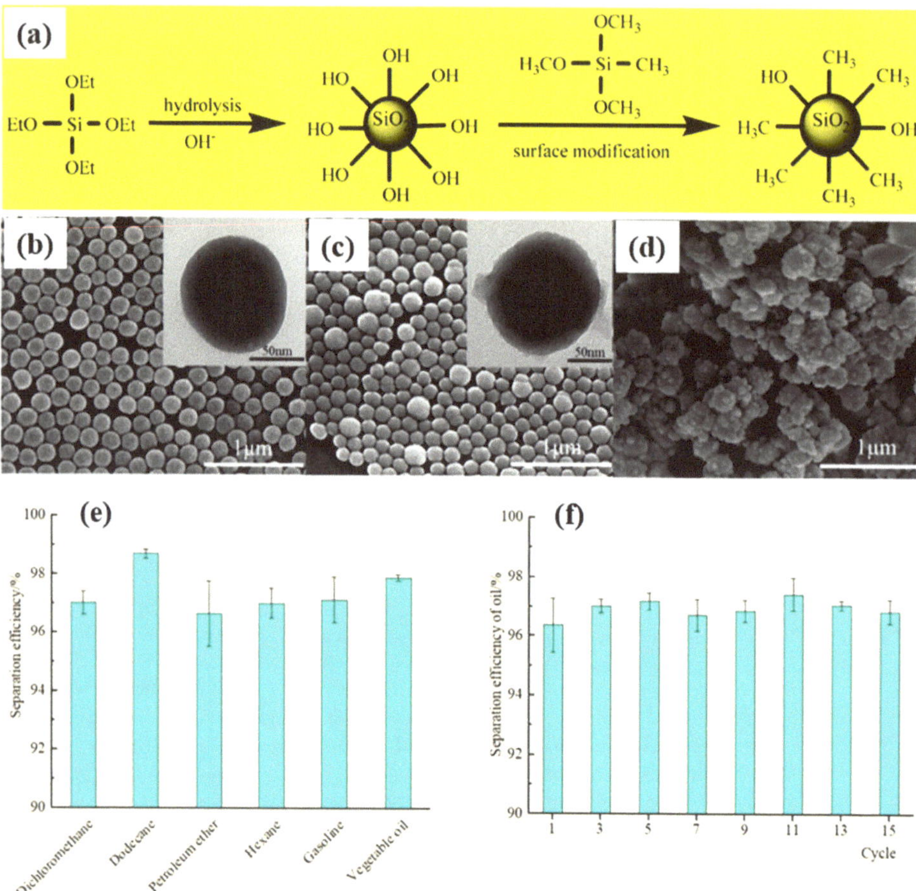

Figure 15. (**a**) Schematic of the preparation of the methylated silica sol; SEM and TEM images of (**b**) pristine silica; (**c**) methylated silica; and (**d**) methylated silica coating. Separation efficiency of the treated PP filter cloth (**e**) for various oil–water mixtures and (**f**) for the dichloromethane–water mixture after 15 cycles [97].

3.2.2. Hydrothermal Technique

The hydrothermal technique stands as a cost-effective, reproducible, and environmentally friendly approach for synthesizing and modifying materials in an aqueous environment under precisely controlled high-temperature and high-pressure conditions. In this method, precursor materials, frequently metal salts or metal–organic compounds, are dissolved in a solution. Subsequently, they are exposed, alongside a substrate, to elevated temperatures and pressures within a sealed reaction vessel. The rigorous conditions within the hydrothermal environment facilitate chemical reactions and promote the nucleation and growth of materials on the substrate. Hydrothermal techniques are widely employed in various scientific and industrial applications, including the synthesis of nanomaterials, crystal growth, and the modification of surfaces to achieve specific functionalities. In the context of superhydrophobic surfaces, the hydrothermal technique can be utilized to create nanostructures and enhance surface properties, leading to water-repellent characteristics [99–106]. Lan et al. [104] created superhydrophobic surfaces on Al alloy substrates by controlling hydrothermal etching and subsequent 1H,1H,2H,2H-perfluorodecyltriethoxysilane (PFDTES) modification. The resulting surface, featuring hierarchical micro/nanostructures, exhibited remarkable self-cleaning capability, anti-corrosion, antibacterial properties against *E. coli* and *S. aureus* bacteria, and anti-icing properties. Tuong et al. [105] prepared the epoxy@ZnO superhydrophobic coating on *Styrax tonkinensis* wood employing a two-step spray coating method aimed at enhancing wood hydrophobicity and color stability. Initially, hydrophobic ZnO particles in micro/nano sizes were synthesized through a hydrothermal process, followed by stearic acid modification, as shown in Figure 16a. Subsequently, the hydrophobic ZnO particles were applied to incompletely cured epoxy pre-coated wood using a spray gun. The hydrophobic ZnO particles contributed to the formation of multiscale roughness, while the incompletely cured epoxy resin pre-coating enhanced the coating's durability, as shown in Figure 16b. Notably, the color stability of wood coated with epoxy@ZnO showed a significant improvement, with the total color changes of coated wood samples surpassing those of uncoated wood samples by over approximately 50%.

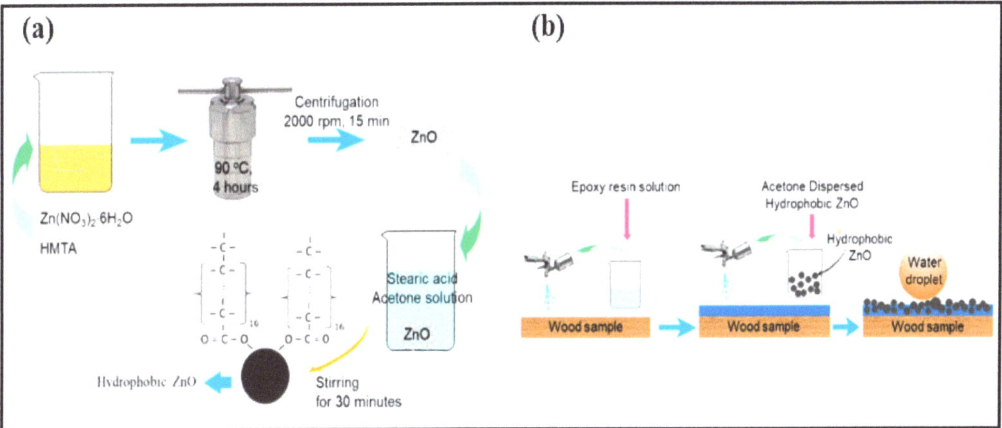

Figure 16. Schematic representation of (**a**) hydrophobic ZnO micro/nanoparticles; (**b**) preparation process of superhydrophobic coatings on wood surfaces [105].

Wan et al. [106] crafted a superhydrophobic copper surface for corrosion protection by employing a combination of etching and hydrothermal methods. Initially, copper underwent a 20 h etching process in an ammonia solution, followed by placement in an autoclave with 60 mL of NaOH solution (0.2 mol/L) for a 2 h reaction at 200 °C. The surface modified with stearic acid exhibited superhydrophobicity with a WCA of $157.7 \pm 1°$. This modified surface demonstrated an impressive corrosion inhibition efficiency of 99.81% in a 3.5 wt% NaCl aqueous solution. Furthermore, the superhydrophobic surface displayed excellent stability in simulated seawater and humid air.

3.2.3. Electrospinning

The electrospinning technique is a versatile method employed to fabricate superhydrophobic surfaces by generating fine fibers through the application of an electric field to a polymer solution or melt. In this process, a syringe containing the polymer solution is subjected to a high voltage, leading to the formation of a charged jet that elongates and eventually solidifies into ultrafine fibers as a result of solvent evaporation or cooling. The resulting fibrous structure, known as a nanofiber mat, creates a high surface area with nanoscale roughness, contributing to the superhydrophobic properties. The morphology of the electrospun fibers is influenced by both the properties of the solution (such as surface tension, viscosity, and conductivity of the polymer solution) and the parameters of the electrospinning process (applied voltage, solution flow rate, and tip-to-collector distance) [107–109].

To enhance hydrophobicity, various hydrophobic agents or surface modifiers can be incorporated into the polymer solution. The electrospinning technique allows for precise control over the fiber morphology and surface characteristics, making it a promising method for developing superhydrophobic surfaces with applications in areas such as water repellency, self-cleaning, and oil–water separation [110–116]. Ding et al. [114] utilized the electrospinning technique to create a superhydrophobic nanofibrous zinc oxide (ZnO) film surface. The process involved electrospinning solutions of poly(vinyl alcohol) and poly(vinyl alcohol)/zinc acetate, followed by a calcination process to produce fibrous zinc oxide superhydrophobic films. By applying a coating of the low-surface-energy material fluoroalkylsilane, the surface wettability of the superhydrophilic ZnO fibrous film (with a WCA~0°) transformed, resulting in a superhydrophobic film (with a WCA~165°) due to the coating of the surface functionalizing agent onto the fibrous films. Similarly, He et al. [115] employed the electrospinning method to prepare a polyvinylidene fluoride (PVDF) nanofibrous membrane featuring superhydrophobic/superoleophilic properties for efficient oil–water separation, coupled with antibacterial capabilities. Modification of the electrospun PVDF nanofibrous surface was achieved through the incorporation of ZnO nanoparticles, tannic acid (TA), and n-dodecyl mercaptan (DT), aimed at reducing surface energy and enhancing surface roughness. The resulting nanofibrous membranes exhibited a water contact angle of 156.5° and a tensile strength of up to 69.25 MPa. The mechanically robust superhydrophobic nanofibrous membrane demonstrated an oil–water separation efficiency exceeding 99%, accompanied by a notable flow flux of 1008.88 $L \cdot m^{-2} h^{-1}$. Additionally, antimicrobial testing revealed a bacteriostatic rate surpassing 98%.

Cai et al. [116] innovatively employed a technique inspired by the traditional Chinese hand-stretched noodle-making process to create a superhydrophobic polyvinylidene fluoride (PVDF) membrane. In this novel method, nanofibers were electrospun onto a substrate covered with (super)hydrophobic nanopowders (candle soot), as shown in Figure 17.

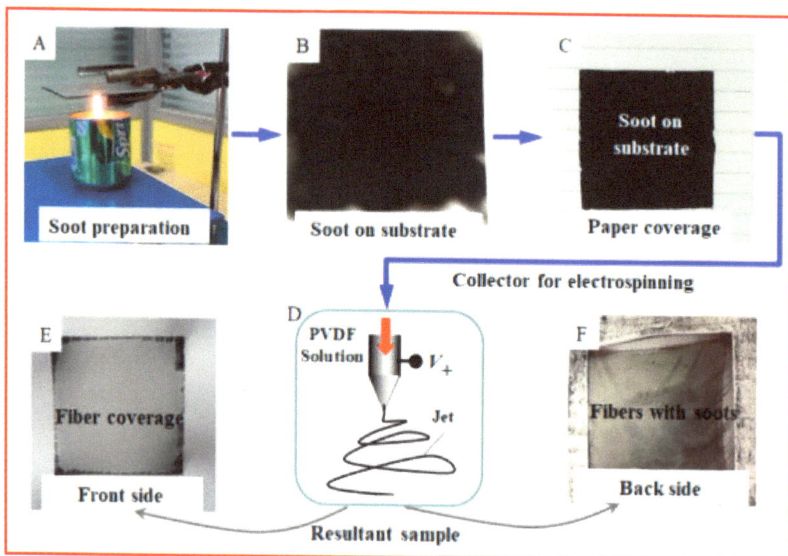

Figure 17. Process for preparing candle soot and superhydrophobic poly(vinylidene fluoride) (PVDF) membranes: (**A,B**) collection of candle soot on sheet metal over the flame; (**C**) paper adhered to the sheet metal; and (**D**) schematic of the electrospinning process using paper and metal as substrates with the deposited fibrous membrane. The soot is pasted to the back side (**F**) of the membrane rather than the front side (**E**) [116].

This approach enhanced the superhydrophobic properties of the electrospun nanofibrous membranes as some nanopowders adhered to the fiber surface. Additionally, heating the substrate enhanced the bonding of nanopowders, allowing them to potentially become embedded within the nanofibers. This advancement in the electrospinning process holds significant promise for various membrane applications.

3.2.4. Chemical Vapor Deposition (CVD)

Chemical vapor deposition (CVD) stands as a vacuum deposition technique based on the chemical reactions of gaseous precursors on a substrate surface. In this method, precursor gases containing the desired elements react on the substrate surface, resulting in the formation of a thin and uniform layer of the intended material. The initiation of the reaction involves heating the substrate, with the chemical reactions unfolding in the gas phase. The volatile by-products are subsequently eliminated, leaving the deposited material in place.

While CVD provides meticulous control over film characteristics such as thickness, composition, and crystal structure, it is a time-consuming and costly process. A notable challenge with CVD lies in its limitation for obtaining large-sized samples. Despite these drawbacks, the versatility of CVD makes it a preferred method for depositing diverse materials, including superhydrophobic coatings [117–124]. This technique proves effective in generating rough surfaces by organizing micro-/nanoparticles into ordered structures or depositing a thin layer of low surface energy materials onto a pre-existing rough surface.

Cai et al. [122] demonstrated the fabrication of transparent superhydrophobic hollow films by using candle soot as a template, followed by CVD of methyltrimethoxysilane (MTMS) and calcination at 450 °C. They optimized the deposition time concerning the transparency and superhydrophobicity of the films. The resulting superhydrophobic films, when coated on a glass substrate exhibited a high transmittance of approximately 90%, with a WCA of more than 165.7° and a SA of about 2.1°. Furthermore, these films displayed remarkable thermal stability and effective moisture resistance even after calcination at temperatures as high as 500 °C. Yang et al. [123] introduced an economical approach to enhance the hydrophobicity of wood using a simple low-temperature CVD technique. The process involved using dichlorodimethylsilane as the CVD chemical source to coat the wood with polydimethylsiloxane (PDMS@wood), resulting in a hydrophobic surface, as depicted in Figure 18a. SEM images reveal that on the tangential section, the wood vessel wall of untreated wood was smooth and distinct (Figure 18b(1,2)). The PDMS@wood cell wall exhibited a particulate morphology on the surface (Figure 18b(3,4)), providing high roughness crucial for forming the hydrophobic surface. The granular and particulate materials were silicon from PDMS, confirmed by EDS analysis of the PDMS@wood samples. The prepared PDMS@wood demonstrated a WCA of 157.28° along with notable thermal stability. Significantly, even after extended storage (30 days) and subjected to a sandpaper abrasion test (Figure 18c), PDMS@wood retained its excellent hydrophobic properties, indicating considerable potential for large-scale industrial production.

Recently, Huang et al. [124] introduced a facile one-step method for fabricating transparent superhydrophobic coatings with customized nanocone array structures using initiated chemical vapor deposition (iCVD). This innovative iCVD process employs condensed nanosized monomer droplets as nucleation centers, leading to the vertical growth of polymer nanocones through a proposed "vapor–liquid-solid" mechanism. The deposition conditions can be tailored to control both the height and density of the nanocones. The optimized nanocone array coating exhibits outstanding water repellency, along with superior anti-icing and anti-frosting capabilities, all while maintaining high light transmittance.

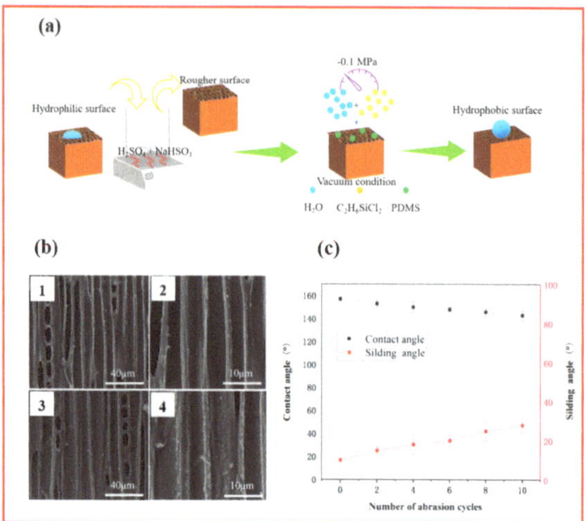

Figure 18. (a) Schematic representation outlining the complete process for fabricating a PDMS@wood coating; (b) SEM images of wood samples on tangential section: (1,2) untreated wood; (3,4) PDMS@wood; and (c) WCA and SA variations with the number of abrasion cycles for the hydrophobic wood surface [123].

3.2.5. Electrochemical Deposition

Electrochemical deposition serves as a versatile and controlled approach for creating superhydrophobic surfaces by electrodepositing materials onto conductive substrates. In this method, the substrate acts as the cathode, and the material to be deposited dissolves in an electrolyte solution. Upon applying an electric current, metal ions from the solution are reduced at the cathode, resulting in the formation of a solid layer on the substrate. Precise adjustment of deposition parameters, such as current density, deposition time, and bath composition, enables the creation of hierarchical micro/nanostructures contributing to surface superhydrophobicity. The technique is not only fast, reproducible, and scalable but also allows the fabrication of diverse surface morphologies like rods, sheets, tubes, fibers, cones, and needles by modifying deposition conditions [125]. Superhydrophobic surfaces produced through electrochemical deposition possess potential advantages, offering ease of large-scale morphology creation and ensuring strong bonding between materials, resulting in durable coatings.

This method provides precise control over surface morphology, composition, and roughness, offering flexibility in tailoring superhydrophobic coatings for applications such as self-cleaning, anti-fouling, and water-resistant materials [126–132]. The electrochemical deposition technique stands as an advantageous avenue for developing customized superhydrophobic surfaces with enhanced functionalities. Li et al. [131] successfully generated a superhydrophobic film with hierarchical porous structures using Fe–myristic acid on a copper substrate through a straightforward one-step electrochemical deposition process. In this electrodeposition procedure, two copper plates served as the cathode and anode, with the electrolyte being an ethanol solution containing Fe–myristic acid. The resulting coating, obtained with an electrodeposition time of 10 min at 20 V, exhibited a maximum contact angle of 159.2° and a minimum sliding angle of 1.7°. Even after undergoing 105 cm abrasion tests, the contact angle of the superhydrophobic film remained at 153.17°, showcasing robust mechanical properties. The superhydrophobic matrix significantly enhanced the corrosion resistance of copper in various environments, including atmospheric conditions, natural seawater, and salt spray. This highlights that the superhydrophobic matrix provides substantial abiotic corrosion inhibition to the underlying copper metal, with an impressive inhibition efficiency of 99.85% observed after 28 days of immersion in seawater.

Prado et al. [132] developed a superhydrophobic coating using composite electrodeposition, incorporating MoS_2 particles into a copper matrix. AISI 316L stainless steel and N80 carbon steel were chosen as substrates, with a thin electrodeposited Ni layer to improve coating adhesion. The resulting coating exhibited a lotus hierarchical coral-like structure, composed of composite Cu and MoS_2 protuberances. The electrodeposition process enabled control over surface roughness and surface energy, with the latter dependent on the quantity of MoS_2 particles in the coating. The Cu–MoS_2 composite coating achieved advancing contact angle values of up to 158.2° with a contact angle hysteresis of 1.8°, as depicted in Figure 19.

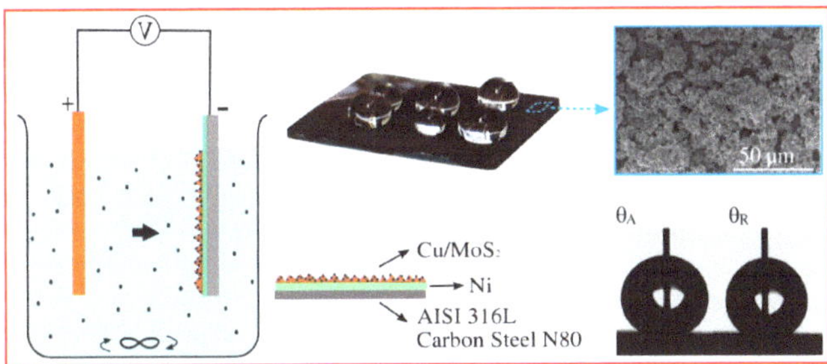

Figure 19. The preparation process of the superhydrophobic Cu–MoS_2 composite coating on the substrate [132].

3.2.6. Layer-by-Layer Deposition

Layer-by-layer (LbL) deposition is a simple, low-cost, and substrate-independent technique employed in the fabrication of superhydrophobic surfaces by sequentially adsorbing layers of oppositely charged materials onto a substrate. In this process, the substrate is alternately dipped into solutions containing positively and negatively charged species, creating a multilayered thin film. The coated film is then rinsed in water and dried after each dip [133]. These films can consist of various materials, such as polymers or nanoparticles, depending on the desired properties of the superhydrophobic surface. The layering process allows precise control over the thickness, composition, and surface morphology of the deposited films. By carefully selecting materials with appropriate surface energies and roughness, the resulting surface can exhibit exceptional water repellency with high contact angles and low sliding angles.

The layer-by-layer deposition technique provides a flexible and scalable approach for tailoring superhydrophobic surfaces, making it applicable in various industries, including self-cleaning, anti-fouling, and water-resistant coatings [134–140]. However, the major disadvantage of this technique is that it is time-consuming and applicable to limited materials. Shao et al. [139] utilized a layer-by-layer assembly method to create a superhydrophobic coating on a wood surface. The process involved the sequential auto-deposition of polydopamine (PDA) and 1H,1H,2H,2H-perfluoro-decyl trichlorosilane (PFDTS). Initially, PDA was deposited on the wood surface, forming a highly cross-linked granular structure that enhanced the surface roughness of the wood. Subsequently, the wood surface underwent further modification with the low-surface-energy PFDTS, resulting in a superhydrophobic wood surface with a water contact angle (WCA) of 154° after the PDA deposition. The coating introduced protrusions and microgrooves through cell walls and cavities in the wood, avoiding the use of solid particles for surface roughening. The deposition of PDA on the wood surface led to the formation of natural micro/nanoparticles with a uniform distribution.

Syed et al. [140] introduced a layer-by-layer (LbL) spin-assembled coating composed of polyaniline–silica composite and tetramethylsilane functionalized silica nanoparticles ($PSC/TMS\text{-}SiO_2$) to achieve a synergistic effect of superhydrophobicity and enhanced anti-corrosion properties on stainless steel surfaces. Notably, the hierarchical integration of these two coating materials with distinct surface roughness and energy in a multilayer structure allowed the wetting feature to transition from a hydrophobic to a hydrophilic state by modulating the layer count (n) with decreasing hydrophilicity. Surfaces with an odd n (TMS–SiO$_2$ surface) exhibited hydrophobic characteristics, while those with an even n (PSC surface) displayed hydrophilic traits. The coating demonstrated superhydrophobicity and self-cleaning abilities with a high water contact angle (CA) of $153° \pm 2°$ and a minimal sliding angle (SA) of $6° \pm 2°$, attributed to the synergy of surface composition and roughness. Beyond its self-cleaning behavior, the coating exhibited significantly enhanced corrosion resistance against aggressive media, maintaining stability even after 240 h of exposure, attributed to superhydrophobicity and an anodic shift in corrosion potential. Table 3 shows the advantages and disadvantages of the superhydrophobic preparation process during practical production.

Table 3. Advantages and disadvantages of the superhydrophobic preparation process during practical production.

Fabrication Method	Principle	Advantages	Disadvantages
Chemical etching	Dissolution of surface layers using etchant solutions	• Simple, fast, cost-efficient, and scalable • Create durable complex micro/nanostructures	• Limited control over surface morphology • Applicable to metals and alloys mainly • Use of toxic solutions • Poor film uniformity
Lithography	Involves the transfer of a pattern from a mask to a substrate using light, radiation, or other forms of energy	• Simple and relatively fast • Precise control over surface features • Ability to create complex patterns • Environmentally friendly • Applicable to wide range of materials • Reusability of templates	• High initial equipment and setup costs, not scalable • Multi-step process; requires a flat substrate • May require cleanroom • Potential waste generation from resist and developer chemicals
Template	Using templates to create structured surfaces	• Low-equipment demand and harmless • Durable, well-defined structures • Scalable	• Template removal may be challenging • Limited flexibility
Plasma	Alteration of surface properties using plasma	• Simple, high aspect ratio structures • Applicable to a wide range of materials • Enhanced adhesion and versatile	• Decrease mechanical properties • Costly; potential toxic gas formation • Limited control over surface roughness
Sol–gel	Condensation polymerization reactions under colloidal liquid systems	• Cost-efficient and applicable to various substrates • Tunability of size and morphology of particles in film, Scalable • Synthesis at normal temperature • Provides high quality films; less deterioration	• Slow process; typically requires multiple processing steps • Limited durability • May require multiple steps
Hydrothermal	Dissolution; recrystallisation processes at high temperature and pressure	• Simple, cost-effective process • Environmentally friendly • Durable and scalable • Allows for the growth of hierarchical structures	• High equipment requirements • Limited to certain materials • Requires precise control of reaction parameters
Electrospinning	Droplet spraying and stretching in electric field	• Inexpensive and environmental-friendly • High superhydrophobic performance • High surface area; fine structures • Film homogeneity	• Non-durable; low fiber strength • Limited control of porosity; relatively slow • Limited to polymer fibers • Requires specialized equipment • Limited scalability
Chemical vapor deposition	Vapor-phase deposition of precursor chemicals	• Uniform coating; residue-free • Film homogeneity • Control over coating thickness and composition	• Complex equipment and high cost • Limited to specific substrates • Requires high temperatures and controlled environments
Electrochemical deposition	Deposition of material through electrochemical reaction	• Cost-efficient and scalable • Uniform coating; control over surface morphology • Tunability of texture morphology	• Limited material compatibility • Limited durability; complex process optimization • Possible toxicity
Layer-by-layer	Inter-particle electrostatic interaction	• Precise control of layer thickness • Cost-efficient and versatile • Multifunctionality; tunable surface properties	• Time-consuming, complex processes • May not be suitable for large surfaces • Limited scalability; potential for delamination • Sensitivity to environmental conditions

4. Applications

Superhydrophobic surfaces have made great progress in the past two decades. The global superhydrophobic coating market has been experiencing significant growth due to the increasing demand across various industries and is expected to witness tremendous revenue opportunities in the upcoming years. These coatings find applications in sectors like automotive, aerospace, electronics, textiles, and construction, among others. As of the most recent market analysis published by Straits Research, the global superhydrophobic coatings market was valued at approximately USD 19.5 million in 2021 and is expected to reach around USD 120 million in 2030, at a compound annual growth rate (CAGR) of 25.6% between 2022 and 2030, as shown in Figure 20a [141].

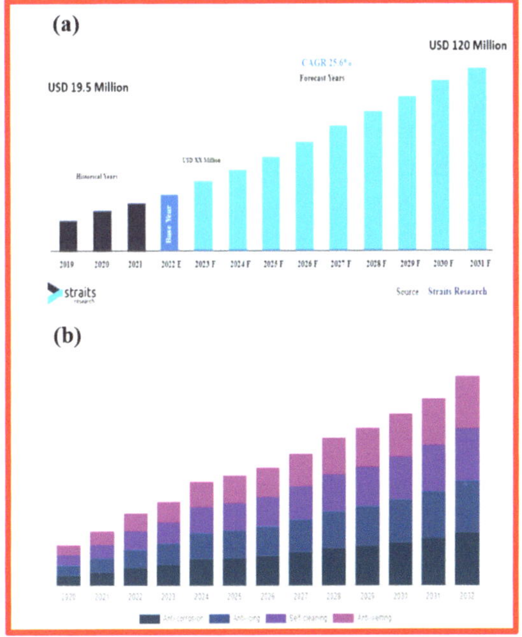

Figure 20. Commercialization status of superhydrophobic coatings: (a) global market value of superhydrophobic surfaces; (b) by property.

Superhydrophobic coatings are formulated to repel water and various liquids, providing attributes like self-cleaning, anti-corrosion, and anti-icing functionalities. In terms of product classification, the global market for superhydrophobic coatings is categorized into anti-corrosion, anti-icing, self-cleaning, and anti-wetting. The segment related to anti-wetting demonstrates the most significant market share and is projected to experience a CAGR of 26.1% throughout the forecast period, as illustrated in Figure 20b [141].

Much research has been carried out to discover methods for developing superhydrophobic surfaces that can be used in a variety of industries for commercialization [142–146]. Therefore, we primarily discuss the uses of superhydrophobic coatings for self-cleaning, anti-icing, anti-fouling, oil–water separation, anti-fogging, anti-corrosion, the medical industry, and other domains, as shown in Figure 21.

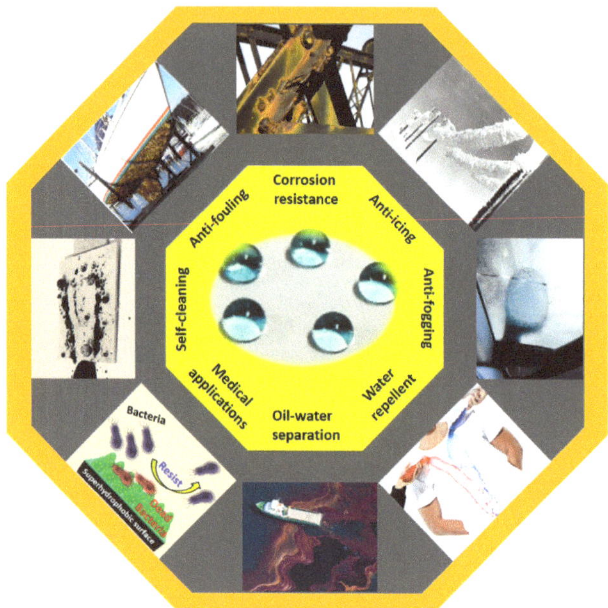

Figure 21. Nature-inspired artificial superhydrophobic surfaces with diverse applications.

4.1. Self-Cleaning

Self-cleaning materials are designed to be effortlessly cleansed with water, which offers diverse applications while reducing maintenance costs and having an environmental impact. As a result, self-cleaning surfaces are being extensively researched and commercialized for use in solar energy panels, paints, windows, satellite dishes, and windscreens for automobiles [108,147–151]. Textile materials that include self-cleaning properties become wash-free or require fewer washings, which makes them convenient for daily usage and adds value to the product. This phenomenon is known as the lotus effect, which was first identified by Neinhaus and Barthlott on the surface of lotus leaves and was later patented in 1998.

Superhydrophobic surfaces exhibit a distinctive set of characteristics, including high water contact angles (WCA > 150°) and low roll-off angles, facilitating the efficient self-cleaning mechanism, as shown in Figure 22a [151,152]. This self-cleaning ability is attributed to the surface's microstructures and nanostructures, which form a unique morphology. Upon contact with water droplets, the surface induces the formation of nearly spherical shapes, minimizing the contact area and enabling easy rolling off of the droplets. This process effectively removes dust, dirt, and contaminants from the surface. The microscale and nanoscale features not only contribute to reduced water adhesion but also enhance the self-cleaning effect. The Cassie–Baxter wetting state, characterized by trapped air pockets between the water droplet and the surface, further aids in minimizing contact and adhesion, solidifying the surface's impressive self-cleaning capabilities. Additionally, these surfaces exhibit stain-resistant properties due to their small contact areas and high contact angles [153].

Wang et al. [154] developed A-SiO$_2$/N-TiO$_2$@HDTMS coatings onto the substrate surface with a water contact angle (WCA) of 157.2° and contact angle hysteresis (CAH) of 2.7°, demonstrating prolonged outdoor self-cleaning functionality. Similarly, Xu et al. [155] employed a sustainable approach using plant-derived phytic acid to etch cotton fabric to enhance its surface roughness. Coated with thermosets derived from epoxidized soybean oil and then covered with stearic acid (STA), the modified fabric achieved a water contact angle of 156.3° and exhibited exceptional superhydrophobic qualities against liquid pollutants and solid dust.

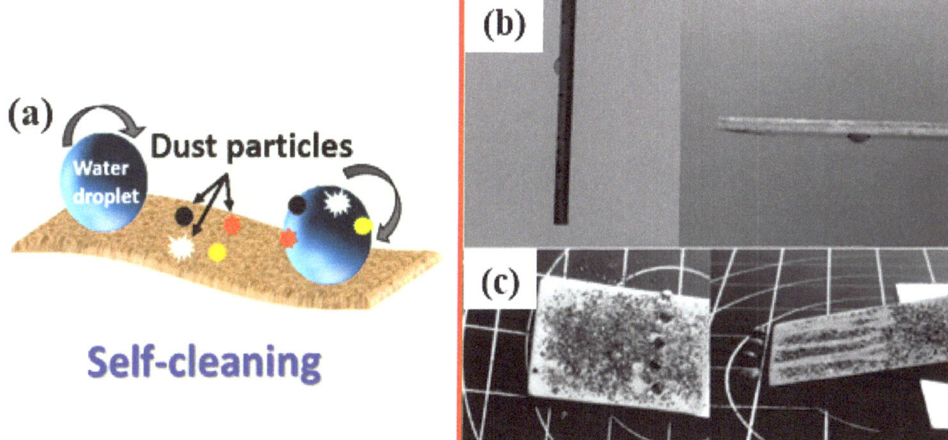

Figure 22. (**a**) Diagram illustrating the self-cleaning functionality of artificial superhydrophobic surfaces. A water droplet, featuring a notably high water contact angle (WCA), effortlessly rolls across the surface, effectively carrying away any adhered dirt or contaminants present on the superhydrophobic surface. The self-cleaning capability of aluminum alloy samples with (**b**) original sample (OS) and (**c**) binary structure (BS) obtained via a multi-step modification process [156].

Shao et al. [157] investigated a novel water-based superhydrophobic coating using polytetrafluoroethylene (PTFE) doped with graphene. The study involved mixing PTFE with graphene nanoparticles to create a PTFE–graphene composite, which was uniformly dispersed in aqueous fluorocarbon resin (FEM-101A-2-SFT) and applied to the substrate. The results revealed that graphene was evenly distributed on the PTFE surface, and the graphene–PTFE composite was immobilized on the coating surface through self-cross-linking fluorocarbon resin. This process generated numerous nanopulled structures on the coating surface, achieving a water contact angle of 153°. The resulting superhydrophobic coating exhibited excellent adhesion and corrosion resistance properties. After 40 cycles of mechanical wear and 4 h of water impact, the water static contact angle remained at 120° and 110°, respectively. The coating demonstrated 5B-level adhesive force according to ASTM standards and exhibited outstanding self-cleaning abilities. The self-cleaning effect was evident when dust was sprinkled on the surface, as water droplets rolled off easily, carrying away the dust and leaving the coating surface clean. This self-cleaning property is attributed to the nanoprotruding structure and low surface energy of the coating surface, preventing water permeation.

Similarly, the self-cleaning performance of aluminum alloy samples was investigated, revealing enhanced capabilities with micro/nanoscale binary structures [154]. As shown in Figure 22b, a water droplet was sprayed on the surface of the untreated sample. The water droplet could not slide down, not even if the sample was tilted to 90° or 180°. Whereas, water droplets on the sample with micro/nanoscale binary structures (BS) were found to slide down smoothly when the sample was tilted at an angle of 2.5° ± 0.7° and carried away brown alumina powders with them (Figure 22c). Table 4 summarizes past research on self-cleaning applications [158–162].

Table 4. Self-cleaning performance of some coated materials on different substrate.

Fabrication Technique	Coating Material	Substrate	Findings	Reference
Modification with HDTMS	Zn-MOF	Cotton fabric	The fabric coating exhibited a WCA of 160° with a hysteresis of 7°, showcasing stable superhydrophobic and self-cleaning characteristics.	[158]
Solvothermal and chemical modification method	Nano zinc sulfide (ZnS)	Zinc	Superhydrophobic ZnS coating has excellent chemical and physical self-cleaning properties.	[159]
Atmospheric pressure plasma polymerization	Hexamethyldisiloxane (HMDSO)	Glass	The thin films on a glass exhibit outstanding superhydrophobic and self-cleaning properties, featuring a WCA of 165° and an SA of about 2°.	[160]
One-step sol. immersion process in Mn (II) aqueous solution and post-modification by stearic acid	Manganese dioxide (MnO_2) microspheres	Mg alloy	Fabricated superhydrophobic magnesium alloy exhibits remarkable self-cleaning properties in both oil and air.	[161]
Dip coating method	Titanium dioxide nanomaterial	Glass	Fabricated superhydrophobic glass shows an excellent potential for self-cleaning action against contaminants.	[162]

4.2. Anti-Icing

The issue of icing on various surfaces poses significant challenges in several technological domains, leading to problems such as increased weight affecting electrical equipment, wind turbine blades, and airplane wings [163,164]. Even on hot days, air conditioners and refrigerators may encounter reduced efficiency when cooling and evaporation coils ice up. The electrothermal method has proven effective in accelerating ice melting. While salt is commonly used to melt snow and ice on roads during the winter, its high solubility poses environmental concerns, as approximately 90% of the salt may runoff, causing dehydration to aquatic life and vegetation along roadsides.

Physical mechanical deicing is the easiest method, but it is limited to easily accessible equipment. This approach becomes challenging when the ice-covered surface is not easily reachable [35]. Moreover, mechanical deicing is considered unsafe and unreliable due to the risk of damaging equipment during the deicing process. In recent times, researchers have directed their efforts towards creating materials characterized by low ice-adhesion strength. This property allows ice formed on these materials to be effortlessly removed either by its own weight or through the action of natural wind.

Zheng et al. [165] presented a method for surface processing titanium alloy using femtosecond laser technology, followed by dual-layer coating functionalization to create superhydrophobic surfaces resembling lotus leaves with enhanced anti-icing properties. Femtosecond laser treatment produced periodic microstructures on the titanium alloy surface. Subsequently, a dual-layer coating was sprayed on the surface, creating a low-surface-energy layer and nanostructure. The formed surface exhibited excellent superhydrophobicity, with an enhanced contact angle of 165° and a reduced sliding angle of 1.2°. Xiao et al. [166] propose a quick and easy method for creating anti-icing/deicing coatings using a two-step deposition process involving a Fe_3O_4@PDMS mixed liquor. This approach quickly produced a photothermal superhydrophobic coating, leveraging the excellent photothermal conversion properties of Fe_3O_4 particles to impart a micro/nanoscale rough structure to the coating surface. PDMS is used as a binder to increase the coating's durability while giving it low surface energy. Experimental results demonstrate that when exposed to simulated sunshine for 10 min, the surface temperature of the coating can reach 85 °C, effectively melt the frozen droplets within 270 s. With a water contact angle of 160°, the resulting coating exhibits outstanding superhydrophobicity and can delay icing time by up to 660 s. Whereas, by employing a one-step spraying method, superhydrophobic coatings composed of PTFE, Al_2O_3, and SiO_2 were successfully applied to glass slides [167]. Evaluation of experiments carried out in an artificial climate chamber (Figure 23) revealed that these three superhydrophobic coatings outperformed glass in resisting glaze icing. This enhanced performance is attributed to their exceptional superhydrophobicity, which facilitates the easy rolling away of water droplets from the surface, significantly reducing contact duration and minimizing the likelihood of freezing. Table 5 provides an overview of past research on anti-icing applications [168–172].

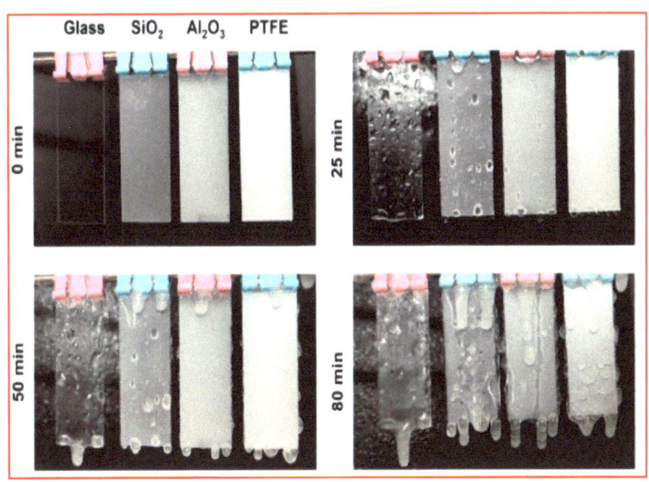

Figure 23. The glaze icing process on bare glass and three different coatings on a glass surface [167].

Table 5. Anti-icing performance of some coated material on different substrate.

Fabrication Technique	Coating Material	Substrate	Findings	Ref.
Laser ablation, followed by modification with HDTMS	Hexadecyltrimethoxysilane (HDTMS)	45 steel	• Superior anti-/de-icing properties; ice on surface thaws in 60 s under 0.5 sun illumination. • Exhibits a 2547s longer freezing delay time and at least ~3 times lower de-icing force than steel substrate.	[168]
Combination of simple chemical etching and anodization, along with modification with poly(dimethylsiloxane) (PDMS)	Silicone oil-infused PDMS (SOIP) coating	Aluminum (Al)	• SOIP coating displayed lower ice-adhesion strength of 22 ± 5 kPa compared to superhydrophobic coatings. • Surface has minimal ice-adhesion (108 kPa) after 20 icing/de-icing cycles, and long-term icephobicity (55 ± 13 kPa) after 4 months of exposure to ambient environment.	[169]
Simple spraying and curing process	Nanosilica co-modified with fluoroalkyl silane and aminosilane	Polyurethane	• Excellent anti-icing efficiency • Delaying water-freezing (700 s) at $-15°$. • Ice-adhesion strength lower than 13 KPa after 25 icing-deicing cycles	[170]
Chemical etching and anodization, later modified with PDMS via thermal vapor deposition.	PDMS	Aluminum	• WCA of fabricated surface more than 160 °C. • Superhydrophobic surface showed exceptional anti-icing properties • Ice formation on superhydrophobic surface was delayed by 45 and 80 min at -10 °C and -5 °C, respectively, at a relative humidity of $80\% \pm 5\%$.	[171]
Crystal growth method	Hollow micro-/nano-structured ZnO (HMN)	Silicon	• HMN ensures a significant thermal resistance between the base and liquid droplet, resulting in enduring anti-icing performance at lower temperatures.	[172]

Wang et al. [173] introduced a two-step spraying process for the preparation of plasmonic photothermal superhydrophobic coatings, incorporating MXene@Au hybrids with waterborne polyurethane. MXene (Ti3C2Tx) exhibits a localized surface plasmon (LSPR) enhancement effect and a broad absorption band [174,175]. Additionally, its good electrical conductivity allows applications in flexible electronic devices and soft robots [176]. Due to its exceptional optical properties and rapid heat transfer capabilities, MXene finds extensive use in light absorption and light-to-heat conversion devices [177,178]. To attain superhydrophobicity, chemically modified SiO_2 nanoparticles were applied to the MXene@Au-WPU layer, resulting in a fSiO_2/MXene@Au-WPU (fluoroalkyl silanes-SiO_2/MXene@Au-WPU) superhydrophobic photothermal coating with a contact angle of 153°. The composite coating proved effective for anti-icing and deicing applications, displaying an ultra-long anti-icing duration of 1053 s under low-temperature and high-humidity conditions (-20 °C, relative humidity 68%). Compared to previous studies, the coating also demonstrated an exceptionally high photothermal deicing efficiency of 73.1%. The coating's high photothermal properties facilitate rapid ice melting in the irradiated area, aided by the superhydrophobic characteristics that guide melted water to slide off. Additionally, the results show that the coating's resistance to corrosive liquids is within a pH range of 1 to 13.

4.3. Anti-Fogging

Fogging is the term used to describe the phenomenon wherein the temperature differential between the surface and the humid environment causes humid air to condense as discrete, small drops of water on optical surfaces. The dispersion of incident light by these water droplets restricts its transmission or reflection on solid surfaces, leading to blurred vision. Fog can accumulate on optical surfaces, including camera lenses, binoculars, swimming goggles, eyeglass lenses, and bathroom mirrors. Variables such as temperature, humidity, and airflow influence the occurrence of fogging. In addition to being annoying, fogging may cause additional problems in a variety of applications, including safety. Fogging, for example, might impair vision during endoscopic surgery and raise the possibility of a failure of the operation [179]. The formation of fog on moving car windscreens and motorcycle helmet visors is intimately associated with road safety. Additionally, fogging on greenhouse cladding materials reduces the amount of light that reaches the crops, which has an impact on agricultural productivity. Furthermore, fogging lowers solar panel efficiency. For various optical applications, it became highly desirable to eliminate or reduce the fogging phenomenon. To mitigate or eliminate fogging in optical applications, researchers commonly employ cost-effective and durable techniques such as applying coating layers to surfaces or modifying them chemically or physically [180,181].

Huang et al. [182] reported the use of diamond as an optical coating for challenging applications, like air force optical windows and offshore oil exploitation. After being treated with oxygen plasma, the diamond films became superhydrophilic, which caused the contact angle to decrease from 87° to less than 5° and resulted in antifogging activity. On the other hand, an ultrathin diamond coating on a quartz slide showed an oil contact angle and a di-chloromethane droplet contact angle of 153° and 157°, respectively, on the O_2-plasma-treated surface. Steaming and freezing tests demonstrated that the coated samples maintained sufficient transparency and anti-fogging activity. Notably, the uncoated sample took 3.5 min to evaporate, while the fog on the diamond thin film evaporated in just 4 s. Varshney and Mohapatra [183] discussed the fabrication of superhydrophobic coatings on brass surfaces using both two-step (chemical etching with a hydrochloric and nitric acid mixture, following a treatment with lauric acid) and one-step (treatment with lauric acid) methods. Treated brass surfaces exhibited rough microstructures, resulting in superhydrophobicity with water contact angles exceeding 173° and sliding angles below 4°. In addition to that, the formed coatings demonstrated anti-fogging and self-cleaning properties. In another approach, combined sol–gel and biotemplating techniques were used to fabricate bio-inspired anti-fogging and anti-reflection surfaces (BFRSs) with multiscale hierarchical columnar structures (MHCS) [184]. The formed surface exhibited rapid elimination of dispersed fog droplets within 6 s (Figure 24). The effective anti-fogging performance of the surface was attributed to the capillary force imbalance of the MHCS acting on the liquid film.

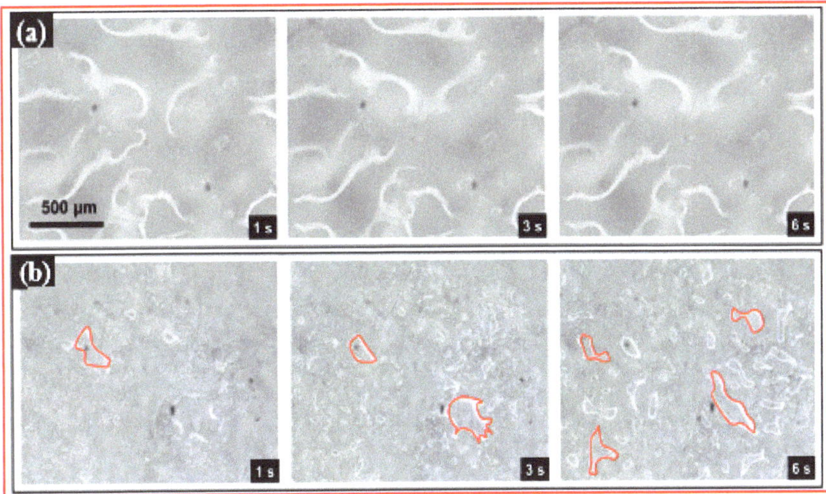

Figure 24. Anti-fog test on the surface of a (**a**) flat plate and (**b**) BFRSs [184].

4.4. Oil–Water Separation

The rapid pace of industrialization, global population growth, and the expansion of the worldwide economy have resulted in the contamination of water, which has emerged as a significant and escalating environmental issue. Oily wastewater is a critical form of water pollution and is commonly produced by industries, oil spills, automotive transportation, and domestic sewage [185,186]. The excess presence of oils poses a severe threat to the environment, which adversely affects aquatic life such as fish, animals, and birds, making them more susceptible to hypothermia and negatively impacting plant growth. Recently, on 25 July 2020, the Japanese-operated MV Wakashio bulk carrier ship collided with a coral reef on the island of Mauritius, causing severe damage to the rich marine ecosystem.

Various methods were reported for the removal of oil and grease from wastewater, including dispersion [187,188], adsorption [189,190], flotation [191,192], biological treatment [193,194], and burning [195]. However, these approaches may come with drawbacks such as low stability, inefficiency, high energy consumption, expensive chemicals, and the potential for secondary pollution. Desirable oil–water emulsion separation technologies should possess characteristics such as excellent stability, cost-effectiveness, easy operation, high separation efficiency, substantial permeation, and minimal secondary contamination throughout the synthesis and separation processes [196].

In response to such challenges, various functional materials with superhydrophobic and superoleophilic properties have undergone extensive study for efficient oil–water separation filtration. These materials include meshes, membranes, sponges, foams, and fabrics. The operational principle of superhydrophobic surfaces in oil–water separation encompasses the synergistic effects of superhydrophobicity, oleophilicity, and distinctive surface structures. This combination enables the selective repulsion of water, the attraction and retention of oil, and the facilitation of a self-cleaning mechanism, ensuring a highly effective and environmentally sustainable separation process, as shown in Figure 25a. Cai et al. [197] utilized natural balsa to create a highly flexible and durable superhydrophobic polydivinylbenzene (PDVB)-wood membrane. The membrane features hydrophobic nanopores achieved by coating porous wood with cross-linked PDVB. In air, the PDVB-wood membrane exhibits unique oil wettability with a $0°$ oil contact angle. The membrane also demonstrates exceptional WCA exceeding 160° and a low WSA of 3.5°. Remarkably, water droplets exhibit bouncing behavior when quickly dropped onto the PDVB–wood membrane. The superhydrophobic properties endure rigorous mechanical and chemical tests, including sandpaper rubbings, tape sticking, and exposure to acid, alkali, and salt solutions, highlighting their durability and stability. With a separation efficiency exceeding 99.98% for surfactant-stabilized water-in-oil emulsions and a high flux of up to 8829.4 L m^{-2} h^{-1}bar^{-1}, the membrane maintains high efficiency even after 20 separation cycles. The remarkable separation performance of the PDVB-wood membrane is attributed to the synergistic effects of superhydrophobicity and nanopores. By combining superhydrophobicity/superoleophilicity with nanoscale pores, the membrane prevents micro/nano-scale water droplets in water-in-oil emulsions from penetrating while allowing free penetration of oil due to its superoleophilicity. This achievement fulfills the demulsification and separation objectives.

Jie et al. [198] used an electrodeposition approach to prepare micro/nano superhydrophobic structures on a copper mesh surface, employing choline chloride/ethylene glycol ionic liquid as an electrolyte. The contact angle of the electrodeposited copper mesh reached 152°, showing superior performance compared to pure copper mesh. Demonstrating a separation efficiency exceeding 95%, the superhydrophobic copper mesh exhibited exceptional oil–water separation capabilities across various oil types. In another study, a simple spraying method was used to prepare a superhydrophobic TiO$_2$/SiO$_2$ nanoparticle coating on sponge, loofah, and metal mesh surfaces. The coating exhibited a surface with convex nano-nipple and nano-scale pore structures, providing microstructural evidence of its superhydrophobic properties, with a water contact angle (WCA) of $\geq$155° and an oil contact angle (OCA) close to 0°. Experimental investigations involving different oil–water mixtures revealed an oil–water separation effectiveness of approximately 95% for the superhydrophobic coating on various substrates. Remarkably, even after 60 separation cycles, the coating maintained a separation efficiency above 90%, indicating its enhanced durability [199]. Kao et al. [200] used a simple painting process to produce a new micro/nanostructure by coating ethylenediaminetetraacetic acid/poly(dimethylsiloxane)/fluorinated SiO$_2$ (EDTA/PDMS/F-SiO$_2$) on a textile surface. The result shows the high oil–water separation efficiency of the superhydrophobic EPS@textile, even in the absence of external force (Figure 25b). Table 6 summarizes past research on oil–water separation applications [158,160,201–203].

Table 6. Oil–water separation performance of some coated material on different substrates.

Fabrication Technique	Coating Material	Substrate	Findings	Ref.
Modification with HDTMS	Zn-MOF	Cotton fabric	Superhydrophobic surface showed exceptional efficiency of 93, 95, 97, 98%, and 100% in separating engine oil, crude oil, n-hexane, chloroform from water, and viscose oils (sunflower and coconut), and 90% for stabilized emulsion.	[158]
Three simple processes: (1) synthesis of MOF-5 nanoparticles; (2) modification using PFOTS; and (3) dip-coating method	Zinc-based metal–organic frameworks (MOF-5)	Sponge	Continuous separation of a variety of oil–organic solvent–water mixtures with a separation efficiency >98%.	[201]
Plasma polymerization	Hexamethyldisiloxane (HMDSO)	Fabric and offset printing paper	Superhydrophobic coating has an excellent separation efficiency with the oil contact angle (OCA) of about 0° under the optimal working conditions.	[160]
Combining chemical etching and hydrothermal processes	PDMS	Aluminum	Superhydrophobic/superoleophilic mesh shows high separation efficiency (94%) and outstanding reusability.	[202]
Dip-coating method	MWCNTs/ZnO composite	Copper mesh	Superhydrophobic and superoleophilic mesh exhibits remarkable reusability and excellent separation efficiency of over 95% for various oils.	[203]

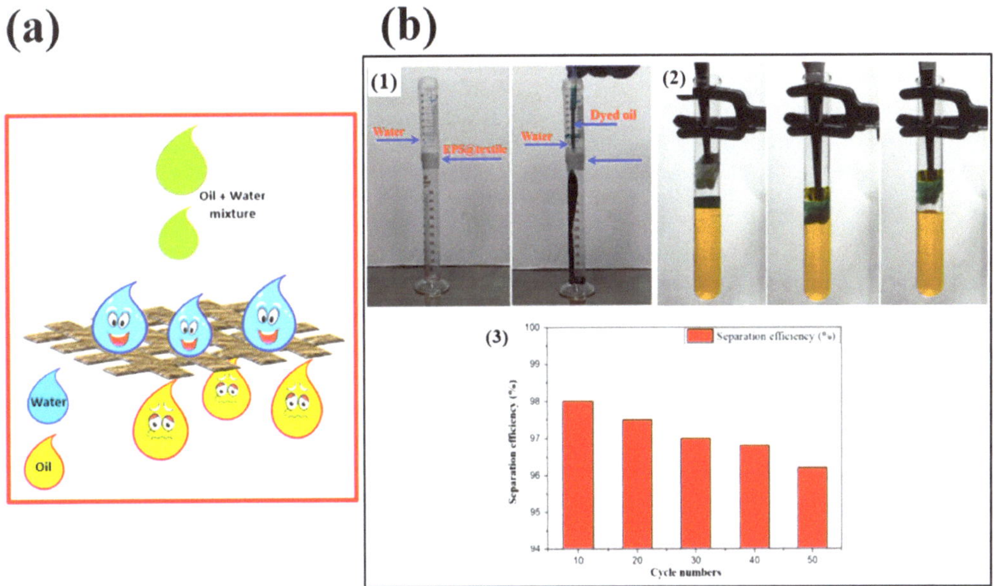

Figure 25. (**a**) Oil–water separation mechanism of superhydrophobic surfaces; (**b**) picture of oil–water separation using an EPS@textile-coated fabric as a (1) filtering membrane; (2) adsorbent; and (3) separation efficiency of the coated fabric [200].

4.5. Anti-Fouling

The accumulation of undesired substances on different surfaces submerged in water is known as fouling. Both inorganic and organic substances, as well as living organisms (biofouling), can be considered fouling materials. Biofouling is a major problem that causes a lot of problems in the maritime environment. Seawater can become contaminated by coatings formed by non-target species and fouling, which provide severe toxicity concerns for marine life. It is now necessary to develop environmentally friendly anti-fouling solutions due to the increasing significance of environmental protection. Researchers are actively exploring biomimetic antifouling coatings, which mimic the micro-structured surfaces found in marine life, as a way to solve the environmental issues related to traditional coatings [204]. Microbial biofouling poses a significant threat to numerous other environmentally related systems, such as water treatment and distribution systems, heat exchangers, and many more. These systems are susceptible to several issues caused by biofouling, including reduced heat transfer efficiency, blockages in pipes, power failures, high maintenance costs, and serious mishaps. As a result, it is preferable to create surfaces with anti-fouling properties using an easy and affordable technique.

Yin et al. [205] introduced an economical and fluorine-free technique for coating superhydrophobic Ni_3S_2 on 304 stainless steel. The resulting superhydrophobic coating significantly impeded water intrusion and exhibited remarkable resilience in maintaining its superhydrophobicity even after exposure to conditions such as heating up to 300 °C or prolonged submersion in ethanol and undergoing five cycles of O_2 plasma etching with heating treatment. Anti-fouling tests validated the effectiveness of the superhydrophobic coating in forming a protective barrier against contamination on the steel surface. In a separate study, Shi et al. [206] developed a superhydrophobic and antibacterial membrane using underwater adhesion technology for directly immobilizing it on substrates in seawater. The membranes displayed excellent hydrophobicity, mechanical strength, and flexibility, boasting a water contact angle (WCA) of 161.3° and a sliding angle (SA) of 4.9°. These qualities remained stable even after 30 days of submersion in seawater, affirming their prolonged service life.

This new approach has general relevance in ocean engineering and may be applied to different substrates to generate antifouling coatings in seawater. Furthermore, research was conducted on the coated fabric surface's ability to inhibit the growth of rhodamine B liquid pollutants [200]. Figure 26a illustrates that the superhydrophobic textile remained remarkably clean, contrasting with the evident pollution of the non-superhydrophobic textile, as shown in Figure 26b. This finding shows the outstanding antifouling property of fabricated superhydrophobic EPS@textile. The unique combination of superhydrophobic characteristics and high hygroscopicity of the uncoated fabric prevented the colored water from permeating the coated superhydrophobic fabric by trapping air.

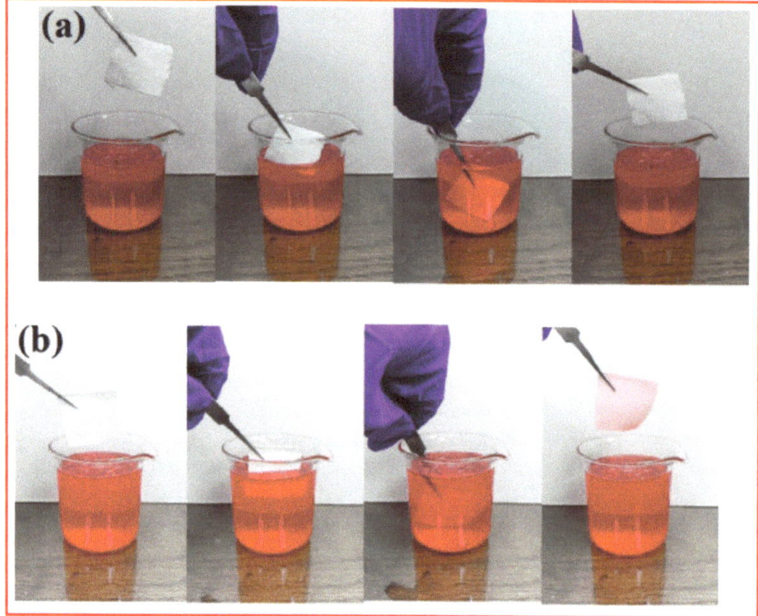

Figure 26. Picture showing the antifouling performance of (**a**) EPS@textile and (**b**) bare textile against water [200].

4.6. Anti-Corrosion

Corrosion refers to the deterioration of a metal surface due to chemical or electro-chemical reactions with the environment. The practical utility of metals and metal alloys is constrained by their susceptibility to corrosion, resulting in substantial financial losses and environmental pollution. Consequently, extensive research efforts, supported by significant funding, are underway to develop anti-corrosive materials. Although chromium-containing compounds are conventionally used for corrosion prevention, their negative impact on both human health and the environment prompts the exploration of alternative methods. One such approach involves directly creating superhydrophobic coatings on metal surfaces to enhance their anti-corrosion properties [207]. This approach utilizes the creation of an air layer between the structured superhydrophobic surface and the solution. These trapped air pockets act as a protective barrier, effectively preventing the corrosive medium from reaching the surface, as depicted in Figure 27.

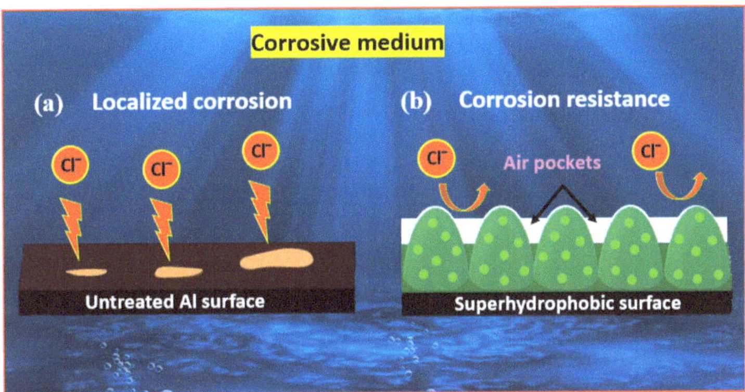

Figure 27. The anti−corrosion mechanism of the superhydrophobic surfaces.

Superhydrophobic coatings have become widely used to enhance the corrosion resistance of various surfaces, including steel, alloys (Al, Cu, Zn, and Fe), and titanium (Ti). Huang et al. [208] introduced an environmentally friendly surfactant-free nanoprecipitation method to fabricate a nanocomposite coating based on polyaniline-titanium oxide (PANI-TiO$_2$). Incorporating two sizes of TiO$_2$ nanoparticles enhanced the coating's hydrophobicity and conductivity. Compared to traditional epoxy coatings, the resulting coating with hierarchical micro/nanostructures demonstrated higher water contact angles (>150°) and superior anti-corrosion properties.

Microplastics are commonly found in saline water, emphasizing the need to study the anticorrosion properties of surfaces in such environments. Rius-Ayra et al. [209] introduced a robust, non-fluorinated superhydrophobic surface through an anodizing process and liquid-phase deposition (LPD) of lauric acid for surface functionalization and superwettable properties. This combined approach resulted in outstanding anticorrosion performance of the metallic substrate in NaCl aqueous solution. The hierarchically structured superhydrophobic surface achieved a water contact angle (WCA) of 154°, a sliding angle (SA) of 1°, and a contact angle hysteresis (CAH) of 1°, effectively removing microplastics from saline water. The anodized aluminum surface demonstrated exceptional anticorrosion capabilities in an aqueous solution with 3.5 wt% NaCl, further enhanced by superhydrophobic characteristics after 60 min of anodization. The functionalized surface exhibited superoleophilicity (0°) and superhydrophobicity (154°), successfully extracting microplastics from the NaCl aqueous solution with an efficiency exceeding 99%.

Similarly, Zhang et al. [210] employed a chemical grafting process to synthesize superhydrophobic SiC, where fluoroalkyl silane (FAS) was chemically bonded to the SiC surface. This superhydrophobic SiC was then integrated into epoxy resin (EP) to create a SiC/EP composite material. The organic molecule grafting on the SiC surface significantly improved the compatibility between nano-SiC and EP. The surface wettability of the coating changed with the addition of F–SiC; without F–SiC, the contact angle was less than 90°, but it increased significantly with F–SiC addition, as shown in Figure 28b. A superhydrophobic composite coating was achieved at 5 wt% F–SiC, reaching a static water contact angle of 150.1° with a roll-off angle of 5.5°. The corrosion resistance was optimal at a 3 wt% addition of F–SiC. The corrosion current of the composite coating was 2–3 orders of magnitude lower than that of the pure EP coating, signifying substantial improvement in corrosion resistance (Figure 28c). Additionally, a 240 h salt spray test evaluated various compositions of composite coatings (EP, 1–5 wt% F-SiC/EP), as depicted in Figure 28d. Results indicated that incorporating an appropriate amount of F-SiC enhanced the coating's insulation capability, with optimal anti-corrosion performance observed in the F-SiC/EP composite coating at approximately 3 wt%. Table 7 summarizes previous research on anti-corrosion applications [168,211–214].

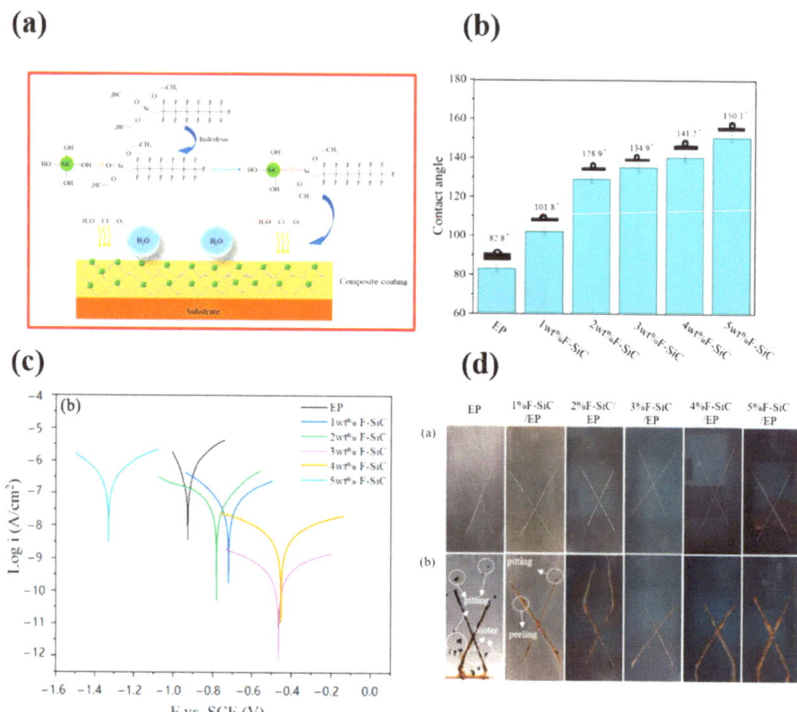

Figure 28. (**a**) Fabrication process for F−SiC and composite coating; (**b**) variation in the WCA of composite coatings with different F−SiC contents; (**c**) potential polarization test comparing different coatings; and (**d**) images depicting salt spray tests on different coatings (**a**) before the test and (**b**) after 240 h of test duration [210].

Table 7. Anti-corrosion performance of some coated materials on different substrates.

Fabrication Technique	Coating Material	Substrate	Findings	Ref.
Pulse laser ablation, followed by modification with HDTMS	Hexadecyltrimethoxysilane (HDTMS)	45 steel	Surface shows superior corrosion resistance compared to steel, with 73.81 times higher charge transfer resistance and 64.78 times lower corrosion current density.	[168]
Combination of chemical etching with hydrothermal process, followed by PDMS coating via a simple vapor deposition method	Polydimethylsiloxane (PDMS)	Aluminum alloy	Corrosion resistance of bare surface is significantly enhanced by a magnitude of three after becoming superhydrophobic surface.	[211]
Hydrothermal method, with modification by sodium laurate (SL) and sodium dodecylbenzene sulfonate (SDBS).	MgAl-LDH laminates	AZ31 alloy	In a 3.5 wt.% NaCl solution, functional coatings demonstrated remarkable anti-corrosion performance.	[212]
Sandblasting and acid treatment, followed by electrodeposition and hydrophobic modification.	Ni-W-TiO$_2$ coating	Steel	Superhydrophobic composite coating can attain a corrosion inhibition rate of 99.63% in 3.5% NaCl solution and at 25 °C.	[213]
Eco-friendly green method	Lauric acid	Concrete	Superhydrophobic concrete (WCA > 153° and WSA < 10°) has better corrosion resistance to internal rebars than ordinary concrete.	[214]

4.7. Anti-Bacterial Property

Bacterial colonization on surfaces poses significant challenges in various industries, healthcare settings, and everyday life. The proliferation of bacteria on surfaces can lead to the formation of biofilms, which are resilient and difficult to remove. In healthcare, bacterial contamination in hospitals can result in healthcare-associated infections (HAIs), posing a serious threat to patients with compromised immune systems. Similarly, in food processing and preparation areas, bacterial contamination can lead to foodborne illnesses. The growth of bacteria on surfaces is not only a hygiene concern but also impacts the durability and performance of materials, especially in outdoor settings where weathering and microbial attack can deteriorate surfaces over time. Consequently, finding effective solutions to mitigate bacterial growth on surfaces is crucial for maintaining public health, ensuring product integrity, and extending the lifespan of materials.

One promising approach to addressing the bacteria problem is the development of superhydrophobic anti-bacterial coatings. Superhydrophobic surfaces exhibit exceptional water repellency, preventing the adhesion and proliferation of bacteria and other microorganisms. These coatings combine the benefits of superhydrophobicity, which repels water and contaminants, with anti-bacterial properties to actively inhibit bacterial growth [215–219]. By incorporating antimicrobial agents or materials with inherent antibacterial properties into the coating formulations, researchers aim to create surfaces that not only repel water but also prevent bacterial colonization. This dual functionality makes superhydrophobic anti-bacterial coatings highly desirable for a wide range of applications, from healthcare settings and food processing facilities to everyday surfaces, ultimately contributing to improved public health, reduced maintenance costs, and enhanced material durability. Ye et al. [220] presented an environmentally friendly nanofibrillated cellulose-based multifunctional superhydrophobic coating (NMSC) through a silylation process involving cellulose, tetraethyl orthosilicate, and cetyl trimethoxysilane. Ethyl orthosilicate hydrolyzed into organosilicon, facilitated by ammonia water, imparted surface roughness to NFC. Hexadecyltrimethoxysilane hydrolysis lowered NFC surface energy. Post-silylation, the water contact angle significantly increased from 42° to 169°, showcasing remarkable hydrophobic, thermal stability, and self-cleaning properties. The NMSC demonstrated prolonged effectiveness against various fluids, including strong acid (pH 1) and alkali (pH 13), alcohols, alkanes, esters, and organic solvents. It maintained static contact angles above 155° during acid/alkali treatments, highlighting exceptional acid and alkali resistance. The NMSC exhibited antibacterial performance, attributed to its hydrophobic surface, which readily penetrated the phospholipid bilayers of bacteria, disrupting bacterial cell membranes and causing structural damage and cytoplasmic leakage.

Agbe et al. [221] illustrated a straightforward two-step procedure for creating a superhydrophobic and antibacterial aluminum surface. The process involves chemical etching in HCl, followed by immersion in an ethanolic solution of octyltriethoxysilane (OTES) and the addition of quaternary ammonium solution (QUATs) through drop-wise deposition. The chemical etching introduces micro- and nano-features with topological terraces, resulting in a surface root mean square (rms) roughness and contact angle (CA) of 6.2 ± 1.5 µm and $16° \pm 0.2°$, respectively. Post-modification, the OTES-QUATs/Al samples exhibited a reduced roughness of 5.8 ± 0.5 µm and an increased CA of $153° \pm 3.7°$, as depicted in Figure 29a–d. The antibacterial activity of the OTES-QUATs solution was noteworthy, displaying a zone of inhibition (ZOI) of 34 ± 1.6, 22 ± 1.4, and 25 ± 0.9 against *Staphylococcus aureus*, *Pseudomonas aeruginosa*, and *Escherichia coli*, respectively. The ZOI indicates the region around the antimicrobial agent where microbial growth is hindered, with a larger ZOI signifying more effective antimicrobial action. Furthermore, the OTES-QUATs-coated aluminum surface demonstrated excellent anti-biofouling properties, achieving a 99.9% reduction in bacterial adhesion for *Staphylococcus aureus*, 99% for *Pseudomonas aeruginosa*, and 99% for *E. coli* bacteria, attributed to the synergistic effects of low-energy OTES, micro/nano roughness, and the presence of QUATs, as depicted in Figure 29.

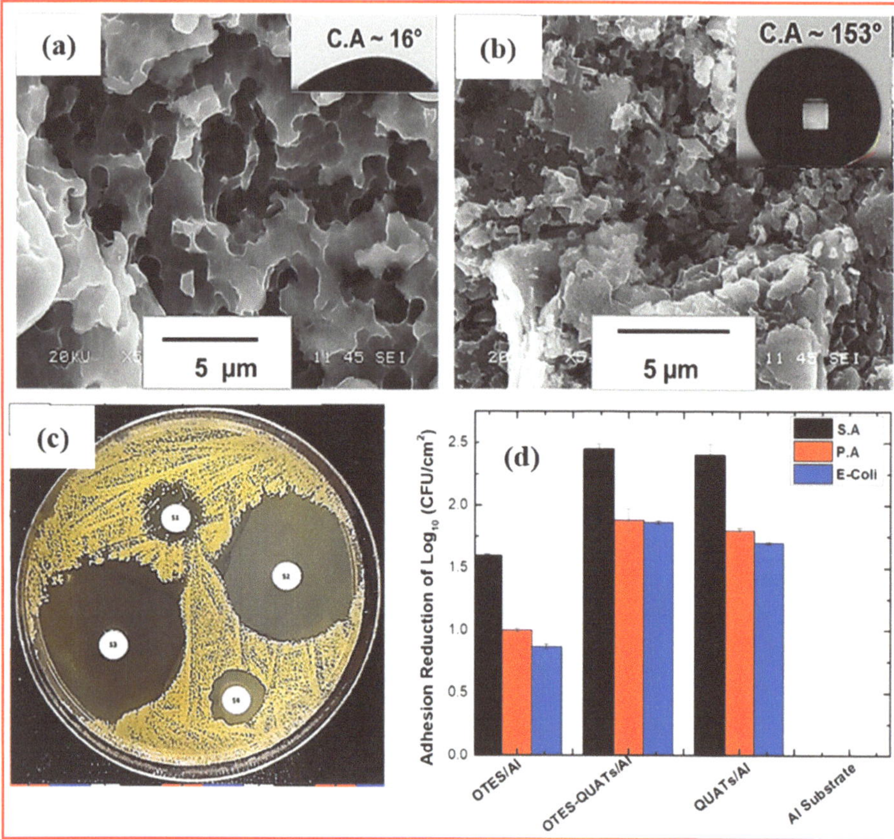

Figure 29. The surface morphology of the etched aluminum substrate (**a**) and OTES-QUATs/Al (**b**). Ethanoic solutions of OTES, OTES-QUATs, QUATs, and ethanol were subjected to disk diffusion assay against *Staphylococcus aureus* bacteria, with distinct regions labeled as S1, S2, S3, and S4, respectively (**c**). A graphical representation depicts the adhesion reduction of *Staphylococcus aureus*, *Pseudomonas aeruginosa*, and *Escherichia coli* on various surfaces, including OTES/Al, OTES-QUATs/Al, QUATs/Al, and etched Al substrate (**d**) [221].

4.8. Water Harvesting

Water is a vital resource for all living organisms, and effective water harvesting plays a crucial role in sustainable water management, especially in regions dealing with water scarcity. Inspired by nature, harvesting water directly from the atmosphere emerges as a promising alternative, particularly in dry areas and regions with abundant fog. Nature provides various examples of efficient water harvesting mechanisms in different organisms and ecosystems, such as cactus spines, spider silk, the elytra of the Namib Desert beetle, and the inner wall of nepenthes [222–225].

The Namib Desert beetle exhibits remarkable water collection and self-transportation capabilities attributed to the surface energy gradient resulting from dual wettability on its elytra surface-wax-coated hydrophobic valleys and hydrophilic bumps. Fog-derived water droplets preferentially condense on hydrophilic bumps, and a surface energy gradient facilitates the transportation of water droplets from hydrophobic valleys to hydrophilic bumps. This nature-inspired mechanism demonstrates a potential solution for efficient water harvesting and transport [226–230].

Taking inspiration from the water-harvesting properties of the Namib Desert beetle, Zhai et al. [231] developed hydrophilic spots, each measuring 750 μm, on a superhydrophobic surface using a poly(acrylic acid) (PAA) water/2-propanol solution. In the patterned region, the advancing water contact angle was 144°, while the receding contact angle was 12°. These hydrophilic patterns emulate the wax-free areas on the Stenocara beetle's back, where tiny water droplets from fog collect. As most droplets roll on the superhydrophobic regions, they eventually adhere to the hydrophilic patterns, forming larger water droplets. This mimics the water-capturing ability of the Stenocara beetle's back, providing a method for capturing small water droplets and converting them into more substantial droplets.

Similarly, Park et al. [232] explored fog harvesting using a flexible hybrid surface featuring a 3D superhydrophilic copper oxide (CuO) pattern on a hydrophobic, rough polydimethylsiloxane (PDMS) background. This design drew inspiration from the bumps found on the curved dorsal surface of Namib desert beetles, as shown in Figure 30a. The process involved transferring a copper (Cu) layer from a silicon (Si) or glass donor substrate to a PDMS receiving substrate, molded on, and then peeled from the donor substrate. Following the transfer, the Cu layer was patterned and oxidized, forming a superhydrophilic CuO pattern on the hydrophobic, rough PDMS substrate. Consequently, the transferred and oxidized Cu layer on the PDMS substrate could exhibit 2D or 3D patterns, depending on the initial morphology of the donor substrate, as depicted in Figure 30b. Evaluation of the water collection rates for the prepared surface indicated that the 3D bumpy structures on a curved surface showed over 16 times higher water collection rates than the flat 2D hybrid surface when subjected to a fog stream.

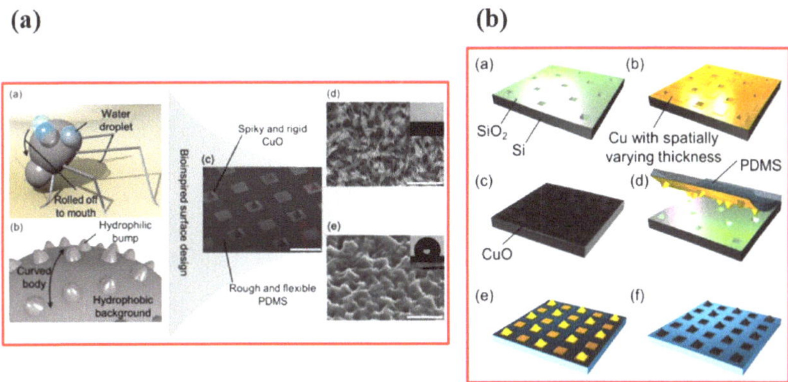

Figure 30. (**a**) A schematic overview of the fog-harvesting Namib Desert beetle and an SEM image of the fabricated flexible hybrid surface; (**b**) a schematic illustration of the fabrication process of the hybrid surface with a 3D superhydrophilic copper oxide pattern on a hydrophobic PDMS surface [232].

Recently, Choi et al. [233] presented a hierarchically structured surface with microgrooved patterns achieved through FFF-type 3D printing. The design aimed to replicate the morphology of cactus spines and the dual wettability observed in the Namib Desert beetle. Microgrooved patterns resembling those of cactus spines were incorporated onto the cactus-shaped PDMS surface using the staircase effect inherent in FFF-type 3D printing. To mimic the dual wettability of Namib Desert beetles, a partial Pt deposition method was introduced, employing a paraffin wax-based masking method for mass-producing complex morphologies. Capitalizing on the surface energy gradient, the hydrophobic region could condense more water droplets than the hydrophilic region, yet the condensed water in the hydrophobic region could autonomously transport itself to the hydrophilic region.

Additionally, the cactus spine morphology facilitated the self-transportation of condensed water droplets from the tip to the base region using the Laplace pressure gradient.

The resulting surface demonstrated superior fog harvesting performance (average weight of 7.85 g for 10 min), benefiting from the synergistic interplay between the Laplace pressure gradient and surface energy gradient.

These findings offer valuable insights for the future design and deployment of hybrid superhydrophobic/superhydrophilic surfaces for cost-efficient atmospheric water harvesting, supporting effective fog harvesting to obtain freshwater under harsh conditions, including dry and polluted water environments.

4.9. Medical Industry

Superhydrophobic polymeric nanocoatings find diverse applications in the medical field, including dentistry, self-cleaning, and drug delivery [234]. Sun et al. [235] innovatively designed a cardiopulmonary bypass tube with superior blood repellency and superhydrophobicity. The results revealed that the superhydrophobic-treated tube clotted in 36 min in terms of coagulation time, biotoxicity, platelet adsorption, and protein, in contrast to a clinical Bioline heparin-coated tube (21 min) and a bare PVC tube (14 min). The superhydrophobic-treated tube demonstrated a remarkable 157% increase in clotting time compared to the bare PVC tube. Additionally, protein and platelet adsorption on the superhydrophobic-treated tube witnessed reductions of 32% and 74%, respectively. Figure 31 illustrates the application of superhydrophobic surfaces in medical devices to prevent blood stickiness [29]. In addressing microleakage in dental composite restorations, superhydrophobic coatings were developed using photo-crosslinked polyurethane (PU) and SiO_2 nanoparticles functionalized with organic fluoro groups (F-SiO$_2$ NPs). The study revealed that a low concentration ratio of PU/F-SiO$_2$ (1:3) in superhydrophobic coatings exhibited desirable attributes, including good transparency, a high contact angle (160.1°), a low sliding angle (<1°), and an excellent hierarchical papillae structure [236].

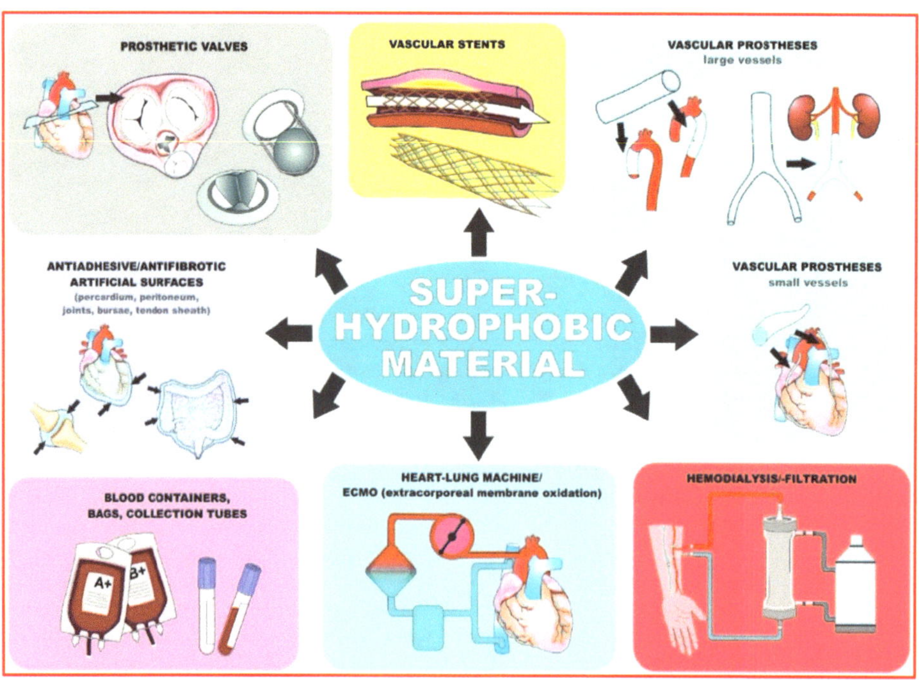

Figure 31. Superhydrophobic coatings for blood-repellent applications in the medical field [29].

While the potential applications are promising, challenges such as long-term durability, scalability, and cost-effectiveness need to be addressed for successful integration into various industries. Ongoing research in materials science is likely to uncover new possibilities, and staying abreast of the latest developments in the field is essential for a comprehensive understanding of the future of superhydrophobic materials.

5. Challenges and Constraints in Superhydrophobic Surfaces

While superhydrophobic surfaces present promising opportunities across various applications, they also encounter several challenges that need to be addressed for their widespread adoption and practical implementation. These challenges encompass both fundamental aspects and practical considerations and are crucial for advancing the field and realizing the full potential of superhydrophobic materials. Several key challenges are worth highlighting:

5.1. Durability and Stability

To enhance the practical usability of artificial superhydrophobic coatings, they must exhibit strong adhesion as well as robust mechanical and chemical stability. Generally, the long-term stability and durability of superhydrophobic surfaces pose significant challenges. External factors such as abrasion, chemical exposure, and environmental conditions can impact the integrity of these surfaces over time. Developing robust materials and fabrication techniques that withstand prolonged usage remains a critical challenge.

5.2. Scalability and Cost-Effectiveness

Many fabrication methods for superhydrophobic surfaces are complex and may not be easily scalable for large-scale industrial applications. The cost of materials and manufacturing processes can pose economic challenges, hindering the widespread adoption of superhydrophobic technologies. Ensuring cost-effectiveness and scalability while maintaining the desired properties is essential for widespread adoption in various industries.

5.3. Biocompatibility and Health Concerns

In biomedical applications, ensuring the biocompatibility of superhydrophobic materials emerges as a critical consideration. Biocompatibility involves a comprehensive assessment of the interactions between living organisms and these superhydrophobic materials. Particularly in medical applications, such as implants or medical devices, it becomes imperative to scrutinize how these superhydrophobic surfaces interact with biological tissues and fluids. The incorporation of certain materials in superhydrophobic coatings, especially those containing nanoparticles or specific chemical compounds, raises potential health concerns. The release of nanoparticles into the body may lead to long-term health impacts. Achieving a delicate equilibrium between attaining superhydrophobic properties and ensuring biocompatibility stands as a crucial factor for the responsible and successful integration of these surfaces across various fields without compromising the well-being of both humans and the environment.

5.4. Multifunctional Superhydrophobic Coating/Surfaces

Multifunctional superhydrophobic surfaces represent a cutting-edge area of research that aims to develop materials with a range of enhanced properties beyond water repellency. These surfaces combine superhydrophobicity with diverse functionalities, broadening their potential applications across various fields, as shown in Figure 32. Despite the advantages of multifunctional superhydrophobic surfaces in many applications, they do present certain limitations. For instance, achieving robust self-cleaning capabilities may compromise other aspects, such as mechanical durability or chemical stability. Additionally, the integration of multifunctionality often involves complex surface structures and coatings, making fabrication and scalability more intricate. Concerns also arise regarding the durability of these surfaces under harsh environmental conditions or during prolonged use. Addressing these limitations is crucial to propelling the practical and widespread adoption of superhydrophobic surfaces across diverse applications.

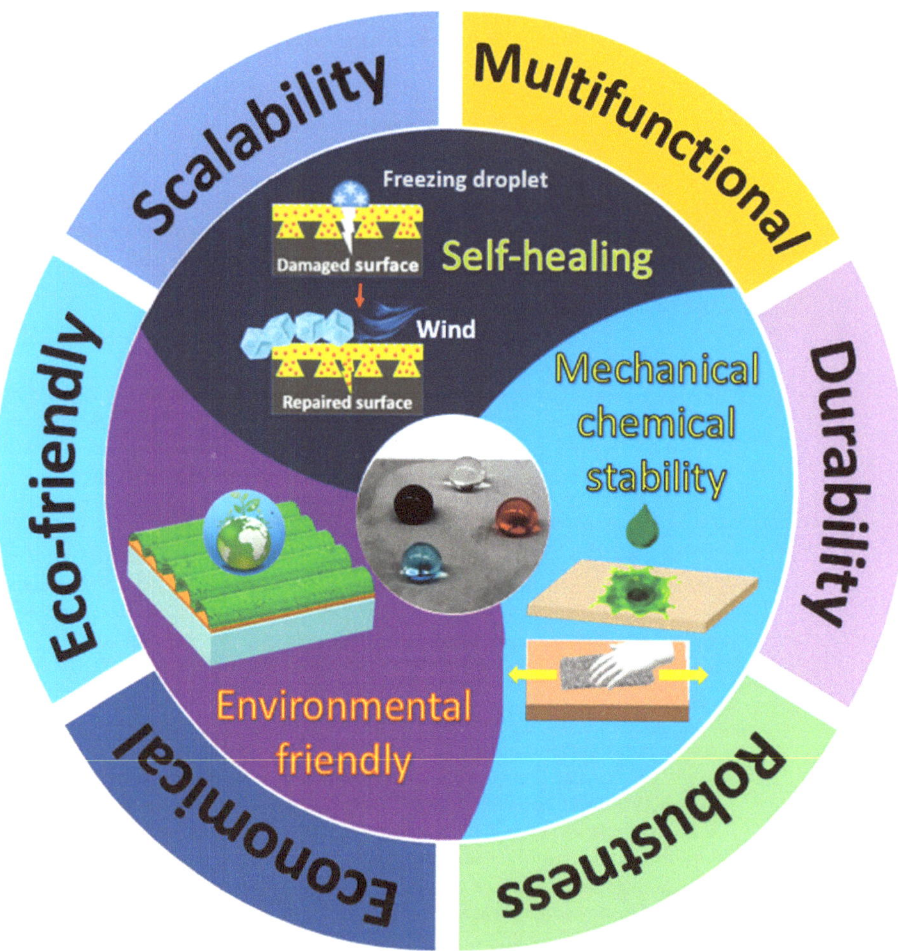

Figure 32. Bio-inspired artificial superhydrophobic surfaces with multifunctional properties.

5.5. Environmental Impact

The environmental implications of superhydrophobic materials, particularly concerning their production and the potential release of nanostructures into the environment, have raised significant concerns. It is essential to develop environmentally friendly fabrication methods and gain a comprehensive understanding of the ecological consequences to ensure responsible use. A key concern is the use of certain chemicals and materials in the manufacturing of superhydrophobic coatings. Reports have confirmed that some solvent-based 'water-repellent' coatings pose health risks, causing harm to the lungs, kidneys, nerves, teeth, and promoting bone decay. The utilization of fluorine-based coatings, particularly in fine spray forms, increases the risk of inhalation and irritation to the eyes and nose.

Over the last 30 years, several hundred cases of serious respiratory inflammation have been reported among consumers using hydrophobic coating 'spray-on' products in Europe and the USA. Some countries have banned the use of fluorinated compounds due to their adverse side effects; for instance, polyfluorooctyl-triethoxysilane (1H,1H,2H,2Hperfluorooctyl triethoxysilane) is banned in Denmark, with restricted use in Canada. Therefore, sustainable and eco-friendly approaches to material selection, manufacturing processes, and end-of-life management are crucial for minimizing any adverse effects on the environment.

As researchers and engineers continue to investigate these challenges, overcoming them will unlock the full potential of superhydrophobic surfaces and drive innovation across diverse fields. Overcoming these obstacles will not only refine the fundamental understanding of superhydrophobicity but also lay the foundation for practical, real-world applications with enduring significance.

6. Progress in Wetting Research through AI and Machine Learning

The incorporation of artificial intelligence (AI) algorithms into materials design is transforming the landscape of materials engineering. This is attributed to their capability to forecast material properties, create novel materials with improved characteristics, and unveil previously unrecognized mechanisms, surpassing traditional intuition in the process. Despite this, there has been limited in-depth exploration into the AI-assisted design of material textures.

The swift progress in artificial intelligence (AI) and machine learning (ML) presents significant possibilities for transforming and accelerating the laborious and expensive material development process. Over the past few decades, AI and ML have inaugurated a fresh era in materials science, employing computer algorithms to assist in exploration, comprehension, experimentation, modeling, and simulation [237,238]. Collaborating with human ingenuity and creativity, these algorithms play a crucial role in uncovering and enhancing innovative materials for upcoming technologies.

Lu et al. [239] introduced a graph-focused deep learning technique specifically designed to capture the intricate nuances inherent in spider web architectures. The utilization of this technique extends beyond understanding these complexities, serving the dual purpose of facilitating the generation of a diverse range of innovative bioinspired structural designs. The authors have laid the foundation for a groundbreaking framework for the generation of spider webs, delving into the realm of bioinspired design guided by rigorous principles. Not confined to spider web emulation, this method demonstrates versatility in addressing various heterogeneous hierarchical structures, spanning a wide spectrum of architected materials. Positioned as a valuable toolset in the domain of generative AI for material applications, this approach bridges the gap between theoretical exploration and the practical actualization of designs, offering profound insights into biology and diverse design possibilities.

Traditionally, the study and design of bioinspired structures have relied on empirical, extensive, and time-consuming top-down strategies devoid of AI algorithms. Scientists painstakingly observe and analyze natural organisms to discern the fundamental principles and structures contributing to their exceptional properties [240–243]. Designing and fabricating materials using advanced manufacturing techniques, conducting experiments to validate performance, and refining designs based on outcomes are integral to this resource-intensive and time-consuming approach, often requiring significant trial and error.

To overcome these challenges and expedite the bioinspired material design process, the integration of AI algorithms offers significant potential to enhance efficiency and effectiveness. Leveraging supervised AI, researchers can expedite material development by swiftly exploring a vast design space, reducing the pool of potential solutions, and pinpointing optimal material compositions and structures for a chosen application [244]. However, it is crucial to note that supervised learning demands high-quality data with accurate labeling about behaviors and characteristics to function correctly. Yu et al. [245] employed reinforcement learning to achieve a refined design within an unfamiliar design space, developing a model capable of learning a biological design strategy through multiple training iterations. In their methodology, they utilized a finite element method (FEM) to compute mechanical properties, treating them as "reward values" for the algorithm to produce materials with enhanced fracture toughness. To ensure highly optimized solutions and bolster confidence, the AI initially analyzed a small set of systems, gradually increasing the size with each successful convergence of results. This iterative approach aimed to minimize calculation time for obtaining the optimal design. Throughout the training process, the model acquired the capability to understand the biological design strategy, enabling its extension to more complex structural optimization problems by incorporating different mechanical properties as reward values. This innovative design framework, adaptable to modifications in system intricacy or the inclusion of additional variables as rewards, demonstrates versatile utility across various applications.

In contrast, Lantada et al. [246] used artificial neural networks (ANNs) to forecast the wettability of bioinspired and biological structures according to their hierarchical surface characteristics. The ANN model was trained using a vast collection of bio-surfaces that the authors had assembled with well-known wettability characteristics. New, distinct topographies were then added in order to assess the algorithm's effectiveness. The wettability findings were compared with those predicted by the AI model once the actual structures were constructed. The convergence of results demonstrated how well artificial intelligence (AI) may help discover new bio-interfaces with hierarchical tribology, allowing for fine control over wettability. This method may take into account a large number of factors that would provide several controlled characteristics and behaviors, increasing its flexibility for particular applications.

Finally, using AI can save on expenses related to developing new materials. Indeed, less intensive laboratory experimentation is required when the design process is streamlined. Gu et al. [247] stated the mechanical characteristics of 100,000 microstructures with respect to computing cost. The procedure took about five days when using FEM, but with their designed ML technique, the same amount of data could be solved in less than a minute during the prediction phase and between 30 s and 10 h during the training phase.

In this way, scientists may focus on the most promising ideas and expedite the material design process, increasing its effectiveness and economy, by utilizing AI's computing capability. Even though this trip is still in its early stages, we are convinced that AI will support materials scientists' research rather than work against it. This will greatly broaden the field's perspectives, creating new opportunities and quickening the development of material design.

7. Practical Implications of the Present Review

This article provides a comprehensive review of superhydrophobic surfaces, encompassing fundamental understanding, natural occurrences, design principles of bio-inspired coatings, potential applications, challenges, and future research directions. The practical implications of this review are multifaceted, offering valuable insights for various stakeholders:

1. Innovative material design: This review identifies key principles inspired by nature that can guide the development of advanced superhydrophobic nano-coating materials. Researchers and industry professionals can leverage these insights for innovative material design, paving the way for the creation of superior products with enhanced functionalities.

2. Functional applications: By drawing inspiration from natural structures, this review proposes practical applications across various domains. These insights could result in the development of coatings customized for specific functions such as self-cleaning, anti-fouling, anti-icing, and more. Implementing nature-inspired coatings with tailored functionalities could significantly impact industries dealing with surface protection and maintenance challenges, ultimately reducing maintenance expenses and prolonging the life of diverse substrates.

3. Environmental sustainability: The exploration of environmentally friendly coatings inspired by nature aligns with global efforts for sustainable practices. Understanding the principles of natural structures allows the development of coatings that exhibit su-

perior performance while contributing to eco-friendly coating technologies, reducing the environmental impact of conventional materials.

4. Advancements in biomedical coatings: This review highlights the relevance of nature-inspired coatings in biomedical applications, offering potential advancements in medical devices, implants, and drug delivery systems. Coatings designed to interact favorably with biological systems can lead to improved biocompatibility and reduced adverse effects, fostering progress in healthcare technologies.

5. Commercialization opportunities: Synthesizing the state-of-the-art in nature-inspired nano-coating materials, this review identifies promising avenues for commercialization and technology transfer. These insights inform the development of marketable products, driving growth in the coatings industry and contributing to economic development.

6. In conclusion, the practical implications of this review extend to fostering innovation, tailoring functionalities in coatings, promoting sustainability, advancing biomedical applications, identifying commercialization opportunities, and facilitating interdisciplinary collaboration. The comprehensive insights provided pave the way for real-world applications and transformative advancements in the field of nano-coating materials.

8. Conclusions

In conclusion, this review highlights the significant advancements in the field of superhydrophobic surfaces and their diverse applications. From the fundamental principles governing their unique wetting properties to the various fabrication techniques employed, a comprehensive understanding of these surfaces has been presented. The exploration of natural examples, such as lotus leaves and butterfly wings, has inspired innovative designs for artificial superhydrophobic surfaces. The applications across different industries, including self-cleaning materials, oil–water separation, and anti-icing technologies, underscore the immense potential of superhydrophobic surfaces in addressing real-world challenges. Moreover, this review emphasizes the importance of continued research to overcome challenges such as durability and scalability and to unlock new opportunities for this technology. As we look toward the future, the development and integration of superhydrophobic surfaces are composed to make a transformative impact, offering sustainable and efficient solutions in various fields.

Author Contributions: S.B. (Subodh Barthwal) and S.U. contributed to the methodology, background literature, writing draft, and formatting of this review article. S.B. (Sumit Barthwal) contributed to conceptualization, writing—review, editing, validation, supervision, and writing. All authors have read and agreed to the published version of the manuscript.

Funding: This research received no external funding.

Data Availability Statement: Not applicable.

Conflicts of Interest: The authors declare no conflict of interest.

References

1. Gerasopoulos, K.; Luedeman, W.L.; Ölçeroglu, E.; McCarthy, M.; Benkoski, J.J. Effects of engineered wettability on the efficiency of dew collection. *ACS Appl. Mater. Interfaces* **2018**, *10*, 4066–4076. [CrossRef]
2. Mondal, B.; Eain, M.M.G.; Xu, Q.; Egan, V.M.; Punch, J.; Lyons, A.M. Design and fabrication of a hybrid superhydrophobic–hydrophilic surface that exhibits stable dropwise condensation. *ACS Appl. Mater. Interfaces* **2015**, *7*, 23575–23588. [CrossRef]
3. Camaiti, M.; Brizi, L.; Bortolotti, V.; Papacchini, A.; Salvini, A.; Fantazzini, P. An environmental friendly fluorinated oligoamide for producing nonwetting coatings with high performance on porous surfaces. *ACS Appl. Mater. Interfaces* **2017**, *9*, 37279–37288. [CrossRef]
4. Zhang, X.; Shi, F.; Nia, J.; Jiang, Y.; Wang, Z. Superhydrophobic surfaces: From structural control to functional applications. *J. Mater. Chem.* **2008**, *18*, 621–633. [CrossRef]
5. Roach, P.; Shirtcliff, N.J.; Newton, M.I. Progress in superhydrophobic surface development. *Soft Matter* **2008**, *4*, 224–240. [CrossRef]
6. Lafuma, A.; Quere, D. Superhydrophobic states. *Nat. Mater.* **2003**, *2*, 457–460. [CrossRef]
7. Wenzel, R.N. Resistance of solid surfaces to wetting by water. *Ind. Eng. Chem.* **1936**, *28*, 988–994. [CrossRef]

8.   Cassie, A.B.D.; Baxter, S. Wettability of porous surfaces. *Trans. Faraday Soc.* **1944**, *40*, 546–551. [CrossRef]
9.   Onda, T.; Shibuichi, S.; Satoh, N.; Tsujii, K. Super-Water-Repellent Fractal Surfaces. *Langmuir* **1996**, *12*, 2125–2127. [CrossRef]
10. Ma, M.; Hill, R.M. Superhydrophobic surfaces. *Curr. Opin. Colloid Interface Sci.* **2006**, *11*, 193–202. [CrossRef]
11. Neinhuis, C.; Barthlott, W. Characterization and distribution of self-cleaning plant surfaces. *Ann. Bot.* **1997**, *79*, 667–677. [CrossRef]
12. Chen, W.; Fadeev, A.Y.; Hsieh, M.C.; Öner, D.; Youngblood, J.; McCarthy, T.J. Ultrahydrophobic and ultralyophobic surfaces: Some comments and examples. *Langmuir* **1999**, *15*, 3395–3399. [CrossRef]
13. Barthlott, W.; Ehler, N. Raster-Elektronenmikroskopie der Epidermis-Oberflächen von Spermatophyten. *Trop. Subtrop. Pflanzenwelt* **1977**, *19*, 110.
14. Bixler, G.D.; Bhushan, B. Rice-and butterfly-wing effect inspired self-cleaning and low drag micro/nanopatterned surfaces in water, oil, and air flow. *Nanoscale* **2014**, *6*, 76–96. [CrossRef] [PubMed]
15. Kumar, M.; Bhardwaj, R. Wetting characteristics of Colocasiaesculenta (Taro) leaf and a bioinspired surface thereof. *Sci. Rep.* **2020**, *10*, 935. [CrossRef]
16. Gao, X.; Yan, X.; Yao, X.; Xu, L.; Zhang, K.; Zhang, J.; Yang, B.; Jiang, L. The dry-style antifogging properties of mosquito compound eyes and artificial analogues prepared by soft lithography. *Adv. Mater.* **2007**, *19*, 2213–2217. [CrossRef]
17. Zheng, Y.; Gao, X.; Jiang, L. Directional adhesion of superhydrophobic butterfly wings. *Soft Matter* **2007**, *3*, 178–182. [CrossRef]
18. Zhu, H.; Huang, Y.; Lou, X.; Xia, F. Beetle-inspired wettable materials: From fabrications to applications. *Mater Today Nano* **2019**, *6*, 100034. [CrossRef]
19. Stark, A.Y.; Subarajan, S.; Jain, D.; Niewiarowski, P.H.; Dhinojwala, A. Superhydrophobicity of the gecko toe pad: Biological optimization versus laboratory maximization. *Philos. Trans. Royal Soc. A* **2016**, *374*, 20160184. [CrossRef]
20. Gao, X.; Jiang, L. Water-repellent legs of water striders. *Nature* **2004**, *432*, 36. [CrossRef]
21. Liu, M.; Wang, S.; Jiang, L. Bioinspired multiscale surfaces with special wettability. *MRS Bull.* **2013**, *38*, 375–382. [CrossRef]
22. Liu, Y.; Wu, M.; Guo, C.; Zhou, D.; Wu, Y.; Wu, Z.; Lu, H.; Zhang, H.; Zhang, Z. A Review on Preparation of Superhydrophobic and Superoleophobic Surface by Laser Micromachining and Its Hybrid Methods. *Crystals* **2023**, *13*, 20. [CrossRef]
23. Guo, Z.; Liu, W. Biomimic from the superhydrophobic plant leaves in nature: Binary structure and unitary structure. *Plant Sci.* **2007**, *172*, 1103–1112. [CrossRef]
24. Barthlott, W.; Neinhuis, C. Purity of the sacred lotus, or escape from contamination in biological surfaces. *Planta* **1997**, *202*, 1–8. [CrossRef]
25. Otten, A.; Herminghaus, S. How plants keep dry: A physicist's point of view. *Langmuir* **2004**, *20*, 2405–2408. [CrossRef] [PubMed]
26. Liu, Y.; Li, G. A new method for producing "Lotus Effect" on a biomimetic shark skin. *J. Colloid Interface Sci.* **2012**, *388*, 235–242. [CrossRef] [PubMed]
27. Kostal, E.; Stroj, S.; Kasemann, S.; Matylitsky, V.; Domke, M. Fabrication of biomimetic fog-collecting superhydrophilic–superhydrophobic surface micropatterns using femtosecond lasers. *Langmuir* **2018**, *34*, 2933–2941. [CrossRef] [PubMed]
28. Feng, L.; Zhang, Y.; Xi, J.; Zhu, Y.; Wang, N.; Xia, F.; Jiang, L. Petal effect: A superhydrophobic state with high adhesive force. *Langmuir* **2008**, *24*, 4114–4119. [CrossRef]
29. Wang, L.; Guo, X.; Zhang, H.; Liu, Y.; Wang, Y.; Liu, K.; Liang, H.; Ming, W. Recent advances in superhydrophobic and antibacterial coatings for biomedical materials. *Coatings* **2022**, *12*, 1469. [CrossRef]
30. Li, J.; Li, M.; Koh, J.J.; Wang, J.; Lyu, Z. 3D-printed biomimetic structures for energy and environmental applications. *DeCarbon* **2024**, *3*, 100026. [CrossRef]
31. Nomeir, B.; Lakhouil, S.; Boukheir, S.; Ali, M.; Naamane, S. Recent progress on transparent and self-cleaning surfaces by superhydrophobic coatings deposition to optimize the cleaning process of solar panels. *Sol. Energy Mater. Sol. Cells* **2023**, *257*, 112347. [CrossRef]
32. Li, Z.; Liu, P.; Chen, S.; Liu, X.; Yu, Y.; Li, T.; Wan, Y.; Tang, N.; Liu, Y.; Gu, Y. Bioinspired marine antifouling coatings: Antifouling mechanisms, design strategies and application feasibility studies. *Eur. Polym. J.* **2023**, *190*, 111997. [CrossRef]
33. Mishra, S.; Yılmaz-Serçinoğlu, Z.; Moradi, H.; Bhatt, D.; Kuru, C.İ.; Ulucan-Karnak, F. Recent advances in bioinspired sustainable sensing technologies. *Nano-Struct. Nano-Objects* **2023**, *34*, 100974. [CrossRef]
34. George, J.S.; Vijayan, P.P.; Hoang, A.T.; Kalarikkal, N.; Nguyen-Tri, P.; Thomas, S. Recent advances in bio-inspired multifunctional coatings for corrosion protection. *Prog. Polym. Sci.* **2022**, *168*, 106858. [CrossRef]
35. Huang, W.; Huang, J.; Guo, Z.; Liu, W. Icephobic/anti-icing properties of superhydrophobic surfaces. *Adv. Colloid Interface Sci.* **2022**, *304*, 102658. [CrossRef] [PubMed]
36. Naikoo, G.A.; Mustaqeem, M.; Hassan, I.U.; Awan, T.; Arshad, F.; Salim, H.; Qurashi, A. Bioinspired and green synthesis of nanoparticles from plant extracts with antiviral and antimicrobial properties: A critical review. *J. Saudi Chem. Soc.* **2021**, *25*, 101304. [CrossRef]
37. Rasouli, S.; Rezaei, N.; Hamedi, H.; Zendehboudi, S.; Duan, X. Superhydrophobic and superoleophilic membranes for oil-water separation application: A comprehensive review. *Mater. Des.* **2021**, *204*, 109599. [CrossRef]
38. Zhang, F.; Guo, Z. Bioinspired materials for water-harvesting: Focusing on microstructure designs and the improvement of sustainability. *Mater. Adv.* **2020**, *1*, 2592–2613. [CrossRef]
39. Adibnia, V.; Mirbagheri, M.; Faivre, J.; Robert, J.; Lee, J.; Matyjaszewski, K.; Lee, D.W.; Banquy, X. Bioinspired polymers for lubrication and wear resistance. *Prog. Polym. Sci.* **2020**, *110*, 101298. [CrossRef]

40. Gao, L.; McCarthy, T.J. Contact angle hysteresis explained. *Langmuir* **2006**, *22*, 6234–6237. [CrossRef]
41. Deng, Y.; Peng, C.; Dai, M.; Lin, D.; Ali, I.; Alhewairini, S.S.; Zheng, X.; Chen, G.; Li, J.; Naz, I. Recent development of super-wettable materials and their applications in oil-water separation. *J. Clean. Prod.* **2020**, *266*, 121624. [CrossRef]
42. Bell, M.S.; Borhan, A. A Volume-Corrected Wenzel Model. *ACS Omega* **2020**, *5*, 8875–8884. [CrossRef]
43. Quéré, D. Rough ideas on wetting. *Phys. A Stat. Mech. Appl.* **2002**, *313*, 32–46. [CrossRef]
44. Quéré, D. Non-sticking drops. *Rep. Prog. Phys.* **2005**, *68*, 2495. [CrossRef]
45. Öner, D.; McCarthy, T.J. Ultrahydrophobic surfaces. Effects of topography length scales on wettability. *Langmuir* **2000**, *16*, 7777–7782. [CrossRef]
46. Kusumaatmaja, H.; Yeomans, J.M. Modeling contact angle hysteresis on chemically patterned and superhydrophobic surfaces. *Langmuir* **2007**, *23*, 6019–6032. [CrossRef] [PubMed]
47. Schwartz, L.W.; Garoff, S. Contact angle hysteresis on heterogeneous surfaces. *Langmuir* **1985**, *1*, 219–230. [CrossRef]
48. Schwartz, L.W.; Garoff, S. Contact angle hysteresis and the shape of the three-phase line. *J. Colloid Interface Sci.* **1985**, *106*, 422–437. [CrossRef]
49. Oliver, J.F.; Huh, C.; Mason, S.G. Resistance to spreading of liquids by sharp edges. *J. Colloid Interface Sci.* **1977**, *59*, 568–581. [CrossRef]
50. Law, K.Y. Contact Angle Hysteresis on Smooth/Flat and Rough Surfaces. Interpretation, Mechanism, and Origin. *Acc. Mater. Res.* **2022**, *3*, 1–7. [CrossRef]
51. Dorrer, C.; Rühe, J. Advancing and receding motion of droplets on ultrahydrophobic post surfaces. *Langmuir* **2006**, *22*, 7652–7657. [CrossRef]
52. Young, T. An essay on the cohesion of fluids. *Philos. Trans. R Soc. Lond.* **1805**, *95*, 65–87.
53. Wenzel, R.N. Surface roughness and contact angle. *J. Phys. Chem.* **1949**, *53*, 1466–1467. [CrossRef]
54. Milne, A.J.B.; Amirfazli, A. The Cassie equation: How it is meant to be used. *Adv. Colloid Interface Sci.* **2012**, *170*, 48–55. [CrossRef] [PubMed]
55. Briant, A.J.; Wagner, A.J.; Yeomans, J.M. Lattice Boltzmann simulations of contact line motion. I. Liquid-gas systems. *Phys. Rev. E* **2004**, *69*, 031602. [CrossRef] [PubMed]
56. Cahn, J.W. Critical point wetting. *J. Chem. Phys.* **1977**, *66*, 3667–3672. [CrossRef]
57. Dupuis, A.; Yeomans, J.M. Modeling droplets on superhydrophobic surfaces: Equilibrium states and transitions. *Langmuir* **2005**, *21*, 2624–2629. [CrossRef] [PubMed]
58. De Gennes, P.G. Wetting: Statics and dynamics. *Rev. Mod. Phys.* **1985**, *57*, 827. [CrossRef]
59. Joanny, J.F.; De Gennes, P.G. A model for contact angle hysteresis. *J. Chem. Phys.* **1984**, *81*, 552–562. [CrossRef]
60. Kumar, A.; Gogoi, B. Development of durable self-cleaning superhydrophobic coatings for aluminium surfaces via chemical etching method. *Tribol. Int.* **2018**, *122*, 114–118. [CrossRef]
61. Kim, J.H.; Mirzaei, A.; Kim, H.W.; Kim, S.S. Facile fabrication of superhydrophobic surfaces from austenitic stainless steel (AISI 304) by chemical etching. *Appl. Surf. Sci.* **2018**, *439*, 598–604. [CrossRef]
62. Zhu, J.; Duan, Y. Facilely etching of superhydrophobic surface with regular mulriple hierarchical micro-nano structures for crowning wettability. *Appl. Surf. Sci.* **2024**, *648*, 159009. [CrossRef]
63. Rodič, P.; Kapun, B.; Panjan, M.; Milošev, I. Easy and Fast Fabrication of Self-Cleaning and Anti-Icing Perfluoroalkyl Silane Film on Aluminium. *Coatings* **2020**, *10*, 234. [CrossRef]
64. Lo, T.N.H.; Lee, J.; Hwang, H.S.; Park, I. Nanoscale Coatings Derived from Fluoroalkyl and PDMS Alkoxysilanes on Rough Aluminum Surfaces for Improved Durability and Anti-Icing Properties. *ACS Appl. Nano Mater.* **2021**, *4*, 7493–7501. [CrossRef]
65. Wei, D.; Wang, J.; Liu, Y.; Wang, D.; Li, S.; Wang, H. Controllable superhydrophobic surfaces with tunable adhesion on Mg alloys by a simple etching method and its corrosion inhibition performance. *Chem. Eng. J.* **2021**, *404*, 126444. [CrossRef]
66. Shirtcliffe, N.J.; Aqil, S.; Evans, C.; McHale, G.; Newton, M.I.; Perry, C.C.; Roach, P. The use of high aspect ratio photoresist (SU-8) for super-hydrophobic pattern prototyping. *J. Micromech. Microeng.* **2004**, *14*, 1384–1389. [CrossRef]
67. Velasco, J.M.; Vlachopoulou, M.E.; Tserepi, A.; Gogolides, E. Stable superhydrophobic surfaces induced by dual-scale topography on SU-8. *Microelectron. Eng.* **2010**, *87*, 782–785. [CrossRef]
68. Yang, Y.; He, H.; Li, Y.; Qiu, J. Using nanoimprint lithography to create robust, buoyant, superhydrophobic PVB/SiO$_2$ coatings on wood surfaces inspired by red roses petal. *Sci. Rep.* **2019**, *9*, 9961. [CrossRef] [PubMed]
69. Feng, J.; Tuominen, M.T.; Rothstein, J.P. Hierarchical superhydrophobic surfaces fabricated by dual-scale electron-beam lithography with well-ordered secondary nanostructures. *Adv. Funct. Mater.* **2011**, *21*, 3715–3722. [CrossRef]
70. Dev, A.; Choudhury, B.D.; Abedin, A.; Anand, S. Fabrication of periodic nanostructure assemblies by interfacial energy driven colloidal lithography. *Adv. Funct. Mater.* **2014**, *24*, 4577–4583. [CrossRef]
71. Wu, R.; Gilavan, M.T.; Akbar, M.A.; Fan, L.; Selvaganapathy, P.R. Direct transformation of polycarbonate to highly conductive and superhydrophobic graphene/graphitic composite by laser writing and its applications. *Carbon* **2024**, *216*, 118597. [CrossRef]
72. Ghasemlou, M.; Le, P.H.; Daver, F.; Murdoch, B.J.; Ivanova, E.P.; Adhikari, B. Robust and Eco-Friendly Superhydrophobic Starch Nanohybrid Materials with Engineered Lotus Leaf Mimetic Multiscale Hierarchical Structures. *ACS Appl. Mater. Interfaces* **2021**, *13*, 36558–36573. [CrossRef]
73. Guo, C.; Liu, K.; Zhang, T.; Sun, P.; Liang, L. Development of flexible photothermal superhydrophobic microarray by photolithography technology for anti-icing and deicing. *Prog. Org. Coat.* **2023**, *182*, 107675. [CrossRef]

74. Wang, H.; He, M.; Liu, H.; Guan, Y. One-Step Fabrication of Robust Superhydrophobic Steel Surfaces with Mechanical Durability, Thermal Stability, and Anti-icing Function. *ACS Appl. Mater. Interfaces* **2019**, *11*, 25586–25594. [CrossRef]
75. Wang, J.; Zhang, Y.; He, Q. Durable and robust superhydrophobic fluororubber surface fabricated by template method with exceptional thermostability and mechanical stability. *Sep. Purif. Technol.* **2023**, *306*, 122423. [CrossRef]
76. Xu, W.; Yi, P.; Gao, J.; Deng, Y.; Peng, L.; Lai, X. Large-Area Stable Superhydrophobic Poly(dimethylsiloxane) Films Fabricated by Thermal Curing via a Chemically Etched Template. *ACS Appl. Mater. Interfaces* **2020**, *12*, 3042–3050. [CrossRef] [PubMed]
77. Wang, J.; Wu, Y.; Zhang, D.; Li, L.; Wang, T.; Duan, S. Preparation of superhydrophobic flexible tubes with water and blood repellency based on template method. *Colloids Surf. A Physicochem. Eng. Asp.* **2020**, *587*, 124331. [CrossRef]
78. Dai, Z.; Guo, H.; Huang, Q.; Ding, S.; Liu, Y.; Gao, Y.; Zhou, Y.; Sun, G.; Zhou, B. Mechanically robust and superhydrophobic concrete based on sacrificial template approach. *Cem. Concr. Compos.* **2022**, *134*, 104796. [CrossRef]
79. Wu, W.; Guijt, R.M.; Silina, Y.E.; Koch, M.; Manz, A. Plant leaves as templates for soft lithography. *RSC Adv.* **2016**, *6*, 22469–22475. [CrossRef]
80. Zhang, T.; Li, M.; Su, B.; Ye, C.; Li, K.; Shen, W.; Chen, L.; Xue, Z.; Wang, S.; Jiang, L. Bio-inspired anisotropic micro/nano-surface from a natural stamp: Grasshopper wings. *Soft Matter* **2011**, *7*, 7973–7975. [CrossRef]
81. Jafari, R.; Asadollahi, S.; Farzaneh, M. Applications of plasma technology in development of superhydrophobic surfaces. *Plasma Chem. Plasma Process.* **2013**, *33*, 177–200. [CrossRef]
82. Li, M.; Wang, G.C.; Min, H.G. Effect of surface roughness on magnetic properties of Co films on plasma-etched Si(100) substrates. *J. Appl. Phys.* **1998**, *83*, 5313–5320. [CrossRef]
83. Notsu, H.; Yagi, I.; Tatsuma, T.; Tryk, D.A.; Fujishima, A. Introduction of oxygen-containing functional groups onto diamond electrode surfaces by oxygen plasma and anodic polarization. *Electrochem. Solid-State Lett.* **1999**, *2*, 522–524. [CrossRef]
84. Ebert, D.; Bhushan, B. Transparent, superhydrophobic, and wear-resistant surfaces using deep reactive ion etching on PDMS substrates. *J. Colloid Interface Sci.* **2016**, *481*, 82–90. [CrossRef]
85. Nguyen-Tri, P.; Altiparmak, F.; Nguyen, N.; Tuduri, L.; Ouellet-Plamondon, C.M.; Prud'homme, R.E. Robust Superhydrophobic Cotton Fibers Prepared by Simple Dip-Coating Approach Using Chemical and Plasma-Etching Pretreatments. *ACS Omega* **2019**, *4*, 7829–7837. [CrossRef]
86. Somrang, W.; Denchitcharoen, S.; Eiamchai, P.; Horprathum, M.; Chananonnawathorn, C. Superhydrophobic and antireflective surface of nanostructures fabricated by CF4 plasma etching. *Mater. Today Proc.* **2018**, *5*, 13879–13885. [CrossRef]
87. Becker, C.; Petersen, J.; Mertz, G.; Ruch, D.; Dinia, A. High Superhydrophobicity Achieved on Poly(ethylene terephthalate) by Innovative Laser-Assisted Magnetron Sputtering. *J. Phys. Chem. C* **2011**, *115*, 10675–10681. [CrossRef]
88. Airoudj, A.; Gall, F.B.; Roucoules, V. Textile with Durable Janus Wetting Properties Produced by Plasma Polymerization. *J. Phys. Chem. C* **2016**, *120*, 29162–29172. [CrossRef]
89. Molina, R.; Teixidó, J.M.; Kan, C.-W.; Jovančić, P. Hydrophobic Coatings on Cotton Obtained by in Situ Plasma Polymerization of a Fluorinated Monomer in Ethanol Solutions. *ACS Appl. Mater. Interfaces* **2017**, *9*, 5513–5521. [CrossRef] [PubMed]
90. Siddig, E.A.; Zhang, Y.; Yang, B.; Wang, T.; Shi, J.; Guo, Y.; Xu, Y.; Zhang, J. Durable Flame-Resistant and Ultra-Hydrophobic Aramid Fabrics via Plasma-Induced Graft Polymerization. *Coatings* **2020**, *10*, 1257. [CrossRef]
91. Ussenkhan, S.S.; Kyrykbay, B.A.; Yerlanuly, Y.; Zhunisbekov, A.T.; Gabdullin, M.T.; Ramazanov, T.S.; Orazbayev, S.A.; Utegenov, A.U. Fabricating durable and stable superhydrophobic coatings by the atmospheric pressure plasma polymerisation of hexamethyldisiloxane. *Heliyon* **2024**, *10*, e23844. [CrossRef]
92. Latthe, S.S.; Nadargi, D.Y.; Rao, A.V. TMOS based water repellent silica thin films by co-precursor method using TMES as a hydrophobic agent. *Appl. Surf. Sci.* **2009**, *255*, 3600–3604. [CrossRef]
93. Yao, X.; Zhao, W.; Zhang, H.; Zhang, Y.; Zhong, L.; Wu, Y. Performance study of a superhydrophobic nanocellulose membrane on the surface of a wood-based panel prepared via the sol-gel method. *Colloid Interface Sci. Commun.* **2023**, *57*, 100758. [CrossRef]
94. Xiong, J.; Sarkar, D.K.; Chen, X.G. Ultravioletdurable superhydrophobic nanocomposite thin films based on cobalt stearate-coated TiO2 nanoparticles combined with polymethylhydrosiloxane. *ACS Omega* **2017**, *2*, 8198–8204. [CrossRef] [PubMed]
95. Ke, C.; Zhang, C.; Wu, X.; Jiang, Y. Highly transparent and robust superhydrophobic coatings fabricated via a facile sol-gel process. *Thin Solid Films* **2021**, *723*, 138583. [CrossRef]
96. Liang, Z.; Geng, M.; Dong, B.; Zhao, L.; Wang, S. Transparent and robust SiO2/PDMS composite coatings with self-cleaning. *Surf. Eng.* **2020**, *36*, 643–650. [CrossRef]
97. Li, J.; Ding, H.; Zhang, H.; Guo, C.; Hong, X.; Sun, L.; Ding, F. Superhydrophobic Methylated Silica Sol for Effective Oil–Water Separation. *Materials* **2020**, *13*, 842. [CrossRef]
98. Liu, L.; Kong, G.; Zhu, Y.; Che, C. Superhydrophobic graphene aerogel beads for adsorption of oil and organic solvents via a convenient in situ sol-gel method. *Colloid Interface Sci. Commun.* **2021**, *45*, 100518. [CrossRef]
99. Gao, L.; Lu, Y. A robust, anti-acid, and high-temperature–humidity-resistant superhydrophobic surface of wood based on a modified TiO2 film by fluoroalkyl silane. *Surf. Coat. Technol.* **2015**, *262*, 33–39. [CrossRef]
100. Wang, K.; Liu, X. Highly fluorinated and hierarchical HNTs/SiO2 hybrid particles for substrate-independent superamphiphobic coatings. *Chem. Eng. J.* **2019**, *359*, 626–640. [CrossRef]
101. Lu, Q.; Cheng, R. Superhydrophobic wood fabricated by epoxy/Cu2(OH)3Cl NPs/stearic acid with performance of desirable self-cleaning, anti-mold, dimensional stability, mechanical and chemical durability. *Colloids Surf. A Physicochem. Eng. Asp.* **2022**, *647*, 129162. [CrossRef]

102. Yuan, J.; Wang, J.; Zhang, K.; Hu, W. Fabrication and properties of a superhydrophobic film on an electroless plated magnesium alloy. *RSC Adv.* **2017**, *7*, 28909–28917. [CrossRef]
103. Wang, J.; Han, F.; Liang, B.; Geng, G. Hydrothermal fabrication of robustly superhydrophobic cotton fibers for efficient separation of oil/water mixtures and oil-in-water emulsions. *J. Ind. Eng. Chem.* **2017**, *54*, 174–183. [CrossRef]
104. Lan, X.; Zhang, B.; Wang, J.; Fan, X.; Zhang, J. Hydrothermally structured superhydrophobic surface with superior anti-corrosion, anti-bacterial and anti-icing behaviors. *Colloids Surf. A Physicochem. Eng. Asp.* **2021**, *624*, 126820. [CrossRef]
105. Tuong, V.M.; Huyen, N.V.; Kien, N.T.; Dien, N.V. Durable Epoxy@ZnO Coating for Improvement of Hydrophobicity and Color Stability of Wood. *Polymers* **2019**, *11*, 1388. [CrossRef] [PubMed]
106. Wan, Y.; Chen, M.; Liu, W.; Shen, X.; Min, Y.; Xu, Q. The research on preparation of superhydrophobic surfaces of pure copper by hydrothermal method and its corrosion resistance. *Electrochim. Acta.* **2018**, *270*, 310–318. [CrossRef]
107. Kulkarni, A.; Bambole, V.A.; Mahanwar, P.A. Electrospinning of polymers, their modeling and applications. *Polym. Plast. Technol. Eng.* **2010**, *49*, 427–441. [CrossRef]
108. Sas, I.; Gorga, R.E.; Joines, J.A.; Thoney, K.A. Literature Review on Superhydrophobic Self-Cleaning Surfaces Produced by Electrospinning. *J. Polym. Sci. B Polym. Phys.* **2012**, *50*, 824–845. [CrossRef]
109. Si, Y.; Guo, Z. Superhydrophobic nanocoatings: From materials to fabrications and to applications. *Nanoscale* **2015**, *7*, 5922–5946. [CrossRef]
110. Cui, M.; Xu, C.; Shen, Y.; Tian, H.; Feng, H.; Li, J. Electrospinning superhydrophobic nanofibrouspoly(vinylidenefluoride)/stearic acid coatings with excellent corrosion resistance. *Thin Solid Film.* **2018**, *657*, 88–94. [CrossRef]
111. Liu, Z.; Tang, Y.; Zhao, K.; Zhang, Q. Superhydrophobic SiO$_2$ micro/nanofibrous membranes with porous surface prepared by freeze electrospinning for oil adsorption. *Colloids Surf. A Physicochem. Eng. Asp.* **2019**, *568*, 356–361. [CrossRef]
112. Raman, A.; Jayan, J.S.; Deeraj, B.D.S.; Saritha, A.; Joseph, K. Electrospun Nanofibers as Effective Superhydrophobic Surfaces: A Brief review. *Surf. Interfaces* **2021**, *24*, 101140. [CrossRef]
113. Pardo-Figuerez, M.; Lopez-Cordoba, A.; Torres-Giner, S.; Lagaron, J.M. Superhydrophobic bio-coating made by co-continuous electrospinning and electrospraying on polyethylene terephthalate films proposed as easy emptying transparent food packaging. *Coatings* **2018**, *8*, 364. [CrossRef]
114. Ding, B.; Ogawa, T.; Kim, J.; Fujimoto, K.; Shiratori, S. Fabrication of a super-hydrophobic nanofibrous zinc oxide film surface by electrospinning. *Thin Solid Films* **2008**, *516*, 2495–2501. [CrossRef]
115. He, G.; Wan, M.; Wang, Z.; Zhao, Y.; Sun, L. Fabrication of firm, superhydrophobic and antimicrobial PVDF@ZnO@TA@DT electrospun nanofibrous membranes for emulsion separation. *Colloids Surf. A Physicochem. Eng. Asp.* **2023**, *662*, 130962. [CrossRef]
116. Cai, X.; Huang, J.; Lu, X.; Yang, L.; Lin, T.; Lei, T. Facile Preparation of Superhydrophobic Membrane Inspired by Chinese Traditional Hand-Stretched Noodles. *Coatings* **2021**, *11*, 228. [CrossRef]
117. Xiong, J.; Das, S.N.; Shin, B.; Kar, J.P.; Choi, J.H.; Myoung, J.M. Biomimetic hierarchical ZnO structure with superhydrophobic and antireflective properties. *J. Colloid Interface Sci.* **2010**, *350*, 344–347. [CrossRef] [PubMed]
118. Pakdel, A.; Zhi, C.; Bando, Y.; Nakayama, T.; Golberg, D. Boron nitride nanosheet coatings with controllable water repellency. *ACS Nano* **2011**, *5*, 6507–6515. [CrossRef]
119. Rezaei, S.; Manoucheri, I.; Moradian, R.; Pourabbas, B. One-step chemical vapor deposition and modification of silica nanoparticles at the lowest possible temperature and superhydrophobic surface fabrication. *Chem. Eng. J.* **2014**, *252*, 11–16. [CrossRef]
120. Zhang, M.; Zhou, T.; Li, H.; Liu, Q. UV-durable superhydrophobic ZnO/SiO$_2$ nanorod arrays on an aluminum substrate using catalyst-free chemical vapor deposition and their corrosion performance. *Appl. Surf. Sci.* **2023**, *623*, 157085. [CrossRef]
121. Sun, W.; Wang, L.; Yang, Z.; Li, S.; Wu, T.; Liu, G. Fabrication of polydimethylsiloxane-derived superhydrophobic surface on aluminium via chemical vapour deposition technique for corrosion protection. *Corros. Sci.* **2017**, *128*, 176–185. [CrossRef]
122. Cai, Z.; Lin, J.; Hong, X. Transparent superhydrophobic hollow films (TSHFs) with superior thermal stability and moisture resistance. *RSC Adv.* **2018**, *8*, 491–498. [CrossRef]
123. Yang, R.; Liang, Y.; Hong, S.; Zuo, S.; Wu, Y.; Shi, J.; Cai, L.; Li, J.; Mao, H.; Ge, S.; et al. Novel Low-Temperature Chemical Vapor Deposition of Hydrothermal Delignified Wood for Hydrophobic Property. *Polymers* **2020**, *12*, 1757. [CrossRef]
124. Huang, X.; Sun, M.; Shi, X.; Shao, J.; Jin, M.; Liu, W.; Zhang, R.; Huang, S.; Ye, Y. Chemical vapor deposition of transparent superhydrophobic anti-Icing coatings with tailored polymer nanoarray architecture. *Chem. Eng. J.* **2023**, *454*, 139981. [CrossRef]
125. Darband, G.B.; Aliofkhazraei, M.; Khorsand, S.; Sokhanvar, S.; Kaboli, A. Science and engineering of superhydrophobic surfaces: Review of corrosion resistance, chemical and mechanical stability. *Arab. J. Chem.* **2020**, *13*, 1763–1802. [CrossRef]
126. Liu, Q.; Chen, D.; Kang, Z. One-Step Electrodeposition Process To Fabricate Corrosion-Resistant Superhydrophobic Surface on Magnesium Alloy. *ACS Appl. Mater. Interfaces* **2015**, *7*, 1859–1867. [CrossRef]
127. Zhang, B.; Zhao, X.; Li, Y.; Hou, B. Fabrication of durable anticorrosion superhydrophobic surfaces on aluminum substrates via a facile one-step electrodeposition approach. *RSC Adv.* **2016**, *6*, 35455–35465. [CrossRef]
128. Meng, H.; Wang, S.; Xi, J.; Tang, Z.; Jiang, L. Facile means of preparing superamphiphobic surfaces on common engineering metals. *J. Phys. Chem. C* **2008**, *112*, 11454–11458. [CrossRef]
129. Tian, J.; Bao, J.; Li, L.; Sha, J.; Duan, W.; Qiao, M.; Cui, J.; Zhang, Z. Facile fabrication of superhydrophobic coatings with superior corrosion resistance on LA103Z alloy by one-step electrochemical synthesis. *Surf. Coat. Technol.* **2023**, *452*, 129090. [CrossRef]
130. Chen, Z.; Hu, Y.; He, X.; Xu, Y.; Liu, X.; Zhou, Y.; Hao, L.; Ruan, Y. One-step fabrication of soft calcium superhydrophobic surfaces by a simple electrodeposition process. *RSC Adv.* **2022**, *12*, 297–308. [CrossRef]

131. Li, B.; Ouyang, Y.; Haider, Z.; Zhu, Y.; Qiu, R.; Hu, S.; Niu, H.; Zhang, Y.; Chen, M. One-step electrochemical deposition leading to superhydrophobic matrix for inhibiting abiotic and microbiologically influenced corrosion of Cu in seawater environment. *Colloids Surf. A Physicochem. Eng. Asp.* **2021**, *616*, 126337. [CrossRef]

132. Prado, L.H.; Virtanen, S. Cu–MoS$_2$ Superhydrophobic Coating by Composite Electrodeposition. *Coatings* **2020**, *10*, 238. [CrossRef]

133. Richardson, J.J.; Cui, J.; Bjo, M.; Braunger, J.A.; Ejima, H.; Caruso, F. Innovation in layer-by-layer assembly. *Chem. Rev.* **2016**, *116*, 14828–14867. [CrossRef] [PubMed]

134. Li, X.; Du, X.; He, J. Self-cleaning antireflective coatings assembled from peculiar mesoporous silica nanoparticles. *Langmuir* **2010**, *26*, 3528–13534. [CrossRef]

135. Wu, M.; An, N.; Li, Y.; Sun, J. Layer-by-layer assembly of fluorine-free polyelectrolyte–surfactant complexes for the fabrication of self-healing superhydrophobic films. *Langmuir* **2016**, *32*, 12361–12369. [CrossRef] [PubMed]

136. Guo, X.J.; Xue, C.H.; Li, M.; Li, X.; Ma, J.Z. Fabrication of robust, superhydrophobic, electrically conductive and UV-blocking fabrics via layer-by-layer assembly of carbon nanotubes. *RSC Adv.* **2017**, *7*, 25560–25565. [CrossRef]

137. Li, Y.; Liu, F.; Sun, J. A facile layer-by-layer deposition process for the fabrication of highly transparent superhydrophobic coatings. *Chem. Commun.* **2009**, *19*, 2730–2732. [CrossRef] [PubMed]

138. Zheng, L.; Su, X.; Lai, X.; Chen, W.; Li, H.; Zeng, X. Conductive superhydrophobic cotton fabrics via layer-by-layer assembly of carbon nanotubes for oil-water separation and human motion detection. *Mater. Lett.* **2019**, *253*, 230–233. [CrossRef]

139. Shao, C.; Jiang, M.; Zhang, J.; Zhang, Q.; Han, L.; Wu, Y. Construction of a superhydrophobic wood surface coating by layer-by-layer assembly: Self-adhesive properties of polydopamine. *Appl. Surf. Sci.* **2023**, *609*, 155259. [CrossRef]

140. Syed, J.A.; Tang, S.; Meng, X. Super-hydrophobic multilayer coatings with layer number tuned swapping in surface wettability and redox catalytic anti-corrosion application. *Sci. Rep.* **2017**, *7*, 4403. [CrossRef]

141. Strait Research. Available online: https://straitsresearch.com/report/superhydrophobic-coatings-market (accessed on 21 February 2024).

142. Liu, K.; Jiang, L. Multifunctional Integration: From Biological to Bio-inspired Materials. *ACS Nano* **2011**, *5*, 6786–6790. [CrossRef]

143. Barthwal, S.; Barthwal, S.; Singh, N.B. Bioinspired Superhydrophobic Nanocomposite Materials: Introduction, Design, and Applications. In *Nanocomposites*; Jenny Stanford Publishing: Dubai, United Arab Emirates, 2022; pp. 429–481.

144. Parkin, I.P.; Palgrave, R.G. Self-cleaning Coatings. *J. Mater. Chem.* **2015**, *15*, 1689–1695. [CrossRef]

145. Blossey, R. Self-cleaning Surfaces-virtual Realities. *Nat. Mater.* **2003**, *2*, 301–306. [CrossRef] [PubMed]

146. Liu, K.; Yao, X.; Jiang, L. Recent developments in bio-inspired special wettability. *Chem. Soc. Rev.* **2010**, *39*, 3240–3255. [CrossRef]

147. Nakajima, A. Design of Transparent Hydrophobic Coating. *J. Ceram. Soc. Jpn.* **2004**, *112*, 533–540. [CrossRef]

148. Latthe, S.S.; Sutar, R.S.; Kodag, V.S.; Bhosale, A.K.; Kumar, A.M.; Sadasivuni, K.K.; Xing, R.; Liu, S. Self-cleaning superhydrophobic coatings: Potential industrial applications. *Prog. Org. Coat.* **2019**, *128*, 52–58. [CrossRef]

149. Nguyen-Tri, P.; Tran, H.N.; Plamondon, C.O.; Tuduri, L.; Vo, D.V.N.; Nanda, S.; Mishra, A.; Chao, H.P.; Bajpai, A.K. Recent progress in the preparation, properties and applications of superhydrophobic nano-based coatings and surfaces: A review. *Prog. Org. Coat.* **2019**, *132*, 235–256. [CrossRef]

150. Yu, S.; Guo, Z.; Liu, W. Biomimetic transparent and superhydrophobic coatings: From nature and beyond nature. *Chem. Commun.* **2015**, *51*, 1775–1794. [CrossRef] [PubMed]

151. Lee, H.J.; Michielsen, S.J. Lotus effect: Superhydrophobicity. *J. Text. Inst.* **2006**, *97*, 455–462. [CrossRef]

152. Rossbach, V.; Patanathabutr, P.; Wichitwechkarn, J. Copying and manipulating nature: Innovation for textile materials. *Fibers Polym.* **2003**, *4*, 8–14. [CrossRef]

153. Li, X.M.; Reinhoudt, D.; Crego-Calama, M. What do we need for a superhydrophobic surface? A review on the recent progress in the preparation of superhydrophobic surfaces. *Chem. Soc. Rev.* **2007**, *36*, 1350–1368. [CrossRef]

154. Wang, X.; Ding, H.; Wang, C.; Zhou, R.; Li, Y.; Ao, W. Self-healing superhydrophobic A-SiO$_2$/N-TiO$_2$@ HDTMS coating with self-cleaning property. *Appl. Surf. Sci.* **2021**, *567*, 150808. [CrossRef]

155. Xu, Q.; Wang, X.; Zhang, Y. Green and sustainable fabrication of a durable superhydrophobic cotton fabric with self-cleaning properties. *Int. J. Biol. Macromol.* **2023**, *242*, 124731. [CrossRef] [PubMed]

156. Dong, X.; Meng, J.; Hu, Y.; Wei, X.; Luan, X.; Zhou, H. Fabrication of self-cleaning superhydrophobic surfaces with improved corrosion resistance on 6061 aluminum alloys. *Micromachines* **2020**, *11*, 159. [CrossRef] [PubMed]

157. Shao, W.; Liu, D.; Cao, T.; Cheng, H.; Kuang, J.; Deng, Y.; Xie, W. Study on Favorable Comprehensive Properties of Superhydrophobic Coating Fabricated by Polytetrafluoroethylene Doped with Graphene. *Adv. Compos. Hybrid Mater.* **2021**, *4*, 521–533. [CrossRef]

158. Nodoushan, R.M.; Shekarriz, S.; Shariatinia, Z.; Montazer, M.; Heydari, A. Multifunctional carbonized Zn-MOF coatings for cotton fabric: Unveiling synergistic effects of superhydrophobic, oil-water separation, self-cleaning, and UV protection features. *Surf. Coat. Technol.* **2023**, *25*, 130194. [CrossRef]

159. Wang, K.; Yu, S.; Li, W.; Song, Y.; Gong, P.; Zhang, M.; Li, H.; Sun, D.; Yang, X.; Wang, X. Superhydrophobic and photocatalytic synergistic Self-Cleaning ZnS coating. *Appl. Surf. Sci.* **2022**, *595*, 153565. [CrossRef]

160. Hossain, M.M.; Wu, S.; Nasir, A.; Mohotti, D.; Yuan, Y.; Agyekum-Oduro, E.; Akter, A.; Bhuiyan, K.A.; Ahmed, R.; Nguyen, V.T.; et al. Superhydrophobic and superoleophilic surfaces prepared by one-step plasma polymerization for oil-water separation and self-cleaning function. *Surf. Interfaces* **2022**, *35*, 102462. [CrossRef]

161. Zang, D.; Xun, X.; Gu, Z.; Dong, J.; Pan, T.; Liu, M. Fabrication of superhydrophobic self-cleaning manganese dioxide coatings on Mg alloys inspired by lotus flower. *Ceram. Int.* **2020**, *46*, 20328–20334. [CrossRef]
162. Khairudin, A.; Shaharuddin, A.S.; Hanafi, M.F.; Sapawe, N. Study of self-cleaning superhydrophobic surface based on titanium dioxide nanomaterial. *Mater. Today Proc.* **2020**, *31*, A63–A66. [CrossRef]
163. He, H.; Guo, Z. Superhydrophobic materials used for anti-icing Theory, application, and development. *iScience* **2021**, *24*, 103357. [CrossRef]
164. Kabardin, I.; Dvoynishnikov, S.; Gordienko, M.; Kakaulin, S.; Ledovsky, V.; Gusev, G.; Zuev, V.; Okulov, V. Optical methods for measuring icing of wind turbine blades. *Energies* **2021**, *14*, 6485. [CrossRef]
165. Zheng, H.; Chang, S.; Ma, G.; Wang, S. Anti-icing performance of superhydrophobic surface fabricated by femtosecond laser composited dual-layers coating. *Energy Build.* **2020**, *223*, 110175. [CrossRef]
166. Xiao, X.; Wei, X.; Wei, J.; Wang, J. Multifunctional Fe_3O_4-based photothermal superhydrophobic composite coating for efficient anti-icing/deicing. *Sol. Energy Mater. Sol. Cells* **2023**, *256*, 112313. [CrossRef]
167. Fan, L.; Xia, M.; Liu, J.; Li, B.; Zhu, T.; Zhao, Y.; Song, L.; Yuan, Y. Fabrication of Superhydrophobic Coatings by Using Spraying and Analysis of Their Anti-Icing Properties. *Coatings* **2023**, *13*, 1792. [CrossRef]
168. Li, X.; Su, H.; Tan, X.; Lin, H.; Wu, Y.; Xiong, X.; Li, Z.; Jiang, L.; Xiao, T.; Chen, W. Photothermal superhydrophobic surface with good corrosion resistance, anti-/de-icing property and mechanical robustness fabricated via multiple-pulse laser ablation. *Appl. Surf. Sci.* **2023**, *646*, 158944. [CrossRef]
169. Barthwal, S.; Lee, B.; Lim, S.H. Fabrication of robust and durable slippery anti-icing coating on textured superhydrophobic aluminum surfaces with infused silicone oil. *Appl. Surf. Sci.* **2019**, *496*, 143677. [CrossRef]
170. Lei, Y.; Jiang, B.; Liu, H.; Zhang, F.; An, Y.; Zhang, Y.; Xu, J.; Li, X.; Liu, T. Mechanically robust superhydrophobic polyurethane coating for anti-icing application. *Prog. Org. Coat.* **2023**, *183*, 107795. [CrossRef]
171. Barthwal, S.; Lim, S.H. Rapid fabrication of a dual-scale micro-nanostructured superhydrophobic aluminum surface with delayed condensation and ice formation properties. *Soft Matter* **2019**, *15*, 7945–7955. [CrossRef] [PubMed]
172. Wang, L.; Teng, C.; Liu, J.; Wang, M.; Liu, G.; Kim, J.Y.; Mei, Q.; Lee, J.K.; Wang, J. Robust anti-icing performance of silicon wafer with hollow micro-/nano-structured ZnO. *J. Ind. Eng. Chem.* **2018**, *62*, 46–51. [CrossRef]
173. Wang, J.; Li, P.; Yu, P. Efficient photothermal deicing employing superhydrophobic plasmonic MXene composites. *Adv. Compos. Hybrid Mater.* **2022**, *5*, 3035–3044. [CrossRef]
174. Lin, H.; Wang, X.; Yu, L.; Chen, Y.; Shi, J. Two-dimensional ultrathin MXene ceramic nanosheets for photothermal conversion. *Nano Lett.* **2017**, *17*, 384–391. [CrossRef] [PubMed]
175. Liu, G.; Zou, J.; Tang, Q.; Yang, X.; Zhang, Y.; Zhang, Q.; Huang, W.; Chen, P.; Shao, J.; Dong, X. Surface modified Ti_3C_2 MXene nanosheets for tumor targeting photothermal/photodynamic/chemo synergistic therapy. *ACS Appl. Mater. Inter.* **2017**, *9*, 40077–40086. [CrossRef]
176. Wu, H.; Sun, H.; Han, F.; Xie, P.; Zhong, Y.; Quan, B.; Zhao, Y.; Liu, C.; Fan, R.; Guo, Z. Negative permittivity behavior in flexible carbon nanofibers-polydimethylsiloxane films. *Eng. Sci.* **2021**, *17*, 113–120. [CrossRef]
177. Chaudhuri, K.; Alhabeb, M.; Wang, Z.; Shalaev, V.M.; Gogotsi, Y.; Boltasseva, A. Highly broadband absorber using plasmonic titanium carbide (MXene). *ACS Photonics* **2018**, *5*, 1115–1122. [CrossRef]
178. Li, R.; Zhang, L.; Shi, L.; Wang, P. MXene Ti_3C_2: An effective 2D light-to-heat conversion material. *ACS Nano* **2017**, *11*, 3752–3759. [CrossRef]
179. Herbots, N.; Watson, C.F.; Culbertson, E.J.; Acharya, A.J.; Thilmany, P.R.; Marsh, S.; Marsh, R.T.; Martins, I.P.; Watson, G.P.; Mascareno, A.M.; et al. Super-Hydrophilic, Bio-compatible Anti-Fog Coating for Lenses in Closed Body Cavity Surgery: VitreOxTM–Scientific Model, In Vitro Experiments and In Vivo Animal Trials. *MRS Adv.* **2016**, *1*, 2141–2146. [CrossRef]
180. Durán, I.R.; Laroche, G. Water drop-surface interactions as the basis for the design of anti-fogging surfaces: Theory, practice, and applications trends. *Adv. Colloid Interface Sci.* **2019**, *263*, 68–94. [CrossRef]
181. Wahab, I.F.; Razak, B.A.; Teck, S.W.; Azmi, T.T.; Ibrahim, M.Z.; Lee, J.W. Fundamentals of antifogging strategies, coating techniques and properties of Inorganic materials; a comprehensive review. *J. Mater. Res. Technol.* **2023**, *23*, 687–714. [CrossRef]
182. Huang, L.; Wang, T.; Li, X.; Wang, X.; Zhang, W.; Yang, Y.; Tang, Y. UV-to-IR highly transparent ultrathin diamond nanofilms with intriguing performances: Anti-fogging, self-cleaning and self-lubricating. *Appl. Surf. Sci.* **2020**, *527*, 146733. [CrossRef]
183. Varshney, P.; Mohapatra, S.S. Durable and regenerable superhydrophobic coatings for brass surfaces with excellent self-cleaning and anti-fogging properties prepared by immersion technique. *Tribol. Int.* **2018**, *123*, 17–25. [CrossRef]
184. Li, W.; Chen, Y.; Jiao, Z. Efficient Anti-Fog and Anti-Reflection Functions of the Bio-Inspired, Hierarchically-Architectured Surfaces of Multiscale Columnar Structures. *Nanomaterials* **2023**, *13*, 1570. [CrossRef] [PubMed]
185. Yu, L.; Han, M.; He, F. A review of treating oily wastewater. *Arab. J. Chem.* **2017**, *10*, S1913–S1922. [CrossRef]
186. Medeiros, A.D.L.M.D.; Junior, C.J.G.D.S.; Amorim, J.D.P.D.; Durval, I.J.B.; Costa, A.F.D.S.; Sarubbo, L.A. Oily wastewater treatment: Methods, challenges, and trends. *Processes* **2022**, *10*, 743. [CrossRef]
187. Croce, D.; Pereyra, E. Study of oil/water flow and emulsion formation in electrical submersible pumps. *SPE Prod. Oper.* **2020**, *35*, 026–036. [CrossRef]
188. Hansen, B.H.; Salaberria, I.; Read, K.E.; Wold, P.A.; Hammer, K.M.; Olsen, A.J.; Altin, D.; Øverjordet, I.B.; Nordtug, T.; Bardal, T.; et al. Developmental effects in fish embryos exposed to oil dispersions–The impact of crude oil micro-droplets. *Mar. Environ. Res.* **2019**, *150*, 104753. [CrossRef] [PubMed]

189. Rabbani, Y.; Shariaty-Niassar, M.; Ebrahimi, S.S. The effect of superhydrophobicity of prickly shape carbonyl iron particles on the oil-water adsorption. *Ceram. Int.* **2021**, *47*, 28400–28410. [CrossRef]

190. Bai, L.; Greca, L.G.; Xiang, W.; Lehtonen, J.; Huan, S.; Nugroho, R.W.N.; Tardy, B.L.; Rojas, O.J. Adsorption and assembly of cellulosic and lignin colloids at oil/water interfaces. *Langmuir* **2018**, *35*, 571–588. [CrossRef]

191. Saththasivam, J.; Ogunbiyi, O.; Lawler, J.; Al-Rewaily, R.; Liu, Z. Evaluating dissolved air flotation for oil/water separation using a hybridized coagulant of ferric chloride and chitosan. *J. Water Process. Eng.* **2022**, *47*, 102836. [CrossRef]

192. Lee, J.; Cho, W.C.; Poo, K.M.; Choi, S.; Kim, T.N.; Son, E.B.; Choi, Y.J.; Kim, Y.M.; Chae, K.J. Refractory oil wastewater treatment by dissolved air flotation, electrochemical advanced oxidation process, and magnetic biochar integrated system. *J. Water Process. Eng.* **2020**, *36*, 101358. [CrossRef]

193. Almaraz, N.; Regnery, J.; Vanzin, G.F.; Riley, S.M.; Ahoor, D.C.; Cath, T.Y. Emergence and fate of volatile iodinated organic compounds during biological treatment of oil and gas produced water. *Sci. Total Environ.* **2020**, *699*, 134202. [CrossRef]

194. Benamar, A.; Mahjoubi, F.Z.; Barka, N.; Kzaiber, F.; Boutoial, K.; Ali, G.A.; Oussama, A. Olive mill wastewater treatment using infiltration percolation in column followed by aerobic biological treatment. *SN Appl. Sci.* **2020**, *2*, 1–12. [CrossRef]

195. Kong, D.; He, X.; Khan, F.; Chen, G.; Ping, P.; Yang, H.; Peng, R. Small scale experiment study on burning characteristics for in-situ burning of crude oil on open water. *J. Loss Prev. Process Ind.* **2019**, *60*, 46–52. [CrossRef]

196. Dansawad, P.; Yang, Y.; Li, X.; Shang, X.; Li, Y.; Guo, Z.; Qing, Y.; Zhao, S.; You, S.; Li, W. Smart membranes for oil/water emulsions separation: A review. *Adv. Membr.* **2022**, *2*, 100039. [CrossRef]

197. Cai, Y.; Yu, Y.; Wu, J.; Wang, K.; Dong, Y.; Qu, J.; Hu, J.; Zhang, L.; Fu, Q.; Li, J.; et al. Durable, flexible, and super-hydrophobic wood membrane with nanopore by molecular cross-linking for efficient separation of stabilized water/oil emulsions. *EcoMat* **2022**, *4*, e12255. [CrossRef]

198. Jie, P.; Leilei, Y.; Wenheng, H.; Kun, C. Preparation of Superhydrophobic Copper Mesh for Highly Efficient Oil-Water Separation. *Int. J. Electrochem. Sci.* **2023**, *19*, 100417. [CrossRef]

199. Xu, P.; Li, X. Fabrication of TiO2/SiO2 superhydrophobic coating for efficient oil/water separation. *J. Environ. Chem. Eng.* **2021**, *9*, 105538. [CrossRef]

200. Kao, L.H.; Lin, W.C.; Huang, C.W.; Tsai, P.S. Fabrication of Robust and Effective Oil/Water Separating Superhydrophobic Textile Coatings. *Membranes* **2023**, *13*, 401. [CrossRef] [PubMed]

201. Barthwal, S.; Jeon, Y.; Lim, S.H. Superhydrophobic sponge decorated with hydrophobic MOF-5 nanocoating for efficient oil-water separation and antibacterial applications. *Sustain. Mater. Technol.* **2022**, *33*, 00492. [CrossRef]

202. Barthwal, S.; Lim, S.H. A durable, fluorine-free, and repairable superhydrophobic aluminum surface with hierarchical micro/nanostructures and its application for continuous oil-water separation. *J. Membr. Sci.* **2021**, *618*, 118716. [CrossRef]

203. Barthwal, S.; Barthwal, S.; Singh, B.; Singh, N.B. Multifunctional and fluorine-free superhydrophobic composite coating based on PDMS modified MWCNTs/ZnO with self-cleaning, oil-water separation, and flame retardant properties. *Colloids Surf. A Physicochem. Eng. Asp.* **2020**, *597*, 124776. [CrossRef]

204. Jin, H.; Tian, L.; Bing, W.; Zhao, J.; Ren, L. Bioinspired marine antifouling coatings: Status, prospects, and future. *Prog. Mater. Sci.* **2022**, *124*, 100889. [CrossRef]

205. Yin, X.; Yu, S.; Wang, K.; Cheng, R.; Lv, Z. Fluorine-free preparation of self-healing and anti-fouling superhydrophobic Ni3S2 coating on 304 stainless steel. *Chem. Eng. J.* **2020**, *394*, 124925. [CrossRef]

206. Shi, S.; Meng, S.; Zhao, P.; Xiao, G.; Yuan, Y.; Wang, H.; Liu, T.; Wang, N. Underwater adhesion and curing of superhydrophobic coatings for facile antifouling applications in seawater. *Compos. Commun.* **2023**, *38*, 101511. [CrossRef]

207. Liu, K.; Jiang, L. Metallic surfaces with special wettability. *Nanoscale* **2011**, *3*, 825–838. [CrossRef] [PubMed]

208. Huang, W.F.; Xiao, Y.L.; Huang, Z.J.; Tsui, G.C.; Yeung, K.W.; Tang, C.Y.; Liu, Q. Super-hydrophobic polyaniline-TiO2 hierarchical nanocomposite as anticorrosion coating. *Mater. Lett.* **2020**, *258*, 126822. [CrossRef]

209. Rius-Ayra, O.; Llorca-Isern, N. A robust and anticorrosion non-fluorinated superhydrophobic aluminium surface for microplastic removal. *Sci. Total Environ.* **2021**, *760*, 144090. [CrossRef]

210. Zhang, Z.; Zhao, N.; Qi, F.; Zhang, B.; Liao, B.; Ouyang, X. Reinforced superhydrophobic anti-corrosion epoxy resin coating by fluorine–silicon–carbide composites. *Coatings* **2020**, *10*, 1244. [CrossRef]

211. Barthwal, S.; Lim, S.H. Robust and chemically stable superhydrophobic aluminum-alloy surface with enhanced corrosion-resistance properties. *Int. J. Precis. Eng. Manuf.-Green Technol.* **2020**, *7*, 481–492. [CrossRef]

212. Huang, M.; Lu, G.; Pu, J.; Qiang, Y. Superhydrophobic and smart MgAl-LDH anti-corrosion coating on AZ31 Mg surface. *J. Ind. Eng. Chem.* **2021**, *103*, 154–164. [CrossRef]

213. Xiang, Y.; Tang, W.; Li, H.; Zhang, Y.; Song, R.; Liu, B.; He, Y.; Guo, X.; He, Z. Fabrication of robust Ni-based TiO2 composite@TTOS superhydrophobic coating for wear resistance and anti-corrosion. *Colloids Surf. A Physicochem. Eng. Asp.* **2021**, *629*, 127394. [CrossRef]

214. Xu, S.; Wang, Q.; Wang, N.; Qu, L.; Song, Q. Study of corrosion property and mechanical strength of eco-friendly fabricated superhydrophobic concrete. *J. Clean. Prod.* **2021**, *323*, 129267. [CrossRef]

215. Wu, M.; Ma, B.; Pan, T.; Chen, S.; Sun, J. Silver-nanoparticle colored cotton fabrics with tunable colors and durable antibacterial and self-healing superhydrophobic properties. *Adv. Funct. Mater.* **2016**, *26*, 569–576. [CrossRef]

216. Ashok, D.; Cheeseman, S.; Wang, Y.; Funnell, B.; Leung, S.F.; Tricoli, A.; Nisbet, D. Superhydrophobic surfaces to combat bacterial surface colonization. *Adv. Mater. Interfaces* **2023**, *10*, 2300324. [CrossRef]

217. Yang, K.; Shi, J.; Wang, L.; Chen, Y.; Liang, C.; Yang, L.; Wang, L. Bacterial anti-adhesion surface design: Surface patterning, roughness and wettability: A review. *J. Mater. Sci. Technol.* **2022**, *99*, 82–100. [CrossRef]
218. Zhang, X.; Wang, L.; Lev, E. Superhydrophobic surfaces for the reduction of bacterial adhesion. *RSC Adv* **2013**, *3*, 12003–12020. [CrossRef]
219. Barthwal, S.; Barthwal, S.; Robust, E.A. Multifunctional Superhydrophobic/Oleophobic Microporous Aluminum Surface via a Two-Step Chemical Etching Process. *Surf. Interfaces* **2024**, *46*, 103933. [CrossRef]
220. Ye, M.; Wang, S.; Ji, X.; Tian, Z.; Dai, L.; Si, C. Nanofibrillated cellulose-based superhydrophobic coating with antimicrobial performance. *Adv. Compos. Hybrid Mater.* **2023**, *6*, 30. [CrossRef]
221. Agbe, H.; Sarkar, D.K.; Chen, X.-G. Tunable superhydrophobic aluminum surfaces with anti-biofouling and antibacterial properties. *Coatings* **2020**, *10*, 982. [CrossRef]
222. Ju, J.; Bai, H.; Zheng, Y.; Zhao, T.; Fang, R.; Jiang, L. A multi-structural and multi-functional integrated fog collection system in cactus. *Nat. Commun.* **2012**, *3*, 1247. [CrossRef]
223. Zheng, Y.; Bai, H.; Huang, Z.; Tian, X.; Nie, F.Q.; Zhao, Y.; Zhai, J.; Jiang, L. Directional water collection on wetted spider silk. *Nature* **2010**, *463*, 640–643. [CrossRef]
224. Chen, H.; Zhang, P.; Zhang, L.; Liu, H.; Jiang, Y.; Zhang, D.; Han, Z.; Jiang, L. Continuous directional water transport on the peristome surface of Nepenthes alata. *Nature* **2016**, *532*, 85–89. [CrossRef] [PubMed]
225. Parker, A.R.; Lawrence, C.R. Water capture by a desert beetle. *Nature* **2001**, *414*, 33–34. [CrossRef]
226. Gao, Y.; Wang, J.; Xia, W.; Mou, X.; Cai, Z. Reusable hydrophilic-superhydrophobic patterned weft backed woven fabric for high-efficiency water-harvesting application. *ACS Sustain. Chem. Eng.* **2018**, *6*, 7216–7220. [CrossRef]
227. Kim, H.; Yang, S.; Rao, S.R.; Narayanan, S.; Kapustin, E.A.; Furukawa, H.; Umans, A.S.; Yaghi, O.M.; Wang, E.N. Water harvesting from air with metal-organic frameworks powered by natural sunlight. *Science* **2017**, *356*, 430–434. [CrossRef]
228. Do, V.; Chun, D. Facile fabrication of extreme-wettability contrast surfaces for efficient water harvesting using hydrophilic and hydrophobic silica nanoparticles. *Colloids Surf. A Physicochem. Eng. Asp.* **2023**, *671*, 131664. [CrossRef]
229. Yu, Z.; Zhang, H.; Huang, J.; Li, S.; Zhang, S.; Cheng, Y.; Mao, J.; Dong, X.; Gao, S.; Wang, S.; et al. Namib desert beetle inspired special patterned fabric with programmable and gradient wettability for efficient fog harvesting. *J. Mater. Sci. Technol.* **2021**, *61*, 85–92. [CrossRef]
230. Bai, H.; Zhao, T.; Wang, X.; Wu, Y.; Li, K.; Yu, C.; Jiang, L.; Cao, M. Cactus kirigami for efficient fog harvesting: Simplifying a 3D cactus into 2D paper art. *J. Mater. Chem. A* **2020**, *8*, 13452–13458. [CrossRef]
231. Zhai, L.; Berg, M.C.; Cebeci, F.C.; Kim, Y.; Milwid, J.M.; Rubner, M.F.; Cohen, R.E. Patterned Superhydrophobic Surfaces: Toward a Synthetic Mimic of the Namib Desert Beetle. *Nano Lett.* **2006**, *6*, 1213–1217. [CrossRef] [PubMed]
232. Park, J.K.; Kim, S. Three-Dimensionally Structured Flexible Fog Harvesting Surfaces Inspired by Namib Desert Beetles. *Micromachines* **2019**, *10*, 201. [CrossRef] [PubMed]
233. Choi, Y.; Baek, K.; So, H. 3D-printing-assisted fabrication of hierarchically structured biomimetic surfaces with dual-wettability for water harvesting. *Sci. Rep.* **2023**, *13*, 10691. [CrossRef]
234. Das, S.; Kumar, S.; Samal, S.K.; Mohanty, S.; Nayak, S.K. A review on superhydrophobic polymer nanocoatings: Recent development and applications. *J. Ind. Eng. Chem.* **2018**, *57*, 2727–2745. [CrossRef]
235. Sun, Z.; Ding, L.; Tong, W.; Ma, C.; Yang, D.; Guan, X.; Xiao, Y.; Xu, K.; Li, Q.; Lv, C. Superhydrophobic blood-repellent tubes for clinical cardiac surgery. *Mater. Des.* **2023**, *232*, 112148. [CrossRef]
236. Cao, D.; Zhang, Y.; Shi, X.; Gong, H.; Feng, D.; Guo, X.; Shi, Z.; Zhu, S.; Cui, Z. Fabrication of superhydrophobic coating for preventing microleakage in a dental composite restoration. *Mater. Sci. Eng. C* **2017**, *78*, 333–340. [CrossRef] [PubMed]
237. Pyzer-Knapp, E.O.; Pitera, J.W.; Staar, P.W.J.; Takeda, S.; Laino, T.; Sanders, D.P.; Sexton, J.; Smith, J.R.; Curioni, A. Accelerating materials discovery using artificial intelligence, high performance computing and robotics. *NPJ Comput. Mater.* **2022**, *8*, 84. [CrossRef]
238. Li, J.; Lim, K.; Yang, H.; Ren, Z.; Raghavan, S.; Chen, P.-O.; Buonassisi, T.; Wang, X. AI Applications through the Whole Life Cycle of Material Discovery. *Matter* **2020**, *3*, 393–432. [CrossRef]
239. Lu, W.; Lee, N.A.; Buehler, M.J. Modeling and design of heterogeneous hierarchical bioinspired spider web structures using deep learning and additive manufacturing. *Proc. Natl. Acad. Sci. USA* **2023**, *120*, e2305273120. [CrossRef] [PubMed]
240. Studart, A.R. Biologically Inspired Dynamic Material Systems. *Angew. Chem. Int. Ed.* **2015**, *54*, 3400–3416. [CrossRef] [PubMed]
241. Sun, J.; Bhushan, B. Nanomanufacturing of bioinspired surfaces. *Tribol. Int.* **2019**, *129*, 67–74. [CrossRef]
242. Li, Y.; Zhang, Q.; Zhang, J.; Jin, L.; Zhao, X.; Xu, T. A top-down approach for fabricating free-standing bio-carbon supercapacitor electrodes with a hierarchical structure. *Sci. Rep.* **2015**, *5*, 14155. [CrossRef]
243. Aziz, M.S. Biomimicry as an approach for bio-inspired structure with the aid of computation. *Alex. Eng. J.* **2016**, *55*, 707–714. [CrossRef]
244. Bonfanti, S.; Guerra, R.; Zaiser, M.; Zapperi, S. Digital strategies for structured and architected materials design. *APL Mater.* **2021**, *9*, 020904. [CrossRef]
245. Yu, C.H.; Tseng, B.Y.; Yang, Z.; Tung, C.C.; Zhao, E.; Ren, Z.F.; Yu, S.S.; Chen, P.Y.; Chen, C.S.; Buehler, M.J. Hierarchical multiresolution design of bioinspired structural composites using progressive reinforcement learning. *Adv. Theory Simul.* **2022**, *5*, 2200459. [CrossRef]

246. Lantada, A.D.; Franco-Martínez, F.; Hengsbach, S.; Rupp, F.; Thelen, R.; Bade, K. Artificial intelligence aided design of microtextured surfaces: Application to controlling wettability. *Nanomaterials* **2020**, *10*, 2287. [CrossRef] [PubMed]
247. Gu, G.X.; Chen, C.-T.; Richmond, D.J.; Buehler, M.J. Bioinspired hierarchical composite design using machine learning: Simulation, additive manufacturing, and experiment. *Mater. Horiz.* **2018**, *5*, 939–945. [CrossRef]

Article

The Role of APTES as a Primer for Polystyrene Coated AA2024-T3

John Halford IV and Cheng-fu Chen *

Department of Mechanical Engineering, University of Alaska Fairbanks, Fairbanks, AK 99775-5905, USA; jhhalfordiv@alaska.edu
* Correspondence: cchen4@alaska.edu; Tel.: +1-(907)-474-7265

Abstract: (3-Aminopropyl)triethoxysilane (APTES) silane possesses one terminal amine group and three ethoxy groups extending from each silicon atom, acting as a crucial interface between organic and inorganic materials. In this study, after APTES was deposited on the aluminum alloy AA2024-T3 as a primer for an optional top coating with polystyrene (PS), its role with regard to stability as a protection layer and interaction with the topcoat were studied via combinatorial experimentation. The aluminum alloy samples primed with APTES under various durations of concentrated vapor deposition (20, 40, or 60 min) with an optional post heat treatment and/or PS topcoat were comparatively characterized via electrochemical impedance spectroscopy (EIS) and surface energy. The samples top-coated with PS on an APTES layer primed for 40 min with a post heat treatment revealed excellent performance regarding corrosion impedance. A primed APTES surface with higher surface energy accounted for this higher corrosion impedance. Based on the SEM images and the surface energy calculated from the measured contact angles on the APTES-primed surfaces, four mechanisms are suggested to explain that the good protection performance of the APTES/PS coating system can be attributed to the enhanced wettability of PS on the cured APTES primer with higher surface energy. The results also suggest that, in the early stages of exposure to the corrosion solution, a thinner APTES primer (deposited for 20 min) enhances protection against corrosion, which can be attributed to the hydrolytic stability and hydrolyzation/condensation of the soaked APTES and the dissolution of the naturally formed aluminum oxide pre-existing in the bare samples. An APTES primer subjected to additional heat treatment will increase the impedance of the coating system significantly. APTES, and silanes, in general, used as adherent agents or surface modifiers, have a wide range of potential applications in micro devices, as projected in the Discussion section.

Keywords: vapor deposition; coating; primer; polystyrene; APTES; silanes; AA2024-T3; corrosion; electrochemical impedance spectroscopy; surface energy

Citation: Halford, J., IV; Chen, C.-f. The Role of APTES as a Primer for Polystyrene Coated AA2024-T3. *Micromachines* **2024**, *15*, 93. https://doi.org/10.3390/mi15010093

Academic Editor: Amir Hussain Idrisi

Received: 13 December 2023
Revised: 26 December 2023
Accepted: 29 December 2023
Published: 31 December 2023

1. Introduction

Silanes and silane esters have demonstrated considerable promise as corrosion-resistant adhesion promoters [1] and undercoats [2] or for use in paint emulsions [3]. APTES, a reactive silane ester (trialkoxysilane) known for its grafting capabilities, enhances compatibility between minerals and organic polymers. APTES can be applied for surface functionalization as a primer or for the surface modification of nanoparticles to form nanocomposites; both applications involve using APTES as a surface-priming agent for top coating with functionalized polymers. Including additional compounds, including magnetite nanoparticles for their magnetic properties, silicate nanoparticles for enhanced strength, titanium dioxide nanoparticles for UV resistance, polyaniline for conductivity, and graphene/carbon nanotubes for reinforcement, further tailors nanocomposites' functionalities [4]. Determining how APTES can be grafted for surface modification or priming is crucial to making a stable and durable coating system.

APTES has been examined for use as a primer compound or as a compatibilizer between organic and inorganic materials [5]. As a primer or adhesion promoter, APTES has a terminal amine group that is less sterically constrained to the silanols, as can be demonstrated by the chair conformational isomer nitrogen–oxygen distance of 0.245 nm in hydrolyzed APTES [5]. The spatial relationship of these functional groups facilitates pH-dependent hydrolysis and the formation of denser networks of oligomers [6]. Tri-alkoxysilanes hydrolyze quickly (in minutes), forming free silanols, but tend to condense much more slowly (in the order of hours). The networking stability of APTES is governed by the competition between the hydrolysis and condensation mechanisms, which can be optimized at a pH value of around 6, given that the rates of condensation and hydrolyzation have minima at pH 4 and 7.5, respectively, for a typical silane [5,7]. Accessibility to water is a key factor in determining the performance of APTES as a primer for the bonding of organic/inorganic materials. Although silane coatings often enhance adhesion, silanes such as γ-aminopropyl silane (γ-APS) perform poorly when immersed in water and may fail wet adhesion tests [2]. Silanes like APTES do not bond to the Cu-rich sites on the surface of aluminum alloys, which are highly susceptible to local pitting [1]. Therefore, silane coatings are typically used as a primer for additional top coatings. In this context, silane compounds are recognized for their potential as effective adhesion promoters between organic coatings and oxidized metal surfaces.

Applications of silanes can be classified based on their grafting functionality on inorganic surfaces [8,9] or their integrity with regard to top-coating polymers for resistance to coating rupture and disbonding [10–12]. A TiO_2-APTES nanocomposite was scrutinized for its interfacial stability and failure [8]. PS has emerged as a potential topcoat due to its efficient water repellency and relatively low surface energy of 31 mJ/m^2 [13]. The non-polar and chemically inert characteristics of PS, owing to its aromatic benzene ring, make it resistant to reactions with acids and bases, simplifying corrosion analysis [13]. PS in a solution form can be directly spin-coated onto the surface of metals like aluminum alloys. However, its effectiveness as a protective coating is limited by its poor adhesion to metals and susceptibility to thin-film rupture when present as a dried film. The dewetting of PS on monolayered APTES was addressed in [12]. Zhang recently addressed the thin-film porosity and integrity issues relating to PS [14] by developing a densely compacted polystyrene/TiO_2 nanocomposite coating. A PS layer can be applied as a topcoat over a silane [15]. This approach effectively mitigates the individual drawbacks of silane and PS coatings. The top-coated PS acts as a barrier, safeguarding the underlying silane primer from water attacks. Simultaneously, the silane primer plays a dual role as an additional protective layer for the metal substrate and an adhesive agent, indirectly binding the PS to the substrate.

However, the role the stacked silane/polymer coating plays after coating failure has not been well characterized, except in a few works [12,15,16]. Chen [15] applied an APTES/toluene solution to a hydroxylated AA2024-T3 surface. The sample surface had been grafted with the hydroxyl group before being soaked in a diluted APTES/anhydrous toluene solution to enable the formation of siloxane bonds or hydrogen bonds with APTES; however, through soaking, the primed APTES layer became thicker and less uniform than a monolayer [6,17], adding undesired complexity in the networked structure. An APTES primer prepared via solvent-based soaking also exhibits a physisorption behavior, yielding a less-stable networked structure because the un-bonded/loosened ethoxy groups hidden in the cured APTES can migrate to the network surface and then react with water for hydrolyzation and subsequent condensation, both of which will change the thickness of the cured APTES layer [17].

In this work, the aluminum alloy AA2024-T3 was chosen for priming with APTES and the subsequent topcoat with PS to characterize the role of the APTES primer in the stacked coating system. AA2024-T3 has been extensively addressed in the literature regarding surface treatments [18]. In this study, the method of concentrated vapor deposition in low-vacuum conditions was used to prime APTES for coating onto the naturally formed aluminum oxide layer of the AA2024-T3 substrate to minimize the accessibility to humidity during deposition. The APTES-primed AA2024 samples, with an optional additional heat treatment, were then top-coated with PS to form a stacked coating system, which is characterized using EIS in a 3.5% NaCl solution. The surface energy of the coating was estimated from measured static contact angles of two working liquids on the coating. The characterization, together with SEM images of the coating, was analyzed for the causes of failures and explained by four hypothetical mechanisms.

2. Materials and Methods

2.1. Materials

Bare samples measuring 1 in. × 1 in. were machined from 5 mm thick wrought AA2024-T3 sheets bought from Online Metals. The cut samples underwent sonication in 97% isopropyl alcohol and then in deionized water, each lasting 5 min. Subsequently, the samples were air-dried, baked at 95 °C for 120 min, and left in an oven overnight. The cleaned samples were then stored in a petri dish sealed with parafilm for later use.

Low-molecular-weight, narrowly dispersed PS flakes (Mn 8000, Mw 8800) were purchased from Polymer Source Inc. (product ID 8096-S, Dorval, Canada). Technical-grade acetone of 94% purity was purchased from Sunnyside Corp. (Wheeling, IL, USA). A PS/acetone solution comprising 200 mg of the PS flakes dissolved in 2.28 mL of technical-grade acetone was prepared, resulting in a nominal PS concentration of 10%. APTES with a purity of $\geq$ 98% was purchased from Millipore Sigma and used without further purification. Milli-Q deionized water and anhydrous ethylene glycol with 99.8% purity (Sigma Aldrich, St. Louis, MO, USA) were used for surface contact angle measurements. For EIS measurements and subsequent in-cell corrosion, a fresh 3.5% NaCl solution was prepared from in-house deionized water and oven-dried, non-iodized, non-fluorinated food-grade NaCl.

2.2. APTES Primed on AA2024-T3 Samples

APTES was applied to the clean AA2024-T3 samples using concentrated vapor deposition. In this process, we employed a heated vapor deposition chamber (Figure 1a) pumped by a dual-stage rotary vacuum pump with the valving system depicted in Figure 2. The chamber undergoes evacuation with valves A and C closed. Once the desired vacuum is attained, valve B is closed to maintain the vacuum in the chamber. To cease pump operation, valve A alleviates backpressure on a rotary vane pump. Additionally, valve C can be optionally used to introduce a dry sparging gas for storage.

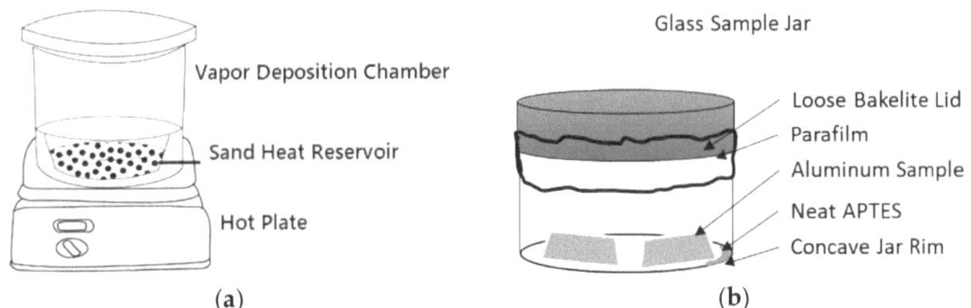

Figure 1. (**a**) Vacuum chamber used as vapor deposition chamber with a metal bowl filled with washed quartz sand; (**b**) 100 mL sample jar for vapor deposition set up for the vacuum chamber.

The vapor deposition chamber was bedded with industrially crushed quartz sand preheated to 150 °C under a vacuum of 10 mbar. Temperature was monitored with a thermocouple inserted into the vapor deposition chamber, while the chamber itself was heated through a hot plate set at 330 °C under ambient conditions. Afterward, a preheated and dried 100 mL glass sample jar (Figure 1b), holding cleaned aluminum samples and APTES reagent was loosely capped (parafilm was applied to the loose cap to reduce the speed of gas transfer in and out of the container). This jar was then loaded into the heated sand bath in the preheated vapor deposition chamber. While maintaining a temperature of 100 °C throughout the process, the vapor deposition chamber was restored to vacuum conditions under 100 mbar. Under vacuum and heated conditions, the APTES solution was gradually deposited onto the samples from the vapor phase. Three groups of samples placed inside the jar were exposed to the neat APTES within the vacuum chamber for 20, 40, or 60 min, respectively.

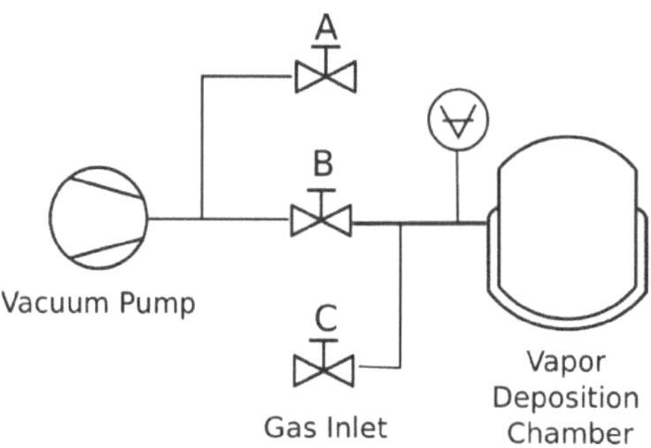

Figure 2. Schematic of vacuum system control.

Vapor reaction times for this method of coating silanes have been characterized for durations as short as 10 min elsewhere [19]. After completion, the aluminum samples were covered and left undisturbed overnight under ambient conditions. One-third of the samples was reserved for contact angle analysis. The remaining samples were used for the subsequent top-coating with the PS/acetone solution.

2.3. Polystyrene Top-Coated on APTES

Half of the remaining APTES-primed samples were spin-coated with 100 µL of acetone containing a 10% PS/acetone solution. The other half of the samples were prepared with further heat treatment and allowed to incubate at an elevated temperature under a pressure of 200 mbar for an additional 24 h to cure the APTES network further before being spin-coated with the 10% PS/acetone solution.

An Ossila brand E441 Spin Coater was programmed to apply 1500 rpm for 5 s, followed by 2000 rpm for another 5 s, and ending with 3000 rpm for 15 s. The coating solution was applied before the conclusion of the first step of the spin-coating program. After the spin-coating process, all the samples underwent 24 h of drying.

2.4. Sample Labeling

The samples in this experiment were given a three-digit code: "1-Y-Z". For samples primed with APTES vapor deposition for 20, 40, or 60 min, the "Y" values are 20, 40, and 60, respectively. Samples are further categorized based on the presence of polystyrene according to the label "Z", where 0 denotes those only primed with APTES, 1 denotes those with polystyrene top-coated on the "as-is" primed APTES, and 2 denotes those with polystyrene top-coated on the primed APTES that underwent further heat treatment.

2.5. Surface Characterization

The sample surfaces were characterized through contact angle measurements with a Rame-Hart goniometer using ethylene glycol and Milli-Q water. The sample lay flat on the surface of the goniometer stage, and 10 µL droplets were manually dispensed out of a 100 µL microsyringe onto the sample surface. The initial contact angle was recorded using DROPimage software (Standard Edition) (Ramé-Hart Instrument Co., Succasunna, NJ, USA). This measurement was replicated across the sample to determine a statistically representative sample.

Exploratory Scanning Electron Microscope (SEM) images were taken using an FEI Quanta200 system in low-vacuum mode. Image processing and scaling were carried out using the freely distributed GNU Image Manipulation Program (GIMP) software (gimp.org, version 2.10.36). All scaling was completed using the pixel length of the SEM Scaling bar.

2.6. Electrochemical Impedance Measurement

EIS was employed using a Gamry G300 potentiostat and the Gamry Echem Analyst 6.33 software (Gamry Instruments Inc., Warminster, PA, USA). Measurements were taken using a conventional three-electrode setup in a horizontal flat cell. A graphite counter electrode and Ag/AgCl/saturated KCl reference electrode were used. The sample was mounted in a 1 cm diameter, 0.78 cm^2 area window. The EIS data were imported into the Python module "impedance.py." The details of the data-fitting procedure are beyond the scope of this paper and will be reported separately, but a summary is given in the Supplementary Materials.

The EIS data of the coated samples were analyzed using a Randles circuit (Figure 3), which contains one time constant featuring two parallel passive elements, C_1 and R_1. Here, C_1 is a constant phase element (CPE) of the form $\frac{1}{Qs^n}$ used to account for non-Faradaic charge transfer, where Q and n are the CPE parameters; the resistance R_1 represents the impedance of the Faradaic current flows. A Warburg element was connected to R_1 to account for any diffusion-limiting reactions.

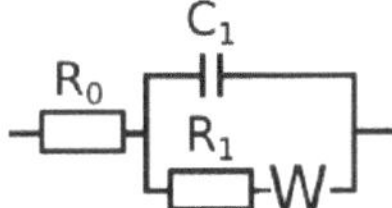

Figure 3. Equivalent circuits for analyzing EIS data: Randles model.

3. Results and Discussion
3.1. APTES Primer

Figure 4 compares the impedance between the bare sample and three types of samples primed with APTES vapor deposition for 20, 40, and 60 min, respectively. In the initial corrosion stage, roughly within the first 100 h, the impedance of the bare sample exceeded that of all APTES-primed samples. This behavior can be attributed to the presence of aluminum oxide and its interaction with the vapor-deposited APTES primer when exposed to the aqueous electrolyte solution, as elaborated in the Discussion section. Among the three tested APTES-primed layers, samples -20- and -40- exhibited similar and superior performance compared to the -60- sample.

In the later stage (after the 100 hr mark), the impedance of the bare sample decreased quickly and persistently, indicating that its alumina protection layer was damaged by corrosion. However, all the APTES-primed samples, regardless of their type, enabled larger impedance than the bare sample in this stage. The impedance of the APTES coating per -60- persistently increased over time; in contrast, the -20- and -40- coating layers began to lose their impedance in the middle of the corrosion period tested. The data points distributed in the Nyquist plots of all the APTES-primed samples exhibit one time constant, evidenced by the semicircle arc fitted to each set of the points. Some curves ended up with a feature of Warburg impedance in the lower-frequency regime, signaling the onset of a reaction under partial or complete mass transport control via diffusion [20].

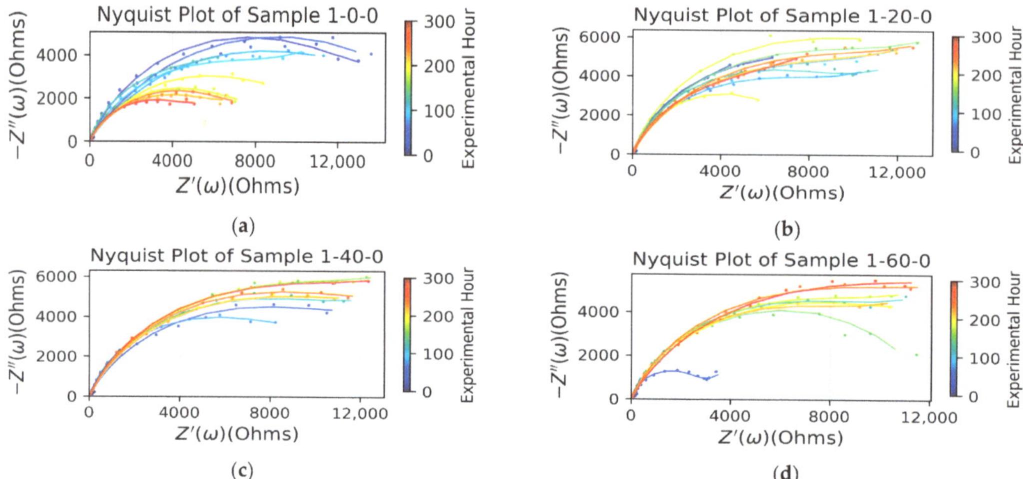

Figure 4. Nyquist plot of measured impedance data (dot) and EIS-fitted trendline over the duration of testing. (**a**) Bare sample and samples coated with a thin layer of APTES vapor-deposited for (**b**) 20 min, (**c**) 40 min, and (**d**) 60 min. All figures share an identical colormap legend regarding timing (in hours), starting from the beginning of an experimental measurement.

Figure 5 compares the fitted parameters of the Randles circuit model. The model shows that the resistance of the bare sample decreased monotonically over time, indicating the dissolution of the naturally formed oxide layer in the corrosion solution. For the APTES-coated samples, the following was revealed:

- In the first 100 h of corrosion, the 1-20-0 APTES-coated sample exhibited the largest resistance, while the 1-60-0 APTES coating had the least. The effective capacitance (as calculated via $Q\omega^n$) of the 1-20-0 sample remained relatively constant from the beginning of corrosion; in contrast, that of the 1-40-0 and 1-60-0 samples decreased over time. (Refer to the individual plots of Q and n in Figure S1 in the Supplementary Materials.) Together with the trendlines in the resistance and capacitance curves, it can be deduced that the change in the modes of electrons and or ion exchange involved in any chemical reactions involving the corrosion and dissolution of the 1-20-0 APTES layer is dictated by the change in resistance, i.e., the Faradaic process.

- After four days of exposure to the corrosion solution, the resistance in the 1-40-0 and 1-60-0 samples increased slightly, which may have been because the coating remained intact while the accumulation of the corrosion products narrowed its porous channels. The resistance of the 1-20-0 sample slightly decreased, implying that the APTES coating in the 1-20-0 sample started to degrade.

- Over the entire corrosion period, the 1-20-0 sample also exhibited a significantly greater Warburg impedance than the other samples shown in Figure 5c. This suggests that the APTES coating prepared as per the -20- protocol enabled a local, interfacial environment that could effectively impede the diffusion of reactive species for charge transfer.

- Figure 6 shows the SEM images of a cleaned bare sample and the samples primed with APTES in the three conditions outlined herein. Note that the SEM imaging of the samples was completed in an exploratory manner to highlight that (1) the commercially wrought surface was not flat but instead exhibited a micro texture caused by the milling process, whereas analytically flat samples (silica or silica oxide) were used for the compared surfaces, and (2) the APTES-primed layer was too thin to cover the manufacturing-induced surface defects of the aluminum samples (e.g., Figure 6d).

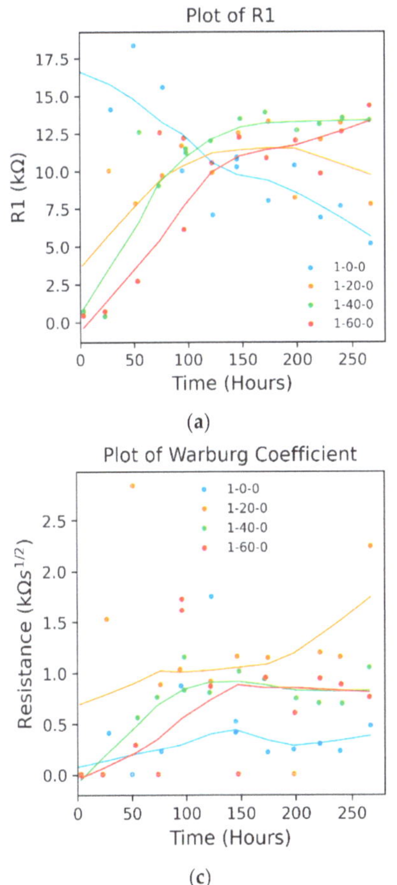

(a)

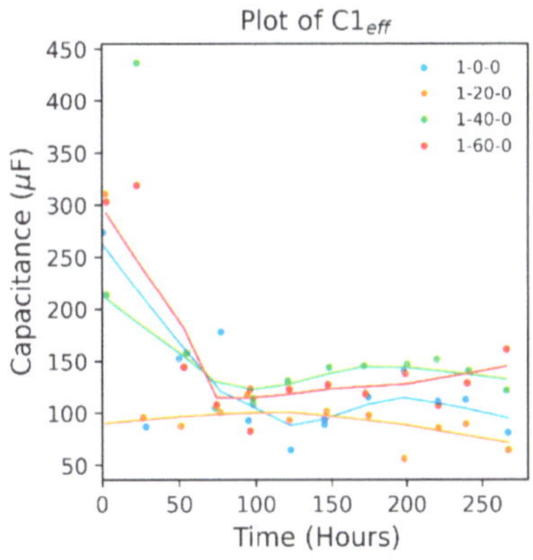

(b)

(c)

Figure 5. The Randles circuit model's fitted parameters (dots) and their trendline for APTES-primed samples. (**a**) Coating resistance R_1. (**b**) Effective capacitance of CPE. (**c**) Warburg impedance. Trendlines were determined using locally weighted scatterplot smoothing (LOWESS) via the statsmodel.py package.

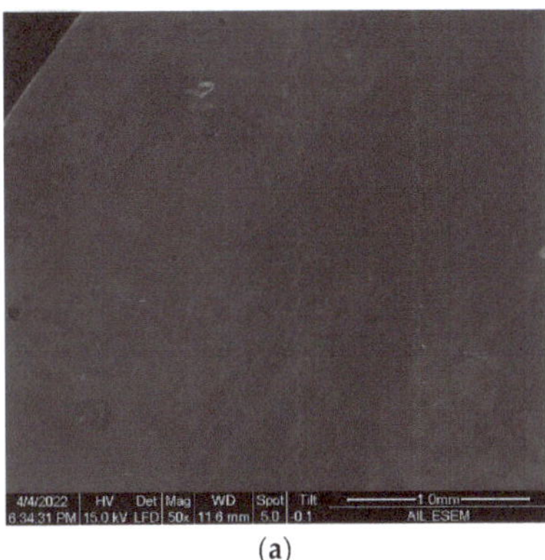

(a)

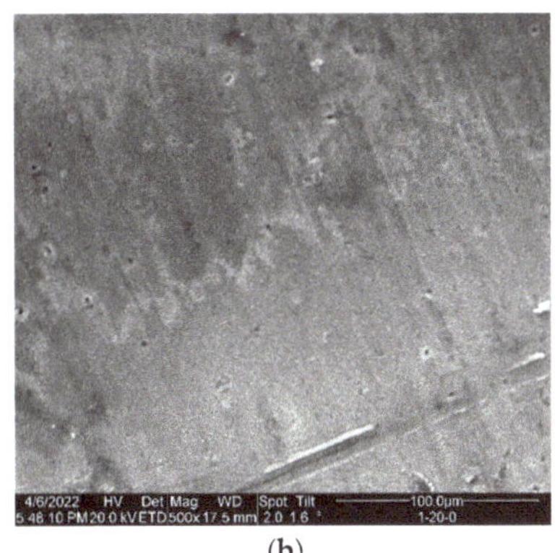

(b)

Figure 6. *Cont.*

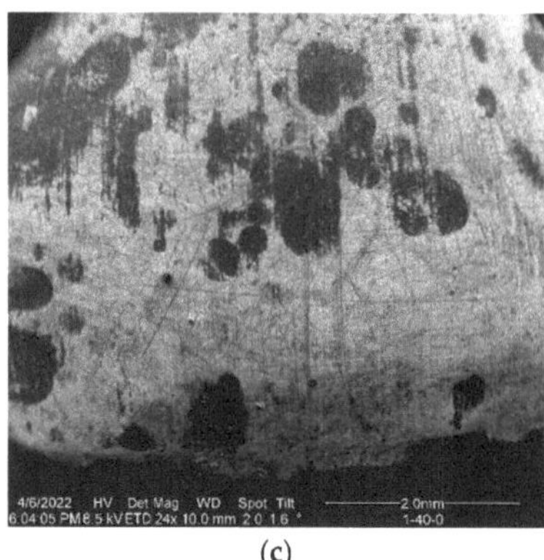
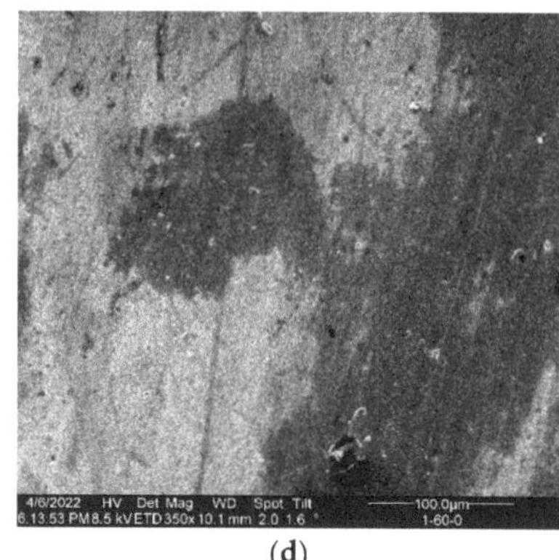

Figure 6. SEM imaging of (**a**) bare sample and APTES-primed samples: (**b**) 1-20-0. (**c**) 1-40-0. (**d**) 1-60-0.

3.2. Polystyrene Top-Coated on APTES

The influence of the primed APTES layer on the impedance of the PS/APTES-coated samples can be seen by comparing the evolving behaviors of the impedance curves of the coated samples in Figure 7.

The samples labeled as 1-Y-1 (Y is 20, 40, or 60) (Figure 7a,c,e) were primed with APTES that had only been cured under the ambient conditions without additional heat treatment. Although the APTES layer primed via vapor deposition for 20 min or 60 min had more significant impedance than the 1-40-1 sample, it degraded over time and thus became hydrolytically unstable, an effect related to partially networked silanols [21].

In the other group of the samples (Figure 7b,d,f), labeled as 1-Y-2 (Y is 20, 40, or 60), the APTES layer had been further heat-treated before the PS layer was top-coated. This group of samples exhibited higher impedance and better hydrolytic stability than the first group (Figure 7a,c,e), which did not undergo additional heat treatment after their APTES layer was polymerized. These results suggest that the post-silanization heat treatment of the APTES layer, following its curing, enhanced networking tightness and hydrolytic stability. This improvement likely occurred due to the removal of any residual water from silanized structures [22]. Heat treatments (post-annealing) may also reduce the adsorbed solvent [4], facilitating condensation and resulting in denser silanol bonds within the networked structure. This group samples all showed increasing impedance over time. For each sample in this group, the diffusion-limiting transport of reactive specimens was evidenced by the upright tail of the impedance curves associated with the 100 h (or earlier) legend. This indicates that over the first 100 h, the electrolyte solution permeated the coating layer, reaching the coating/substrate interface, giving rise to a local environment and thus allowing the reactive species to react with the metal substrate; such species-exchange passages will become narrowed once the corrosion products are stuck at the reaction sites because their removal (likely via diffusion) is impeded by the insoluble coating, despite the porous structure of the coating. Once the corrosion products had accumulated locally at the coating/substrate interface, the sample became less conductive, as shown by the curves of increasing impedance after the 100 h mark. The APTES/PS coating in sample 1-40-2 performed significantly better in these samples because of its relatively higher impedance.

The surface energy of the APTES-primed surface that underwent an additional heat treatment was interpolated and listed in Table 2. Assuming surface energy conservation to be a linear combination of the weighted energies of constituent parts and considering PS characteristics as being independent of the coated surface, the heat-treated APTES without PS can be described as having an energy similar to the 1-20-Z samples plus the difference between the 1-Y-1 and 1-Y-2 samples.

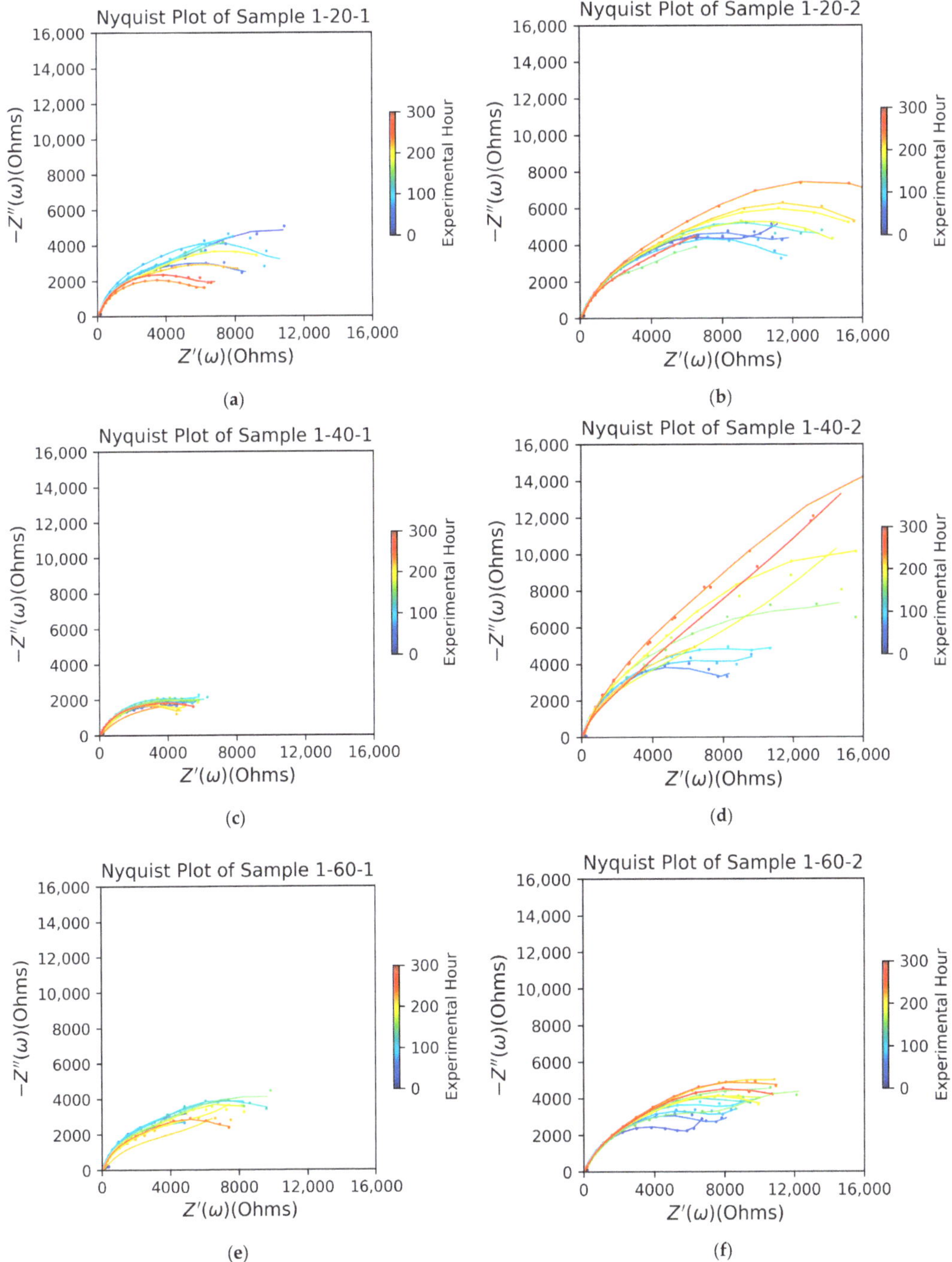

Figure 7. Impedance of PS top-coated samples with the APTES primer cured without further treatment (**a**,**c**,**e**) and with a further heat treatment (**b**,**d**,**f**). Each APTES primer was vapor-deposited for 20, 40, or 60 min, as indicated by the labels of -20-, -40-, or -60-, respectively.

The samples that were primed with APTES via vapor deposition for 40 min (the 1-40-Z samples, where Z is 1 or 2) held their impedance persistently with a minimum extent of degradation, as shown by the monotonically increasing trend of the impedance curves. This implies that the 40 min primed APTES layer cured with the densest network, preventing the loss of its polymerized structure to re-hydrolyzation [23].

3.3. Surface Characterization

The dispersion and polar components of the surface energy on the cured APTES surface were calculated by applying the Fowkes model with the measured contact angles of water and ethylene glycol [24,25]. Table 1 lists the properties of water and ethylene glycol used for the calculations. The properties of acetone are also listed to allow for the determination of the wettability of the PS/acetone solution on the sample surface.

Table 1. Polar and dispersion parts of the surface energy of the working liquid agents used.

Solvent	Mol. Formula	CAS #	Surface Energy/Tension (mJ/m^2)			Reference
			γ_l	Dispersion γ^D	Polar γ^P	
DI Water	H_2O	7732-18-5	72.8	26.85	45.9	[26]
Ethylene glycol	$C_2H_6O_2$	107-21-1	48.0	29.0	19.0	[26]
Acetone	$(CH_3)_2CO$	62-53-3	24.5	--	--	[27]

Table 2 lists the measured static contact angles and the calculated surface energy. Typical images of the measured contact angles are listed in the Supplementary Materials, Figure S2. Despite a preference for using dynamic contact angle measurements for a comprehensive wettability analysis, static contact angle measurements were taken due to equipment limitations. The statistics were developed through multiple tests conducted on fresh areas of the same sample. All statistics were derived from a sample space of more than 10 measurements. The wide statistical variations presented are due to the non-homogeneous, non-uniform nature of the wrought aluminum sample surface. It is crucial to acknowledge that achieving analytical precision in contact angle measurements for coating wettability necessitates perfectly flat surfaces. However, wrought aluminum sheets do not have these characteristics.

Table 2. Measured contact angles and calculated surface energy components.

Sample (1-Y-Z)	Contact Angle (°)		APTES		PS	Surface Energy (mJ/m^2)		
	DI Water	Ethylene Glycol	Vapor Deposition Time (min)	Additional Heat Treatment	Top-Coated	Dispersion γ_s^D	Polar γ_s^P	Total $\gamma_s^P + \gamma_s^D$
1-0-0	82 ± 21	65 ± 12	--	--	--	13.3	11.1	24.4
1-20-0	63 ± 11	47 ± 14	20	NO	NO	9.6	29.6	39.2
1-40-0	63 ± 10	34 ± 8	40	NO	NO	23.1	17.0	40.2
1-60-0	60 ± 12	22 ± 16	60	NO	NO	29.6	15.2	44.7
1-20-0 *	47	56	20	YES	NO	43.3	15.9	59.2
1-40-0 *	48	57	40	YES	NO	46.6	13.9	60.5
1-60-0 *	53	62	60	YES	NO	50.7	10.1	60.8
1-20-1	81 ± 14	66 ± 12	20	NO	YES	10.7	13.8	24.4
1-40-1	83 ± 6	56 ± 5	40	NO	YES	29.6	3.5	33.1
1-60-1	83 ± 6	59 ± 15	60	NO	YES	24.3	5.1	29.4
1-20-2	90 ± 7	57 ± 5	20	YES	YES	44.4	0.1	44.5
1-40-2	94 ± 8	58 ± 7	40	YES	YES	53.1	0.3	53.4
1-60-2	93 ± 8	60 ± 1	60	YES	YES	45.4	0.0	45.4

* Interpolated data.

The results in Table 2 reveal the following:

- The measured contact angles of DI water on the cured APTES surface and dried PS surfaces are statistically indistinguishable. However, the contact angle of ethylene glycol on the cured APTES surface decreased with the increase in the vapor deposition time of coating APTES.
- On the APTES-primed surface, the dispersion energy increased over an increasing duration of vapor deposition of the APTES layer. All the APTES-primed samples except for 1-20-0 were characterized by surface energy with a larger dispersion component than the polar component.
- On the top-coated PS surface, the total surface energy calculated for the 1-Y-2 samples (in which the surface of the cured APTES layer had undergone additional heat treatment before being top-coated with PS) was 50% or higher than that of the 1-Y-1 samples (in which the cured APTES layer did not undergo additional heat treatment). Both groups of samples have much smaller dispersion energy. The polar component of the surface energy was relatively smaller than the dispersion component, except for the 1-20-1 sample surface. Note that among the PS-coated samples, the 1-40-2 samples had the greatest surface energy.

The SEM images of the APTES-primed samples top-coated with PS in Figure 8 show the heterogeneity of the coating, with various dewetted patterns formed on every PS-top-coated sample. Consistent patterns in PS dewetting were difficult to discern because of the high coating heterogeneity. Cracks in the coating surface can be observed in all the types of PS-coated samples. The crazing is consistent with low-weight PS, which does not have the tensile strength of polymers above the molecular entanglement weight. Small PS dewetted patterns such as "microdots" of approximately 2–3 µm can be observed on all the samples (also shown in the Supplementary Materials, Figure S3). The dewetted patterns in the 1-Y-2 samples constitute a less dense distribution of small patterns and larger packs with crazing.

The top-coated PS layer covered the underneath of the APTES-primed layer (see Supplementary Materials, Figure S4). However, the top coating was not uniform, as the dried PS had dewetted into various shapes, wherein crazing is apparent in some large patterns. The 1-40-2 sample had a relatively uniform top coat, as shown in Figure 8d.

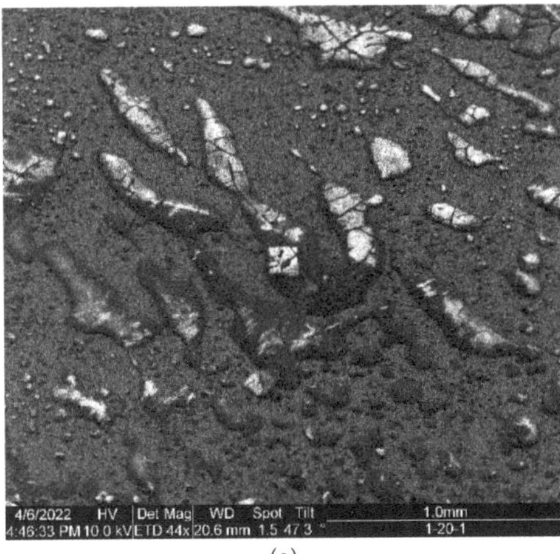

(a)

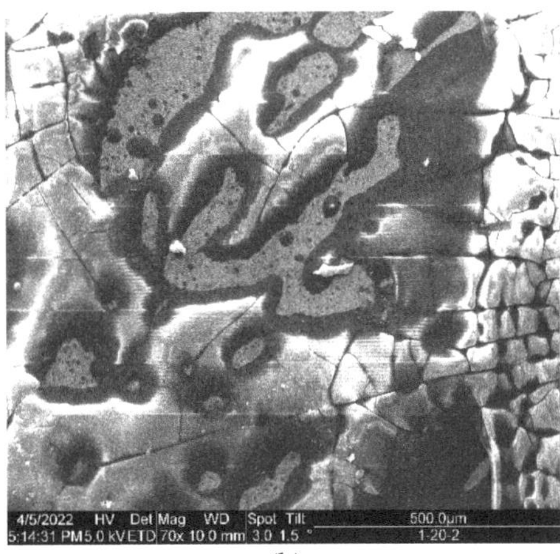

(b)

Figure 8. *Cont.*

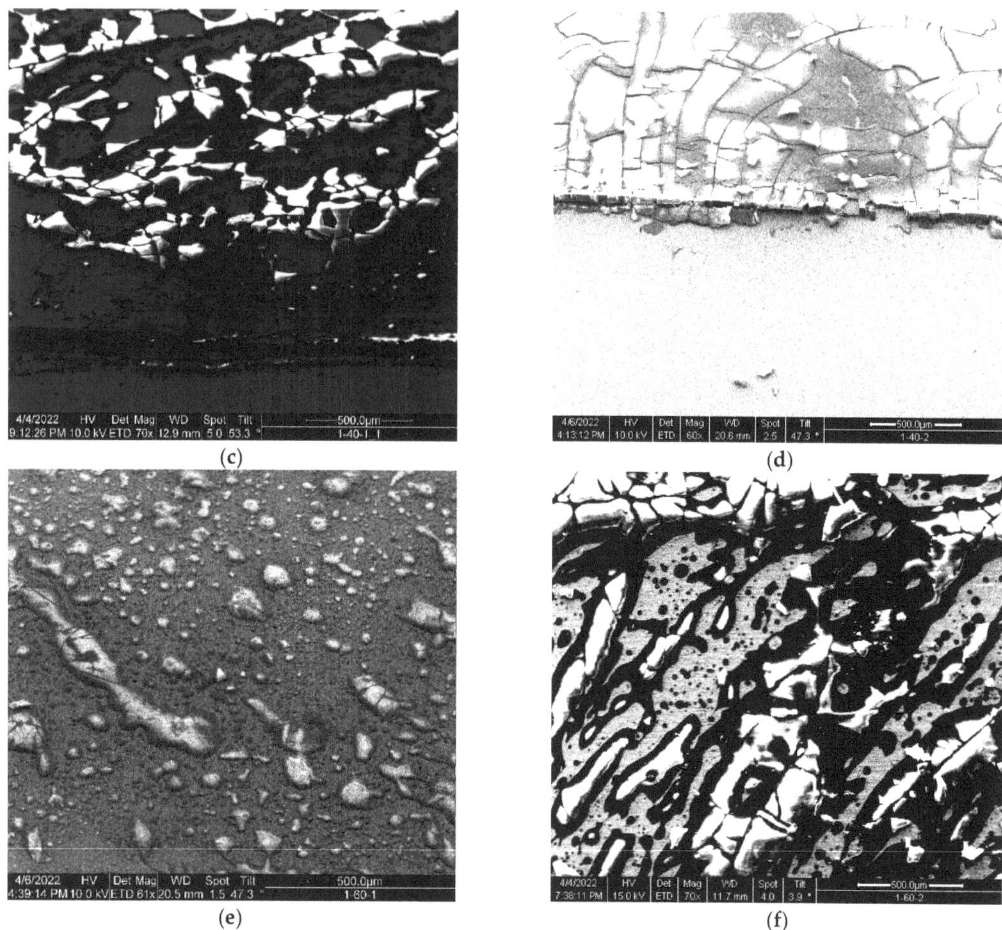

Figure 8. SEM images of dewetted PS patterns on APTES for the samples: (**a**) 1-20-1, (**b**) 1-20-2, (**c**) 1-40-1, (**d**) 1-40-2, (**e**) 1-60-1, and (**f**) 1-60-2.

4. Discussion

In the context of the various failure scenarios, the naturally formed aluminum oxide, silane primer, and PS top-coating interact with one another in the coating system. The interactions may minimize ion exchange during corrosion and generate undesired local environments, potentially compromising the individual effectiveness of each material as a corrosion barrier. A few mechanisms derived from the experimental results are discussed below with regard to their interactions in the related failure modes. Note that these mechanisms may need further adjustments before they are applied to explain the roles of APTES when not prepared using the concentrated vapor deposition method used herein [4,28].

4.1. The Role of APTES in Interaction with the Aluminum Oxide Substrate

APTES was applied to AA2024-T3 through concentrated vapor-phase deposition in a heated, high-vacuum environment with negligible humidity. The hydrolyzation and condensation of ethoxy groups are impeded in the absence of water, resulting in a networked APTES primer that will primarily rely on hydrogen bonds. As a result, the cured APTES primer in the samples was expected to have a structure with a lower density of covalent silanol bonds. However, a more desirable APTES network would feature a higher density of these bonds to grant it better hydrolytic stability as a barrier layer. It can thus be gleaned that during the initial stage (the first 100 h of soaking in the corrosion solution), the naturally formed aluminum oxide layer served as the primary barrier for the corrosion protection of AA2024-T3, because the primed APTES layer is less hydrolytically stable, keeping in mind the three mechanisms related to the reaction of APTES and its interaction with the aluminum oxide substrate developed.

- Mechanism 1: APTES under the effect of water adsorption and permeation

Water access becomes possible when primed samples are later exposed to ambient humidity or immersed in aqueous electrolytes. This leads to the absorption of water molecules on the APTES surface, gradually permeating into the networked APTES layer. This process will then promote hydrolysis and condensation reactions. These water-induced reactions form the desired silanol bonds that can contribute to yielding an improved structural integrity of the APTES layer. However, the permeation of water through the APTES network is a slow process.

- Mechanism 2: APTES reacts with aluminum oxide

Secondly, APTES also reacts with aluminum oxide exclusively by forming stable siloxy bonds (M-O-Si) [29]. This reaction helps reduce the density of unbonded voids or hydrogen bonds susceptible to water attack.

- Mechanism 3: Loss of the naturally formed aluminum oxide to APTES hydrolyzation

However, the APTES-primed layer also facilitates the dissolution of the naturally formed aluminum oxide. The hydrolyzation of the amino groups in APTES makes the solution locally alkaline [30], and aluminum oxide's solubility changes in alkaline solutions [31].

These three mechanisms competed with one another to collectively determine the overall anti-corrosion performance of the APTES-primed AA2024-T3 samples, which were under the stacked barrier of the natural Al_2O_3 layer and the APTES-primed layer. The reason why the 1-20-0 APTES-coated sample exhibited the highest resistance (Figure 4b–d) can be attributed to the following factors:

- The reduced extent of alkaline conditions resulting from hydrolyzation in the APTES layer of the 1-20-0 sample contributed to the increased stability of the aluminum oxide layer.
- The hydrostatic stability of APTES [4]—the APTES primer on the 1-20-0 sample—demonstrated relatively robust hydrolytic stability, as indicated by the consistent evolving behavior of the effective capacitance in Figure 5b. Consequently, the structural integrity of the APTES primer on the 1-20-0 sample remained the most pronounced, resulting in the minimal release of OH- ions and the local solution becoming less alkaline.

This behavior can also clarify why the 1-60-0 APTES coating had the smallest impedance during the initial 100 h of corrosion. The hydrolyzation in the 1-60-0 APTES layer fostered a more potent alkaline environment, leading to increased dissolution of aluminum oxide. Meanwhile, as an APTES layer produced by concentrated vapor deposition is typically only tens of angstroms thick [6] and becomes even thinner after hydrolysis [17], the role of thickness in the APTES layers of the three samples (-20-, -40-, and -60-) in providing barrier protection becomes less significant compared to that of the naturally formed aluminum oxide layer.

4.2. The Role of APTES in Interaction with the Top Coating

The extent and uniformity of PS coverage play a crucial role in determining impedance. A PS/acetone solution was applied as a topcoat over the primer, a cured APTES layer in this study. The following mechanism is suggested to explain the effect of the surface energy of a primed APTES in the extent of PS-coated coverage and its relation to the impedance of a sample with the naturally formed aluminum oxide, APTES primer, and PS topcoat.

- Mechanism 4: The surface energy of primed APTES vs. the surface tension of the PS solution.

The PS/acetone solution partially wets the APTES surface; once the PS layer was dried after the acetone had evaporated, various dewetted patterns of dried PS formed, as shown in Figure 8 (and Figures S3 and S4 in the Supplementary Materials). The extent of the wettability of the APTES-primed surface via the PS/acetone solution was determined by the competition between the surface tension of acetone and the surface energy of the cured APTES. Because the PS/acetone solution has a lower surface tension than the surface energy of the cured APTES (refer to Tables 1 and 2), the PS/acetone solution can wet the APTES-primed surface regardless of the duration of the vapor deposition. The higher the surface energy of the APTES primer surface, the more wetted it will be by the PS/acetone solution. Both observed trends can be related to the fact that a substrate with higher surface energy tends to be more wetted by a liquid with lower surface tension. Enhanced wetting results in fewer dewetting artifacts or patterns. This relationship was illustrated by Ashley et al. [32], who demonstrated that the number density of dewetted polystyrene patterns is proportionate to the surface energy of the cured APTES on which the PS was deposited. The textural roughness and heterogeneity of the wrought sample surface also facilitate wettability. Because the surface tension of the PS/acetone solution is less than the surface energy of the primed APES surface, a higher surface energy of the primed APTES promotes a larger extent of wetting via the PS/acetone solution.

From Mechanism 4, it can be surmised that an APTES primer with a higher surface energy (which can be induced with an additional heat treatment) can be wetted to a greater extent by the PS/acetone solution. The dried PS layer from such an APTES surface with higher energy will form larger patterns [32]; however, given that the chosen PS with low molecular weight has less tension, the dried PS layer can easily be cracked. This can explain why the 1-Y-2 samples exhibited dewetted patterns consisting of larger packs with crazing.

4.3. Potential Applications for Micro Devices

The vapor-phase deposition of silanes is applicable in the photo-lithography-based microfabrication process [33]. As described in this paper, this technique holds significant potential for a wide range of applications in microfabricated devices requiring the bonding or adhesion of dissimilar materials [34]. These applications include hermetic sealing, chemical sensing, and the modification of the mechanical characteristics of microstructures. Several practical applications are enumerated below.

Firstly, a conformal polymeric topcoat on a microdevice can effectively address inherent stress-induced warpage. This intrinsic stress arises from thermal effects caused by the selective removal (etching) of dissimilar materials from each thin film individually deposited at different temperatures on a silicon wafer during fabrication. This thermal stress distorts patterned features, altering their intended dimensions, and can cause warping in micro cantilevered structures [35]. To mitigate this warpage, Kuchiji et al. addressed the issue using a polyurethane coating [36]. Polyurethane, more commonly used than polystyrene, exhibits a relatively strong elasticity modulus, enabling a thin coating to counteract intrinsic thermal stress and reduce warpage in microdevices. The durability of the polyurethane coating on vibratory microstructures is crucial and can be enhanced via the APTES-modification of polyurethane [37]. This modification involves directly mixing a low dose (e.g., 1%) of concentrated APTES with waterborne polyurethane or using APTES as a primer for top coating with 2K polyurethane solutions.

Secondly, an APTES-modified surface exhibits hydrophobic behavior (as indicated in Table 2), making a microsensor chip water/moisture repellent [38].

Thirdly, for microsensors requiring a biocompatible or hermetic seal for chronic implantation, a conformal coating of low-Young's-modulus polymers can preserve the designed sensing capability [36,39]. To ensure strong adhesion and durability of the added coating for a long-term deployment, silanes like APTES can serve as a primer for top-coating with Parylene for in vivo applications [40] or be mixed with polyurethane as a hybrid composite for ex vivo applications [41]. In contrast to using strong polyurethane to modify the mechanical behavior of a microdevice, polymers with a low elasticity modulus like Parylene C can be coated on an APTES-primed surface for encapsulation so that further patterning will not alter the intended mechanical impedance of microdevices [36,39,42]. Another low-elastic-strength polymer, polyimide (Kapton), is commonly used in poly(dimethylsiloxane)-based soft lithography with APTES as the bonding agent [43,44]. These applications are all rooted in the use of silanes, with a specific focus on APTES as an adherent bonding agent or a surface modifier for top coating with a polymer of choice to serve various purposes.

5. Conclusions

The complexity of the arrangement of and interactions among the three layers of the naturally formed aluminum oxide, primed silane, and top-coated PS film make the stacked coating system difficult to manage for predictable anti-corrosion performance.

In this paper, we addressed this challenge by examining the role of the silane primer APTES, commonly used as an adherent agent for inorganic/organic bonding. For a primed layer made via concentrated vapor deposition, the findings indicate that an APTES primer of an optimal thickness (as in the 1-40-2 sample) offers the best corrosion resistance. This result is attributed to the relatively large surface energy in this primed APTES surface that enables better wettability via the PS solution and, consequently, more uniform coverage of the dried PS topcoat. During the early exposure to the corrosion solution, a thinner APTES primer would enhance protection against corrosion. This can be explained by the mechanisms of hydrolytic stability and the interaction between the hydrolyzation/condensation in soaked APTES and the dissolution of the naturally formed aluminum oxide pre-existing in the bare samples. An APTES primer subjected to an additional heat treatment before top coating always performs better. Finally, the applications of APTES and silanes in micro devices were projected.

Supplementary Materials: The following supporting information can be downloaded at https://www.mdpi.com/article/10.3390/mi15010093/s1. 1. Summary of data fitting using the Python Module "impedance.py". Figure S1: Fitted CPE parameters. Figure S2: Images of water contact angles measured on the individual substrate surfaces. Figure S3: SEM images of top-coated PS, dried, on the APTES-primed samples. Figure S4: SEM images of dewetted PS patterns on APTES with an articulately made edge or groove.

Author Contributions: Conceptualization, J.H.IV and C.-f.C.; methodology, J.H.IV and C.-f.C.; software, J.H.IV; validation, J.H.IV and C.-f.C.; formal analysis, J.H.IV; investigation, J.H.IV and C.-f.C.; resources, J.H.IV and C.-f.C.; data curation, J.H.IV and C.-f.C.; writing—original draft preparation, J.H.IV and C.-f.C.; writing—review and editing, J.H.IV and C.-f.C.; visualization, J.H.IV and C.-f.C.; supervision, C.-f.C.; project administration, C.-f.C.; funding acquisition, C.-f.C. All authors have read and agreed to the published version of the manuscript.

Funding: This research was supported by the NASA EPSCoR Program Cooperative Agreement Notice (CAN) (80NSSC20M0137).

Data Availability Statement: Raw EIS data can be found at https://sites.google.com/a/alaska.edu/cf-chen/home/organic-inorganic-coating.

Acknowledgments: The SEM imaging was performed at the Advanced Instrumentation Laboratory (AIL), University of Alaska Fairbanks. The authors thank Junqin Zhang for assisting in the EIS instrumentation setup.

Conflicts of Interest: The authors declare no conflicts of interest.

References

1. Hintze, P.E.; Calle, L.M. Electrochemical Properties and Corrosion Protection of Organosilane Self-Assembled Monolayers on Aluminum 2024-T3. *Electrochim. Acta* **2006**, *51*, 1761–1766. [CrossRef]
2. Lyon, S.B.; Bingham, R.; Mills, D.J. Advances in Corrosion Protection by Organic Coatings: What We Know and What We Would like to Know. *Prog. Org. Coat.* **2017**, *102*, 2–7. [CrossRef]

3. Bera, S.; Rout, T.K.; Udayabhanu, G.; Narayan, R. Comparative Study of Corrosion Protection of Sol–Gel Coatings with Different Organic Functionality on Al-2024 Substrate. *Prog. Org. Coat.* **2015**, *88*, 293–303. [CrossRef]

4. Sypabekova, M.; Hagemann, A.; Rho, D.; Kim, S. Review: 3-Aminopropyltriethoxysilane (APTES) Deposition Methods on Oxide Surfaces in Solution and Vapor Phases for Biosensing Applications. *Biosensors* **2022**, *13*, 36. [CrossRef]

5. Plueddemann, E.P. *Silane Coupling Agents*; Springer US: New York, NY, USA, 1991.

6. Zhu, M.; Lerum, M.Z.; Chen, W. How to Prepare Reproducible, Homogeneous, and Hydrolytically Stable Aminosilane-Derived Layers on Silica. *Langmuir* **2012**, *28*, 416–423. [CrossRef] [PubMed]

7. Brochier Salon, M.C.; Belgacem, M.N. Competition between Hydrolysis and Condensation Reactions of Trialkoxysilanes, as a Function of the Amount of Water and the Nature of the Organic Group. *Colloids Surf. A Physicochem. Eng. Asp.* **2010**, *366*, 147–154. [CrossRef]

8. Meroni, D.; Lo Presti, L.; Di Liberto, G.; Ceotto, M.; Acres, R.G.; Prince, K.C.; Bellani, R.; Soliveri, G.; Ardizzone, S. A Close Look at the Structure of the TiO2-APTES Interface in Hybrid Nanomaterials and Its Degradation Pathway: An Experimental and Theoretical Study. *J. Phys. Chem. C* **2017**, *121*, 430–440. [CrossRef]

9. Simon, A.; Cohen-Bouhacina, T.; Porté, M.C.; Aimé, J.P.; Baquey, C. Study of Two Grafting Methods for Obtaining a 3-Aminopropyltriethoxysilane Monolayer on Silica Surface. *J. Colloid Interface Sci.* **2002**, *251*, 278–283. [CrossRef]

10. Witucki, G.L. A Silane Primer: Chemistry and Applications of Aikoxy Silanes. *J. Coat. Technol.* **1993**, *65*, 57–60.

11. Mahdavian, M.; Ramezanzadeh, B.; Akbarian, M.; Ramezanzadeh, M.; Kardar, P.; Alibakhshi, E.; Farashi, S. Enhancement of Silane Coating Protective Performance by Using a Polydimethylsiloxane Additive. *J. Ind. Eng. Chem.* **2017**, *55*, 244–252. [CrossRef]

12. Choi, S.H.; Newby, B.M.Z. Stability Enhancement of Polystyrene Thin Films on Aminopropyltriethoxysilane Ultrathin Layer Modified Surfaces. In *Silanes and Other Coupling Agents*; Mittal, K.L., Ed.; CRC Press: Boca Raton, FL, USA, 2020; Volume 4, pp. 189–208.

13. Románszki, L.; Datsenko, I.; May, Z.; Telegdi, J.; Nyikos, L.; Sand, W. Polystyrene Films as Barrier Layers for Corrosion Protection of Copper and Copper Alloys. *Bioelectrochemistry* **2014**, *97*, 7–14. [CrossRef] [PubMed]

14. Zhang, J.; Zhang, L. Polystyrene/TiO2 Nanocomposite Coatings to Inhibit Corrosion of Aluminum Alloy 2024-T3. *ACS Appl. Nano Mater.* **2019**, *2*, 6368–6377. [CrossRef]

15. Chen, C.-F. Polystyrene Coating on APTES-Primed Hydroxylated AA2024-T3: Characterization and Failure Mechanism of Corrosion. *Solids* **2023**, *4*, 254–267. [CrossRef]

16. Choi, S.-H. Dewetting of Polystyrene Thin Films on Organosilane Modified Surfaces. Ph.D. Thesis, University of Akron, Akron, OH, USA, 2006.

17. Kim, J.; Seidler, P.; Fill, C.; Wan, L.S. Investigations of the Effect of Curing Conditions on the Structure and Stability of Amino-Functionalized Organic Films on Silicon Substrates by Fourier Transform Infrared Spectroscopy, Ellipsometry, and Fluorescence Microscopy. *Surf. Sci.* **2008**, *602*, 3323–3330. [CrossRef]

18. Chidambaram, D.; Halada, G.P. Infrared Microspectroscopic Studies on the Pitting of AA2024-T3 Induced by Acetone Degreasing. *Surf. Interface Anal.* **2001**, *31*, 1056–1059. [CrossRef]

19. Yadav, R.; Tirumali, M.; Wang, X.; Naebe, M.; Kandasubramanian, B. Polymer Composite for Antistatic Application in Aerospace. *Def. Technol.* **2020**, *16*, 107–118. [CrossRef]

20. Taylor, S.R.; Gileadi, E. Physical Interpretation of the Warburg Impedance. *Corrosion* **1995**, *51*, 664–671. [CrossRef]

21. Etienne, M.; Walcarius, A. Analytical Investigation of the Chemical Reactivity and Stability of Aminopropyl-Grafted Silica in Aqueous Medium. *Talanta* **2003**, *59*, 1173–1188. [CrossRef]

22. Qiao, B.; Wang, T.J.; Gao, H.; Jin, Y. High Density Silanization of Nano-Silica Particles Using γ-Aminopropyltriethoxysilane (APTES). *Appl. Surf. Sci.* **2015**, *351*, 646–654. [CrossRef]

23. Smith, E.; Chen, W. How to Prevent the Loss of Surface Functionality Derived from Aminosilanes. *Langmuir* **2008**, *24*, 12405–12409. [CrossRef]

24. Rudawska, A. Assessment of Surface Preparation for the Bonding/Adhesive Technology. *Surf. Treat. Bond. Technol.* **2019**, 227–275. [CrossRef]

25. Fowkes, F.M. Attractive Forces at Interfaces. *Ind. Eng. Chem.* **2002**, *56*, 40–52. [CrossRef]

26. Zdziennicka, A.; Krawczyk, J.; Szymczyk, K.; Jańczuk, B. Components and Parameters of Liquids and Some Polymers Surface Tension at Different Temperature. *Colloids Surf. A Physicochem. Eng. Asp.* **2017**, *529*, 864–875. [CrossRef]

27. Wu, N.; Li, X.; Liu, S.; Zhang, M.; Ouyang, S. Effect of Hydrogen Bonding on the Surface Tension Properties of Binary Mixture (Acetone-Water) by Raman Spectroscopy. *Appl. Sci.* **2019**, *9*, 1235. [CrossRef]

28. Jothi Prakash, C.G.; Prasanth, R. Approaches to Design a Surface with Tunable Wettability: A Review on Surface Properties. *J. Mater. Sci.* **2021**, *56*, 108–135. [CrossRef]

29. Rozyyev, V.; Murphy, J.G.; Barry, E.; Mane, A.U.; Sibener, S.J.; Elam, J.W. Vapor-Phase Grafting of a Model Aminosilane Compound to Al2O3, ZnO, and TiO2 Surfaces Prepared by Atomic Layer Deposition. *Appl. Surf. Sci.* **2021**, *562*, 149996. [CrossRef]

30. Jing, M.; Zhang, L.; Fan, Z.; Liu, X.; Wang, Y.; Liu, C.; Shen, C. Markedly Improved Hydrophobicity of Cellulose Film via a Simple One-Step Aminosilane-Assisted Ball Milling. *Carbohydr. Polym.* **2022**, *275*, 118701. [CrossRef]

31. Zhang, J.; Klasky, M.; Letellier, B.C. The Aluminum Chemistry and Corrosion in Alkaline Solutions. *J. Nucl. Mater.* **2009**, *384*, 175–189. [CrossRef]

32. Ashley, K.; Sehgal, A.; Amis, E.J.; Raghavan, D.; Karim, A. Combinatorial Mapping of Polymer Film Wettability on Gradient Energy Surfaces. *Mat. Res. Soc. Symp. Proc.* **2001**, *700*, S4.7. [CrossRef]
33. Popat, K.C.; Johnson, R.W.; Desai, T.A. Characterization of Vapor Deposited Thin Silane Films on Silicon Substrates for Biomedical Microdevices. *Surf. Coat. Technol.* **2002**, *154*, 253–261. [CrossRef]
34. Kim, J.B.; Meng, E. Review of Polymer MEMS Micromachining. *J. Micromech. Microeng.* **2016**, *26*, 013001. [CrossRef]
35. Reu, P.L.; Chen, C.-F.; Engelstad, R.L.; Lovell, E.G.; Bayer, T.; Greschner, J.; Kalt, S.; Weiss, H.; Wood, O.R.; Mackay, R.S. Electron Projection Lithography Mask Format Layer Stress Measurement and Simulation of Pattern Transfer Distortion. *J. Vac. Sci. Technol. B Microelectron. Nanometer Struct. Process. Meas. Phenom.* **2002**, *20*, 3053–3057. [CrossRef]
36. Kuchiji, H.; Masumoto, N.; Baba, A. Piezoelectric MEMS Wideband Acoustic Sensor Coated by Organic Film. *Jpn. J. Appl. Phys.* **2023**, *62*, SG1021. [CrossRef]
37. Karna, N.; Joshi, G.M.; Mhaske, S.T. Structure-Property Relationship of Silane-Modified Polyurethane: A Review. *Prog. Org. Coat.* **2023**, *176*, 107377. [CrossRef]
38. Baselt, D.R.; Fruhberger, B.; Klaassen, E.; Cemalovic, S.; Britton, C.L.; Patel, S.V.; Mlsna, T.E.; McCorkle, D.; Warmack, B. Design and Performance of a Microcantilever-Based Hydrogen Sensor. *Sens. Actuators B Chem.* **2003**, *88*, 120–131. [CrossRef]
39. You, Z.W.; Wei, L.; Zhang, M.L.; Yang, F.H.; Wang, X.D. Design of a Novel MEMS Implantable Blood Pressure Sensor and Stress Distribution of Parylene-Based Coatings. In Proceedings of the 2022 IEEE 16th International Conference on Solid-State and Integrated Circuit Technology, ICSICT 2022, Nanjing, China, 25–28 October 2022.
40. Sasaki, H.; Onoe, H.; Osaki, T.; Kawano, R.; Takeuchi, S. Parylene-Coating in PDMS Microfluidic Channels Prevents the Absorption of Fluorescent Dyes. *Sens. Actuators B Chem.* **2010**, *150*, 478–482. [CrossRef]
41. Sardon, H.; Irusta, L.; González, A.; Fernández-Berridi, M.J. Waterborne Hybrid Polyurethane Coatings Functionalized with (3-Aminopropyl)Triethoxysilane: Adhesion Properties. *Prog. Org. Coat.* **2013**, *76*, 1230–1235. [CrossRef]
42. Scholten, K.; Meng, E. Materials for Microfabricated Implantable Devices: A Review. *Lab Chip* **2015**, *15*, 4256–4272. [CrossRef]
43. Tang, L.; Lee, N.Y. A Facile Route for Irreversible Bonding of Plastic-PDMS Hybrid Microdevices at Room Temperature. *Lab Chip* **2010**, *10*, 1274–1280. [CrossRef]
44. Borók, A.; Laboda, K.; Bonyár, A. PDMS Bonding Technologies for Microfluidic Applications: A Review. *Biosensors* **2021**, *11*, 292. [CrossRef]

 *micromachines*

Article

Comparison of PDMS and NOA Microfluidic Chips: Deformation, Roughness, Hydrophilicity and Flow Performance

Tatiana Turcitu [†], Curtis J. K. Armstrong [†], Niko Lee-Yow, Maya Salame, Andy Vinh Le and Marianne Fenech *

Department of Mechanical Engineering, University of Ottawa, Ottawa, ON K1N 6N5, Canada; msala029@uottawa.ca (M.S.)
* Correspondence: marianne.fenech@uottawa.ca
[†] These authors contributed equally to this work.

Abstract: Microfluidic devices are frequently manufactured with polydimethylsiloxane (PDMS) due to its affordability, transparency, and simplicity. However, high-pressure flow through PDMS microfluidic channels lead to an increase in channel size due to the compliance of the material. As a result, longer response times are required to reach steady flow rates, which increases the overall time required to complete experiments when using a syringe pump. Due to its excellent optical properties and increased rigidity, Norland Optical Adhesive (NOA) has been proposed as a promising material candidate for microfluidic fabrication. This study compares the compliance and deformation properties of three different characteristic sized (width of parallel channels: 100, 40 and 20 μm) microfluidic devices made of PDMS and NOA. The comparison of the microfluidics devices is made based on the Young's modulus, roughness, contact angle, channel width deformation, flow resistance and compliance. The experimental resistance is estimated through the measurement of the flow at a given pressure and a precision flow meter. The characteristic time of the system is extracted by fitting the two-element resistance-compliance (RC) hydraulic circuit model. The compliance of the microfluidics chips is estimated through the measurement of the characteristic time required for channels to achieve an output flow rate equivalent to that of the input flow rate using a syringe pump and a precision flow meter. The Young modulus was found to be 2 MPa for the PDMS and 1743 MPa for the NOA 63. The surface roughness was found to be higher for the NOA 63 than for the PDMS. The hydrophilicities of materials were found comparable with and without plasma treatment. The results show that NOA devices have lower compliance and deformation than PDMS devices.

Keywords: characteristic time; resistance; compliance; microfluidics; PDMS; NOA 63; contact angle; hydrophilicity; roughness; young modulus; RC hydraulic circuit model

 **check for updates**

Citation: Turcitu, T.; Armstrong, C.J.K.; Lee-Yow, N.; Salame, M.; Le, A.V.; Fenech, M. Comparison of PDMS and NOA Microfluidic Chips: Deformation, Roughness, Hydrophilicity and Flow Performance. *Micromachines* **2023**, *14*, 2033. https://doi.org/10.3390/mi14112033

Academic Editor: Amir Hussain Idrisi

Received: 27 September 2023
Revised: 26 October 2023
Accepted: 27 October 2023
Published: 31 October 2023

1. Introductions

Microfluidic chips are mostly manufactured using Polydimethylsiloxane (PDMS) [1]. PDMS is widely used due to its transparency [2], low costs, and ease of manufacturing [1,3,4]. The prefusion of microfluidic chips can require hundreds of bars to overcome the system's resistance. Under pressure, the dimension of the channels can significantly change depending on the compliance [5]. Constant and precise dimensions of microfluidic channels are crucial for various applications that require precise control of fluid flow and interactions. The dimensions of microfluidic channels directly influence fluid behavior, including flow rate, pressure drop, mixing, and diffusion [6]. Any variation in channel dimensions can lead to inconsistent results and unreliable experimental outcomes. The flow rate of a fluid through a microfluidic channel is directly proportional to the channel dimensions, such as width, height, and length [7]. By precisely controlling these dimensions, researchers can manipulate the flow rate and achieve desired fluid velocities [8]. This is crucial for applications such as drug delivery, where precise control of flow rate is necessary to ensure accurate dosing. In microreactors, where chemical reactions occur within microfluidic channels, precise dimensions ensure efficient mixing and reaction rates [9]. Similarly, in biological applications, such as cell culture and analysis, constant channel dimensions allow for consistent fluid flow and controlled interactions between cells and reagents [10]. Compliance is the change in volume for any given applied pressure. When using a syringe pump as the system's input source, the time it takes for the system to reach steady flow conditions can vary from seconds to hours depending on the fluidic resistance and compliance. This time, between the initial state and when steady flow conditions are reached, is called the response time or the characteristic time.

PDMS has a low elastic modulus (1 to 3 MPa), which allows the deformation of the device under high pressures [3,4,11]. At a low Reynolds number, the law that governs fluid flow is the Hagen Poiseuille model. The hydraulic resistance of a microchannel is inversely proportional to the hydraulic diameter of the channel raised to the power of four (4) [12]. As a result, a small variation in this dimension, due to the deformation of the elastomer resulting from pressure, can cause significant variation in the hydraulic resistance, and thus invalidate the expected flow rate. Additionally, the accurate evaluation of channel dimensions is of utmost significance, as it significantly impacts shear estimation. In microrheology studies, this can directly influence the formulation of viscosity laws and the establishment of relationships between shear and microstructures [13–15].

To overcome the limitations associated with PDMS, other materials have been used for the fabrication of microfluidic devices. One of the promising materials is the optical glue Norland Optical Adhesive (NOA). The manufacturing process used to design the NOA chip is comparable to that of the PDMS gold standard [16–18]. This material is clear and has a high Young modulus (1655 MPa) [17]. This allows NOA to be used in high-pressure flow systems with minimal compliance as shown by Elodie Sollier et al. when comparing PDMS to other forms of polymer-based materials, including NOA, in high-pressure flow systems [19]. Their findings revealed that the maximum pressure (Pmax) at which delamination occurs in NOA is approximately 74–79 PSI, while PDMS bonded to glass experienced delamination at pressures as low as 36 PSI [19]. This study compares the compliance of PDMS microfluidic devices to that of NOA microfluidic devices, while also assessing the material roughness and hydrophilicity. Surface roughness increases the surface area, which leads to absorption effect and the possibility of trapped air bubbles during flow [20]. So, it is important to study the surface roughness of both materials due to the challenge of micro-scale control, the interfacial properties, the complex boundary effects and the lack of theoretical characterization [20]. Hydrophilicity is vital in microfluidics for the precise handling of small liquid volumes. In a "blood-on-a-chip" device, hydrophilic surfaces enable the accurate manipulation of blood samples within microchannels, facilitating precise diagnostic and analytical procedures. Indeed, hydrophilicity is vital in microfluidics for the precise handling of small liquid volumes. In a "blood-on-a-chip" device, hydrophilic surfaces enable the accurate manipulation of blood samples within microchannels, facilitating precise diagnostic and analytical procedures [21].

2. Materials and Methods

2.1. Microfluidic Devices Geometry

The geometries of the microfluidic chips used were previously designed in Niko Lee-Yow's study and consist of two tapered chambers connected with parallel channels (Figure 1) [22]. The chip has an inlet path and two outlet paths. Three chips (A, B and C) of different dimensions were used. The dimensions of the parallel channels, tapered chambers, and inlet and outlet paths for the different chips are presented in Tables 1 and 2. Three chips were designed to present the same order of total resistance.

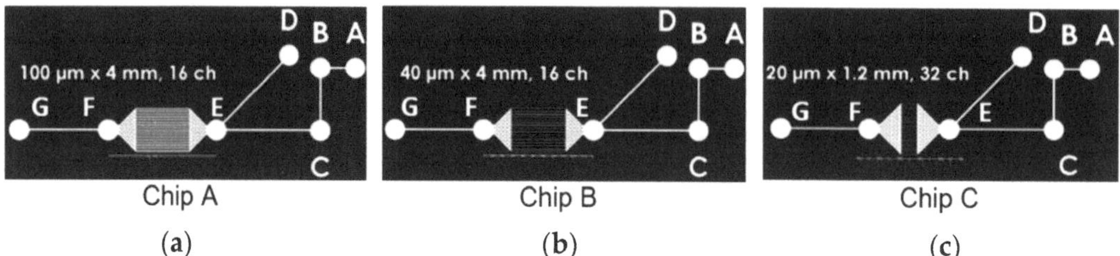

Figure 1. Mask used for SU-8 structures to make the PDMS stamp of (**a**) chip A, (**b**) chip B and (**c**) chip C. The parallel channel widths are 100 µm for chip A, 40 µm for chip B and 20 µm for chip C.

Table 1. Parallel channels and tapered channels dimensions of different chips.

Parallel Channels			
Parameters	**A**	**B**	**C**
Width (d_1) (µm)	100	40	20
Height (d_2) (µm)	100	100	100
Length (L) (µm)	4000	4000	1200
Number of parallel channels	16	16	32
Tapered Channels			
Width of entrance (b_1) (µm)	500	500	500
Width of exit (b_2) (µm)	3100	3100	3740
Height of channel (h) (µm)	100	100	100
Radius of micropillar (r) (µm)	50	50	50
Length of tapered channels (λ) (µm)	1270	1270	1591
Half distance between fibers (Δ) (µm)	50	50	50

Table 2. Inlet and outlet channels dimensions.

Parameters	Inlet Channel (Ports GF)	Outlet Channel (Ports ED)	Outlet Channel (Ports EC)	Outlet Channel (Ports CB)	Outlet Channel (Ports BA)
Width (d_1) (µm)	100	100	100	100	100
Height (d_2) (µm)	150	150	150	150	150
Length (L) (µm)	5000	6000	6129	2925	1104

2.2. Manufacturing

NOA63 and NOA81 were both considered, as they were both used in previous studies as microfluidic device materials [17,18]. NOA63 was chosen due to its increased viscosity and low shrinkage [17,18].

2.2.1. SU-8 Wafer

The molds used for the replica molding of PDMS chips were made with SU-8 photoresist patterned on 3-inch silicon wafer. After Piranha solution cleaning (mixture of H_2SO_4 and H_2O_2) and nitrogen drying, a thin layer of SU-8 50 (MicroChem, Westborough, USA) was manually poured onto the surface of the silicon wafer and spin coated at 500 rpm with an acceleration of 100 rpm/s for 10 s [23]. Immediately after the spread step, a final speed of 1000 rpm was achieved at an acceleration of 300 rpm/s and held for a total of 30 s. After two-step soft baking at 65 °C for 10 min and 95 °C for 30 min, a thin layer of glycerol was used between the mask and wafer to improve the contact [23]. Glycerol was used to ensure a similar refractive index so the light will not refract in the gap between the wafer and mask [24]. The mask/glycerol/wafer was exposed to UV at 500–650 mJ/cm^2 at 350–400 nm [23]. The mask used for the SU-8 structure imprint is negative, as shown in Figure 1. A post-exposure bake at 65 °C for 1 min and at 95 °C for 10 min was completed on the SU-8 wafer [23]. Once the SU-8 wafer was cooled to room temperature, it was submerged into the SU-8 developer for 10 min, rinsed with isopropyl alcohol (99%) and dried with nitrogen gas.

2.2.2. NOA Device

NOA63 was chosen due to its high viscosity (2500 cps) at 25 °C and low shrinkage of approximately 1.5% [16,17]. NOA devices are manufactured using a patterned PDMS stamp. The details of the NOA device manufacturing process are shown in Figure 2a. Using a silanization process, the PDMS molded from the SU-8 wafer was used to fabricate the PDMS stamp. The PDMS-PDMS replica molding was adapted from Zhuang et al. [25]. The PDMS was created using a PDMS 5:1 mixture of main agent and curing agent, respectively. This creates a harder mold allowing the PDMS to act as a more effective mold during the silanization process [26–28]. The harder PDMS was placed in a vacuum chamber, directly above 2 drops of trichloro (1H 1H 2H 2H-perfluorooctyl) silane (PFOTS). The PDMS master was de-gassed for 15–20 min causing the PFOTS to create a thin layer over the PDMS surface. The PDMS was then placed on a hot plate for 10 min at 150 °C, before being placed in a pool of deionized water for 10 min to remove any excess silane (PFOTS) that may remain on the PDMS. It could then be used as a mold for the PDMS stamp by using a traditional 10:1 ratio of main agent to curing agent, as typically used in microfluidic devices [28]. The 10:1 ratio was then poured on top of the PDMS and heated at 75 °C in the oven for 1 h. The PDMS stamp was finally cut away from the PDMS.

The manufacturing method of the NOA63 microfluidic devices was adapted from the technique presented by Sim, Jae Hwan et al. [16]. A 1.25 cm diameter circle of NOA63 was dispensed onto a standard glass slide. The PDMS stamp generated from the double molding with a classical SU-8 wafer was used to manually stamp the uncured NOA63. The stamped NOA63 was then placed under 365 nm wavelength UV light, of intensity 40.25 mW/cm^2, for 10 s causing the glue to stay in a precured state for a total energy exposure of 402.5 mJ/cm^2. Therefore, the stamped channels were retained by the NOA. The PDMS stamp was removed from the NOA and a glass cover slip was placed, and aligned, on top of the precured NOA. The cover slip was prepared in advance with laser cut holes for the appropriate inlet and outlet ports using a CO_2 laser etching machine (*Epilog Laser*). The etching machine was set to a speed of 85% and a power of 90%. The coverslip was secured to a wooden surface and 10 laser shots were used to create the inlet and outlet holes. Once aligned atop the precured NOA63, the entire NOA63 device was placed under UV light at an intensity of 40.25 mW/cm^2 and a wavelength of 365 nm for 20 min for a total energy exposure of 241.5 J/cm^2. An airtight adhesive bond formed between the glass and the precured NOA. Pieces of PDMS were pre-punched at a 19-gauge diameter and plasma-bonded onto the glass coverslip, completing the NOA device.

2.2.3. PDMS Device

The PDMS devices were manufactured using the SU-8 wafer with a patterned mask as presented in Figure 1. The manufacturing process for the PDMS microfluidic devices is detailed in Figure 2.

The PDMS waqs made by using a traditional 10:1 ratio of main agent to curing agent, typically used in microfluidic devices [28]. This mixture was de-gassed for 1 h until no air bubbles were present. The 10:1 ratio was then poured on top of the SU-8 wafer and heated at 75 °C in the oven for 1 h. The PDMS device was finally cut away from the SU-8 wafer.

Inlet and outlet holes were pre-punched at a 19-gauge diameter into the device, and it was plasma-bonded to a glass slide.

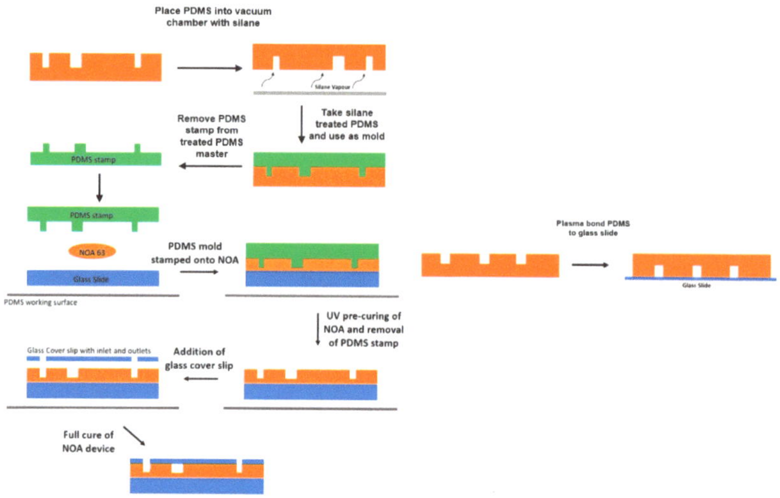

Figure 2. (**a**) Schematic layout of the NOA device manufacturing method using NOA 63. NOA microfluidic chips require a PDMS of the channels to makes a stamp, which is taken from the patterned wafer. A PDMS device is created using a 5:1 ratio of the main agent and the curing agent thus creating a harder PDMS mold [26–28]. The mold is treated by vaporizing trichloro (1H, 1H, 2H, 2H-perfluorooctyl) silane (PFOTS) in a vacuum [25]. This leaves a small film of PFOTS on the surface of the PDMS, which can then be used as its own mold to fabricate PDMS devices with correctly oriented channels from the original PDMS device. The PDMS stamp is then slowly placed down on top of the NOA. The NOA is precured using UV light of 365 nm wavelength and 40.25 mW/cm^3 intensity. The PDMS is then peeled off and the remaining NOA becomes an inverse replica of the PDMS stamp. A glass cover slip with the appropriate inlet and outlet holes is placed on top of the precured NOA chip. The NOA chip is then fully cured for 20 min under UV light. (**b**) Schematic layout of the PDMS device manufacturing method using a PDMS device. The PDMS device is taken from the patterned wafer and plasma bonded to a glass cover slip.

2.3. Surface Roughness

The surface roughness was measured using the *DektakXT profilometer* (Bruker Corporation, Billerica, MA, USA). A PDMS sample of 34 mm length, width of 14 mm and thickness of 2 mm was used. A sample of NOA 63 of 30 mm length, 14 mm width and thickness of 2 mm was used. The PDMS samples were made on a clean SU-8 wafer to assure the surface did not have imperfections. The NOA sample was made on the PDMS surface in contact with the wafer. Three tests were performed for each sample. For the NOA, the tests were performed on a length of 20 mm with leveling. The PDMS tests were performed on a length of 25 mm with no leveling.

2.4. Tensile Strength Test

The tensile strengths of the PDMS and NOA63 was measured using the *Instron 4482 materials testing machine* (Instron Corporation, Norwood, MA, USA). Three PDMS samples measuring 74 mm in length, 26 mm in width and 3 mm in height were tested with tensile grips of 10 N. Similarly, three NOA63 samples of 65 mm length, 20 mm in width and 1.6 mm in height were tested using tensile grips of 10 kN. All samples were tested at a rate of 15 mm/min until failure.

2.5. Channels Width Deformation

The channels' deformation under flow conditions was measured to better understand the compliance in the chips. For this purpose, images of the channels were captured using a microscope featuring $40\times$ objective while maintaining a constant flow. Deionized water was pushed from a 500 µL Hamilton syringe by a syringe pump at rates of 25, 50 and 100 µL/h. The channel width was then measured for each flow using ImageJ 1.53.

2.6. Contact Angle

The contact angle measurement was used to observe the hydrophilic and hydrophobic properties of the materials. The same samples of PDMS and NOA63 were used for surface roughness measurements. Both PDMS and NOA samples were cleaned with tape to remove any particles. Three tests were performed on each sample with and without plasma treatment. The tests of the samples without the plasma treatment were first performed followed by the tests with plasma treatment. A plastic syringe was utilized to distribute 1 microliter of water on each sample. A digital microscope captured side-view images, and each sample underwent three tests. The contact angle was then measured using ImageJ ROI manager. The plasma treatment was made using the *Laboratory Corona Treater* (Electro-Technic Products inc., Chicago, IL, USA) device for about 1 min on each sample. This device distributes the plasma using an electric current. Plasma was applied to the surface of the material, causing the chips to become hydrophilic.

2.7. Experimental Setup

2.7.1. Pressure Controlled Setup for the Resistance of Microfluidic Devices

Resistance can be defined as the slope of the linear relationship between pressure and flow. A pressure-controlled system (Flow EZ, Fluigent, France) coupled with a microflow sensor (Flow Unit S, Fluigent, France) was used to test the resistance for all devices. The fluidic circuit consisted of rigid components with low compliant properties, which include 20-gauge metal tubes and small but rigid polymer tubing. Additionally, a 0.2 µm filter was placed at the top of the reservoir to filter the deionized water to prevent the flow meter from clogging. A detailed flow chart of the experimental setup can be seen in Figure 3a. Pressure was reduced by 0.5 mbar every 20 s from around 100 µL/h until no flow rate was detected. Pressure-flow data were collected with and without the chip and analyzed to calculate the resistance. The process was repeated three times for all three chips for both materials. The average of the flow for each applied pressure was used to plot the pressure flow. The chip's resistance was then calculated by deducting the resistance of the tubing and filter alone from the resistance of the whole circuit, which included the chip. A total of 6 microfluidic devices were tested: 3 PDMS devices and 3 NOA devices.

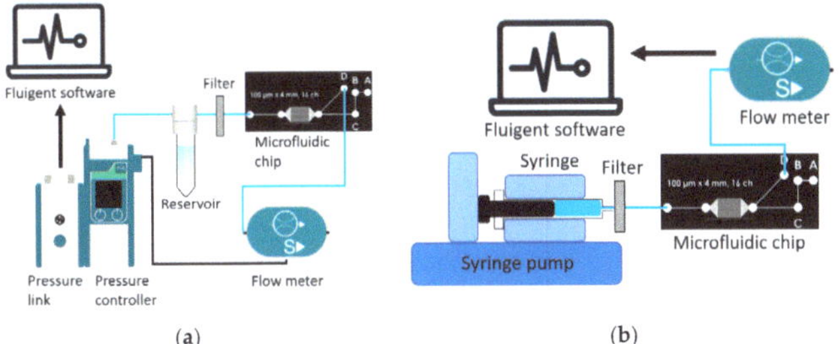

Figure 3. (**a**) Pressure control setup for the resistance testing of the microfluidic devices. Setup includes a pressure controller. All fluid runs through rigid polymer tubing. After exiting the microfluidic chip, the flow is directed into a precision flow meter size S (small). (**b**) Flow set up for the characteristic time testing of microfluidic devices. Setup includes a precision syringe pump with 500 µL glass syringes with 20-gauge metal tubes. All fluid runs through rigid polymer tubing. After exiting the microfluidic chip, the flow is directed into a precision flow meter size S (small).

2.7.2. Flow Setup for the Characteristic Time of the Microfluidic Devices

A flow-control system was used to test the characteristic time of the devices by analyzing the time delay of the output flow rate versus the flow rate imposed by a pump. The setup consisted of a glass syringe (500 µL *Hamilton*) mounted in a syringe pump (*Nexus 3000*, Chemyx, Stafford, TX, USA) couple with a microflow sensor (Flow unit S, Fluigent, France). The fluidic circuit was composed of 20-gauge metal tubes, and small but rigid polymer tubing. As for the pressure control setup, a filter was placed at the tip of the syringe to filter the deionized water. A detailed flow chart of the experimental setup can be seen in Figure 3b. The syringe pump was set to accommodate flow rates of 25 µL/h, 50 µL/h, and 100 µL/h through the microfluidic device. The microflow sensor data were compared to the syringe pump's flow rate settings to calculate the characteristic time, in accordance with Section 2.8.3. The process was repeated 3 times and the average results at each point were used to plot flow over time for each chip and flow rate. A total of 6 microfluidic devices were tested: 3 PDMS devices and 3 NOA devices.

2.8. Analysis

The fluidic system was modeled using the two-element Resistance Compliance (RC) hydraulic circuit model in Section 2.8.2 [29]. The data were collected from the Fluigent flow software and were processed using MATLAB software. This was done by fitting an exponential function, derived from the RC hydraulic circuit model to the flow rate over time plot as shown in Section 2.8.3.

2.8.1. Theoretical and Hydraulic Resistance Estimation

The hydrodynamic resistance is the slope of the line of the pressure over the flow. The hydrodynamic resistance can be expressed as follows [30]:

$$R_{hyd} = \frac{\Delta P}{Q} \tag{1}$$

where ΔP is the difference of pressure, R_{hyd} is the hydrodynamic resistance of the microfluidic device and Q is the flow.

The theoretical hydrodynamic resistance of the devices is computed by summing the individual resistances in series or parallel as with electrical circuits using the following equations [22]:

$$R_{eq_{serie}} = R_{hyd_1} + R_{hyd_2} + \cdots + R_{hyd_n} \tag{2}$$

$$R_{eq_{parallel}}{}^{-1} = R_{hyd_1}{}^{-1} + R_{hyd_2}{}^{-1} + \cdots + R_{hyd_n}{}^{-1} \tag{3}$$

The hydrodynamic resistance of the rectangular cross-section channels can be calculated using the Hagen–Poiseuille equation as follows [22]:

$$R = \frac{12\,\mu L}{wh^3 \left(1 - \frac{0.63h}{w}\right)} \tag{4}$$

where h is the height of the channel, w is the width of the channel, μ is the viscosity of the fluid, and L is the length of the channel in question. The hydrodynamic resistance of the tapered channels can be calculated using the Darcy law, as follows [22]:

$$R = \frac{9\lambda\pi\mu\sqrt{2}}{8r^2 h \left(\frac{\Delta}{r}\right)^{5/2}(b_1 + b_2)} \tag{5}$$

where b_1 is the width of entrance, b_2 the width of exit, h the height of the channel, r the radius of the micropillar, λ the length of the tapered channels and Δ the half distance between fibers.

The experimental resistance of the chip can be obtained by plotting the pressure over the mean flow. The regression line is then modeled as follows:

$$\Delta P = R_{exp} \times Q + P \tag{6}$$

where R_{exp}, the slope, is the experimental resistance of the microfluidic device, Q is the average flow at each pressure and P is the initial pressure. To get the experimental resistance of only the chip, the slope of the regression line with and without the chip must be modeled. Then, the slope of the system without the chip can be subtracted from the slope of the full system (with the chip) as follows:

$$R_{chip_{exp}} = R_{full\ system} - R_{no\ chip} \tag{7}$$

2.8.2. Two-Element Resistance Compliance (RC) Hydraulic Circuit Model

The flow of the system can be described by:

$$Q_{in} - Q_{out} = Q_c \tag{8}$$

where Q_{in} is the inflow to the system, controlled by the syringe pump. Q_{out} is the outflow of the system measured by the flow meter, and Q_c is the rate of storage of the system itself. Q_c can be further described by:

$$Q_c = C\frac{dP}{dt} \tag{9}$$

where C is the compliance of the system and $\frac{dP}{dt}$ is the pressure change over time inside the system. This means the volumetric rate of storage of the system is directly related to the compliance of the system. Q_{out} can be defined, assuming Hagen–Poiseuille flow, as:

$$Q_{out} = \frac{\Delta P}{R_s} \tag{10}$$

where ΔP is the drop-in pressure of the system, and R_s is the peripheral resistance. When the pressure at the outflow is assumed to be close to zero, it is reduced to the pressure within the storage chamber. Thus Equations (8)–(10) can be re-written as:

$$Q_{in} - \frac{p}{R_s} = C\frac{dp}{dt} \tag{11}$$

Integrating to solve for $p(t)$ using initial conditions of $p = p_0$ (initial pressure) and $t = 0$ (time) the pressure can be written as:

$$p(t) = R_s Q_{in} - (R_s Q_{in} - p_0)e^{-\left(\frac{t}{R_s C}\right)} \tag{12}$$

where p_0 is the initial pressure and t is the time. From Equation (12), dp/dt can be rewritten as:

$$\frac{dp}{dt} = \left(\frac{R_s Q_{in} - p_0}{R_s C}\right)e^{-\left(\frac{t}{R_s C}\right)} \tag{13}$$

Combining Equations (13), (9) and (8), the system equation can be rewritten as:

$$Q_{out} = Q_{in} - \left(\frac{R_s Q_{in} - p_0}{R_s}\right)e^{\left(-\frac{t}{R_s C}\right)} \tag{14}$$

where p_0 is the initial pressure of the system, t is the time, and $R_s C$ represents the characteristic time of the system.

2.8.3. Characteristic Time and Compliance

The characteristic time of each trial is calculated and used to compare the compliance of the system. This is assuming that the resistance of the external system is constant and the only change in compliance from the system comes from the microfluidic device.

The outflow of the system, Q_{out} (which is a function of time t), can then be modeled using the following equation:

$$Q_{out} = A - Be^{-(\frac{t}{D})} \tag{15}$$

The three constants are used to fit the raw data, acquired experimentally, from Equation (14). The constant A represents the Q_{in}, the constant B represents the maximum rate of storage, and the constant D is the characteristic time of the system.

With the characteristic time, the compliance can be calculated using the following equation:

$$C = \frac{\tau}{R_e} \tag{16}$$

where C is compliance, τ is the characteristic time experimentally found for PDMS or NOA using Equations (8)–(15), and R_e is the total experimental resistance of the microfluidic device found using the pressure controller.

2.9. Statistical Analysis

Unpaired Student's t test was used for comparisons between the two groups. Graph-Pad Prism 9 was used to perform the statistical tests and the graphical representations. A p value less than 0.05 was considered to be statistically significant. The data were considered a normal distribution, due to the nature of the data describing a physical property of a material [31].

3. Results

3.1. Mechanical Properties

The surface roughness of NOA63 and PDMS was found in terms of the arithmetic mean height of primary profile (Pa) and the root mean square height of the primary profile (Pq). The results are presented in Table 3. The samples of NOA63 and PDMS underwent three tensile strength tests each with the Instron machine. The average Young's modulus found for each material is compared to the theoretical values in Table 3.

Table 3. Arithmetic mean height of primary profile (Pa) and root mean square height of the primary profile (Pq) of the surface samples and experimental Young's modulus of NOA63 and PDMS, as well as those found in the literature.

Sample	Pa (μm)	Pq (μm)	Experimental Young's Modulus (MPa)	Literature Young's Modulus (MPa)
NOA63	1559 ± 77	1867 ± 98	1743 ± 173	1655 [32]
PDMS	147 ± 4	169 ± 4	2 ± 0.2	1–3 [3,4,6]

It was observed that NOA63 has an arithmetic mean height of the primary profile and root mean square height about 10.6 times higher and 11.0 times higher than the PDMS, respectively.

3.2. Channels Width Deformation

The channels of the NOA63 and PDMS microfluidic devices were measured using microscopy and ImageJ at the different flow rates (25, 50 and 100 μL/h). The average values of these measurements are presented in Figure 4.

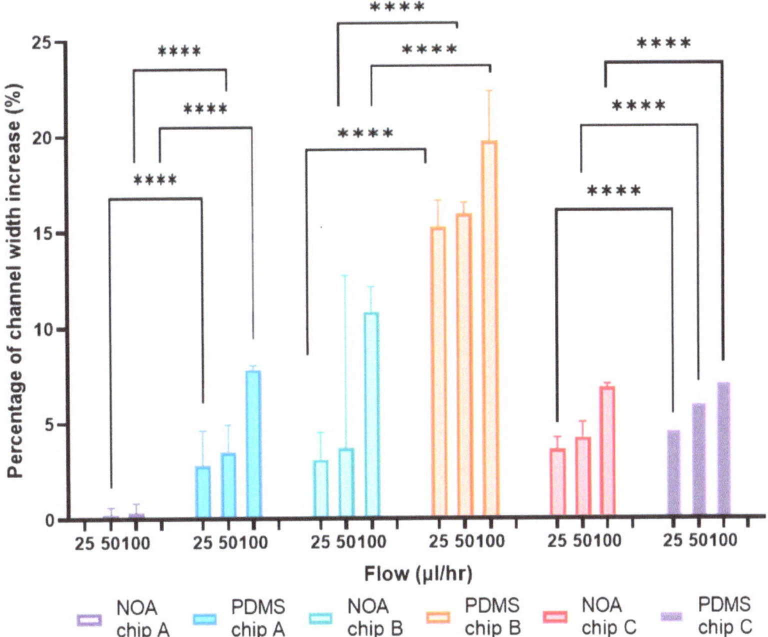

Figure 4. Comparison of NOA and PDMS channels' width deformation of chip A, B and C at the different flow rates (25, 50 and at 100 μL/h). The initial channel width of the PDMS chip A is 95.4 μm, that of chip B is 36 μm and that of chip C is 17.7 μm. The initial channel width of NOA chip A is 101.7 μm, chip B is 38.2 μm and chip C is 25.6 μm. (**** *p*-value < 0.0001).

It was observed that all PDMS devices have a lower initial channel width than NOA devices: 95.4 μm vs. 101.7 μm, 36.0 μm vs. 38.2 μm and 17.7 μm vs. 25.6 μm for PDMS vs. NOA in chip A, B and C, respectively. For all three chips, an increase in the channel width can be observed for the PDMS device as the flow rate increases. For the NOA devices, chip B and C show an increase in the channel width as the flow increases, but chip A displays a constant channel width. For chip A, at 100 μL/h, there is an increase of 7.8% for the PDMS device and of 0.2% for the NOA device. For chip B, at 100 μL/h, there is an increase of 19.7% for the PDMS and of 10.8% for the NOA device. Chip B has the highest increase in the percentage of channel width. For chip C, at 100 μL/h, there is an increase of 7.1% for the PDMS device and of 6.8% for the NOA device. PDMS shows a higher channel increase as the flow rate increases for chip A, B and C than NOA.

3.3. Contact Angle

The contact angle of the NOA63 and PDMS was obtained with and without portable plasma treatment. Figure 5 illustrates the contact angle of a water droplet on the surfaces.

The results for the contact angle without and with plasma treatment for NOA and PDMS are, respectively, 81.4° ± 4.9, 44.9° ± 3.1, 91.4° ± 1.9 and 32.1° ± 4.4. A higher contact angle means the surface is more hydrophobic, while a lower contact angle means a more hydrophilic surface. By treating the materials with the plasma device, it can be observed that the contact angle decreases for both materials. This means that the microfluidic devices become more hydrophilic after plasma treatment. Without plasma treatment, PDMS is more hydrophobic than NOA, and with plasma treatment, NOA is more hydrophobic than PDMS.

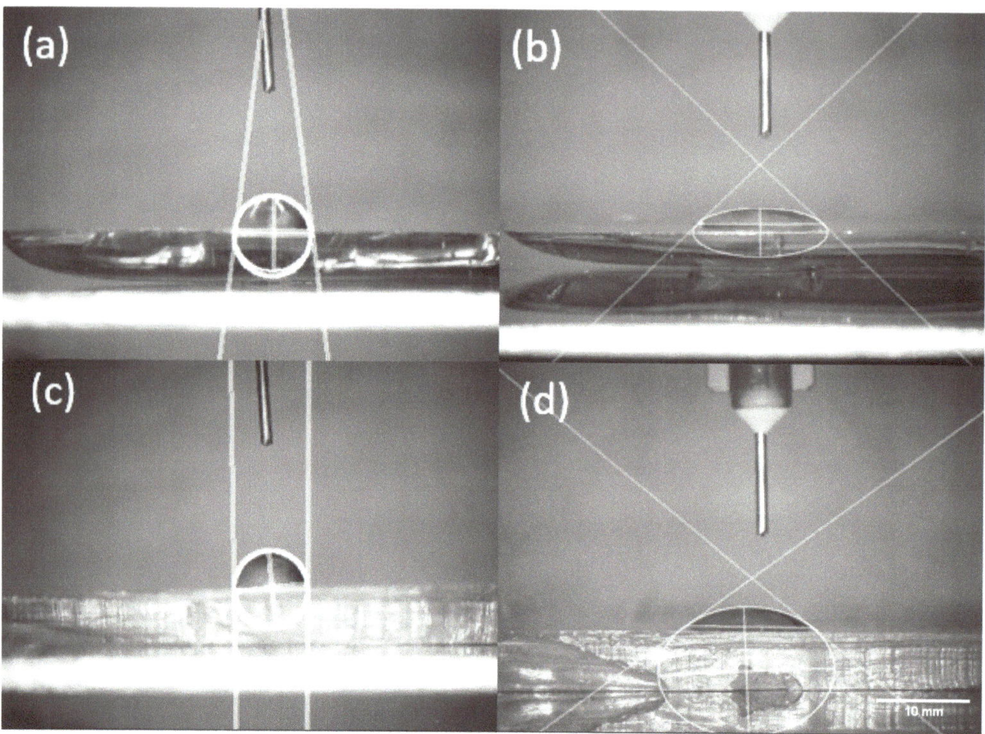

Figure 5. Contact angle for (**a**) NOA63 without plasma treatment (81.4° ± 4.9), (**b**) NOA63 with plasma treatment (44.9° ± 3.1), (**c**) PDMS without plasma treatment (91.4° ± 1.9) and (**d**) PDMS with plasma treatment (32.1° ± 4.4).

3.4. Resistance Estimation

The hydraulic resistances of the chips were found by following the procedure in 2.7.1 and analyzed using the procedure described in 2.8.1. Examples of the pressure in function of the flow are presented in Figure 6 and the hydraulic resistances are presented in Table 4.

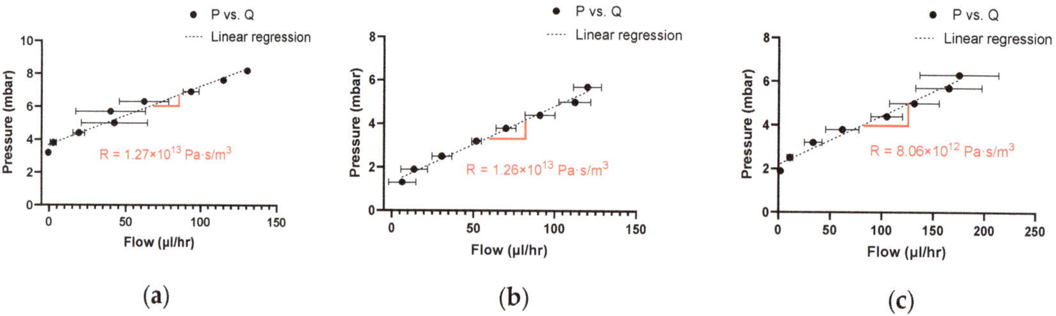

Figure 6. Examples of the pressure in function of the flow for the full hydrodynamic system of (**a**) NOA63 chip A, (**b**) PDMS chip A and (**c**) without the chip.

By subtracting the experimental resistance without a chip from the experimental resistance of the NOA63 or PDMS, the experimental resistance of the chip can be found.

The experimental resistances of the microfluidic devices are compared to the theoretical values in Table 4.

Table 4. Experimental resistance of all 3 chips of NOA63 and PDMS and the theoretical resistance. The theoretical resistance range is calculated with the initial channel width dimension from 0 and the larger width with deformation.

Chip	Theoretical Resistance (10^{12} Pa·s/m^3)	NOA63 Experimental Resistance (10^{12} Pa·s/m^3)	PDMS Experimental Resistance (10^{12} Pa·s/m^3)
A	4.78–4.79	4.64	4.54
B	5.34–5.66	5.44	5.65
C	5.09–5.75	5.44	5.04

3.5. Characteristic Time Estimation

Examples of the flow rate as a function of the time for both NOA and PDMS chip A are presented in Figure 7.

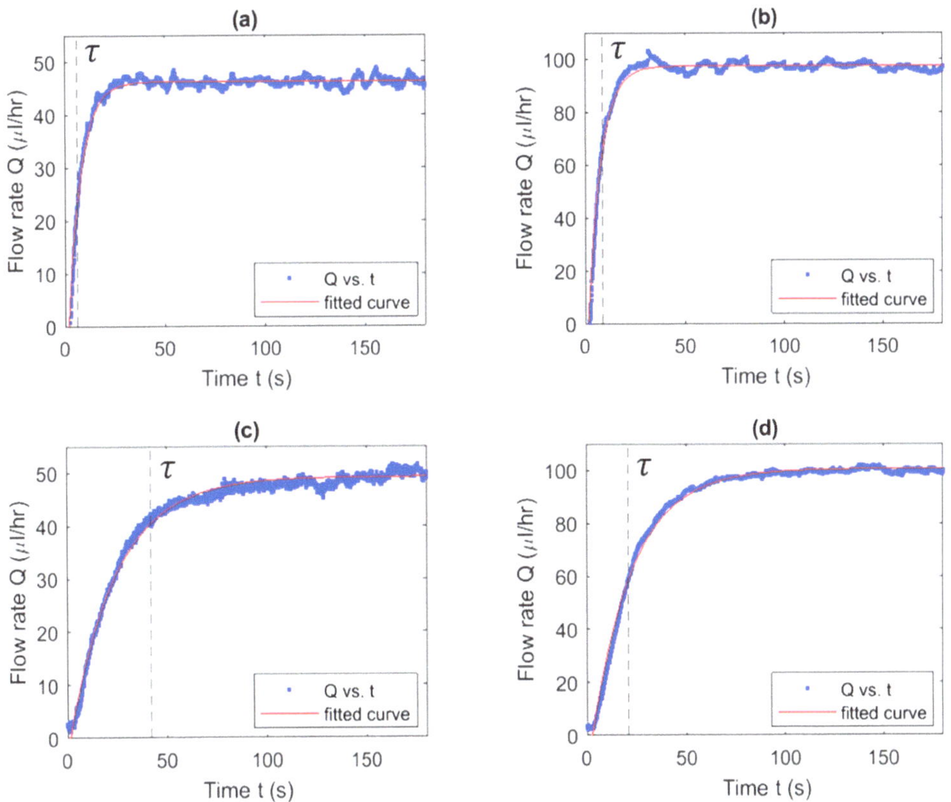

Figure 7. Examples of flow rate versus time graphs for chip A (**a**) at flow rate of 50 µL/h for NOA63, (**b**) at flow rate of 100 µL/h for NOA 63, (**c**) at flow rate of 50 µL/h for PDMS, and (**d**) at flow rate of 100 µL/h for PDMS. Each device shows a significant initial rise from zero to the plateau value (Qout) measured from the Fluigent Flow Meter (S).

All the PDMS and NOA63 trials' characteristic times were averaged and are graphed in Figure 8.

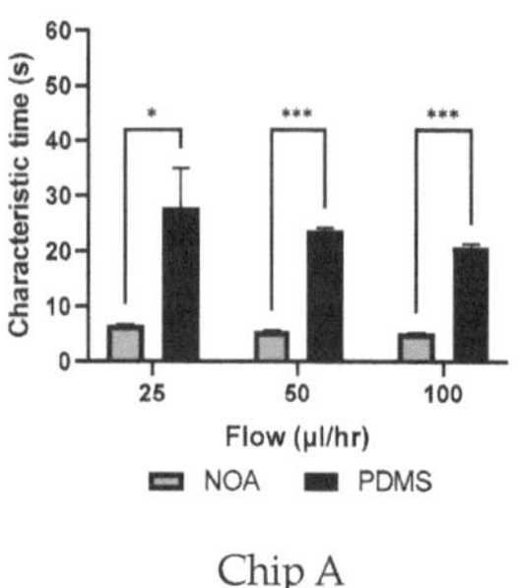

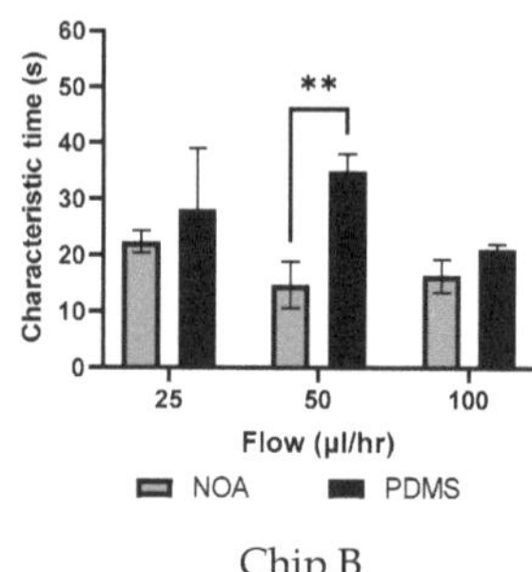

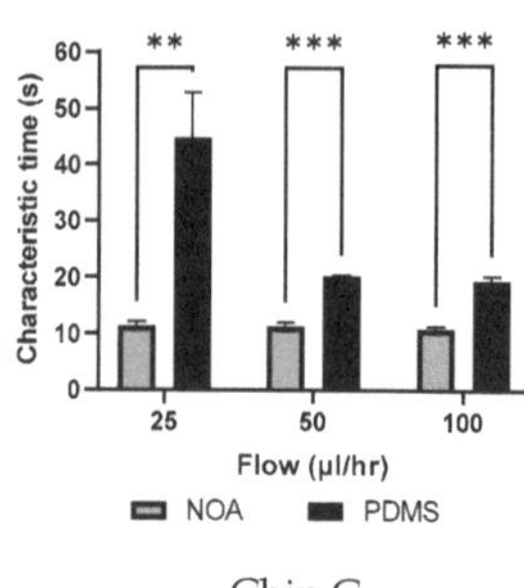

Figure 8. Comparing NOA and PDMS devices of chip A, B and C, average characteristic times at three different flow rates (25, 50, 100 in µL/h) of chip A (* p-value < 0.05, ** p-value < 0.01, *** p-value < 0.001).

As shown in Figure 8, there is a significant difference between the characteristic times of the NOA device versus the PDMS device for all chips. For chip A, it was found that the PDMS devices had a characteristic time around 4 times longer than that of the NOA devices. At a flow rate of 25 µL/h, the PDMS device exhibited significantly longer characteristic times compared to the NOA device, which recorded times of 32.2 s and 7.7 s, respectively. For chip B, a significant difference was observed at a flow rate of 50 µL/h. At this flow rate, PDMS had a higher characteristic time than NOA of 34.9 s and 14.8 s, respectively. For chip C, a significant difference can be observed between PDMS and NOA devices at all flow rates. At a flow rate of 25 µL/h, PDMS had a higher characteristic time than NOA, of 44.7 s and 11.4 s, respectively. PDMS has on average a higher characteristic time than NOA devices of 4 times longer for chip A, 1.6 times longer for chip B and 2.5 times longer for chip C. The results presented in Figure 8 suggest a decay of characteristic time as the flow rates increased for all chips.

The statistical analysis revealed that the NOA devices provided more consistent results than the PDMS devices. Chip A showed standard deviations of ±0.38 µL/h, ±0.27 µL/h, and ±0.14 µL/h for flow rates of 25 to 100 µL/h. In comparison, the PDMS devices showed more inconsistency, showing standard deviations of ±12.2 µL/h, ±1.0 µL/h, and ±1.2 µL/h for the same flow rates of 25 to 100 µL/h. Similar trends can be observed for chips B and C.

3.6. Compliance

The compliance of all devices was found using the time characteristic extracted from Figure 7 and the experimental resistance from Table 4. The compliance found for each device at the different flow rates is presented in Figure 9.

For both PDMS and NOA devices, chip A and chip C showed a decrease in the compliance as the flow rate increased. For chip A, PDMS showed a decrease in the compliance of 36%, while NOA showed a decrease of 18% from 25 µL/h to 100 µL/h. For chip C, PDMS showed a decrease in the compliance of 57%, while NOA underwent a decrease of 4% from 25 µL/h to 100 µL/h. For chip B, a decrease in the compliance could be observed as the flow rate increased for both NOA and PDMS devices, but at some flow rates there was an increase in the compliance.

An estimation of the compliance using the dimension of the channels can also be given as $C_{l=}\frac{V_{in} \cdot a^2}{P}$, where V_{in} is the initial volume, a is the percentage by which the width increased, and P is the pressure applied to flow in the channel. The relation of the compliance calculated from the volume in function of the compliance found from the characteristic time is presented in Figure 10. All experimental outliers were removed.

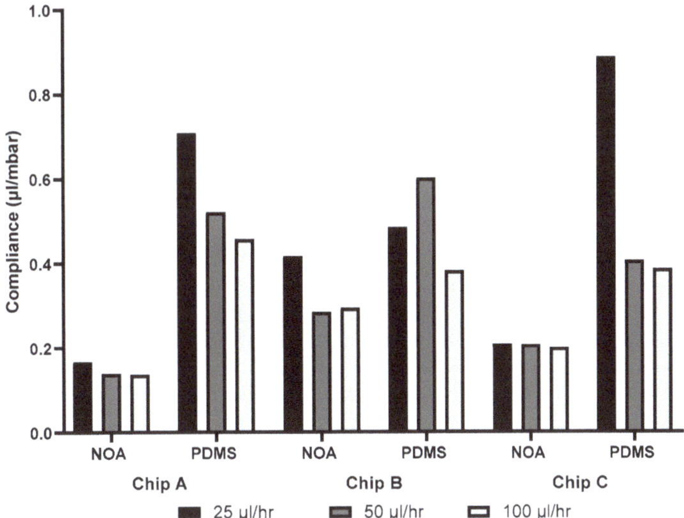

Figure 9. Compliance from characteristic time of both NOA and PDMS chips A, B and C at the different flow rates (25, 50 and 100 μL/h).

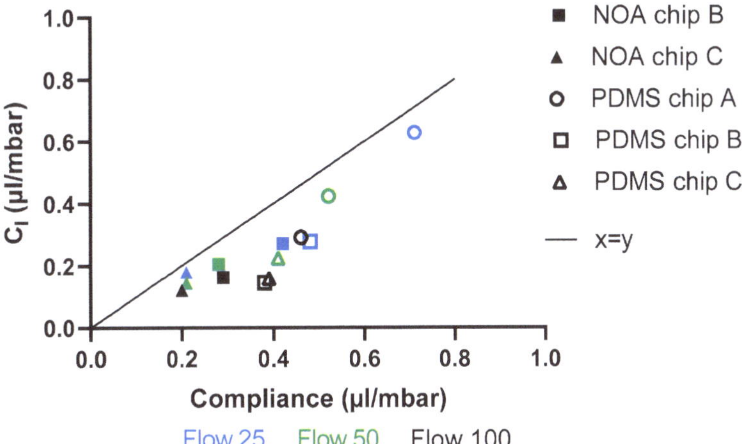

Figure 10. Compliance from the channel deformation in function of the compliance from the characteristic time (τ) for all chips at flow rates 25, 50 and 100 μL/h.

From Figure 10, it can be observed that for all flow rates there is a correlation between the compliance from the channels deformation and the compliance from the characteristic time, while C_l seems underestimated.

4. Discussion

4.1. Material Properties

4.1.1. Surface Roughness and Young's Modulus

The surface roughness can be found in Table 4. The mean arithmetic height and root mean square height are about 11 times higher for NOA63 than PDMS. This means that NOA63 has a rougher surface than PDMS. The Young's modulus of the materials was found in Table 4. The PDMS has a Young's modulus of 2.1 ± 0.2 MPa and the

NOA63 has a Young's modulus of 1743 ± 173 MPa. These values agree with those in the literature [3,4,11,17]. NOA63 is more rigid than PDMS, which will allow less deformation of the channels under a constant flow, while it is a well-established fact that preparing PDMS with a 1:5 ratio increases its stiffness, resulting in a Young's modulus of approximately 2.7 MPa [33]. This value remains significantly inferior to that of NOA, which is 800 times higher. This will lead to less compliance in the microfluidic devices. If a different type of Norland Optical Adhesive is used the Young's modulus changes considerably (325 MPa for NOA81 compared to 1655 MPa for NOA63) [19].

4.1.2. Hydrophilicity/Hydrophobicity of Surface

In Figure 6, a higher intercept, which is the initial pressure, can be observed for the NOA63 device than the PDMS. A higher initial pressure means that the chip is more hydrophobic. The contact angle measurement determines the hydrophilicity/hydrophobicity of the material surface. Both NOA63 and PDMS microfluidic devices received a plasma treatment. NOA63 presented a higher contact angle with plasma treatment ($44.9° \pm 3.1$) and had a higher initial pressure (3.7 mbar) than the PDMS ($32.1° \pm 4.4$ and 1.3 mbar). Therefore, NOA63 microfluidic devices are more hydrophobic than the PDMS.

4.2. Experimental Resistance

4.2.1. Channels Width Deformation

A difference between the channel widths of NOA63 and PDMS was observed in Figure 4. This difference can be explained by the double molding. The double molding decreases the average dimension, which explains the narrower width of the PDMS than the NOA [34].

Also, the difference between the resistance values of PDMS and NOA63 for each chip can be explained by the channel's width deformation. The propagation of error of the channel width was calculated to understand the effect of a difference in the channel's dimensions. Using a precision error of 0.5% for the width of the channels, the propagation error for chip A can be estimated at 1.24×10^{13} Pa·s/m^3. Due to the channel's deformation measurement and error propagation, high sensitivity is achieved in the channel's width measurement for the calculation of the theoretical resistance. This can explain the difference between the experimental and theoretical values in Table 4.

4.2.2. Dependency of Characteristic Time on Flow Rate

The rate of storage for the system, Q_c, is described by Equation (9) as the compliance (C) multiplied by the change in pressure over time $\left(\frac{dP}{dt} \right)$. As the flow rate increases, the system's pressure increases as well, which ultimately increases the volume of the rate of storage of the system. By decreasing the resistance, the characteristic time ($R_e C$) also decreases. PDMS microfluidic devices had a higher characteristic time for all chips and all flow rates compared to the NOA63 devices. Also, a decrease in the characteristic time can observed as the flow rate increases for all chips. This explains the decay, which becomes evident as the flow rate increases.

4.3. Flow Rate Measurement Uncertainty

The Fluigent flow meter S, with a range of ± 420 μL/h, was used to accommodate the low flow rates used in the protocol. The accuracy of the device was $\pm 5\%$ of the measured value for all flow rates of 25 μL/h and higher, and ± 1.26 μL/h for all measurements below 25 μL/h. The repeatability of the device was within 0.5% for all measurements taken above 42 μL/h and ± 0.21 μL/h below 42 μL/h. Thus, all measurements taken were repeatable within 0.5% of the measured value except those taken at 25 μL/h, which were repeatable within 0.84% and 2.1% of the measured value, respectively. This can explain why there are more peaks at lower flow rates for both NOA and PDMS.

The high standard deviations for the PDMS and NOA devices can be explained by the inaccuracy of the Fluigent flow meter at low flow rates (25 μL/h). Although the standard deviations are consistently high for the PDMS devices, the inconsistency in characteristic time, compared to NOA devices, suggests differences in the PDMS vs. NOA devices. In PDMS devices, small variations in material properties can occur with changes in temperature, holding time, and altered mass ratios of pre-polymer to curing agent [26,35]. Furthermore, the PDMS thickness can affect flow conditions and can display significant bulging or deformation during use [4,14,36]. Slight alterations to the PDMS mechanical properties, through uncontrolled parameters, coupled with the channel's deformation may also contribute to the large standard deviations with PDMS devices. Consistency with the NOA results suggests a more repeatable system than PDMS equivalents. This increase is expected due to the 2% disparity in the Fluigent flow meter at the aforementioned flow rate.

4.4. Compliance

Since the flow is modeled using Equation (11), the characteristic time is calculated as system resistance (R_e) multiplied by the compliance (C). The system setup used was consistent throughout all trials. Therefore, the same rigid tubing and glass syringes were used throughout the entire experimental process. Thus, the only change to the resistance and compliance of the system can be directly associated with the change in the microfluidic device being tested. Due to the low compliance of the materials, the system outside of the microfluidic device was assumed to be negligible when comparing characteristic time values. It is also important to note that the resistance of the system outside of the microfluidic device (syringes and tubing, etc.) is constant. The value was measured to be 8.06×10^{12} Pa·s/m^3. Thus, the characteristic times seen in Figure 8 are directly related to the compliance (presented in Figure 9) of each microfluidic device, therefore suggesting that the PDMS devices have, on average, a higher compliance than that of the NOA devices. A negative correlation between compliance and flow rate has been observed. This indicates that compliance decreases as the flow rate increases. This phenomenon can be explained by the association of reduced deformability with an increase in channel width.

4.5. Anticipated Periodic Instability of the Output Flow

The results presented in Figure 7, show the flow rate change over time of PDMS and NOA, respectively. The data demonstrate a clear rise and plateau, following the derived model presented. For both NOA and PDMS devices, the raw data plateaued with a sinusoidal behaviour. This behaviour was analyzed and found to be, as anticipated, caused by the syringe pump's (*Nexus 3000*, Chemyx, Stafford, TX, USA) stepper motor [37,38].

5. Conclusions

The compliance of three different chips of NOA63 microfluidic devices and PDMS microfluidic devices has been quantified under repeatable flow conditions using a pressure-controlled setup and a flow. On average, the characteristic times for chip A were found to be 4 times longer for PDMS devices than NOA devices, 1.6 times for chip B and 2.5 times for chip C. Also, the experimental resistance of the microfluidic device was found for all three chips of both PDMS and NOA. PDMS has a higher percentage of error in the hydraulic resistance due to a higher channel deformation. This leads to a higher compliance. These results concur with the material properties of both NOA and PDMS, having elastic moduli of 1655 MPa and 2 MPa, respectively. The presented results demonstrate that NOA microfluidic devices are less compliant than PDMS microfluidic devices. It also suggests that NOA microfluidic devices could increase consistency in microfluidic research due to their significantly lower standard deviations. This could potentially encourage NOA devices to become the more prevalent option in microfluidic research and in high-pressure microfluidic flow systems. These findings could be used to better understand the use of NOA as a microfluidic material, as well as its properties of compliance and corresponding benefits to microfluidic research.

Author Contributions: Conceptualization: T.T., C.J.K.A., A.V.L. and M.F. Formal analysis: T.T., C.J.K.A., A.V.L. and M.F. Funding acquisition: M.F. Investigation: T.T., C.J.K.A., M.S. and N.L.-Y. Methodology: T.T., C.J.K.A. and A.V.L. Resources: M.F. Visualization: T.T., C.J.K.A., A.V.L. and M.F. Supervision: M.F. Writing—original draft: T.T., C.J.K.A. and M.F. Writing—review and editing: A.V.L., N.L.-Y. and M.S. All authors have read and agreed to the published version of the manuscript.

Funding: This research was funded by National Science and Engineering Research Council of Canada, garnt #RGPIN-2020-07020.

Data Availability Statement: All data that support the findings of this study are included within the article.

Acknowledgments: The authors want to thank Graham Jerrey Rivers Killaire and Leo Denner for their help in measuring the roughness and young modulus.

Conflicts of Interest: The authors declare no conflict of interest.

References

1. McDonald, J.C.; Whitesides, G.M. Poly(dimethylsiloxane) as a Material for Fabricating Microfluidic Devices. *ChemInform* **2002**, *33*, 265. [CrossRef]
2. Toepke, M.W.; Beebe, D.J. PDMS absorption of small molecules and consequences in microfluidic applications. *Lab Chip* **2006**, *6*, 1484–1486. [CrossRef]
3. Hardy, B.S.; Uechi, K.; Zhen, J.; Kavehpour, H.P. The deformation of flexible PDMS microchannels under a pressure driven flow. *Lab Chip* **2009**, *9*, 935–938. [CrossRef]
4. Gervais, T.; El-Ali, J.; Günther, A.; Jensen, K.F. Flow-induced deformation of shallow microfluidic channels. *Lab Chip* **2006**, *6*, 500–507. [CrossRef]
5. Lee, K.S.; Ram, R.J. Plastic-PDMS bonding for high pressure hydrolytically stable active microfluidics. *Lab Chip* **2009**, *9*, 1618–1624. [CrossRef]
6. Squires, T.M.; Quake, S.R. Microfluidics: Fluid physics at the nanoliter scale. *Rev. Mod. Phys.* **2005**, *77*, 977–1026. [CrossRef]
7. Kirby, B.J. *Micro- and Nanoscale Fluid Mechanics: Transport in Microfluidic Devices*, 1st ed.; Cambridge University Press: Cambridge, UK, 2012; ISBN 978-0-521-11903-0.
8. Beebe, D.J.; Mensing, G.A.; Walker, G.M. Physics and Applications of Microfluidics in Biology. *Annu. Rev. Biomed. Eng.* **2002**, *4*, 261–286. [CrossRef] [PubMed]
9. Stone, H.; Stroock, A.; Ajdari, A. Engineering Flows in Small Devices: Microfluidics Toward a Lab-on-a-Chip. *Annu. Rev. Fluid Mech.* **2004**, *36*, 381–411. [CrossRef]
10. El-Ali, J.; Sorger, P.K.; Jensen, K.F. Cells on chips. *Nature* **2006**, *442*, 403–411. [CrossRef]
11. Liu, M.; Sun, J.; Sun, Y.; Bock, C.; Chen, Q. Thickness-dependent mechanical properties of polydimethylsiloxane membranes. *J. Micromech. Microeng.* **2009**, *19*, 035028. [CrossRef]
12. Oertel, H. *Biofluid Mechanics*; Springer: New York, NY, USA, 2010; Volume 158, ISBN 978-1-4398-4518-9.
13. Chayer, B.; Pitts, K.L.; Cloutier, G.; Fenech, M. Velocity measurement accuracy in optical microhemodynamics: Experiment and simulation. *Physiol. Meas.* **2012**, *33*, 1585. [CrossRef]
14. Mehri, R.; Mavriplis, C.; Fenech, M. Design of a microfluidic system for red blood cell aggregation investigation. *J. Biomech. Eng.* **2014**, *136*, 064501. [CrossRef]
15. Pitts, K.L.; Fenech, M. Micro-particle Image Velocimetry for Velocity Profile Measurements of Micro Blood Flows. *J. Vis. Exp.* **2013**, *74*, e50314. [CrossRef]
16. Sim, J.H.; Moon, H.J.; Roh, Y.H.; Jung, H.W.; Bong, K.W. Fabrication of NOA microfluidic devices based on sequential replica molding. *Korean J. Chem. Eng.* **2017**, *34*, 1495–1499. [CrossRef]
17. Dupont, E.P.; Luisier, R.; Gijs, M.A.M. NOA 63 as a UV-curable material for fabrication of microfluidic channels with native hydrophilicity. *Microelectron. Eng.* **2010**, *87*, 1253–1255. [CrossRef]
18. Kim, H.; Yang, Y.; Kim, M.; Nam, S.W.; Lee, K.M.; Lee, N.Y.; Kim, Y.S.; Park, S. Simple Route to Hydrophilic Microfluidic Chip Fabrication Using an Ultraviolet (UV)-Cured Polymer. *Adv. Funct. Mater.* **2007**, *17*, 3493–3498. [CrossRef]
19. Sollier, E.; Murray, C.; Maoddi, P.; Carlo, D.D. Rapid prototyping polymers for microfluidic devices and high pressure injections. *Lab Chip* **2011**, *11*, 3752–3765. [CrossRef]
20. Rakesh, T.G.; Girisha, C.; Satish, V.T. A Review on the Effect of Surface Roughness and Liquid Slip on fluid flow in PDMS Microchannels. *IOP Conf. Ser. Mater. Sci. Eng.* **2021**, *1189*, 012021. [CrossRef]
21. Pitts, K.L.; Abu-Mallouh, S.; Fenech, M. Contact angle study of blood dilutions on common microchip materials. *J. Mech. Behav. Biomed. Mater.* **2013**, *17*, 333–336. [CrossRef]
22. Lee-Yow, N.; Pitts, K.L.; Fenech, M. Optically Clear Biomicroviscometer with Modular Geometry Using Disposable PDMS Chips. In Proceedings of the 2017 IEEE International Symposium on Medical Measurements and Applications (MeMeA), Rochester, MN, USA, 7–10 May 2017; IEEE: Rochester, MN, USA, 2017; pp. 72–77.

23. MicroChem. "SU8_2035-2100.pdf". Available online: https://www.seas.upenn.edu/~nanosop/documents/SU8_2035-2100.pdf (accessed on 26 October 2023).

24. Chuang, Y.J.; Tseng, F.G.; Lin, W.K. Reduction of diffraction effect of UV exposure on SU-8 negative thick photoresist by air gap elimination. *Microsyst. Technol.* **2002**, *8*, 308–313. [CrossRef]

25. Zhuang, G.; Kutter, J.P. Anti-stiction coating of PDMS moulds for rapid microchannel fabrication by double replica moulding. *J. Micromech. Microeng.* **2011**, *21*, 105020. [CrossRef]

26. Khanafer, K.; Duprey, A.; Schlicht, M.; Berguer, R. Effects of strain rate, mixing ratio, and stress-strain definition on the mechanical behavior of the polydimethylsiloxane (PDMS) material as related to its biological applications. *Biomed. Microdevices* **2009**, *11*, 503–508. [CrossRef] [PubMed]

27. Ye, X.; Liu, H.; Ding, Y.; Li, H.; Lu, B. Research on the cast molding process for high quality PDMS molds. *Microelectron. Eng.* **2009**, *86*, 310–313. [CrossRef]

28. Johnston, I.D.; McCluskey, D.K.; Tan, C.K.L.; Tracey, M.C. Mechanical characterization of bulk Sylgard 184 for microfluidics and microengineering. *J. Micromech. Microeng.* **2014**, *24*, 035017. [CrossRef]

29. Roh, C.; Lee, J.; Kang, C. The Deformation of Polydimethylsiloxane (PDMS) Microfluidic Channels Filled with Embedded Circular Obstacles under Certain Circumstances. *Molecules* **2016**, *21*, 798. [CrossRef]

30. Tabeling, P. *Introduction to Microfluidics*; Oxford University Press (OUP): Oxford, UK, 2023.

31. Tavoularis, S. Measurement in Fluid Mechanics. *J. Fluids Struct.* **2007**, *23*, 159–160. [CrossRef]

32. NOA63. Available online: https://www.norlandprod.com/adhesives/noa%2063.html (accessed on 20 October 2023).

33. Sotiri, I.; Tajik, A.; Lai, Y.; Zhang, C.; Kovalenko, E.; Nemr, C.; Ledoux, H.; Alvarenga, J.; Johnson, E.; Patanwala, H.; et al. Tunability of liquid-infused silicone materials for biointerfaces. *Biointerphases* **2018**, *13*, 06D401. [CrossRef]

34. Gitlin, L.; Schulze, P.; Belder, D. Rapid replication of master structures by double casting with PDMS. *Lab Chip* **2009**, *9*, 3000–3002. [CrossRef] [PubMed]

35. Xie, S.; Wu, J.; Tang, B.; Zhou, G.; Jin, M.; Shui, L. Large-Area and High-Throughput PDMS Microfluidic Chip Fabrication Assisted by Vacuum Airbag Laminator. *Micromachines* **2017**, *8*, 218. [CrossRef]

36. Kang, C.; Roh, C.; Overfelt, R.A. Pressure-driven deformation with soft polydimethylsiloxane (PDMS) by a regular syringe pump: Challenge to the classical fluid dynamics by comparison of experimental and theoretical results. *RSC Adv.* **2014**, *4*, 3102–3112. [CrossRef]

37. Li, Z.; Mak, S.Y.; Sauret, A.; Shum, H.C. Syringe-pump-induced fluctuation in all-aqueous microfluidic system implications for flow rate accuracy. *Lab Chip* **2014**, *14*, 744. [CrossRef] [PubMed]

38. Zeng, W.; Jacobi, I.; Beck, D.J.; Li, S.; Stone, H.A. Characterization of syringe-pump-driven induced pressure fluctuations in elastic microchannels. *Lab Chip* **2014**, *15*, 1110. [CrossRef] [PubMed]

 *micromachines*

Article

Commercially Accessible High-Performance Aluminum-Air Battery Cathodes through Electrodeposition of Mn and Ni Species on Fuel Cell Cathodes

Paloma Almodóvar [1,*], Belén Sotillo [2], David Giraldo [1,3], Joaquín Chacón [1], Inmaculada Álvarez-Serrano [3] and María Luisa López [3]

1   Albufera Energy Storage, 28001 Madrid, Spain; dgiraldo@ucm.es (D.G.); joaquin.chacon@albufera-energystorage.com (J.C.)
2   Departamento de Física de Materiales, Facultad de Física, Universidad Complutense de Madrid, 28040 Madrid, Spain; bsotillo@ucm.es
3   Departamento de Química Inorgánica, Facultad de Química, Universidad Complutense de Madrid, 28040 Madrid, Spain; ias@ucm.es (I.Á.-S.); marisal@ucm.es (M.L.L.)
*   Correspondence: palmodov@ucm.es

Abstract: This study presents a cost-effective method for producing high-performance cathodes for aluminum-air batteries. Commercial fuel cell cathodes are modified through electrodeposition of nickel and manganese species. The optimal conditions for electrodeposition are determined using a combination of structural (Raman, SEM, TEM) and electrochemical (LSV, EI, discharge curves) characterization techniques. The structural analysis confirms successful incorporation of nickel and manganese species onto the cathode surface. Electrochemical tests demonstrate enhanced electrochemical activity compared to unmodified cathodes. By combining the favorable properties of electrodeposited manganese species with nickel species, a high-performance cathode is obtained. The developed cathode exhibits capacities of 50 mA h cm^{-2} in aluminum-air batteries across a wide range of current densities. The electrodeposition method proves effective in improving electrochemical performance. A key advantage of this method is its simplicity and cost-effectiveness. The use of commercially available materials and well-established electrodeposition techniques allows for easy scalability and commercialization. This makes it a viable option for large-scale production of high-performance cathodes for the next-generation energy storage devices.

Keywords: metal-air batteries; Al-air; electrodeposition

check for updates

Citation: Almodóvar, P.; Sotillo, B.; Giraldo, D.; Chacón, J.; Álvarez-Serrano, I.; López, M.L. Commercially Accessible High-Performance Aluminum-Air Battery Cathodes through Electrodeposition of Mn and Ni Species on Fuel Cell Cathodes. *Micromachines* **2023**, *14*, 1930. https://doi.org/10.3390/mi14101930

Academic Editor: Amir Hussain Idrisi

Received: 26 September 2023
Revised: 12 October 2023
Accepted: 13 October 2023
Published: 14 October 2023

1. Introduction

Metal-air batteries such as Li-air, Zn-air, Mg-air, and Al-air batteries are promising for the future generation of energy for mobility and stationary applications because they use oxygen from the air as one of the battery's main reactants, reducing the weight of the battery and freeing up more space devoted to energy storage. Among all these metal-air batteries, Li-air shows the highest theoretical gravimetric capacity (3860 A h kg^{-1}). However, this technology poses significant safety problems related to the presence of Li. Al-air technology presents the second highest gravimetric capacity (2980 A h kg^{-1}), just after Li-air, and the highest volumetric capacity. In addition, these systems do not exhibit many of the environmental and safety problems related to Li based batteries, and their disposable components are fully recyclable [1,2].

Metal-air cathodes are composed, usually, of four parts: a current collector (to ensure the good electrical conductivity), a catalyst layer (to ensure the oxygen reduction reaction, ORR), a carbonaceous gas diffusion layer—GDL (responsible of an efficient reduction of oxygen by the contact between oxygen, the electrolyte and the catalyst)—and a O$_2$ permeable membrane (to ensure the O$_2$ flow, while preventing electrolyte leakage to the outside).

There are different methods and patents describing the manufacture and production of these air cathodes. However, in most cases, they describe how a paste/slurry/suspension of the catalytic material is initially added to the GDL carbon by using binders and solvents (which are then removed with high temperature treatments). Afterward, a metallic mesh is placed in the catalyst containing face and a Polytetrafluoroethylene (PTFE) film in the contrary face, and is finally pressed all together in a hot roll press machine [3]. However, despite the advancements in research and development, these methods are often challenging to implement on an industrial scale. Scaling up from laboratory processes to commercial production poses significant difficulties. Additionally, the cost of the components used in metal-air cathodes, such as the catalytic materials and specialized membranes, can also hinder their commercialization. Due to these factors, commercially available metal-air cathodes are currently limited, making it challenging to bring metal-air batteries to the market at a large scale. Researchers and manufacturers are actively working towards developing more scalable and cost-effective production methods to address these challenges and enable widespread adoption of metal-air batteries.

The most widely used catalysts in metal-air batteries are manganese oxides, in particular MnO_2, since they have a very good relationship in terms of cost and electrochemical activity [4,5]. However, due to their low conductivity, they result in cathodes that present low voltages and low power and need to be doped with different materials, such as cobalt nickel or form composites with carbon to improve these shortcomings [6,7].

On the other hand, nickel hydroxide has been extensively studied in various energy storage systems, such as Ni-Fe and Ni-Cd, or in supercapacitors [8,9]. However, in metal-air batteries, its electrochemical activity has rarely been studied, and it has only been used as a dopant to improve the conductive properties of other materials or as a component of the current collector in different systems [10,11].

Specifically, aluminum-air batteries consist of an air cathode, an aqueous electrolyte, and an aluminum anode. Within aqueous electrolytes, there are two dominant lines of research, alkaline electrolytes based on KOH or NaOH solutions and neutral electrolytes based on NaCl. The main problem with alkaline electrolytes is the spontaneous self-corrosion of the aluminum anode that occurs when it makes contact with them, which greatly limits the lifetime of the battery. In the case of neutral electrolytes, self-corrosion of the aluminum is avoided, and a passivation layer appears on the aluminum surface which reduces the reactivity of the anode and generates parasitic reactions during the electrochemical activity of the battery [1,12]. Different ways of suppressing these effects have been studied, such as the search for commercial anti-corrosive alloys or the coating of the aluminum anodes with a thin layer of carbon that avoids the passivation of the anode when it meets the electrolyte [13]. The right combination of these can avoid these problems and lead to promising energy storage systems. This work presents a novel method that offers a quick, simple, and low-cost approach to transform commercially available fuel cell cathodes into high-performance cathodes for aluminum-air batteries. The transformation involves the electrodeposition of catalytically active nickel and manganese species onto the surface of the cathodes. By utilizing this method, it becomes possible to compare and investigate the electrochemical activity of various electrodes based on nickel and manganese oxides in primary aluminum-air batteries with a neutral electrolyte.

The results obtained from this study provide valuable insights into the combination of nickel and manganese oxides as an electrode with exceptional electrochemical properties in aluminum-air batteries. Moreover, this approach offers the advantage of using commercially available materials and well-established electrodeposition techniques. By employing this simple method, it becomes feasible to produce cathodes for metal-air batteries that can be easily commercialized. Furthermore, the findings from this work have the potential to be applied to other batteries within the metal-air family, thereby broadening the scope of its practical applications.

2. Materials and Methods

2.1. Cathode Synthesis

A 2M solution of the corresponding metal nitrate in each case ($Mn(NO_3)_2 \cdot 4H_2O$ or $Ni(NO_3)_2 \cdot 6H_2O$, both from Sigma–Aldrich, St. Louis, MO, USA) was prepared in 100 mL distilled water. Moreover, 2M solutions of 9:1, 7:3, 3:2, and 1:1 of Ni:Mn ratio were also prepared. A glassy carbon electrode—positive electrode—and different commercial fuel cell electrodes (Freudenberg H23I2)—negative electrode—of 2.5 × 2.5 cm (surface immersed in the solution) were placed in each solution at a constant distance. Between the electrodes, currents between 20 and 100 mA and varying times between 1 and 10 min were applied with an 8-channel Arbin Instruments BT2143 workstation (Munich, Germany) to obtain the electrodeposition of different manganese and nickel species as a result. Once the electrodeposition was completed, the cathodes were dried at 40 °C for 10 min. It is worth noting that despite the different electrodeposition conditions applied to the samples, the variation in the amount of electrodeposited material between them was practically negligible, ranging from 10 to 13 mg in all cases. This is why all the electrochemical results obtained from these samples would be expressed in cm^{-2}.

2.2. Characterization

Scanning electron microscopy (SEM) images were obtained in a FEI Inspect-S SEM instrument. Energy dispersive X-ray microanalysis (EDS) elemental mappings were acquired with a QUANTAX 70 detector (Bruker, Berlin, Germany) attached to a Hitachi TM3000 SEM microscope working at 15 kV. Micro-Raman measurements were carried out at room temperature in a Horiba Jobin-Ybon LabRAM HR800 on an Olympus BX 41 confocal microscope system with a 633 nm He-Ne laser. High-resolution transmission electron microscopy (HRTEM) images, electron diffraction (ED) patterns, EDS spectra, and mapping in STEM mode were acquired in a JEOL 300FEG electron microscope (Nieuw-Vennep, The Netherlands). The samples were prepared by crushing the powders under n-butanol and dispersing them over copper grids covered with a holey carbon film.

2.3. Electrochemical Characterization

To perform the electrochemical characterization, a laboratory-scale metal-air cell prototype was designed with a 2×2 cm^2 open window. Different electrodeposited cathodes, a 2M NaCl solution-based electrolyte, and a 2×2 cm^2 commercial 7475 aluminum anode coated with a 0.025 mm layer of carbon black (Cabot) and PVDF (Sigma–Aldrich, St. Louis, MO, USA) in a 3:2 ratio were placed on it. E-type glass fibber was placed on top of the aluminum, which was utilized as a separator to avoid a short-circuit between the anode and cathode. The electrochemical tests were carried out using an 8-channel Arbin Instruments BT2143 workstation at room temperature. Once the cell was assembled, it was left for 1 h to allow for good impregnation of the electrodes, and the open-circuit voltage (OCV) was recorded and the possible self-discharge of the cell was observed. The cut-off potential for all cases was 0.0 V. Constant current density discharges of 5 mA/cm^2 were applied to explore the evolution of the voltage. Discharging times and specific capacities referred to the 2×2 cm^2 open window. Linear sweep voltammetry (LSV) curves were performed on the same device and in the same voltage range at a rate of 1 mV/s.

Electrochemical impedance spectroscopy (EIS) data were obtained by applying an AC voltage of 5 mV in the (0.01–100 kHz) frequency range by using a battery tester Biologic BCS—810.

All measurements were repeated at least three times to ensure reproducibility.

3. Results and Discussion

Initially, different commercial cathodes were electrodeposited with nickel and manganese nitrate solutions as described in the Materials and Methods section. We can distinguish different electrodeposition times and currents for the two solutions as detailed in Table 1.

Table 1. Description of currents and times applied for the different electrodepositions.

Current (mA)	Time (min)
20	5
	10
30	2
50	1
	5
100	1

The different electrodeposited cathodes can be observed in Figure 1. In Figure 1a, we find the cathodes electrodeposited with the manganese nitrate solution, where a layer of slightly brown color uniformly covers the commercial fuel cell cathode, and are the cathodes that have been subjected to longer electrodeposition times and present a more intense color. Figure 1b shows the evolution of the voltage during the electrodeposition time. In all cases, it can be clearly observed how it increases above 2 V, even reaching values higher than 3 V when we work in the range of higher currents. The same occurs with the cathodes electrodeposited with the nickel solution (Figure 1c), which, in this case, show the same layer but with a pale green color (characteristic color in the electrodeposition of nickel nitrate [14]) and again it can be seen that, in all cases, values higher than 2 V are reached in electrodeposition (Figure 1d).

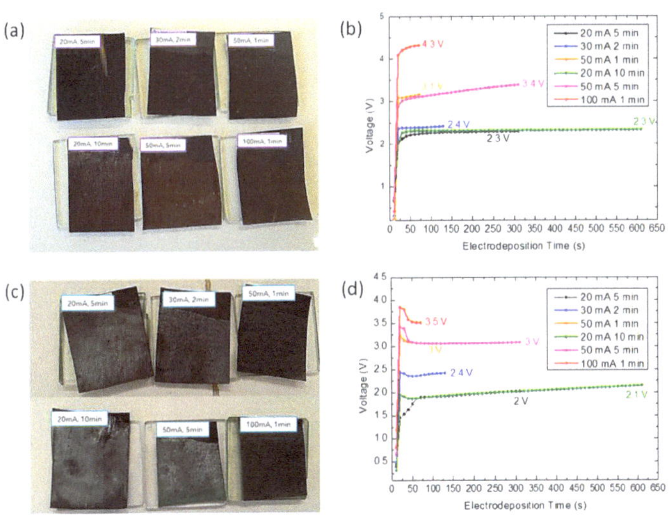

Figure 1. Images of the fuel cell cathodes electrodeposited at different currents and times in (**a**) a manganese nitrate solution with corresponding (**b**) V vs. t electrodeposition curves, and (**c**) a nickel nitrate solution with corresponding (**d**) V vs. t electrodeposition curves.

If we now study the morphology of the structures deposited on the surface of the cathodes via SEM (Figures 2 and 3), we can observe that the applied current and time are clearly key factors to be considered when it comes to the electrodeposition of these species. In Figure S1, SEM-EDS images of the fuel cell before being electrodeposited can be observed. It is visible that only PTFE-coated carbon fibers are present in the fuel cell and that no other coating or element signal was detected.

Figure 2. SEM images of commercial fuel cell cathodes electrodeposited in a manganese nitrate solution at different currents and times. Insets show high magnification images of the selected samples.

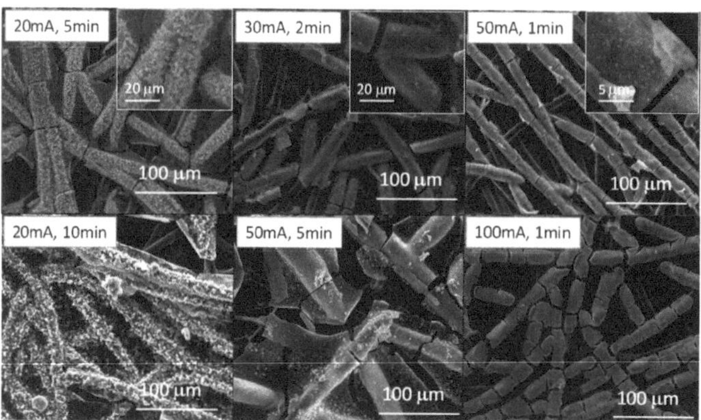

Figure 3. SEM images of commercial fuel cell cathodes electrodeposited in a nickel nitrate solution at different current densities and times. Insets show high magnification images of the selected samples.

Manganese nitrate electrodeposited microstructures can be seen in Figure 2. In the case of the sample electrodeposited at 20 mA for 5 min, the carbon filaments are coated with small platelet-shaped particles that seem to tend to coalesce, along with circular structures with many cavities, similarly to corals. If we increase the current and/or time, we can observe how the small particles disappear and start to form more and more homogeneous coatings on the GDL carbon filaments. In turn, the coral-like structures disappear and seem to merge with this coating, as in the case of the samples subjected to 50 mA for 1 min and 20 mA for 10 min. It can be seen how in the case of the sample subjected to 100 mA for 1 min, the coating of the filaments begins to present a very thick appearance and begins to plug the cavities that allow the circulation of O_2. Therefore, this sample was discarded to analyze its electrochemical behavior in aluminum-air batteries, and it was determined that the maximum current applied could not exceed 50 mA to obtain satisfactory coatings in this type of configuration.

For samples electrodeposited in the nickel nitrate solution (Figure 3), all the samples present similar morphologies, i.e., uniform coatings are observed on the carbon filaments. At low currents (20 mA), these coatings seem to present a more granular appearance, which tends to disappear and become more fused on the surface as the current is increased (>30 mA). Again, at 100 mA, the coating on the filaments becomes very thick and begins to form plates that merge and may plug the O_2 penetration holes. For a better comparison with the manganese samples and due to the observed increase trend in the fusing of the coatings, this sample was also discarded from the process to continue the electrochemical study.

To determine which phases have been electrodeposited on the carbon fibers, Raman micro-spectroscopy was used. Raman spectra of the samples electrodeposited in the manganese nitrate solution are shown in Figure 4. Table 2 summarizes the phases observed in the electrodeposition process, which depend on the current and time applied. These two parameters determine the different species with different oxidation states that are electrodeposited. Specifically, at lower currents and time, Mn (III) species are found on the surface of the carbon fibers. Meanwhile, at higher currents, Mn (IV) species start to appear. The species obtained are directly related to the voltage values reached during electrodeposition, and are the samples that reached higher voltages (see Figure 1b), i.e., the highest oxidation states. The presence of these phases has been previously observed in the electrodeposition of manganese nitrate following the subsequent reaction sequence, in which the following disproportionation and hydrolysis processes take place [15]:

$$2Mn^{2+} \rightarrow 2Mn^{3+} + 2e^- \tag{1}$$

$$Mn^{3+} + 2H_2O \rightarrow MnOOH + 3H^+ \tag{2}$$

$$MnOOH \rightarrow MnO_2 + H^+ + e^- \tag{3}$$

It should be noted that within the 1100–4000 cm^{-1} range, we can observe the signal of the carbon peaks (G, D, and 2D bands) from the fuel cells carbon fibers, superimposed on the signals originating from various manganese phases. In Figure S2, a reference Raman spectrum of the fuel cell cathode, before being subjected to any electrodeposition process, can be found. If we compare this spectrum with the signals referring to electrodeposited carbon fibers, it can be observed how in some cases the carbon signal has suffered an impact after the electrodeposition process, i.e., carbon fibers are altered. Specifically, the 2D band (2700 cm^{-1}) shifts to higher frequencies, and, even in some samples, a new band associated with C-H vibrations (2950 cm^{-1}) appears [16–18]. This fact is accentuated in the samples that have been subjected to longer electrodeposition times, as they are the samples that were electrodeposited for 10 min and the ones showing the highest intensity of the mentioned C-H band (Figure 4b). This damage may be due to the oxidation and/or reaction of the carbon on the surface during electrodeposition when exceeding 1.8 V [19]. The appearance of a band at 1100 cm^{-1} is another indicator of the modification produced in the carbon fibers. This band is directly related to the presence of amorphous carbons and stretching in the C-C bands [20,21]. It is once again more pronounced in the sample that has been electrodeposited for the longest time (20 mA, 10 min). Therefore, this reaction can also interact with the electrodeposited manganese species. Specifically, by measuring the Raman spectrum on the coral-like spheres (inset Figure 4a) in the electrodeposited sample at 20 mA for 5 min, the spectrum obtained is related to $MnCO_3$. In the rest of the samples, it is possible to find bands superimposed on the carbon signals, which may be due to Mn-C vibrations. However, as they are fused with the coatings (as seen in SEM), it is difficult to know their exact nature.

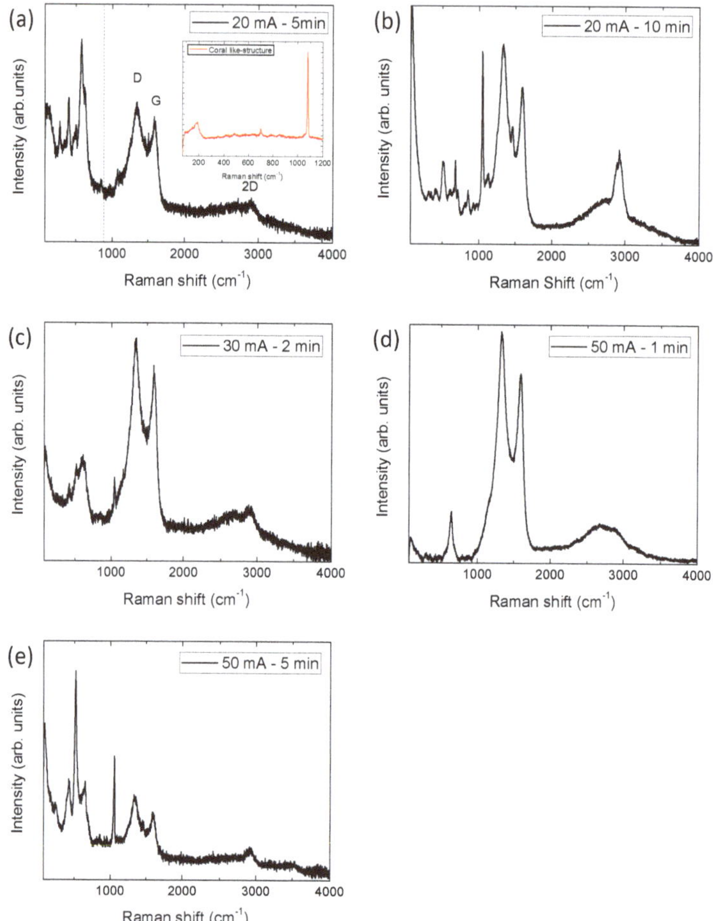

Figure 4. Raman spectra of the different fuel cell electrodeposited in the manganese nitrate solution: (**a**) 20 mA 5 min (inset corresponds to the Raman spectra of the coral-like structures), (**b**) 20 mA, 10 min; (**c**) 30 mA, 2 min; (**d**) 50 mA, 1 min; and (**e**) 50 mA, 5 min.

Table 2. Raman peak assignments of the manganese species electrodeposited on the commercial fuel cell cathodes.

Sample	Raman Peaks (cm^{-1})	Assignment	Reference
20 mA, 5 min	280 403 504 578 631	Mn_2O_3	[22]
	700 1080	$MnCO_3$	[23]

Table 2. *Cont.*

Sample	Raman Peaks (cm^{-1})	Assignment	Reference
20 mA, 10 min	325 408 514 612 681 715	MnO_2, tunnel structure (3 × 3)	[24]
	1042	NO_3, adsorbed	[25]
30 mA, 2 min	380 485 558 624 655	$MnOOH + MnO_2$, mixed forms	[24,26]
	1046	NO_3, adsorbed	[25]
50 mA, 1 min	640	Mn_3O_4	[26,27]
50 mA, 5 min	430 520 650	MnO_2, tunnel structure (2 × 2)	[24]
	1080	$MnCO_3$	[23]

In the case of the Raman spectra of samples electrodeposited with nickel hydroxide (Figure 5), we found the presence of $Ni(OH)_2$ in all cases. Its presence is confirmed by the broad band centered at ~470 cm^{-1} [8]. The appearance of this phase is expected, since the electrodeposition of nickel from nickel nitrate has been extensively studied and is driven by the following reaction [28,29]:

$$NO_3^- + H_2O + 2e^- \rightarrow NO_2^- + 2OH^- \tag{4}$$

The hydroxide ions cause a steep increase in the pH close to the electrode surface and nickel hydroxide precipitation takes place, as follows:

$$Ni^{2+} + 2OH^- \rightarrow Ni(OH)_2 \tag{5}$$

It can be observed that the Ni electrodeposition is more hostile to carbon fibers, since in all cases, even for very low electrodeposition times, a band due to C-H groups (at ~2950 cm^{-1}) appears. Furthermore, in this case, the current increase has a very significant impact on the carbon fibers, which are the samples electrodeposited at 50 mA and show are the most affected by D, G, and 2D band signals (widening, displacement, and decrease of its signal). This is also related to the observation of a high concentration of H^+ generated around the glassy carbon during the nickel deposition. On the one hand, this effect allows the nickel electrodeposition on the fuel cell electrode to generate a rich OH^- environment (nitrate ions (NO_3^-) are reduced to produce hydroxide ions (OH^-), which reacts with Ni^{2+} ions to form metal hydroxides [30]). On the other hand, it produces the partial reaction of nickel with the carbonaceous electrode [31].

The main structural differences between samples are found in the intercalated species inside the nickel hydroxide matrix. It has been previously observed that, due to the laminar nature of $Ni(OH)_2$, it has a high affinity towards the anion intercalation in its structure [8,32]. Predominantly, NO_3^- ion intercalation (1050 cm^{-1}) can be detected; the presence of NO_3^- ions inside its structure is very frequent when the precursor of this phase is nickel nitrate. In addition, carbonate ions are also present inside $Ni(OH)_2$, this process is related to the peaks centered at 870 and 1070 cm^{-1} [8,33]. These last ions probably owe their origin to the structural damage suffered by the carbon fibers during electrodeposition. It can be clearly observed that the samples subjected to 20 mA for 10 min, 50 mA for 1 min, and 50 mA for 5 min show the highest intensity of the peaks related to species' intercalation.

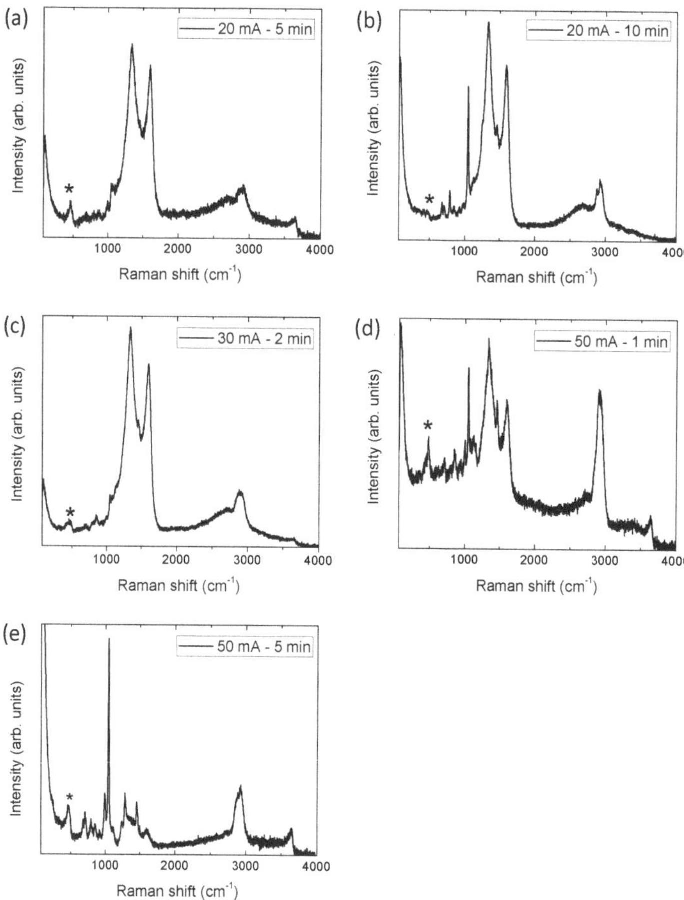

Figure 5. Raman spectra of the different fuel cell electrodeposited in the nickel nitrate solution: (**a**) 20 mA, 5 min; (**b**) 20 mA, 10 min; (**c**) 30 mA, 2 min; (**d**) 50 mA, 1 min; and (**e**) 50 mA, 5 min. Ni(OH)$_2$ signal is indicated by *.

To study the electrochemical behavior of the prepared electrodes in aluminum-air batteries, different experiments were carried out, which prepared the cells as described in the Materials and Methods section and kept the dimensions and the amount of electrolyte constant in all of the cells. On the one hand, LSV curves were performed to analyze the electrochemical activity of the samples in these batteries; on the other hand, discharge protocols at a constant current (20 mA, 5 mA cm^{-2}) were performed to observe which samples present the highest electrochemical activity in these types of batteries.

Based on the LSV curves of the samples electrodeposited in the manganese nitrate solution (Figure 6a), it can be observed how all of the samples start to show activity at a voltage of ~0.6 V, improving, in all cases, the activity of the reference cathode (Reference FC, corresponding to the commercial fuel cell without a catalyst), which does not start showing activity until 0.5 V. By comparing the LSV curves, it can be seen that the samples electrodeposited at 20 mA for 5 min and at 50 mA for 1 min provide the higher ORR kinetic-limiting current densities (>−8 mA, 2 mA cm^{-2}) [34,35]. This is clearly reflected in the discharge curves (Figure 6b) since they are the samples which provide the highest discharge time, i.e., the highest capacity. In particular, the sample electrodeposited at 20 mA for 5 min stands out, reaching discharge times higher than 5.5 h, which is equivalent to capacities higher than 110 mA h–27.5 mA h cm^{-2}. This represents an increase of 2 h of discharge compared to the reference fuel cell electrode (Figure S3). A priori, the phases with the best electrochemical activity in this type of battery are the phases with Mn (IV)-derived oxides in their structure [21]. However, in this case, the presence of phases derived from manganese and carbon ($MnCO_3$) and the optimal coating of the carbon fibers (thin coatings without blocking the entry of O_2 into the cell) greatly favor the electrochemical activity of the battery. The presence of Mn-C derivatives has already been observed to considerably improve the activity of primary metal-air batteries compared to pure species [36].

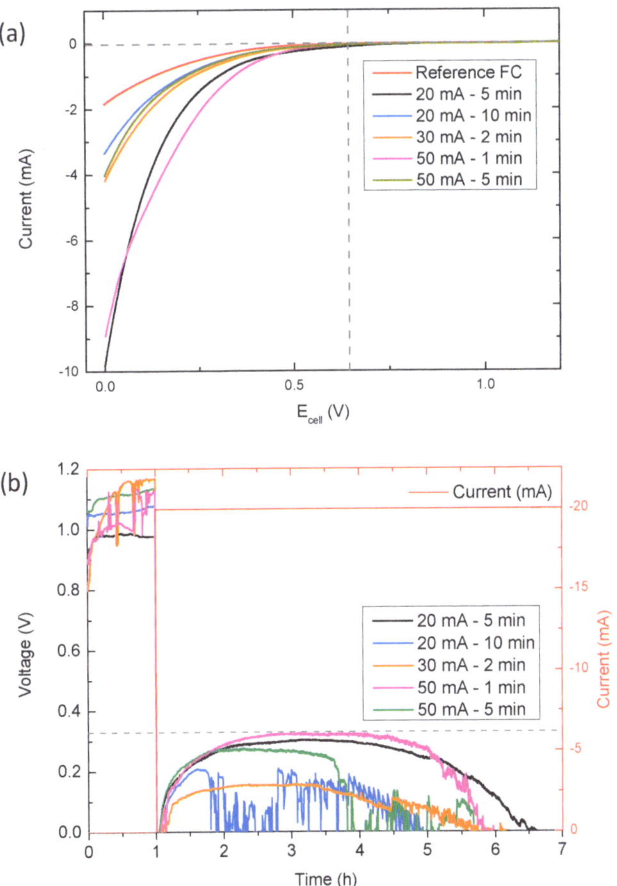

Figure 6. Aluminium-air electrochemical measurements with the electrodeposited manganese nitrate cathodes: (**a**) LSV curves and (**b**) constant current discharge curves.

Nevertheless, it should be noted that the morphology of the electrodeposited oxides also plays an important role. Both of the samples subject to 20 mA for 5 min and the 50 mA for 1 min show the longest discharge times and we can find the coatings with coral-like cavities in both of them. These morphologies, which are not found or are fused with other structures in the other samples, give rise to a better electrocatalytic activity than that of the samples with MnO_2 species [21]. This could be due to the presence of thicker and smoother coatings on the carbon fibers in the samples with MnO_2 species, which may reduce the contact between oxygen, the electrolyte, and the catalyst. To verify this factor, EIS measurements were performed (Figure S4a and Table S1), in which it can be observed how the resistance of the samples increases directly in relation to the thickness of the coatings obtained. With this observation, we can assume that not only the electrocatalytic species that are used are important, but that their morphology and the architecture of the air cathode are also important.

In the case of nickel electrodeposited samples, the LSV curves (Figure 7a) show that the cathodes electrodeposited at 20 mA for 5 min and 50 mA for 1 min show the highest ORR kinetic-limiting current densities (~-6 mA, -1.5 mA cm^{-2}). It should be noted that in this case the $Ni(OH)_2$, cathodes start their activity at $\sim$0.75 V, significantly increasing the voltage with respect to the reference cathode. Once again, these samples with higher ORR-limiting current densities show the longest discharge time (Figure 7b), reaching 6.5 h (130 mA h–32.5 mA h cm^{-2}). In particular, the sample electrodeposited at 20 mA for 5 min is the one which shows the most stable discharge curve. These results are also in agreement with those obtained in EIS (Figure S4b and Table S1), which is that the sample electrodeposited with Ni at 20 mA for 5 min offers the lowest resistance. This cathode exhibits an improvement of more than 3 h in the total discharge time with respect to the reference commercial fuel cell cathode.

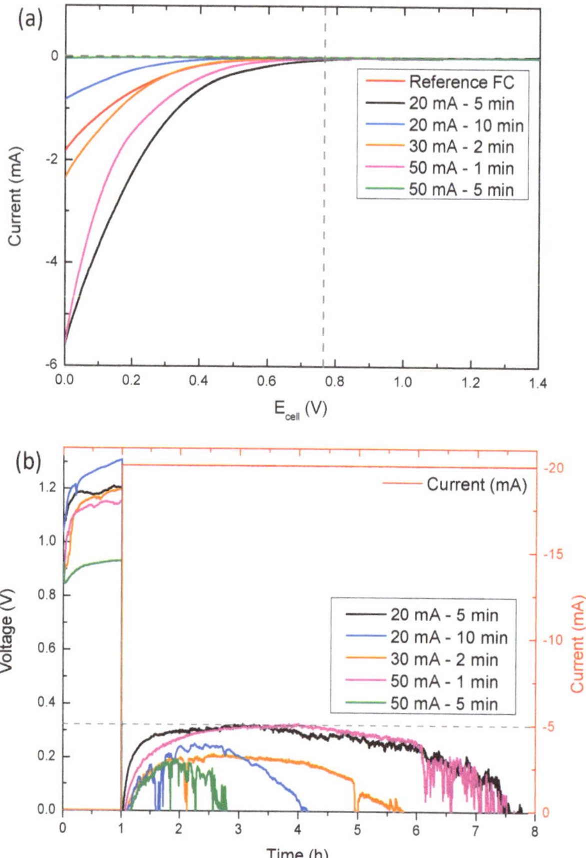

Figure 7. Aluminum-air electrochemical measurements with the electrodeposited nickel nitrate cathodes: (**a**) LSV curves and (**b**) constant current discharge curves.

The results obtained agree with previous observations in relation to the ORR activity of $Ni(OH)_2$ with intercalated species [37,38]. The ORR activity of $Ni(OH)_2$ increases and stabilizes when it has a certain amount of intercalated species in its structure. However, there comes a point that an increase of these species begins to destabilize its structure and its ORR activity begins to decrease again. Taking this into account, and the results observed in Raman, we can infer that the sample electrodeposited at 20 mA for 5 min presents the optimum number of intercalated species to favor and improve its catalytic activity. On the contrary, the samples that present a higher number of intercalated species, electrodeposited at 20 mA for 10 min and at 50 mA for 5 min, destabilize their structure and considerably reduce their activity.

Comparing the results obtained in different electrodeposited samples, we can conclude that, in both manganese nitrate solution and nickel nitrate solution, the best results are obtained from the electrodepositions performed at 20 mA for 5 min.

To compare the morphology and structure of these two samples, an additional TEM analysis was performed (Figures S5 and S6). Results corroborate that both show nanometric and porous coatings on the carbon fibers, promoting ORR activity without affecting the contact between oxygen, the electrolyte, and the catalyst. Moreover, from the LSV curves, it can be determined that, while the sample electrodeposited with the manganese species presents a higher ORR kinetic-limiting current density, the sample electrodeposited with nickel presents a higher voltage at which the activity starts.

Considering the results obtained above, an attempt was made to combine manganese and nickel ORR activities with the electrodeposition of mixed solutions of nickel and manganese nitrates in different ratios (9:1, 7:3, 3:2, and 1:1 of Ni:Mn, respectively). To evaluate the electrochemical activity of these new combined Ni:Mn cathodes, discharge curves at 20 mA (5 mA cm^{-2}) with the previously described configuration were again performed. In Figure 8, it can be observed that the sample with a Ni:Mn ratio in a 3:2 solution shows the best electrochemical activity, reaching discharge times of 9 h (180 mA h–45 mA h cm^{-2}), which corresponds to an increase of more than 100 mA h (25 mA h cm^{-2}) in the battery capacity with respect to the commercial reference fuel cell cathode. If we again perform LSV (Figure S7) on this electrode, at a Ni:Mn ratio in a 3:2 solution, and compare it with the results previously obtained for the single nickel and manganese electrodeposited electrodes, it can be clearly seen that this electrode combines the good ORR properties of both of them. On the one hand, it maintains the voltage at which the ORR activity begins, which is characteristic of nickel hydroxide. On the other hand, the ORR kinetic-limiting current density reaches similar values to those obtained with the manganese electrode; this proves the efficiency of the combination of the two solutions.

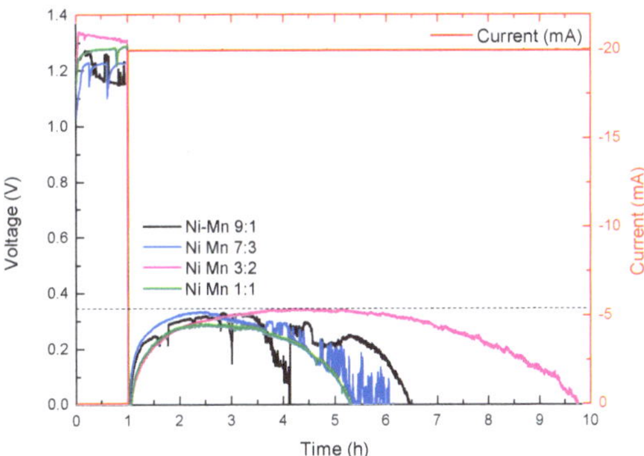

Figure 8. Constant current discharge curves of the different concentrations in electrodeposited Ni:Mn cathodes.

In view of the good results obtained for the electrodeposited Ni:Mn (3:2) cathode, new discharges at different currents were carried out to analyze its response (Figure 9). Despite varying the discharge current between 30 and 5 mA (7.5 mA cm^{-2}–1.25 mA cm^{-2}), we obtain similar capacity values close to 200 mA h (50 mA h cm^{-2}). It is evident from these discharge curves that as we increase the discharge current density, the average discharge voltage plateau decreases: ~0.6 V at 1.25 mA cm^{-2}, ~0.42 V at 2.5 mA cm^{-2}, ~0.33 V at 5 mA cm^{-2}, and ~0.24 V at 7.5 mA cm^{-2}. However, the power density of the batteries exhibits a logarithmic increase, which is the expected behavior in this type of metal-air battery, appearing to reach its maximum at values close to 1.8 mW cm^{-2} (achieved during a discharge at 30 mA: ~0.24 V at 7.5 mA cm^{-2}). These results are depicted in Figure S8. These findings demonstrate how a simple, fast, and inexpensive electrodeposition process can significantly improve the electrochemical activity of commercial cathodes in aluminum-air batteries.

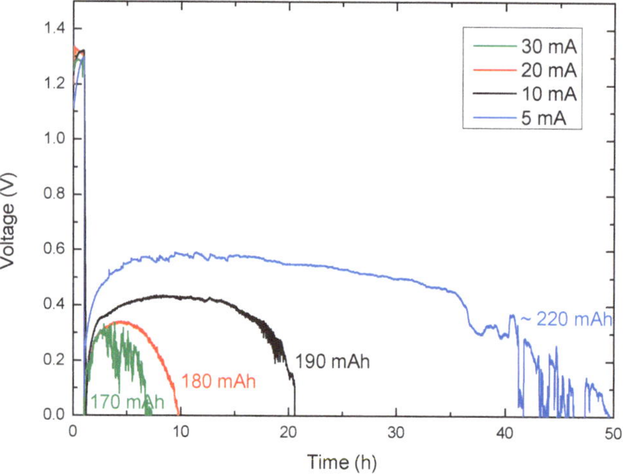

Figure 9. Ni:Mn (3:2) discharge curves at different constant currents: 30 mA (green), 20 mA (red), 10 mA (black), and 5 mA (blue).

To analyze this electrode with improved properties in depth, a structural study was conducted. Figure 10a shows the images obtained with SEM, in which a coating with cavities can be appreciated again, reminding us of the coral-like structures which have previously presented a favorable behavior. In the case of the Raman spectrum of this sample (Figure 10b), we can observe the presence of a highly amorphous signal, which is unable to identify any specific peak, and can only identify a broad band in the area corresponding to the characteristic signals of the nickel and manganese oxides and hydroxides [100–1000 cm^{-1}]. This signal has been observed previously in Ni(OH)$_2$ doped with manganese; the presence of manganese could disorder the structure of the nickel hydroxide by intercalating in its layered structure [39].

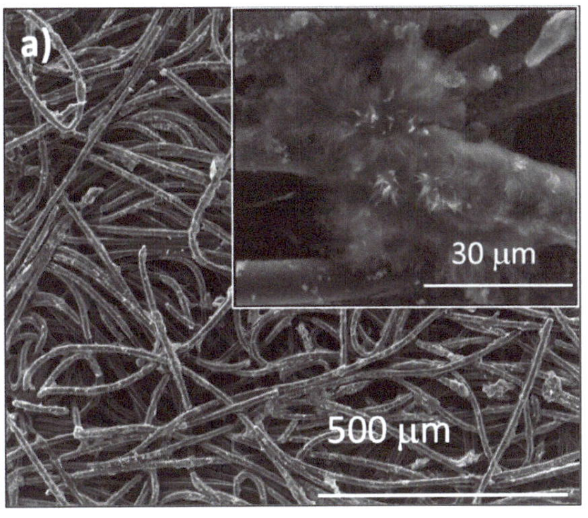

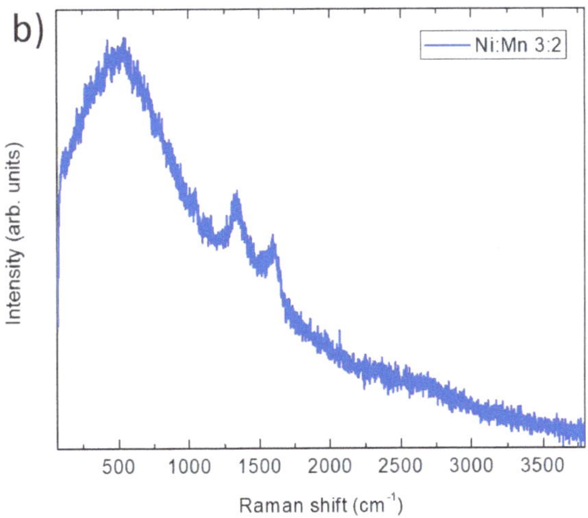

Figure 10. (**a**) SEM images and (**b**) Raman spectra of the Ni:Mn (3:2) electrode.

To understand the origin of this signal, a TEM-EDS study was performed. Figure 11 shows different HRTEM micrographs of the Ni:Mn (3:2) sample. This sample is formed by an interesting nanoarchitecture built from nearly transparent nanosheets (regions I in Figure 11a,b) assembled through stalk-like formations (regions II in Figure 11a,c), which are composed by corrugated nanosheets of about 10–15 nm in width. Both types of morphology (region I and region II) exhibit the same composition of 3Ni:1Mn, as determined from the corresponding EDS spectra (Figure S9). This composition aligns with the initial molar ratios used during electrodeposition and highlights the preference for nickel electrodeposition over manganese.

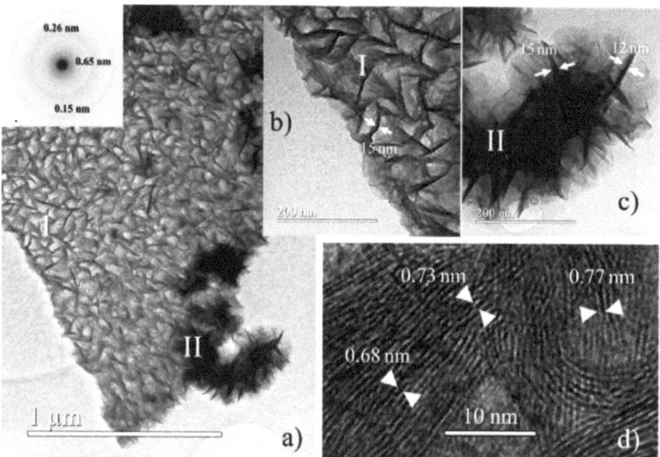

Figure 11. HRTEM images for the Ni:Mn sample at (**a**) low magnification (inset shows the corresponding ED pattern); (**b,c**) gathered images corresponding to regions labelled as I and II in image (**a**), respectively; and at (**d**) high magnification.

The inset of Figure 11a gathers a representative ED pattern, which can be indexed based on the $Ni(OH)_2$-type structure, similar to that obtained for the Ni 20 sample electrodeposited at mA for 5 min. As the sample presents very low crystallinity, the periodic contrasts appear only along small distances and it is difficult to determine precise values, as would be the case in well-grown crystals. Thus, the d-spacing values of 0.65 nm could be indexed to the (003) crystal planes of $Ni(OH)_2$; however, they are probably more representative of the (003) crystal planes of NiOOH. The other two distances indicated in the ED pattern (inset of Figure 11a), 0.26 and 0.15 nm, could also correspond to both phases. Moreover, the distances corresponding to the (003) crystal planes range between 0.65 and 0.78 nm (Figure 11d), pointing to the probable coexistence of $Ni(OH)_2$ and MOOH, with M=Mn, Ni. Furthermore, the EDS mappings (Figure S10) of the Ni:Mn nanosheets indicate compositional homogeneity at a nanoscale. Indeed, Mn doping at the atomic level (up to 6%) has been previously reported as a strategy to induce local contraction of the metal-O/metal bond length, stabilizing the structure in this way [39]. In addition, Mn doping of $Ni(OH)_2$ has been proved to promote the production of high-valence $Ni^{III}OOH$ [40]. Thus, the stabilization of trivalent Ni and Mn cations should be an interesting electronic feature of the Ni:Mn sample that leads to an improved electrochemical response.

Lastly, it is important to highlight that all the samples examined in this study were characterized both before and after the electrochemical measurements. These assessments revealed no significant differences in the electrodes, indicating that the cathode elements (the carbon fibers and the electrodeposited catalysts) do not undergo degradation during battery discharge. This suggests their potential for reuse in another battery of the same type, with the sole requirement of having to replace the consumed aluminum. A representative SEM image, along with its corresponding EDS analysis, of the Ni:Mn 3:2 sample after cycling is presented in Figure S11.

4. Conclusions

In this work, we have shown that commercial fuel cell cathodes can be converted into high-performance cathodes for metal-air batteries through electrodeposition. Two types of sample series have been tested, one with the electrodeposition of nickel species and another with manganese species. We have determined the optimal parameters for the electrodeposition of both species, which is a current of 20 mA and a deposition time of 5 min. Electrodeposition of manganese species results in high ORR kinetic-limiting current density, while the electrodeposition of nickel presents a higher voltage at which the activity starts. Along with the importance of the electrocatalytic species used, the morphology and the architecture of the air cathode play an important role. A porous coating of the carbon fibers, formed by interconnected sheets or platelet-shaped particles, favors the contact between oxygen, the electrolyte, and the catalyst. The combination of Ni and Mn in optimal proportions produces cathodes with good properties that are merged from both species, leading to the construction of 50 mA h cm^{-2} Al-air batteries. In conclusion, this study presents a novel, cost-effective method for producing high-performance cathodes for aluminum-air batteries through electrodeposition of nickel and manganese species. The method in this study offers a readily scalable solution that enhances electrochemical performance. The results highlight its potential for practical applications in emerging energy storage technologies.

Supplementary Materials: The following supporting information can be downloaded at: https://www.mdpi.com/article/10.3390/mi14101930/s1. Figure S1. SEM-EDS compositional mapping of commercial fuel cell cathode: (a) SE signal, (b) carbon signal and (c) F signal. (d) EDS spectra. Figure S2. Raman spectra of the commercial fuel cell cathode. Figure S3. Constant current discharge curve of the reference commercial fuel cell cathode. Figure S4. EIS Nyquist plots – insets show the equivalent circuits: (a) Manganese nitrate electrodeposited samples and (b) Nickel nitrate electrodeposited samples. Figure S5. HRTEM images for the Mn (20 mA-5min) sample (a) at low magnification (inset shows the corresponding ED pattern indexed based on cubic Mn2O3) and at high magnification in different regions showing the presence of (b) Mn2O3 and (c) MnCO3. Figure S6. HRTEM images for the Ni (20 mA-5min) sample (a) at low magnification (inset shows the corresponding ED pattern indexed based on hydrated Ni(OH)2) and (b) at high magnification. Figure S7. LSV curves comparison between the samples electrodeposited at 20 mA 5 min: Ni:Mn 3:2 (black), Mn 20 mA 5 min (pink) and Ni 20 mA 5min (blue). Figure S8. Discharge voltage and power density vs discharge current density for Ni:Mn 3:2 electrode. Figure S9. EDS obtained from the different HRTEM regions for the Ni:Mn sample. Cu signal originates from the sample holder. Figure S10. (a) STEM image of the Ni:Mn sample and (b) Ni, (c) Mn and (d) O corresponding mappings. Figure S11. SEM image (a) and corresponding EDS (b) of the Ni:Mn 3:2 sample after cycling. Table S1. Relevant results obtained from Nyquist plots fitting.

Author Contributions: P.A.: conceptualization, investigation, formal analysis, and writing—original draft preparation; B.S.: investigation, formal analysis, and writing—review and editing; D.G.: investigation; J.C.: supervision and resources; I.Á.-S.: investigation, formal analysis, and writing—review and editing; M.L.L.: supervision, formal analysis, and writing—review and editing. All authors have read and agreed to the published version of the manuscript.

Funding: This work has been partially supported by the Clean Sky 2 joint with the European Union's Horizon 2020 research and innovation programme, under the O2FREE project (grant agreement no. 101008179).

Acknowledgments: Sotillo acknowledges the financial support from the Comunidad de Madrid via the project PR65/19-22464 (R+D projects for young researchers) and the Research Talent Attraction Program (2017-T2/IND-5465).

Conflicts of Interest: The authors declare no conflict of interest.

References

1. Liu, Y.; Sun, Q.; Li, W.; Adair, K.R.; Li, J.; Sun, X. A comprehensive review on recent progress in aluminum–air batteries. *Green Energy Environ.* **2017**, *2*, 246–277. [CrossRef]
2. Elia, G.A.; Kravchyk, K.V.; Kovalenko, M.V.; Chacón, J.; Holland, A.; Wills, R.G. An overview and prospective on Al and Al-ion battery technologies. *J. Power Sources* **2021**, *481*, 228870. [CrossRef]
3. Martin, J.J.; Neburchilov, V.; Wang, H.; Qu, W. Air cathodes for metal-air batteries and fuel cells. In Proceedings of the 2009 IEEE Electrical Power & Energy Conference (EPEC), Montreal, QC, Canada, 22–23 October 2009; pp. 1–6. [CrossRef]
4. Matsuki, K.; Kamada, H. Oxygen reduction electrocatalysis on some manganese oxides. *Electrochim. Acta* **1986**, *31*, 13–18. [CrossRef]
5. Wei, S.; Liu, H.; Wei, R.; Chen, L. Cathodes with MnO$_2$ catalysts for metal fuel battery. *Front. Energy* **2019**, *13*, 9–15. [CrossRef]

6. Roche, I.; Chaînet, E.; Chatenet, M.; Vondrák, J. Durability of carbon-supported manganese oxide nanoparticles for the oxygen reduction reaction (ORR) in alkaline medium. *J. Appl. Electrochem.* **2008**, *38*, 1195–1201. [CrossRef]

7. Xia, Z.; Zhu, Y.; Zhang, W.; Hu, T.; Chen, T.; Zhang, J.; Liu, Y.; Ma, H.; Fang, H.; Li, L. Cobalt ion intercalated MnO_2/C as air cathode catalyst for rechargeable aluminum–air battery. *J. Alloys Compd.* **2020**, *824*, 153950. [CrossRef]

8. Hall, D.S.; Lockwood, D.J.; Bock, C.; MacDougall, B.R. Nickel hydroxides and related materials: A review of their structures, synthesis and properties. *Proc. R. Soc. A Math. Phys. Eng. Sci.* **2015**, *471*, 20140792. [CrossRef] [PubMed]

9. Ash, B.; Nalajala, V.S.; Popuri, A.K.; Subbaiah, T.; Minakshi, M. Perspectives on Nickel Hydroxide Electrodes Suitable for Rechargeable Batteries: Electrolytic vs. Chemical Synthesis Routes. *Nanomaterials* **2020**, *10*, 1878. [CrossRef]

10. Nie, Q.; Xu, N.; Zhou, X.; Qiao, J. Nickel Foam as a New Air Electrode Material to Enhance the Performance in Rechargeable Zn-Air Batteries. *ECS Trans.* **2018**, *85*, 35–40. [CrossRef]

11. Chowdhury, A.; Lee, K.-C.; Lim, M.S.W.; Pan, K.-L.; Chen, J.N.; Chong, S.; Huang, C.-M.; Pan, G.-T.; Yang, T.C.-K. The Zinc-Air Battery Performance with Ni-Doped MnO_2 Electrodes. *Processes* **2021**, *9*, 1087. [CrossRef]

12. Wu, P.; Wu, S.; Sun, D.; Tang, Y.; Wang, H. A Review of Al Alloy Anodes for Al–Air Batteries in Neutral and Alkaline Aqueous Electrolytes. *Acta Met. Sin. Engl. Lett.* **2020**, *34*, 309–320. [CrossRef]

13. Pino, M.; Herranz, D.; Chacón, J.; Fatás, E.; Ocón, P. Carbon treated commercial aluminium alloys as anodes for aluminium-air batteries in sodium chloride electrolyte. *J. Power Sources* **2016**, *326*, 296–302. [CrossRef]

14. Wohlfahrtmehrens, M.; Wohlfahrt-Mehrens, M.; Oesten, R.; Wilde, P.; Huggins, R. The mechanism of electrodeposition and operation of $Ni(OH)_2$ layers. *Solid State Ion.* **1996**, *86–88*, 841–847. [CrossRef]

15. Yi, C.P.; Majid, S.R. *The Electrochemical Performance of Deposited Manganese Oxide-Based Film as Electrode Material for Electrochemical Capacitor Application, Semiconductors—Growth and Characterization*; IntechOpen: London, UK, 2017. [CrossRef]

16. Pawlyta, M.; Rouzaud, J.-N.; Duber, S. Raman microspectroscopy characterization of carbon blacks: Spectral analysis and structural information. *Carbon* **2015**, *84*, 479–490. [CrossRef]

17. Merlen, A.; Buijnsters, J.G.; Pardanaud, C. A Guide to and Review of the Use of Multiwavelength Raman Spectroscopy for Characterizing Defective Aromatic Carbon Solids: From Graphene to Amorphous Carbons. *Coatings* **2017**, *7*, 153. [CrossRef]

18. Zou, J.W.; Schmidt, K.; Reichelt, K.; Dischler, B. The properties of *a*-C:H films deposited by plasma decomposition of C_2H_2. *J. Appl. Phys.* **1990**, *67*, 487–494. [CrossRef]

19. Yi, Y.; Weinberg, G.; Prenzel, M.; Greiner, M.; Heumann, S.; Becker, S.; Schlögl, R. Electrochemical corrosion of a glassy carbon electrode. *Catal. Today* **2017**, *295*, 32–40. [CrossRef]

20. Paillard, V. On the origin of the 1100 cm $^{-1}$ Raman band in amorphous and nanocrystalline sp^3 carbon. *EPL Europhys. Lett.* **2001**, *54*, 194–198. [CrossRef]

21. Dessie, Y.; Tadesse, S.; Eswaramoorthy, R.; Abebe, B. Recent developments in manganese oxide based nanomaterials with oxygen reduction reaction functionalities for energy conversion and storage applications: A review. *J. Sci. Adv. Mater. Devices* **2019**, *4*, 353–369. [CrossRef]

22. Julien, C.; Massot, M.; Poinsignon, C. Lattice vibrations of manganese oxides Part I. Periodic structures. *Spectrochim. Acta Part A Mol. Biomol. Spectrosc.* **2004**, *60*, 689–700. [CrossRef] [PubMed]

23. Zhao, C.; Li, H.; Jiang, J.; He, Y.; Liang, W. Phase Transition and vibration properties of $MnCO_3$ at high pressure and high-temperature by Raman spectroscopy. *High Press. Res.* **2018**, *38*, 212–223. [CrossRef]

24. Post, J.E.; McKeown, D.A.; Heaney, P.J. Raman spectroscopy study of manganese oxides: Tunnel structures. *Am. Miner.* **2020**, *105*, 1175–1190. [CrossRef]

25. Fontana, M.D.; Ben Mabrouk, K.; Kauffmann, T.H. Raman spectroscopic sensors for inorganic salts. In *Spectroscopic Properties of Inorganic and Organometallic Compounds: Techniques, Materials and Applications*; Royal Society of Chemistry: London, UK, 2013; Volume 44, pp. 40–67. [CrossRef]

26. Cheng, S.; Yang, L.; Chen, D.; Ji, X.; Jiang, Z.-J.; Ding, D.; Liu, M. Phase evolution of an alpha MnO_2-based electrode for pseudo-capacitors probed by in operando Raman spectroscopy. *Nano Energy* **2014**, *9*, 161–167. [CrossRef]

27. Shah, H.U.; Wang, F.; Toufiq, A.M.; Ali, S.; Khan, Z.U.H.; Li, Y.; Hu, J.; He, K. Electrochemical Properties of Controlled Size Mn_3O_4 Nanoparticles for Supercapacitor Applications. *J. Nanosci. Nanotechnol.* **2018**, *18*, 719–724. [CrossRef] [PubMed]

28. Jayashree, R.; Kamath, P. Nickel hydroxide electrodeposition from nickel nitrate solutions: Mechanistic studies. *J. Power Sources* **2001**, *93*, 273–278. [CrossRef]

29. Streinz, C.C.; Hartman, A.P.; Motupally, S.; Weidner, J.W. The Effect of Current and Nickel Nitrate Concentration on the Deposition of Nickel Hydroxide Films. *J. Electrochem. Soc.* **1995**, *142*, 1084–1089. [CrossRef]

30. Abebe, E.M.; Ujihara, M. Simultaneous Electrodeposition of Ternary Metal Oxide Nanocomposites for High-Efficiency Supercapacitor Applications. *ACS Omega* **2022**, *7*, 17161–17174. [CrossRef]

31. Liu, D.; Lazenby, R.A.; Sloan, J.; Vidotti, M.; Unwin, P.R.; Macpherson, J.V. Electrodeposition of Nickel Hydroxide Nanoparticles on Carbon Nanotube Electrodes: Correlation of Particle Crystallography with Electrocatalytic Properties. *J. Phys. Chem. C* **2016**, *120*, 16059–16068. [CrossRef]

32. Lee, J.W.; Ko, J.M.; Kim, J.-D. Hierarchical Microspheres Based on α-$Ni(OH)_2$ Nanosheets Intercalated with Different Anions: Synthesis, Anion Exchange, and Effect of Intercalated Anions on Electrochemical Capacitance. *J. Phys. Chem. C* **2011**, *115*, 19445–19454. [CrossRef]

33. Delmas, C.; Faure, C.; Borthomieu, Y. The effect of cobalt on the chemical and electrochemical behaviour of the nickel hydroxide electrode. *Mater. Sci. Eng. B* **1992**, *13*, 89–96. [CrossRef]
34. Wang, Y.-J.; Fang, B.; Zhang, D.; Li, A.; Wilkinson, D.P.; Ignaszak, A.; Zhang, L.; Zhang, J. A Review of Carbon-Composited Materials as Air-Electrode Bifunctional Electrocatalysts for Metal–Air Batteries. *Electrochem. Energy Rev.* **2018**, *1*, 1–34. [CrossRef]
35. Huang, C.; Zhao, X.; Hao, Y.; Yang, Y.; Qian, Y.; Chang, G.; Zhang, Y.; Tang, Q.; Hu, A.; Chen, X. Self-Healing SeO_2 Additives Enable Zinc Metal Reversibility in Aqueous $ZnSO_4$ Electrolytes. *Adv. Funct. Mater.* **2022**, *32*, 2112091. [CrossRef]
36. Flegler, A.; Hartmann, S.; Weinrich, H.; Kapuschinski, M.; Settelein, J.; Lorrmann, H.; Sextl, G. Manganese Oxide Coated Carbon Materials as Hybrid Catalysts for the Application in Primary Aqueous Metal-Air Batteries. *J. Carbon Res.* **2016**, *2*, 4. [CrossRef]
37. Farjami, E.; Deiner, L.J. Evidence for oxygen reduction reaction activity of a $Ni(OH)_2$/graphene oxide catalyst. *J. Electrochem. Soc. J. Mater. Chem. A* **2013**, *1*, 15501–15508. [CrossRef]
38. Godínez-Salomón, F.; Rhodes, C.P.; Alcantara, K.S.; Zhu, Q.; Canton, S.; Calderon, H.; Reyes-Rodríguez, J.; Leyva, M.; Solorza-Feria, O. Tuning the Oxygen Reduction Activity and Stability of $Ni(OH)_2$@Pt/C Catalysts through Controlling Pt Surface Composition, Strain, and Electronic Structure. *Electrochim. Acta* **2017**, *247*, 958–969. [CrossRef]
39. Zhang, Z.; Huo, H.; Wang, L.; Lou, S.; Xiang, L.; Xie, B.; Wang, Q.; Du, C.; Wang, J.; Yin, G. Stacking fault disorder induced by Mn doping in $Ni(OH)_2$ for supercapacitor electrodes. *Chem. Eng. J.* **2021**, *412*, 128617. [CrossRef]
40. Yang, X.; Zhang, H.; Xu, W.; Yu, B.; Liu, Y.; Wu, Z. A doping element improving the properties of catalysis: In situ Raman spectroscopy insights into Mn-doped NiMn layered double hydroxide for the urea oxidation reaction. *Catal. Sci. Technol.* **2022**, *12*, 4471–4485. [CrossRef]

 *micromachines*

Article

A Novel Approach for Identifying Nanoplastics by Assessing Deformation Behavior with Scanning Electron Microscopy

Jared S. Stine [1,2], Nicolas Aziere [3], Bryan J. Harper [2] and Stacey L. Harper [1,2,*]

[1] School of Chemical, Biological and Environmental Engineering, Oregon State University, Corvallis, OR 97331, USA; stinej@oregonstate.edu
[2] Department of Environmental and Molecular Toxicology, Oregon State University, Corvallis, OR 97331, USA; bryan.harper@oregonstate.edu
[3] School of Electrical Engineering and Computer Science, Oregon State University, Corvallis, OR 97331, USA; azieren@oregonstate.edu
[*] Correspondence: stacey.harper@oregonstate.edu; Tel.: +1-541-737-2791

Abstract: As plastic production continues to increase globally, plastic waste accumulates and degrades into smaller plastic particles. Through chemical and biological processes, nanoscale plastic particles (nanoplastics) are formed and are expected to exist in quantities of several orders of magnitude greater than those found for microplastics. Due to their small size and low mass, nanoplastics remain challenging to detect in the environment using most standard analytical methods. The goal of this research is to adapt existing tools to address the analytical challenges posed by the identification of nanoplastics. Given the unique and well-documented properties of anthropogenic plastics, we hypothesized that nanoplastics could be differentiated by polymer type using spatiotemporal deformation data collected through irradiation with scanning electron microscopy (SEM). We selected polyvinyl chloride (PVC), polyethylene terephthalate (PET), and high-density polyethylene (HDPE) to capture a range of thermodynamic properties and molecular structures encompassed by commercially available plastics. Pristine samples of each polymer type were chosen and individually milled to generate micro and nanoscale particles for SEM analysis. To test the hypothesis that polymers could be differentiated from other constituents in complex samples, the polymers were compared against proxy materials common in environmental media, i.e., algae, kaolinite clay, and nanocellulose. Samples for SEM analysis were prepared uncoated to enable observation of polymer deformation under set electron beam parameters. For each sample type, particles approximately 1 μm in diameter were chosen, and videos of particle deformation were recorded and studied. Blinded samples were also prepared with mixtures of the aforementioned materials to test the viability of this method for identifying near-nanoscale plastic particles in environmental media. Based on the evidence collected, deformation patterns between plastic particles and particles present in common environmental media show significant differences. A computer vision algorithm was also developed and tested against manual measurements to improve the usefulness and efficiency of this method further.

Keywords: SEM; detection; microplastics; machine learning; polymers

Citation: Stine, J.S.; Aziere, N.; Harper, B.J.; Harper, S.L. A Novel Approach for Identifying Nanoplastics by Assessing Deformation Behavior with Scanning Electron Microscopy. *Micromachines* **2023**, *14*, 1903. https://doi.org/10.3390/mi14101903

Academic Editor: Amir Hussain Idrisi

Received: 2 September 2023
Revised: 27 September 2023
Accepted: 4 October 2023
Published: 5 October 2023

1. Background

In recent years, studies focusing on the extent of global plastic pollution, specifically micro and nanoplastic (MNP) pollution, have increased exponentially [1]. With this increase in research, mounting evidence of the ubiquity of MNPs in the environment has raised concerns over their potential implications for the health of terrestrial and aquatic ecosystems [2–5]. Microplastics are commonly described as plastic particles less than 5 mm in size, while nanoplastics have been described as having at least one dimension smaller than 1000 nanometers (nm) [6]. The smallest nanoplastics are of particular concern due to their increased capacity for biological interactions [5,7,8]. Given the difficulty of measuring or identifying nanoplastics from environmental samples due to their small size and low mass, there remains a methodological gap in characterizing environmental nanoplastics [6,9]. New, accessible approaches for detecting nanoplastics would greatly improve our understanding of their presence in the environment and provide additional data for regulatory decision-makers.

Following concentration and recovery from environmental samples, analysis using tools like pyrolysis gas chromatography–mass spectrometry (Py-GC-MS) may support the identification of nanoplastics by polymer type, but quickly detecting environmentally relevant concentrations of nanoplastics may prove a challenge as limits of detection are on the order of micrograms per liter range [6,10,11]. In addition to the significant preconcentration needed to enable the effective use of Py-GC-MS [12], it also requires the destruction of a sample, inhibiting the collection of nanoplastic morphology or particle count data once analyzed. The typical non-destructive methods used for collecting chemical fingerprints to identify microplastics by polymer type, including Fourier-transform infrared and Raman spectroscopy, have been reported to have limitations due to small particle size and background interference when used to characterize nanoplastics [9]. A recent study focused on applying scanning transmission X-ray spectromicroscopy (STXM) and near-edge X-ray absorption fine-structure spectroscopy (NEXAFS) to image and characterize spiked nanoplastics recovered from different environmental matrices [13]. While this method seems viable for its intended purpose, it still requires considerable time to image particles and perform spectral analysis. For environmental nanoplastics research to advance more rapidly, simpler and more accessible methods are needed.

Methods targeting unique molecular structures of different polymers may provide an analytical fingerprint associated with individual polymer types, allowing researchers to better characterize nanoplastics. In electron energy loss spectroscopy (EELS), materials are exposed to an electron beam with a known energy input, while energy losses resulting from inelastic scattering of electrons are measured to create spectra unique to a given material [9,14]. Similar techniques commonly coupled with SEM, including X-ray photoelectron spectroscopy (XPS) and energy-dispersive X-ray spectroscopy (EDS), have also been applied while attempting to identify MNPs within a sample. One study investigating Indian beach sediment utilized both SEM and EDS to verify the presence of PVC, PE, and PET microplastics [15]. However, the microplastics identified using this methodology were between 36 μm and 5 mm in size and were identified primarily based on strong carbon signatures in their elemental spectra. This methodology alone would likely prove more challenging for characterizing nanoplastics in samples of mixed environmental media with other high-carbon signature materials. Researchers using these techniques to identify MNPs typically rely on heavier elemental signatures as markers that may not always be present or known to be unique to plastics [15–17].

Due to the intensity of the SEM electron beam, many organic and sensitive samples may degrade during imaging if left untreated [18,19]. Often, it is desirable to coat sensitive samples with thin carbon or gold–palladium coatings to help protect the sample from radiation damage incurred by the electron beam [20]. If left uncoated, organic materials such as polymers with relatively low thermal conductivity values are susceptible to deformation with only moderately elevated temperatures [21]. By maintaining constant electron beam parameters during SEM imaging, it may be possible to identify unique deformation profiles for anthropogenic plastics at the nanoscale that are distinct from other environmental media. Furthermore, utilizing SEM enables the observation of materials down to the nanoscale, which is outside the detection limits for many other analytical methods.

Electron-beam irradiation is commonly used for sterilizing ultra-high molecular weight polyethylene (UHMWPE) materials used in biomedical applications [22,23]. Irradiation of linear hydrocarbon polymers can result in C-C and C-H bond cleavage, radical and hydrogen removal, chain scission, cross-linking, and oxidation (in the presence of oxygen) [22]. Polymer research has shown that chemical cross-linking is an irreversible process that is commonly used to improve the strength, stiffness, and rigidity of polymeric materials [24]. These chemical alterations, occurring concurrently with high enough irradiation doses, can lead to spatial rearrangement and complex physical changes in polymer characteristics [23]. During SEM, this spatial rearrangement of the polymer matrix can result in changes to polymer surface characteristics that are observable in real time. By controlling the conditions of irradiation during SEM, we hypothesized that physical alterations of MNPs of similar sizes from the most common commercial plastics would occur predictably by polymer type and be unique from the deformation patterns seen in non-anthropogenic polymers. This study sought to test the capacity for irradiation of individual particles during SEM to provide a spatiotemporal fingerprint to identify environmental MNPs.

A supporting computer vision (CV) algorithm has been written in Python using the popular PyTorch application processing interface (API) to aid the processing power of the developed methodology [25]. PyTorch contains tools allowing software development to design and train deep neural networks. The dataset of videos collected with SEM, along with their associated masks representing particles to segment, were used as training data. Using a specifically designed objective function, the network was trained to accurately predict pixels belonging to either the foreground or background. The architecture, the objective function, and the training process were all programmed with PyTorch. Once the deep neural network was trained, it was stored in memory and used to infer a segmentation mask for new input data. The prediction quality was quantitatively evaluated using a standard metric for the segmentation algorithm, namely intersection over union (IoU). IoU quantifies the amount of overlap between the predicted and manually annotated masks. A 100% IoU corresponds to a predicted mask perfectly aligned with the manually annotated one. As artificial intelligence (AI) systems continue to develop and increase in sophistication, the accuracy of automated analysis of data collected using the methodology described in this study is expected to improve.

2. Materials and Methods

2.1. Materials and Characterization

Environmentally relevant polymer types were investigated by identifying the most prominent commercial plastics found in environmental waste [26]. After reviewing the fundamental properties of engineered polymers, the degree of crystallinity was noted to be affected during SEM irradiation [23,27] and hypothesized to be the most predictive of polymer deformation during irradiation. Considering the predicted deformation profiles of different polymer types during SEM imaging, the degree of crystallinity values and overall polymeric structure were used to identify polymer types anticipated to capture a wide spectrum of particle deformation behavior. The three plastic materials chosen for this study were polyvinyl chloride (PVC), polyethylene terephthalate (PET), and high-density polyethylene (HDPE). Pristine samples of each chosen polymer type were selected from the Hawaii Pacific University Center for Marine Debris Research Polymer Identification Kit and individually fragmented using a Retsch Cryomill to generate environmentally relevant MNPs for SEM analysis.

Non-plastic materials were also studied for comparison against MNPs to determine if plastics deform differently under SEM irradiation analysis. Non-plastic materials were selected to capture a range of media commonly found in environmental samples. Algae were selected as a proxy for common biological material. Samples of algae (*Raphidocelis subcapitata*) were prepared using specimens cultured within the laboratory. For a non-polymeric material, aluminum silicate (kaolinite) was selected as a proxy for soft silt and sedimentary particles commonly found in environmental samples. Kaolinite materials were obtained through Sigma–Aldrich. Environments are also rich in natural polymeric materials, and a naturally derived polymer would also be needed to compare against the anthropogenic polymers in this study. Cellulose was selected as a proxy for naturally occurring polymer materials common in environmental media and could be mistaken as an MNP. Cellulose used in this study was obtained through Sigma–Aldrich.

2.2. SEM Sample Preparation

All SEM sample preparation occurred within a laminar flow hood with materials obtained through Ted Pella, Inc. (Redding, CA, USA). Aluminum SEM specimen mounts were first prepared by placing a piece of double-stick carbon tape and affixing a 5 × 5 mm silicon wafer to the top of the tape. Compressed air was then used to remove any potential dust or particulate debris that may have been present on the sample. Plastic samples were prepared by suspending milled plastics in ultrapure water and subsequently, using glass pipette tips, drop-casting each sample onto individual specimen mounts. Both kaolinite and nanocrystalline cellulose samples were prepared similarly by suspending the dry powder materials in ultrapure water and subsequently drop casting onto individual SEM specimen mounts. Since algae samples were already suspended in aqueous media, they were diluted 50:50 with ultrapure water to reduce the concentration of algal cells prior to drop casting. Samples were left uncoated to enable observation of deformation during SEM irradiation. All validation samples were prepared similarly using mixtures of plastic materials and environmentally relevant media.

2.3. Experimental Design

While field emission SEM (FE-SEM) may also be suitable for obtaining high-resolution images of MNPs at low voltages, the higher voltages desired and the general ease of accessibility from a typical SEM were preferable. To ensure consistent energy input across multiple particle deformation observations, electron beam voltage and current were maintained at 5 kilovolts (kV) and 33 nanoamperes (nA), respectively, with the beam aperture set to 50 μm. The beam scan rate was held constant at 500 nanoseconds with a horizontal field width (HFW) of 9.95 μm and a working distance of 10.6 mm for each sample. For the purposes of method development, groups of five of the smallest particles of each material type (between 1 and 10 μm in length) were selected, and videos of particle deformation were recorded for at least 40 s for analysis using a Quanta 3D dual beam SEM (FEI Company, Hillsboro, OR, USA). Images were automatically collected every second during recording for particle deformation analysis.

Electron beam settings were held constant throughout all plastic sample observations for PVC, PET, and HDPE. After the initial observations of the selected plastic materials, it was decided that a higher beam current may help further interrogate differences between the deformation behavior of particle materials. All subsequent algae, kaolinite, and cellulose observations occurred with a beam voltage of 5 kV at a current of 37.9 nA, with all other experimental parameters remaining the same. Additional observations of different particles from the PVC, PET, and HDPE samples were also recorded at the higher beam current to determine potential changes in observed deformation patterns resulting from the increased SEM irradiation.

Following the deformation observations of all the materials used in this study, three blinded validation samples were prepared using mixtures of the same materials with the addition of environmentally relevant media. These samples were then observed under the same electron beam parameters with the increased 37.9 nA beam current. Particles present in the validation samples were identified systematically prior to observation. Validation samples were divided into quadrants during SEM analysis, with 10 particles of similar size from each quadrant being selected and recorded under SEM irradiation. In total, 40 randomly selected particles from each of the three blinded samples were recorded for deformation analysis. Deformation profiles from these blinded particles were analyzed and compared against the deformation profiles collected for the six known materials used in this study. The intent of the blinded validation study was to evaluate the utility of this method for detecting MNPs from complex environmental samples.

2.4. Manual Data Evaluation

Analysis of the change in cross-sectional areas of individual particles was the primary focus of this study. Once the SEM irradiation observations were recorded for particle groups for each sample material, particle cross-sections were measured using images taken at specific time points. Particle cross-sectional area measurements were collected using ImageJ Version 1.53i image processing software (NIH, Madison, WI, USA). As the plastic particles were of irregular size and shape due to milling, variability among particle deformation within individual polymer samples was expected. Following the characterization of deformation profiles for known materials, particle cross-section measurements from blinded samples were then collected. After the characterization of deformation profiles for particles of unknown origin, these data were compared against those of the known materials. If the unknown particles exhibited similar deformation profiles to those of known plastics, this would indicate potential MNPs present in the blinded samples.

Once the analysis of particles from blinded samples was complete and deformation profiles were analyzed, the presence of MNPs in blinded samples was proposed and validated by the person who prepared the blinded samples. Successful identification of MNPs present in mixed samples would further validate this method and indicate its potential to help close the methodological gap for identifying environmental nanoplastics. However, the method described herein for manual cross-sectional particle area measurements does not lend itself to the practical and rapid collection of environmental data on MNPs. More automated measurement techniques would be needed to develop this method further.

2.5. Automated Data Evaluation with Computer Vision Analysis

Automated data evaluation was developed to expand the usefulness of the developed methodology. By developing a computational tool that can observe the SEM irradiation of particles and calculate the changes in particle size, particle deformation profiles could be produced in a fraction of the time.

Computer vision (CV) systems are proper candidates to tackle the problem of measuring changing particle sizes within the observational data collected in this study. Given an image representing a particle, the task consists of recognizing which pixels represent it. This is known as the segmentation task, a popular problem that has been widely studied over the years by the CV community. Traditional segmentation algorithms may be used, including watershed [28], grab-cut [29], or image preprocessing followed by thresholding. However, these methods suffer from poor generalization power, are sensitive to noise, and require tedious manual tuning of parameters. They become poor candidates for SEM image processing, which can be highly noisy with varying contrasting occurring. With the recent advance of AI systems, particularly deep learning, a new set of algorithms was developed, leveraging deep neural networks' discrimination power. This family of techniques achieves state-of-the-art performance in various CV tasks, like segmentation.

Deep learning algorithms must be trained on a large set of annotated data to perform well on the desired downstream task. Since the particle data collected were not annotated with the corresponding image masks representing the particle, selecting a database for training the deep network with publicly available annotated data was necessary. The database selected is called PhC-C2DH-U373 [30] for cell segmentation. This dataset is appropriate for training the deep network because the images are annotated with corresponding expert-made segmentation masks, and the cells represented are visually similar to the particles observed in this study. The visual domain is also similar since both databases contain images taken using SEM. The deep network model comprises the popular UNet architecture [31]. UNet is a popular choice for the segmentation algorithm because it was designed to consider the image at multiple scales and is robust to noise perturbations. It achieves state-of-the-art performance on multiple benchmark datasets on the segmentation task.

The first and last frames of all particle deformation videos were manually annotated to fill the gap between the cell and particle deformation datasets. UNet architecture was then trained on a joint set of images with cell and plastic deformation masks, increasing its ability to generalize to unseen images containing plastic particles.

2.6. Statistical Analysis

Statistical analyses were performed using SigmaPlot version 15.0 (Systat Software, San Jose, CA, USA). Differences between sample groups were considered significant when $p \leq 0.05$. Significant differences in deformation behavior between material types were based on cross-sectional measurement data collected over time and analyzed using a two-way repeated-measures analysis of variance (RM-ANOVA) and Bonferroni post hoc analysis. Significant differences in initial deformation rate data across material types were determined using one-way ANOVA and Tukey's post hoc analysis. Normality and equal variance of data were determined using Shapiro–Wilk and Brown–Forsythe tests, respectively. Correlations between variables within the study were also performed using linear regression analysis.

3. Results

3.1. Plastic Particle Deformation

Data collected for PVC, PET, and HDPE particles at a 33 nA beam current showed a variation in deformation patterns between the different polymer types. Figure 1 shows particle deformation over time as a percentage of the initially measured cross-sectional area. The data shown are the mean of five different particle measurements for each polymer type ($n = 5$), with standard error bars showing the deformation variability between measured particles. Trends in the data show that the measured particle cross-sectional area is generally reduced for the lower crystallinity polymer types during SEM irradiation. Although distinct differences between particle deformation seem apparent, the higher variability in PVC particle deformation adds uncertainty to the dataset. Statistical comparisons of the three deformation profiles show that PVC and PET deformation profiles significantly differed from HDPE ($p < 0.001$) but not from each other.

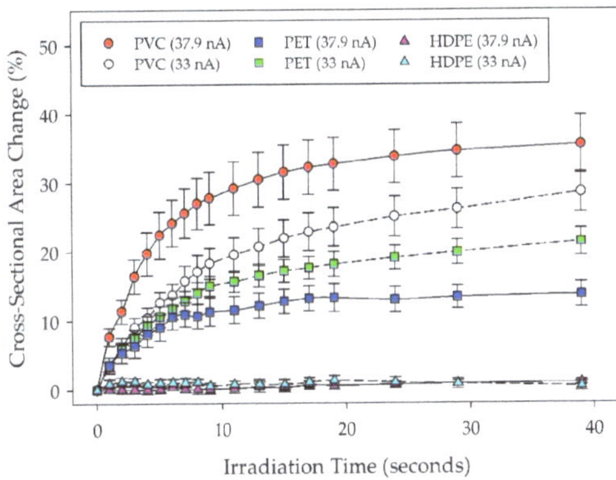

Figure 1. Deformation profiles for PVC, PET, and HDPE under both 33 nA and 37.9 nA beam currents. Particle shrinkage is depicted as a change in manually measured particle cross-sectional area over time, normalized against the initial particle cross-sectional area at the start of irradiation. The plot shows the mean particle measurements for each material type ($n = 5$) with standard error.

Following initial plastic particle deformation observations, additional data on plastic materials were collected at a higher beam current of 37.9 nA to observe changes resulting from increased irradiation. PVC, PET, and HDPE were re-evaluated by selecting five additional particles within each sample for analysis ($n = 5$). Data collected for all plastic materials under both 33 nA and 37.9 nA beam currents are shown in Figure 1. Statistical comparisons of the deformation profiles between the plastic materials at different beam currents indicated significantly increased deformation behavior for PVC at a higher beam current ($p = 0.042$) but no significant change for either PET or HDPE. The apparent reduced average deformation of PET particles observed at the 37.9 nA beam current is likely an artifact of variations in the degree of crystallinity across the particles selected in the different study groups. Comparisons across the three materials at the higher beam current also indicate that differences in deformation behavior between all of the plastic materials were statistically significant from each other. Based on these observations, deformation profiles of lower crystallinity plastics under higher currents appear more likely to display accelerated deformation rates and greater changes in measured cross-sectional area than higher crystallinity plastics. Linear regression analysis also showed that the final deformation measurements for each material ($t = 39$ s) correlated to the reported degree of crystallinity values with an R^2 value of 0.87.

3.2. Plastic vs. Non-Plastic Media Particle Deformation

Generally, data collected for algae, kaolinite, and cellulose particles indicated less deformation than plastic particles. Notable differences in particle morphology were also present during deformation observations. Interestingly, blebbing was observed during algal cell irradiation, which enabled additional qualitative distinction of algal media from other media in this study. Kaolinite was characterized by a markedly different contrasting quality over the other materials and appeared to have more jagged features when compared to the other materials. While cellulose appeared to have a similar morphology to the plastic materials observed in this study, it did not appear to degrade as readily as the lower crystallinity plastics in this study. Even at the higher current of 37.9 nA, all of the non-plastic media tested appeared to display less particle shrinkage when compared to both PVC and PET samples.

Figure 2 shows particle deformation profiles for algae, kaolinite, and cellulose, as well as the plastic materials tested at 37.9 nA beam current. Data plotted in Figure 2 are the mean of five individual particle measurements for each material type ($n = 5$) collected during SEM irradiation with calculated standard error bars to indicate variability within material types. Measurements were then normalized to a percentage of the initially measured cross-sectional area over time. Statistical comparisons across the material types shown indicate that both PVC and PET display deformation behavior significantly different from the rest of the materials tested and from each other ($p < 0.001$). Additional comparisons yielded no significant differences in particle deformation behavior between the non-plastic media or HDPE. These observations further indicate that using this methodology, the particle deformation behavior of low crystallinity plastics is significantly different from the behavior of common non-plastic environmental media and high crystallinity plastics. SEM images showing typical particle deformation for each material type are shown in Figure 3.

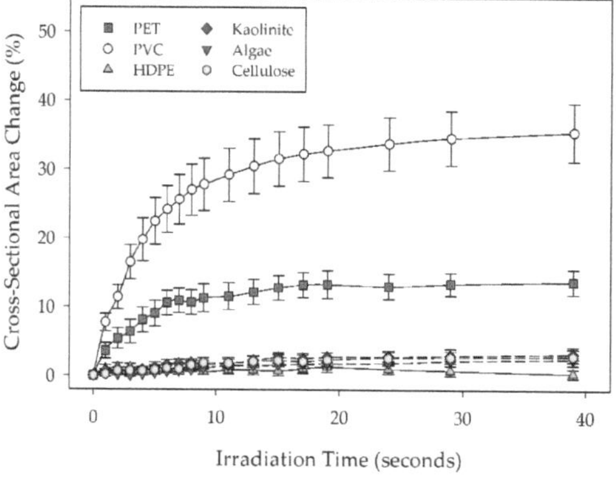

Figure 2. Deformation profiles for all material types under a 37.9 nA beam current. Particle shrinkage is depicted as a change in manually measured particle cross-sectional area over time, normalized against the initial particle cross-sectional area at the start of irradiation. The plot shows the mean particle measurements for each material type ($n = 5$) with standard error.

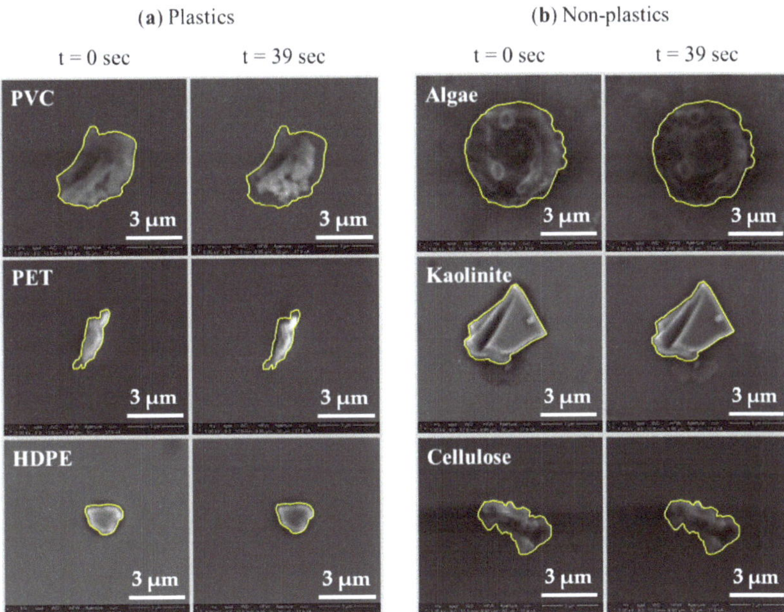

Figure 3. SEM images depicting typical particle deformation behavior for (**a**) plastic and (**b**) non-plastic materials from $t = 0$ s to $t = 39$ s. The highlighted boundary shows the initially measured cross-sectional area to emphasize differences over time.

3.3. Blinded Validation Sample Analysis

Measurements were collected at five time points for 0, 2, 4, 10, 20, and 40 s of irradiation to more rapidly collect the deformation profiles for particles from blinded samples. Deformation profiles were then plotted and visually assessed to determine if any particles of unknown origin exhibited deformation behavior similar to the lower crystallinity plastics in this study (PVC and PET). Similar particle shrinkage behavior to these known plastics was considered indicative of anthropogenic polymers, and those particles were further studied. Unknown particles were deemed suspect when their overall particle deformation at the end of observation ($t = 39$ s) showed deformation greater than the least deforming known PET particles assessed in this study (8.1% at $t = 39$ s).

Based on analyses of the 120 unknown particles studied across the three blinded validation samples, 35 unknown particles exhibited suspect deformation behavior. Of the total suspect particles, 19, 1, and 15 were present in blinded validation samples #1, #2, and #3, respectively. Suspect particles in each validation sample were then grouped and compared for statistical similarity to known deforming plastics PVC and PET. Upon comparison, it was found that PVC was significantly different from the grouped suspect particles in all three validation samples ($p < 0.002$), whereas no significant differences were apparent between PET and any of the three validation samples. These data indicated the possible presence of PET in every blinded validation sample.

To further interrogate the presence of PVC within the validation samples, a new set of suspect unknown particles were grouped based on overall particle deformation at the end of observation ($t = 39$ s) and showed deformation greater than the least deforming known PVC particle assessed in this study (20.0% at $t = 39$ s). Of the total suspected PVC particles, 3, 0, and 2 were present in blinded validation samples #1, #2, and #3, respectively. After a statistical comparison, PVC was no longer found to be statistically different from validation sample #3. This analysis indicated that the suspected particles in validation sample #1 exhibited deformation profiles that were different from those of the suspected particles in validation sample #3. Additionally, this analysis showed no statistical difference between PVC and validation sample #3, indicating the potential presence of PVC. Following these analyses predicting the presence of PET in all validation samples and PVC in validation sample #3, the data were verified with the preparer of the blinded samples for comparison. The results of those predictions are highlighted in Table 1.

Table 1. Summarized findings of blinded validation study testing identification methodology.

Validation Sample	Statistically Suspect *	Verified	Sample Description
#1	PET	Yes	PET, Algae
#2	PET	Yes	PET, HDPE, Kaolinite
#3	PVC, PET	Yes	PVC, PET, HDPE, Silty Soil

* denotes a significant similarity relative to known plastic media ($p \leq 0.05$).

3.4. Analysis of Materials with AI-Assisted Data Processing

The same particle observation data analyzed with manual measurements was also analyzed using the developed machine learning algorithm for comparison to provide enhanced data-generating power. The benefits of machine learning analysis are exceptionally enhanced speed of data collection while also collecting data at additional time points. Figure 4 depicts the computationally generated particle measurements for the plastic materials at the lower 33 nA beam current. Statistical analysis of the three plastic materials at a 33 nA beam current showed PVC to be significantly different from HDPE ($p = 0.005$), but PET was found not to be significantly different from HDPE ($p = 0.088$). PVC and PET were also not found to be significantly different from each other. These results deviate slightly from those derived from the manual measurements, possibly suggesting the need for further training of the machine learning algorithm to improve the accuracy of measurements and reduce the variability of the data collected.

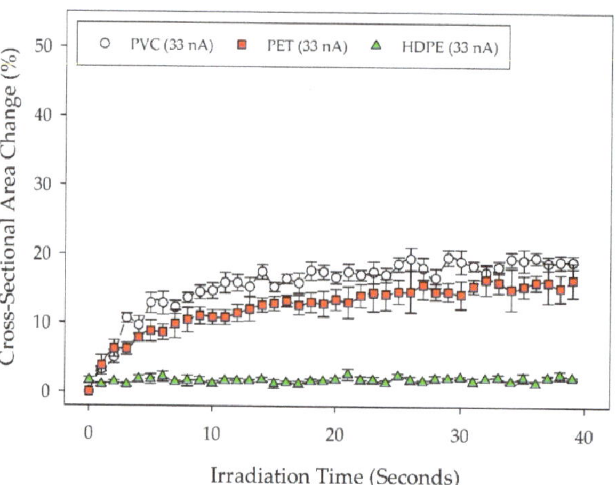

Figure 4. Deformation profiles for PVC, PET, and HDPE under a 33 nA beam current. Particle shrinkage is depicted as a change in computationally measured particle cross-sectional area over time, normalized against the initial particle cross-sectional area at the start of irradiation. The plot shows the mean particle measurements for each material type ($n = 5$) with standard error.

Figure 5 depicts computationally generated particle measurements for all materials tested at the 37.9 nA beam current. Upon visual inspection, the same trends in the data are apparent, albeit with increased variability. Statistical comparisons across the material types indicated that PVC displays deformation behavior significantly different from all the other materials except for PET ($p = 0.07$). Given the high variability of the PET data generated, PET was no longer shown to be significantly different from the other materials, including HDPE ($p = 0.064$). These results also deviate slightly from those derived from the manual measurements. A comparison of manual and computationally generated particle measurements is detailed below in Table 2. Each material type was assessed to determine significant differences between manual and computationally generated data.

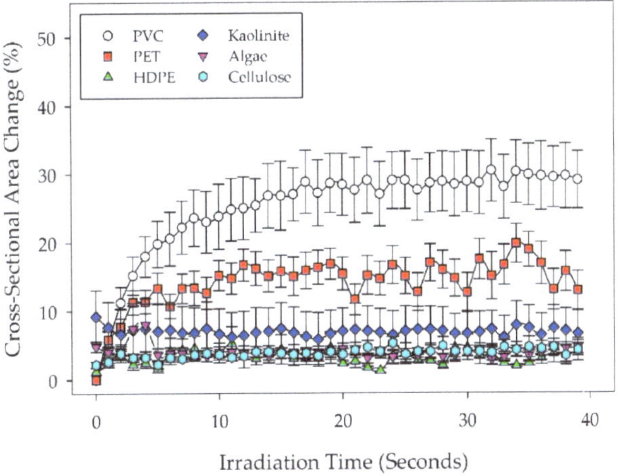

Figure 5. Deformation profiles for all material types under a 37.9 nA beam current. Particle shrinkage is depicted as a change in computationally measured particle cross-sectional area over time, normalized against the initial particle cross-sectional area at the start of irradiation. The plot shows the mean particle measurements for each material type ($n = 5$) with standard error.

Table 2. Statistical comparisons of manual measurements against computationally generated measurements for each material type.

Beam Current	Material Type	Significant Difference *
33 nA	PVC	No
	PET	No
	HDPE	No
37.9 nA	PVC	No
	PET	No
	HDPE	No
	Kaolinite	No
	Algae	No
	Cellulose	No

* denotes a significant difference between measurement methods ($p \leq 0.05$).

The trends between manual and computationally generated measurements are similar, such as no statistically significant differences across computational and manually derived data by material type. However, due to increased variability in the computational data, challenges remain with identifying statistically significant differences between material types using the computational dataset. Higher resolution SEM data from particle observations with increased particle-background contrast for each material type would likely improve machine learning algorithm measurements, allowing better determination of statistically significant differences. Additional training of the machine learning algorithm would also improve the overall discrimination of particles within lower-quality observation data and improve the consistency of computationally generated measurements.

To further verify the feasibility of AI-assisted data processing to identify MNPs from samples of unknown origin, known materials were compared against the same three blinded samples using only computationally generated datasets. With cross-sectional areas generated at one-second intervals, known materials were compared against grouped suspect particles from the same three validation samples following the aforementioned methodology. Based on analyses of the 120 unknown particles studied across the three blinded validation samples, 25 and 11 unknown particles exhibited suspect deformation behavior for PET and PVC, respectively. Of the total suspect PET particles, 9, 1, and 15 were present in blinded validation samples #1, #2, and #3, respectively. Of the total suspect PVC particles, 4, 0, and 7 were present in blinded validation samples #1, #2, and #3, respectively.

Suspect particles in each validation sample were again grouped and compared for statistical similarity to known deforming plastics PVC and PET using two-way RM-ANOVA with the computational datasets. Upon comparison, no statistically significant difference was found between the computational datasets of known PET and the grouped suspect PET particles from all three individual validation samples, accurately suggesting the possible presence of PET MNPs in all three validation samples. The same comparison between the computational datasets of known PVC and grouped suspect PVC particles from individual validation samples showed no statistically significant difference between known PVC particles and suspect PVC particles from both validation samples #1 and #3. As previously tabulated in Table 1, PVC was only present in validation sample #3, meaning that the statistical comparison of computational datasets for known and suspected PVC correctly indicated the presence of PVC in validation sample #3 but incorrectly suggested the likelihood of PVC in validation sample #1. To summarize these comparisons, it was possible to accurately predict the presence of PET particles in all three validation samples using only the computational datasets, but the presence of PVC was inaccurately suggested for validation sample #1. This analysis of computational datasets suggests that it is feasible to use computational datasets solely to predict the presence of MNPs in samples of unknown origin. However, AI detection systems would likely need to be trained with larger datasets to ensure more accurate predictions. A table of these comparisons is provided in Table 3.

Table 3. Summarized statistical analyses between computational datasets comparing known and grouped suspected MNPs from blinded validation samples.

Validation Sample	Statistically Suspect *	Verified	Sample Description
#1	PVC, PET	PET only	PET, Algae
#2	PET	Yes	PET, HDPE, Kaolinite
#3	PVC, PET	Yes	PVC, PET, HDPE, Silty Soil

* denotes no statistical difference relative to known plastic media ($p \leq 0.05$).

4. Discussion

As noted earlier in this study, the random sizes and shapes of particles were expected to introduce variability across particles of similar material types. A regression analysis of the starting cross-sectional area of deforming plastic particles against the percent of total deformation by the end of observation indicated a poor correlation with an R^2 value of 0.05. The effects of irradiation are a well-studied topic in the field of polymeric materials design [27]. Other research has suggested that the degree of deformation from electron beam irradiation also depends on the polymer's structure [32]. Previous research has also shown that irradiation can cause a host of chemical changes, including chain scission and cross-linking in polymers, leading to overall structural changes and spatial rearrangement [23,27]. When considering the degree of crystallinity values of the tested materials, the deformation behaviors of the studied plastics matched predictions made prior to investigation, with PVC deforming the most, followed by PET and HDPE, respectively. Crystallinity is considered a significant property in the design of polymer material and was shown to have a positive correlation to the deformation and shrinkage behavior observed during irradiation in this study.

This same logic may also explain why HDPE was able to withstand the same irradiation as PVC and PET without undergoing the same deformation, which could be attributed to its already high degree of crystallinity. A plastic's crystallinity degree is based on the density of tightly packed folded molecular chains, or crystalline lamellae regions, versus the amorphous regions where molecular chains are loosely or irregularly formed within the polymer structure [33]. Materials with higher degrees of crystallinity are associated with higher melting temperatures and greater strength and rigidity. Ranges noted in the polymer literature list degrees of crystallinity for PVC, PET, and HDPE as approximately 10%, 35%, and 75%, respectively [34,35]. Considering these concepts, increased deformation of MNPs under electron beam irradiation shows an inverse relationship to the degree of crystallinity, wherein highly crystalline MNPs would require more irradiative energy to experience significant observable deformation behavior.

The study of the plastic materials at different beam currents also allowed for the observation of differing levels of deformation, specifically with the PVC particles. By studying the same materials under both 33 nA and 37.9 nA currents, it was possible to observe how an increase in irradiation could increase particle shrinkage. Additionally, observing the non-plastic media under the higher irradiative parameters used in this study further accentuated the differences in their deformation profiles compared to plastics. In future developments of this methodology, observing the deformation of each plastic type under increasing levels of irradiation could further enhance distinctions between the deformation profiles of MNPs and non-plastic environmental media.

While the manual measurement techniques used in this investigation were enough for a proof of concept, the ultimate vision for this methodology would be to apply more computational methods for rapidly assessing the visual data collected during SEM. Using the machine learning algorithm implemented in this study on additional samples of known plastics, developing an MNP deformation database would be feasible. The development of such a tool could allow scientists and researchers to take a given environmental sample and observe particle deformations in real-time with a software package capable of rapidly estimating the presence of MNPs. Although this methodology requires access to SEM equipment, which can be costly to operate, it provides a simple, more accessible methodology to augment existing analytical techniques used in detecting environmental MNPs. SEM also has the added benefit of allowing observation of the size and occurrence of particles.

The fields of AI, and more specifically, deep learning, are rapidly evolving. New neural network architectures and training strategies emerge yearly, achieving enhanced state-of-the-art performance in various tasks. While it is possible to improve accuracy using the latest methods or include more annotated data, accuracy can also be improved in other ways. By analyzing the spatiotemporal changes of the particles observed in this study, it is possible to leverage other well-studied tasks of AI systems, such as video classification. Classifying particle deformation into categories by polymer type and source would simplify the post-processing step of fitting the new sample to the established deformation profile of a given plastic polymer type.

The rationale for performing blinded validation studies in this investigation was to test the concept of using irradiation-induced deformation to detect MNPs in a sample with unknown media, as would likely be the case when studying an environmental sample. The results of applying the methodology described herein can potentially improve the detection of environmental MNPs. Additional studies using this methodology on less pristine and weathered samples of commercial plastics would further test its utility. Weathered or aged MNPs are expected to have varying degrees of UV radiation exposure that may impact the degree of crystallinity of aged materials. Understanding how different levels of UV exposure affect plastic particle deformation is critical in developing this methodology for use on environmental MNPs. Using similar methods to those described in this study, the deformation behavior of pristine and UV-irradiated particles of the same polymer type could be characterized and compared for differences. Under such circumstances, the analytical capabilities of the aforementioned computational tools would be necessary to make this a desirable methodology to the broader environmental research community.

Based on this work, some key aspects of data capture that should be considered include the beam settings of the SEM instrument, the number of particles present in a sample, and the presence of media-obscuring particles. Having the correct SEM settings enables better particle resolution while aiding in observing particle morphology during deformation. Diluting samples before preparation on SEM specimen mounts can prevent particles from being obscured by large piles of microscopic debris, which aids in the more rapid collection of particle deformation data. In addition to environmental debris, the presence of biological media coating particles would likely inhibit the ability to find them using this methodology. Digestion of biological media would be recommended if this methodology is applied to an environmental sample. Guidelines for aiding in the analysis of collected observational data include having clear contrast between particles and their background, identifying isolated particles within the SEM imaging frame, and maintaining a consistent frame rate in exported video files. Following these best practices enhances the quality of observational data, enabling machine learning algorithms to better assess potential deformational changes in irradiated particles. Table 4 summarizes how sample preparation, data collection, and data analysis could be improved to make this methodology more reliable for future researchers.

Table 4. Guidance for data collection and analysis using SEM particle deformation methodology.

Process	Guidance	Description	Benefits
Sample Preparation	Digest	use mild digestion to remove organic matter	reduces organics surrounding MNPs to better observe particle deformation
	Fraction	separate particles of specific size ranges	improves homogeneity of particles, enabling expedited particle selection
	Dilute	reducing concentration of particles in sample	reduces aggregation, enabling expedited particle selection
Data Collection	E-Beam	optimize beam settings for voltage and current	enables observation of more discrete differences in particle deformation
	Contrast	optimize contrast settings of instrument	improves identification of particle boundaries during data analysis
	Resolution	optimize beam scan rate and image resolution	improves identification of particle boundaries during data analysis
	Selection	identification of discrete particles	improves reliability of computational methods for measuring particles
	Sample	increase number of particle observations	improves statistical power of particle deformation characterization
Data Analysis	Materials	collect data on a wide variety of materials	expands library of particle deformation behaviors, improving characterization of particles from different source materials
	Condition	collect data on particles of different condition	expands characterization of particles that have been UV aged or degraded through different processes
	Analytics	pair method with other analytical techniques	other techniques (such as EDS) may help characterize particles
	Training	provide additional training data to AI	improves computational analysis, reducing error and variability
	Software	develop SEM software package and database	enables cataloging of deformation behavior and more rapid analysis

Successful implementation of this method would help highlight the extent to which plastic persists in the environment, paving the way for a more comprehensive understanding of the risks associated with plastic pollution. Identifying nanoplastics by polymer type in complex environmental matrices is the ultimate validation of this methodology, providing a novel approach to help close the methodological gap for studying environmental MNPs.

Author Contributions: S.L.H. acquired funding and contributed to the conceptualization of the study; J.S.S. and B.J.H. designed experiments; N.A. contributed to writing, computational design, and data analysis; J.S.S. conducted experiments, data analysis, and preparation of the final manuscript. All authors have read and agreed to the published version of the manuscript.

Funding: This work was supported by the National Science Foundation (NSF) Growing Convergence Research Grants #1935018 and #1935028.

Institutional Review Board Statement: Not applicable.

Informed Consent Statement: Not applicable.

Data Availability Statement: The data presented in this study are available upon request.

Acknowledgments: The authors would like to thank Teresa Sawyer and Peter Eschbach at the OSU electron microscopy facility for assisting with the SEM data collection of samples analyzed in this study. The authors would also like to thank Maksim Postnov for assisting with data analysis.

Conflicts of Interest: The authors declare no conflict of interest.

References

1. Zhou, M.; Wang, R.; Cheng, S.; Xu, Y.; Luo, S.; Zhang, Y.; Kong, L. Bibliometrics and visualization analysis regarding research on the development of microplastics. *Environ. Sci. Pollut. Res. Int.* **2021**, *28*, 8953–8967. [CrossRef] [PubMed]
2. Jiang, B.; Kauffman, A.E.; Li, L.; McFee, W.; Cai, B.; Weinstein, J.; Lead, J.R.; Chatterjee, S.; Scott, G.I.; Xiao, S. Health impacts of environmental contamination of micro- and nanoplastics: A review. *Environ. Health Prev. Med.* **2020**, *25*, 29. [CrossRef]
3. Yee, M.S.L.; Hii, L.W.; Looi, C.K.; Lim, W.M.; Wong, S.F.; Kok, Y.Y.; Tan, B.K.; Wong, C.Y.; Leong, C.O. Impact of Microplastics and Nanoplastics on Human Health. *Nanomaterials* **2021**, *11*, 496. [CrossRef] [PubMed]
4. Amobonye, A.; Bhagwat, P.; Raveendran, S.; Singh, S.; Pillai, S. Environmental Impacts of Microplastics and Nanoplastics: A Current Overview. *Front. Microbiol.* **2021**, *12*, 768297. [CrossRef] [PubMed]
5. Lai, H.; Liu, X.; Qu, M. Nanoplastics and Human Health: Hazard Identification and Biointerface. *Nanomaterials* **2022**, *12*, 1298. [CrossRef] [PubMed]
6. Schwaferts, C.; Niessner, R.; Elsner, M.; Ivleva, N.P. Methods for the analysis of submicrometer- and nanoplastic particles in the environment. *TrAC Trends Anal. Chem.* **2019**, *112*, 52–65. [CrossRef]
7. Hoshyar, N.; Gray, S.; Han, H.; Bao, G. The effect of nanoparticle size on in vivo pharmacokinetics and cellular interaction. *Nanomedicine* **2016**, *11*, 673–692. [CrossRef]
8. Lehner, R.; Weder, C.; Petri-Fink, A.; Rothen-Rutishauser, B. Emergence of nanoplastic in the environment and possible impact on human health. *Environ. Sci. Technol.* **2019**, *53*, 1748–1765. [CrossRef]
9. Caldwell, J.; Taladriz-Blanco, P.; Lehner, R.; Lubskyy, A.; Ortuso, R.D.; Rothen-Rutishauser, B.; Petri-Fink, A. The micro-, submicron-, and nanoplastic hunt: A review of detection methods for plastic particles. *Chemosphere* **2022**, *293*, 133514. [CrossRef]
10. Mintenig, S.M.; Bäuerlein, P.S.; Koelmans, A.A.; Dekker, S.C.; Van Wezel, A.P. Closing the gap between small and smaller: Towards a framework to analyse nano- and microplastics in aqueous environmental samples. *Environ. Sci. Nano* **2018**, *5*, 1640–1649. [CrossRef]
11. Nguyen, B.; Claveau-Mallet, D.; Hernandez, L.M.; Xu, E.G.; Farner, J.M.; Tufenkji, N. Separation and Analysis of Microplastics and Nanoplastics in Complex Environmental Samples. *Acc. Chem. Res.* **2019**, *52*, 858–866. [CrossRef] [PubMed]
12. Li, Y.; Wang, Z.; Guan, B. Separation and identification of nanoplastics in tap water. *Environ. Res.* **2022**, *204*, 112134. [CrossRef] [PubMed]
13. Foetisch, A.; Filella, M.; Watts, B.; Vinot, L.-H.; Bigalke, M. Identification and characterisation of individual nanoplastics by scanning transmission X-ray microscopy (STXM). *J. Hazard. Mater.* **2021**, *426*, 127804. [CrossRef] [PubMed]
14. Pal, R.; Bourgeois, L.; Weyland, M.; Sikder, A.K.; Saito, K.; Funston, A.M.; Bellare, J.R. Chemical fingerprinting of polymers using electron energy-loss spectroscopy. *ACS Omega* **2021**, *6*, 23934–23942. [CrossRef]
15. Tiwari, M.; Rathod, T.D.; Ajmal, P.Y.; Bhangare, R.C.; Sahu, S.K. Distribution and characterization of microplastics in beach sand from three different Indian coastal environments. *Mar. Pollut. Bull.* **2019**, *140*, 262–273. [CrossRef] [PubMed]
16. Wagner, J.; Wang, Z.M.; Ghosal, S.; Rochman, C.; Gassel, M.; Wall, S. Novel method for the extraction and identification of microplastics in ocean trawl and fish gut matrices. *Anal. Methods* **2017**, *9*, 1479–1490. [CrossRef]
17. Matharu, R.K.; Charani, Z.; Ciric, L.; Illangakoon, U.E.; Edirisinghe, M. Antimicrobial activity of tellurium-loaded polymeric fiber meshes. *J. Appl. Polym. Sci.* **2018**, *135*, 46368. [CrossRef]
18. Okada, T.; Ogura, T. Nanoscale imaging of untreated mammalian cells in a medium with low radiation damage using scanning electron-assisted dielectric microscopy. *Sci. Rep.* **2016**, *6*, 29169. [CrossRef]

19. Beale, E.V.; Warren, A.J.; Trincao, J.; Beilsten-Edmands, J.; Crawshaw, A.; Sutton, G.; Stuart, D.; Evans, G. Scanning electron microscopy as a method for sample visualization in protein X-ray crystallography. *IUCrJ* **2020**, *7*, 500–508. [CrossRef]

20. Fischer, E.R.; Hansen, B.T.; Nair, V.; Hoyt, F.H.; Dorward, D.W. Scanning Electron Microscopy. *Curr. Protoc. Microbiol.* **2012**, *25*, 2B.2.1–2B.2.47. [CrossRef]

21. Egerton, R.F.; Li, P.; Malac, M. Radiation damage in the TEM and SEM. *Micron* **2004**, *35*, 399–409. [CrossRef] [PubMed]

22. Czaja, K.; Sudoł, M. Studies on electron-beam irradiation and plastic deformation of medical-grade ultra-high molecular weight polyethylene. *Radiat. Phys. Chem.* **2011**, *80*, 514–521. [CrossRef]

23. Cybo, J.; Maszybrocka, J.; Duda, P.; Bartczak, Z.; Barylski, A.; Kaptacz, S. Properties of ultra-high-molecular-weight polyethylene with a structure modified by plastic deformation and electron-beam irradiation. *J. Appl. Polym. Sci.* **2012**, *125*, 4197–4208. [CrossRef]

24. Mane, S.; Ponrathnam, S.; Chavan, N. Effect of Chemical Cross-linking on Properties of Polymer Microbeads: A Review. *Can. Chem. Trans.* **2015**, *3*, 473–485. [CrossRef]

25. Paszke, A.; Gross, S.; Massa, F.; Lerer, A.; Bradbury, J.; Chanan, G.; Killeen, T.; Lin, Z.; Gimelshein, N.; Antiga, L.; et al. PyTorch: An Imperative Style, High-Performance Deep Learning Library. *NeurIPS* **2019**, *32*, 1–12.

26. Charles, D.; Kimman, L.; Saran, N. *The Plastic Waste Makers Index*; Minderoo Foundation: Nedlands, Australia, 2021.

27. Ashfaq, A.; Clochard, M.C.; Coqueret, X.; Dispenza, C.; Driscoll, M.S.; Ulański, P.; Al-Sheikhly, M. Polymerization Reactions and Modifications of Polymers by Ionizing Radiation. *Polymers* **2020**, *12*, 2877. [CrossRef]

28. Malpica, N.; Ortiz de Solórzano, C.; Vaquero, J.J.; Santos, A.; Vallcorba, I.; García-Sagredo, J.M.; del Pozo, F. Applying Watershed Algorithms to the Segmentation of Clustered Nuclei. *Cytometry* **1997**, *28*, 289–297. [CrossRef]

29. Rother, C.; Kolmogorov, V.; Blake, A. GrabCut: Interactive Foreground Extraction using Iterated Graph Cuts. *ACM Trans. Graph.* **2004**, *23*, 309–314. [CrossRef]

30. Maška, M.; Ulman, V.; Svoboda, D.; Matula, P.; Matula, P.; Ederra, C.; Urbiola, A.; España, T.; Venkatesan, S.; Balak, D.M.W.; et al. A benchmark for comparison of cell tracking algorithms. *Bioinformatics* **2014**, *30*, 1609–1617. [CrossRef]

31. Ronneberger, O.; Fischer, P.; Brox, T. U-Net: Convolutional Networks for Biomedical Image Segmentation. In *Proceedings of the Medical Image Computing and Computer-Assisted Intervention (MICCAI)*; Springer International Publishing: Munich, Germany, 2015; pp. 234–241.

32. Singh, P.; Venugopal, B.R.; Nandini, D.R. Effect of Electron Beam Irradiation on Polymers. *J. Mod. Mater.* **2018**, *5*, 24–33. [CrossRef]

33. Balani, K.; Verma, V.; Agarwal, A.; Narayan, R. Physical, Thermal, and Mechanical Properties of Polymers. *Biosurfaces* **2015**, 329–344. [CrossRef]

34. Hay, J.N.; Biddlestone, F.; Walker, N. Crystallinity in poly(vinyl chloride). *Polymer* **1980**, *21*, 985–987. [CrossRef]

35. Ehrenstein, G.W.; Theriault, R.P. *Polymeric Materials: Structure, Properties, Applications*; Hanser Publications: Liberty Township, OH, USA, 2001; ISBN 9781569903100.

Review

Bio-Inspired Nanomaterials for Micro/Nanodevices: A New Era in Biomedical Applications

Mohammad Harun-Ur-Rashid [1], Israt Jahan [2], Tahmina Foyez [3] and Abu Bin Imran [4,*]

[1] Department of Chemistry, International University of Business Agriculture and Technology, Dhaka 1230, Bangladesh; mrashid@iubat.edu
[2] Department of Cell Physiology, Graduate School of Medicine, Nagoya University, Nagoya 466-8550, Japan; israt.kyoto@gmail.com
[3] Department of Pharmacy, United International University, Dhaka 1212, Bangladesh; tahmina@pharmacy.uiu.ac.bd
[4] Department of Chemistry, Bangladesh University of Engineering and Technology, Dhaka 1000, Bangladesh
* Correspondence: abimran@chem.buet.ac.bd

Abstract: Exploring bio-inspired nanomaterials (BINMs) and incorporating them into micro/nanodevices represent a significant development in biomedical applications. Nanomaterials, engineered to imitate biological structures and processes, exhibit distinctive attributes such as exceptional biocompatibility, multifunctionality, and unparalleled versatility. The utilization of BINMs demonstrates significant potential in diverse domains of biomedical micro/nanodevices, encompassing biosensors, targeted drug delivery systems, and advanced tissue engineering constructs. This article thoroughly examines the development and distinctive attributes of various BINMs, including those originating from proteins, DNA, and biomimetic polymers. Significant attention is directed toward incorporating these entities into micro/nanodevices and the subsequent biomedical ramifications that arise. This review explores biomimicry's structure–function correlations. Synthesis mosaics include bioprocesses, biomolecules, and natural structures. These nanomaterials' interfaces use biomimetic functionalization and geometric adaptations, transforming drug delivery, nanobiosensing, bio-inspired organ-on-chip systems, cancer-on-chip models, wound healing dressing mats, and antimicrobial surfaces. It provides an in-depth analysis of the existing challenges and proposes prospective strategies to improve the efficiency, performance, and reliability of these devices. Furthermore, this study offers a forward-thinking viewpoint highlighting potential avenues for future exploration and advancement. The objective is to effectively utilize and maximize the application of BINMs in the progression of biomedical micro/nanodevices, thereby propelling this rapidly developing field toward its promising future.

Keywords: bio-inspired nanomaterials; micro/nanodevices; biomedical applications; nanotechnology; biomimetic polymers; microfabrication; nano-biotechnology

Citation: Harun-Ur-Rashid, M.; Jahan, I.; Foyez, T.; Imran, A.B. Bio-Inspired Nanomaterials for Micro/Nanodevices: A New Era in Biomedical Applications. *Micromachines* **2023**, *14*, 1786. https://doi.org/10.3390/mi14091786

Academic Editor: Amir Hussain Idrisi

Received: 21 August 2023
Revised: 14 September 2023
Accepted: 16 September 2023
Published: 18 September 2023

1. Introduction

Bio-inspired nanomaterials (BINMs), alternatively referred to as biomimetic nanomaterials (BNMs), are a class of materials that are intentionally engineered and manufactured to replicate the intricate structures, functionalities, or mechanisms observed in natural biological systems [1–3]. These advancements offer a novel trajectory for the field of material science, facilitating the creation of materials possessing unique characteristics that can effectively tackle a wide range of scientific, technological, and environmental obstacles through successful applications in multidimensional sectors, including medicine and healthcare [4], biotechnology and bioengineering [5], energy [6], environment [7], material science [8], robotics [9,10], and many more [11–13]. Various biological entities, including proteins, DNA [14,15], cells [16], and complete organisms [17], can serve as sources of inspiration for these materials. Using DNA's self-assembling properties has facilitated the construction of shapes and patterns at the nanoscale level [18]. The adhesive characteristics exhibited by gecko feet have served as a source of inspiration for developing sophisticated adhesive materials [19]. Furthermore, the self-cleaning and hydrophobic characteristics exhibited by the lotus leaf have prompted advancements in creating self-cleaning surfaces and coatings with water-repellent properties [20]. The interdisciplinary field of BINMs integrates principles from biology, chemistry, physics, and material science. A bottom-up approach is often employed, commencing at the atomic or molecular level and progressing upward. This is juxtaposed with the conventional top-down methodology, wherein the initial focus is on a larger system that is subsequently deconstructed into smaller constituent parts. Various techniques can be employed to fabricate BINMs, such as molecular self-assembly, in which molecules autonomously organize themselves into desired structures. Another method is biosynthesis, which involves utilizing biological organisms such as bacteria, fungi, or plants to synthesize nanomaterials. The field of BINMs holds significant potential for scientific investigation; however, it is not devoid of inherent obstacles. One of the primary obstacles lies in the capacity to regulate the synthesis and assembly processes of these materials in order to attain the intended properties [21]. There exist additional concerns regarding the potential environmental and health ramifications associated with these nanomaterials, necessitating the need for further examination and comprehensive testing prior to their widespread implementation [22]. Notwithstanding these challenges, BINMs constitute a captivating and burgeoning area of investigation. The advancement of our knowledge in the fields of biology and nanotechnology is expected to enhance the possibilities for the development of novel and influential BINMs.

With the advancement and comprehension of BINMs, an opportunity arises to investigate the pragmatic utilization of these materials in the configuration of micro/nanodevices. Micro/nanodevices, as their nomenclature implies, are miniature devices that operate at the micro- or nanolevel. The significance of micro/nanodevices has increased substantially due to their potential to enhance capabilities in diverse sectors, including medicine, environmental monitoring, electronics, and energy production. These devices provide an unparalleled degree of control and accuracy at a minuscule level, enabling us to devise and develop solutions to previously insurmountable obstacles. The potential for developing novel and influential micro/nanodevices using BINMs is anticipated to grow due to technological advancements and improved comprehension of biological systems. Through the utilization of the distinct characteristics exhibited by these nanomaterials in the form of nanocomposite gels [23–26] and films [27–29], structural colored nanomaterials [30–32], organo-metallic nanomaterials [33], molecular machines [34], and nanobiosensors [35] have found widespread application and have replaced mainly more conventional bulk materials in a variety of sectors [36–41] as well as in theoretical inquiries [42]. Researchers and practitioners can fabricate devices that imitate or draw inspiration from biological systems to execute targeted functions, frequently surpassing the efficiency and efficacy of conventional devices. Medicine and healthcare are highly significant domains for applying micro/nanodevices [43,44]. Environmental monitoring is a field that extensively utilizes micro/nanodevices [45]. These encompass sensors capable of detecting various environmental pollutants, even in exceedingly low concentrations. These devices can continuously monitor air and water quality, thereby offering significant data that can be utilized to safeguard the environment. The electronics and computing sector represents a significant domain in which micro/nanodevices are widely used [46]. Modern electronic and computing devices rely on various components, such as transistors found in computers and sensors present in smartphones, which collectively serve as the fundamental infrastructure for these technologies. By further reducing the size of these devices and enhancing their operational capabilities, it is possible to develop electronic devices that are more potent and consume less energy. Micro/nanodevices are paramount in energy production and storage [47]. Nanostructured materials have been employed to improve the efficiency of solar cells, fuel cells, and batteries, among other applications. These devices have the potential to enhance energy efficiency, mitigate expenses, and foster the adoption of renewable energy sources. Although the prospect of micro/nanodevices is vast, there are still obstacles to overcome in manufacturing, integration, reliability, and safety. Current investigations in BINMs and their utilization in micro/nanodevices are actively tackling these obstacles, thus laying the foundation for a novel epoch in diverse industries.

The convergence of BINMs and micro/nanodevices is driving a transformative shift across various academic fields, with a particular emphasis on biomedicine. The convergence described in this context capitalizes on the distinctive characteristics of BINMs, which are derived from biological systems, to optimize the functionality of micro/nanodevices. These devices, in turn, provide a pragmatic framework for implementing these nanomaterials. The inherent characteristic of BINMs, known as the "bottom-up" approach, is highly compatible with the micro/nanoscale. This compatibility facilitates the formation of intricate structures through the process of self-assembly. The advantageous collaboration between nanomaterials and biological systems in biomedicine is of great significance, as it allows for the customization of nanomaterials to enhance their interaction capabilities with biological entities. One example of improving targeted drug delivery systems involves the integration of nanomaterials into micro/nanodevices, thereby enabling the accurate administration of drugs to particular cells or tissues. Moreover, developing micro/nanosensors with high sensitivity is feasible, thereby improving the capability for early disease detection and accurate environmental monitoring. Although this interdisciplinary field offers significant prospects, it is important to acknowledge the persistent challenges associated with the control of nanomaterial synthesis and assembly, their integration into devices, and the assurance of safety and efficacy in practical applications. However, the potential advantages signify a promising outlook for this convergence.

In this thorough analysis, we set out on a complex trip to investigate the field of BINMs and their significant implications for creating and operating micro/nanodevices, particularly those used in the biomedical industry. We start by delving deeply into the idea of biologically inspired nanomaterials, illuminating the intrinsic functional possibilities they bring, and defining thestructure–function correlations observed in nature (Section 2). The many forms of BINMs and their key properties are then discussed (Section 3). The benefits of these intriguing nanomaterials in improving the performance of micro/nanodevices are underlined as we go along, from their flawless biocompatibility to their adaptability (Section 4). In other nanotechnology fields, a wide range of non-biomedical uses of BINMs are also covered (Section 5). The complex synthesis of BINMs, motivated by natural structures, biomolecules, and processes, is then covered in detail (Section 6). Design guidelines for BINM interfaces, emphasizing functionalization strategies and associated difficulties, significantly deepen our understanding (Section 7). We elaborate on the numerous uses of BINMs in micro/nanodevices, primarily focusing on the biomedical sector, including drug delivery systems, organ-on-chip technologies, wound healing approaches, and antimicrobial surfaces (Section 8). As this analysis draws to a close, we consider the ongoing difficulties associated with using BINMs in biomedical applications (Section 9). Finally, we believe in the prospects for the future and offer a few closing thoughts to summarize our discussion (Section 10). This article offers researchers, academics, and business executives a comprehensive grasp of the state of the art and the projected trajectory of BINMs in micro/nanodevices. Figure 1 represents the table of contents of this review article.

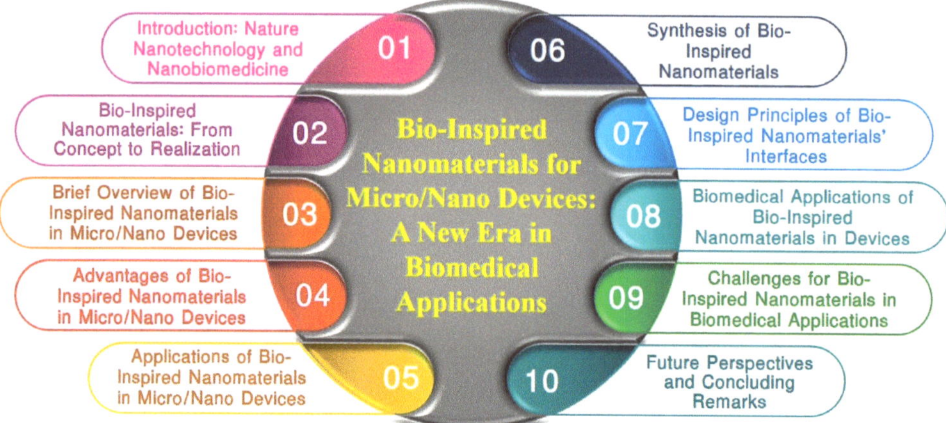

Figure 1. Bio-inspired nanomaterials for micro/nanodevices in biomedical applications.

2. Bio-Inspired Nanomaterials: From Concept to Realization

The progression of BINMs serves as a testament to the notable amalgamation of biology and nanotechnology. They are designed to replicate the structures, functions, or processes observed in nature. This approach enables the development of novel materials that possess distinctive properties. The development of BINMs typically commences with a comprehensive comprehension of the biological system that researchers seek to emulate. The comprehensive understanding and replication of distinct structures, processes, or functions observed in nature at the nanoscale necessitate collaboration among biologists, chemists, and material scientists. The practical implementation of nanomaterials inspired by biological systems can be a multifaceted undertaking requiring meticulous planning and regulation. In certain instances, researchers can employ a biological process directly to synthesize nanomaterials. One illustrative instance involves using bacteria or fungi to generate nanoparticles (NPs), transforming these organisms into miniature factories for nanomaterial production.

In some instances, scientists may be required to employ alternative approaches to accomplish their objectives. The potential application entails the development of artificial structures capable of autonomous assembly, emulating the structural characteristics observed in biological systems. One instance illustrating this phenomenon is the advancement in the creation of synthetic peptides capable of self-assembly into nanofibers, which resemble the nanofibers present in the extracellular matrix of various tissues [48]. Additionally, there exists the challenge of expanding the scale of these processes. Although these nanomaterials can be synthesized in a laboratory setting, scaling up the production process while preserving their intended characteristics poses a greater challenge. Furthermore, the realization of BINM concepts frequently necessitates meticulous optimization. It may be necessary to carefully adjust their properties to optimize the performance of these nanomaterials. This process may entail modifying various parameters, including the dimensions, morphology, or elemental composition of the nanomaterials. Developing BINMs involves a complex and intricate journey, necessitating a comprehensive comprehension of biology and nanotechnology. This research direction holds promise in generating novel materials that can effectively tackle various scientific and technological challenges. The application of biomimetics to multiple domains, such as design, product development, service enhancement, and biomedicine, can be facilitated through a basic research method comprising six distinct steps (Figure 2) [49].

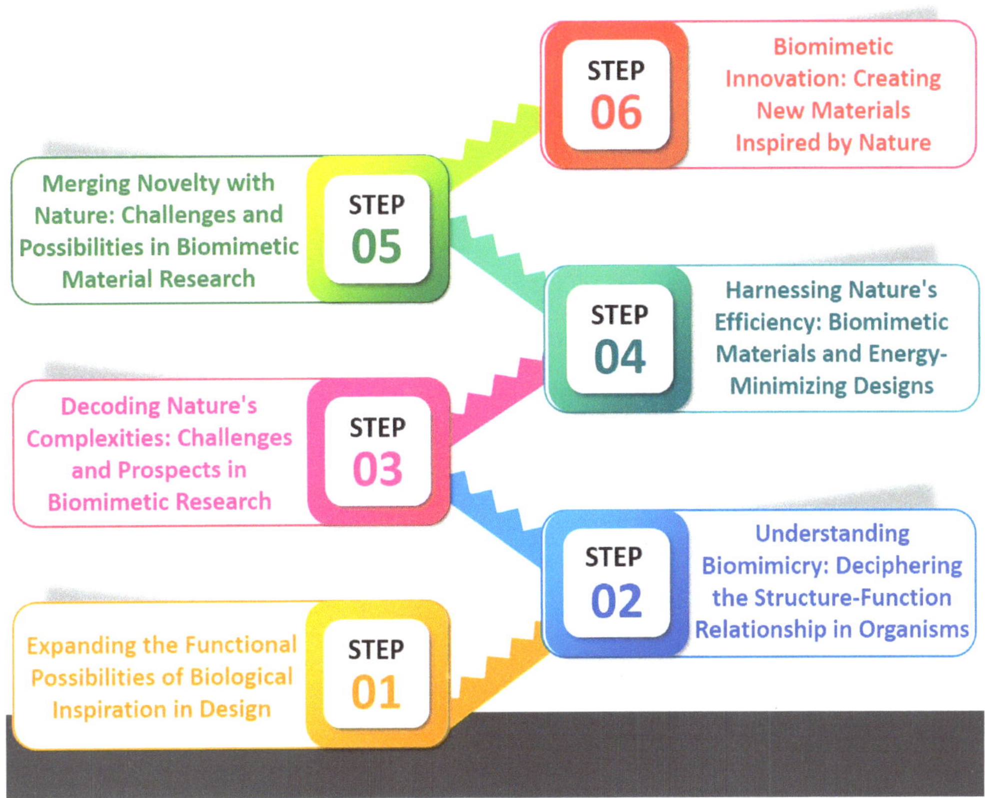

Figure 2. Implementation of BINM-based micro/nanodevices in biomedical applications using intensive research composed of six steps.

2.1. Transforming the Functional Possibilities of Biological Inspiration in Design

Biomimicry, the process of using ideas from the natural world to address problems faced by humans, has long been at the forefront of ground-breaking discoveries. However, it is crucial to realize that biological things' functional potential should not be replicated as it appears in nature. Instead, they should act as a springboard for creative, research-based designs. It has been seen in nature that millions of years of evolution have shaped these animals for certain functions in their settings, from the coordinated flight of birds to the exquisite designs on butterfly wings. It is possible that merely reproducing these capabilities will not satisfy the particular needs and limitations of human cultures and technology surroundings. Determining the basic principles and mechanisms that nature uses and then adapting or improving them for human use are crucial. The Japanese Shinkansen bullet train is one of the examples [50]. When these trains departed tunnels at high speeds, noise pollution posed a huge difficulty to the engineers. The engineers rebuilt the train's nose by taking inspiration from the kingfisher, whose streamlined beak allows it to plunge into water with little splash. This improved speed and fuel economy while also reducing noise. Although nature served as the source of inspiration, the design was an adaption rather than a close match in this case.

The fundamental motivation should be broader despite the obvious economic attractiveness of such inventions. Discoveries or the confluence of disparate concepts can increase profitability and market supremacy. The objective is to improve the comfort and quality of human life, not just make money from a novelty. Designs that are straightforward but innovative, derived from nature but made specifically for people, have the power to change industries as diverse as transportation, healthcare, energy, and architecture. As a result, even while nature offers a priceless store of design solutions, the difficulty for innovators is in interpreting and implementing these solutions. The actual achievement is in developing goods or procedures that skillfully combine the brilliance of nature with the requirements and aspirations of people.

2.2. Understanding Biomimicry: Deciphering the Structure-Function Relationship in Organisms

Understanding the connection between an organism's function and the governing principles of that function is a crucial component of biomimicry, a discipline that aims to mimic nature's time-tested patterns and tactics. The development of bio-inspired designs and technologies is based on this understanding. To this aim, conducting a diligent study and assembling thorough databases to amass knowledge and utilize various materials according to their features is essential. In biomimicry, the interesting relationship between form and function is crucial. Designing self-cleaning materials, for instance, is influenced by the complex surface structure of a lotus leaf, which makes it water-repellent. A thorough knowledge of thisstructure–function link can be gained through cutting-edge scientific methods like scanning electron microscopy. This method makes it possible to notice the little details critical to an organism's ability to operate, which is an essential first step in the biomimicry process.

Scanning electron microscopy provides a thorough image of the surface topography and composition of a material to understand better how an organism functions. For instance, the invention of specifically textured surfaces that limit fouling or microbial growth was motivated by observing the microscopic structure of a shark's skin, which is made up of tiny, tooth-like scales known as denticles [51]. To effectively employ biomimicry, it is imperative to acquire a comprehensive understanding of the relationship between structure and function in living organisms, which should be underpinned by robust research and meticulously curated databases. This approach has the potential to further unlock the vast possibilities of nature-inspired concepts and technologies, thereby facilitating the development of a sustainable future.

2.3. Decoding Nature's Complexities: Challenges and Prospects in Biomimetic Research

Biomimetics, the study and creation of engineering systems and contemporary technologies using biological techniques and systems found in nature, presents special potential and challenges. Understanding the intricate connections between organisms, their micro- and nanostructures, and their environment is perhaps the most important of them. Harnessing the potential of these structures requires understanding how they work, especially for those that have not yet been adequately investigated. These difficulties are multifaceted. For instance, an organism may use a certain structure to perform a given function in a particular environment, yet the same structure may be used otherwise in a different situation. Additionally, there could be less obvious tertiary or even secondary functions. Resolving this complex dance between structure and function that depends on the surrounding environment is like solving a complex puzzle for biomimetics. Understanding biological complexities and reproducing them in synthetic materials are complex tasks in biomimetic research. It takes skill and accuracy in material design and engineering to replicate the structures seen in nature, which frequently exist on the nano- or microscale [52,53].

The merging of biology, natural history, and material science is the next step in biomimetic research to address these issues. Each of these disciplines gives a unique viewpoint and set of instruments that can aid in revealing the mysteries of biological architecture. Understanding living things and how they work is made possible by biology, which serves as the basis for biomimicry. Natural history sheds light on how these processes have changed over time and their contributions to the organism's success and survival. Last but not least, material science provides the skills and knowledge required to mimic these biological structures with artificial materials, enabling the implementation of biomimetic principles in practical settings. This integrative approach to biomimetics has great potential benefits. Medicine, architecture, energy, and manufacturing are just a few industries that might undergo a revolution if we can mimic and exploit the efficiency, adaptability, and sustainability inherent in natural systems. Though the road is difficult, the promise of biomimetic research keeps scientists and engineers motivated to discover the mechanisms behind nature's intricate design.

2.4. Harnessing Nature's Efficiency: Biomimetic Materials and Energy-Minimizing Designs

Exciting research prospects in the developing discipline of biomimicry are focused on identifying unique functional and environmental adaption strategies of organisms. Discovering how these organisms adopt energy-minimizing designs, a concept essential for our sustainable future, is a critical part of this frontier [54–56]. This is about using the creativity and efficiency of nature to inform and advance our designs and technologies. One successful instance of innovation is the development of antireflective coatings, which drew inspiration from an initially unremarkable source, namely the structure of a moth's eye [57]. Despite their tiny size, moth eyes feature complex designs about 200 nm and astonishingly reflect visible light. The efficiency of solar panels can be increased by minimizing light reflection, and the legibility of electronic displays can be improved. Scientists have duplicated these nanostructures to create antireflective coatings, which have a variety of applications.

Remodeling hierarchical structures and the associated functions taken from nature is crucial in creating novel biomimetic materials. This procedure involves copying these structures and comprehending and using the guiding concepts. Then, designs and technologies that impact human society adapt and incorporate these concepts. Advancements in several sectors may result from creating novel materials motivated by the hierarchical structures found in nature [58]. Scientists have developed reusable, residue-free, and temperature-resistant adhesives by comprehending the nanoscale hair-like structures on a gecko's feet.

2.5. Merging Novelty with Nature: Challenges and Possibilities in Biomimetic Material Research

Biomimetics is constantly growing, and new innovations and discoveries are routinely made. Integrating recently found materials with ongoing biomimetic research is crucial to this subject. This integration is believed to be crucial to comprehend the possible uses and constraints of such materials, opening up new avenues in technology and design. However, in order to fully fulfill this potential, it is imperative to develop a thorough grasp of both the advantages and disadvantages of biomimetics. Every newly discovered substance or method has special benefits and drawbacks. In contrast to conventional materials and processes, biomimetic designs, while frequently bringing about enhanced efficiency and sustainability, can also present cost, manufacturing complexity, or durability obstacles. Understanding the morphological and functional applications of novel materials is crucial, in addition to considering the pros and downsides. While the functional features describe how the material functions or interacts under various circumstances, the morphological qualities specify the material's physical and structural properties. By gaining information into these areas, scientists can forecast how the material would perform in various applications and what adjustments might be required to maximize its performance.

Unexpected outcomes may arise from integrating novel materials into biomimetic designs [59–61]. These results need to be carefully examined and comprehended since they can indicate new applications for the materials or unforeseen limitations of the designs. Untangling these findings requires a systematic, step-by-step approach that progressively unveils the essence of the substance and its potential, much like peeling back the layers of an onion. It is difficult to advance in this field, it is true. The complicated and sophisticated nature of the systems being investigated and imitated makes biomimetic material research challenging. Nevertheless, despite the difficulties, there is an intense study going on because of the potential that biomimetics provides for developing future solutions that are more sustainable, effective, and innovative. There are many challenges in realizing the full potential of novel materials in biomimetics. Despite this, there is still a strong commitment to overcoming these obstacles since it is recognized that the benefits, such as advancing our technological skills and promoting sustainable practices, make the effort worthwhile.

2.6. Biomimetic Innovation: Creating New Materials Inspired by Nature

At its essence, biomimicry is about taking inspiration from nature to create new things. Analyzing the structure and function of biological components is a vital step in this process. This knowledge frequently acts as the starting point for creating new materials and the creative application of existing ones. For instance, new high-strength fibers have been developed due to the structural versatility of spider silk, a substance that is both stronger and lighter than steel. Similarly, the development of color-changing materials has been driven by the unique design of butterfly wings, which can reflect light without pigmentation. The fundamental principle is to gain knowledge from the complexity of natural structures and functions and then apply this knowledge to inspire and guide the creation of new materials. It is required to conduct thorough testing and analyses of the structures and operations of biological materials to accomplish this. This gives researchers important insights into these materials' potential by enabling them to comprehend how they behave under diverse circumstances. These revelations can then influence the design and synthesis of novel materials with comparable properties.

Once these novel materials are created, they can be integrated with recent developments in various industries, including chemistry, nanotechnology, and medicine. This interdisciplinary approach can open up a wealth of cutting-edge uses that will considerably enhance human lives. For example, in the field of medicine, materials modeled after gecko feet are being created for their exceptional adhesive characteristics, which have the potential to revolutionize surgical techniques and wound healing [62]. The creation of catalysts based on enzymatic processes has been stimulated by biomimicry in chemistry. The design of water-repellent coverings in nanotechnology results from features like the nanoscale hairs that make lotus leaves self-cleaning [63]. Despite the enormous promise, it is necessary to recognize the difficulties. The intricacy of natural systems differs significantly from that of artificial systems; therefore, merely mimicking nature is not the solution. Instead, it is about comprehending and putting these biological systems' core principles to use to develop novel, long-lasting, and efficient solutions.

3. Brief Overview of Micro/Nanodevices and Types of Bio-Inspired Nanomaterial

Micro/nanodevices frequently exhibit distinct properties and behaviors, which can be attributed to quantum effects and other phenomena that manifest exclusively at these reduced dimensions. They encompass a diverse array of instruments, such as sensors, actuators, and electronic components, among various others. Historically, micro/nanodevices have predominantly employed silicon-based materials due to their exceptional semiconductor characteristics, widespread availability, and well-developed knowledge of silicon processing methodologies. The significant impact of the semiconductor industry on this phenomenon can be attributed to its extensive utilization of silicon in producing microprocessors and various electronic components. Other materials, such as gallium arsenide, silicon carbide, and various polymers, ceramics, and metals, are employed per specific device specifications.

The emergence of micro/nanodevices has led to notable advancements in biomedical applications. The diminutive dimensions of these entities facilitate engagements with biological systems at the cellular and molecular scale, thereby facilitating the development of accurate diagnostics, therapeutics, and research instruments. An illustration of the efficacy of nanoscale drug delivery systems lies in their ability to selectively target afflicted cells with minimal impact on surrounding healthy tissues. In diagnostics, they can swiftly identify disease biomarkers even at highly diluted levels, thereby facilitating timely identification and intervention [64]. Moreover, micro/nanodevices have been employed in tissue engineering and regenerative medicine to manipulate cellular behavior and facilitate tissue proliferation. Despite the considerable potential, several challenges need to be addressed. A notable obstacle lies in the manufacturing process, which necessitates the meticulous and consistent production of these devices at a miniature scale. One additional obstacle pertains to incorporating these devices into broader systems, necessitating the resolution of concerns related to connectivity, compatibility, and power management. There are other apprehensions regarding nanoscale materials' potential health and environmental ramifications, as they exhibit distinct behaviors compared to their macroscale counterparts. Finally, the challenges pertaining to the stability, performance in varying conditions, and durability of micro/nanodevices are of utmost importance and require attention as we further exploit their capabilities. BINMs might be classified as magnetic biomimetic, metal and metal oxide biomimetic, and organic, ceramic, and hybrid biomimetic [65]. Figure 3 lists the types of BINMs and their unique characteristics that have made them potential candidates for multiple biomedical applications.

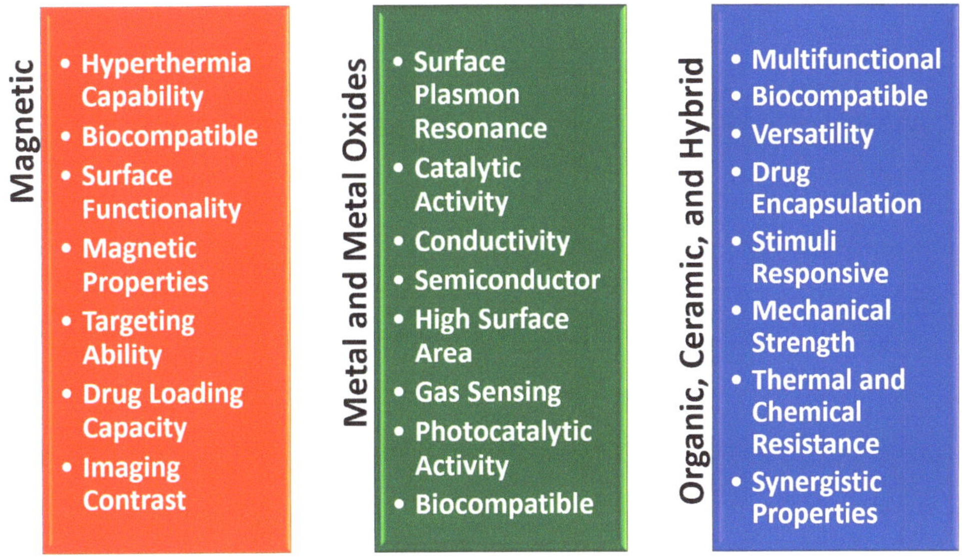

Figure 3. Types of BINMs and their unique characteristics have made them potential candidates for multiple biomedical applications.

3.1. Magnetic BINMs

Magnetic BINMs nanoscale particles are intentionally designed and manipulated to replicate and imitate natural biological processes and structures. These particles leverage their inherent magnetic properties to serve a multitude of applications. The design of these NPs is influenced by biological systems, including cells, proteins, and viruses, to fabricate functional materials with distinctive characteristics. The bio-inspired nature of these NPs entails emulating specific attributes observed in biological systems. For instance, certain magnetic NPs are engineered to replicate the morphology and organization of specific cells, enabling them to engage with biological tissues selectively [66]. Individuals can imitate the actions of biomolecules, such as enzymes or receptors, to carry out specific tasks related to drug delivery or sensing. The manipulation of magnetic NPs by applying external magnetic fields enables precise control over their movement and facilitates targeted interactions within biological systems. The characteristic mentioned above is utilized in various biomedical contexts, including but not limited to drug administration, medical imaging, and the application of magnetic hyperthermia in cancer treatment. In general, magnetic NPs inspired by biological systems integrate the principles of nanotechnology and biomimicry to generate novel materials that hold promise for fields such as medicine, biotechnology, environmental remediation, and other domains.

Numerous magnetic BINMs have been successfully produced thanks to the application of biomimetic synthesis techniques, from the hard magnetic alloys FePt and CoPt to ferrimagnetic Fe_3O_4 found in magnetotactic bacteria (MTB) [67]. Because they use biological structures that can be altered through chemical or genetic engineering for certain functionalities and scalable production, biomimetic approaches have special advantages. Most previous publications on magnetic NP synthesis concentrated on physical and chemical techniques. However, more recent developments in magnetic BINMs have enabled the lab to replicate MTB-like chains of magnetic NPs, showcasing promising biomimetic techniques for biomedical applications [68]. They come in a variety of forms. One method uses magnetosome-associated MTB proteins to biomineralize Fe_3O_4 and produce polymer-coated and non-polymer-coated magnetite NPs. Chemotherapy, magnetic hyperthermia, enzyme immobilization, and photothermia are only a few of the uses for these NPs [69–71]. Purified and functionalized engineered structures derived from bacterial magnetosomes can be used as contrast agents in magnetic resonance imaging, magnetic particle imaging, and magnetic hyperthermia [72]. Hydroxyapatite (HAP)-coated magnetite NPs are a different class of magnetic BINMs created by mixing a liquid HAP precursor with a solution containing magnetite cores. These constructions help deliver genetic material, magnetic scaffold creation for bone tissue repair, and magnetic hyperthermia [73,74]. The design of biocatalysts for targeted enzyme prodrug therapy is made possible by combining magnetic NPs with active ingredients in a biomimetic matrix, such as SiO_2 [75]. Methods for developing magnetic BINMs can be divided into two categories: those that use magnetic NPs that have already been obtained or are available commercially and those that make magnetic NPs from scratch. The latter group has a better chance of producing atomically precise structures that resemble their natural analogs. These techniques include recombinant MamC-based anaerobic biosynthesis [76], PEGylated human ferritin NP-based magnetite biomineralization [77], and encapsulation or biotinylation of isolated bacterial magnetosomes [72]. Despite a long research history, magnetic BINMs are not as commonly used in biomimetics as other materials. Nevertheless, recent developments in biomimetic synthesis and the distinctive characteristics of magnetic BNMs imply they have enormous potential for various biomedical applications.

3.2. Metal and Metal Oxide BINMs

Metal and metal oxide BINMs represent a captivating category of NPs that emulate natural structures and functionalities. Nanomaterials possess distinct characteristics that render them exceptionally well suited for various biomedical applications. Metal NPs, such as gold and silver, are frequently employed in biomedical research owing to their remarkable optical characteristics, substantial surface area, and adjustable surface chemistry. These technologies find utility in cancer treatment, precise administration of pharmaceuticals, and medical imaging. Gold NPs can undergo functionalization through the attachment of antibodies, enabling them to selectively target cancer cells and facilitate the direct administration of therapeutic agents to the tumor site [78]. Silver NPs possess antimicrobial properties, making them highly advantageous for wound dressings and antibacterial therapies [79]. Metal oxides, including iron oxide NPs, exhibit magnetic characteristics that make them well suited for various applications, such as magnetic targeting, hyperthermia-based cancer treatment, and the development of contrast agents for magnetic resonance imaging (MRI) [80]. External magnetic fields enable the precise localization of iron oxide NPs within the body, facilitating targeted drug delivery and localized therapeutic interventions. BINMs present a highly promising avenue for advancing various disciplines, including medicine, diagnostics, and therapeutics. The valuable attributes of NPs, such as their biocompatibility, functionality, and capacity to interact with biological systems, render them instrumental in addressing a wide range of health challenges and enhancing patient outcomes. Nevertheless, additional investigation is required to comprehensively comprehend their conduct within intricate biological settings and guarantee their reliability and effectiveness in clinical contexts.

By choosing particular micro- or nanoenvironments during synthesis, biomimetic synthesis techniques have been utilized to regulate the physical and chemical properties of metal and metal oxide nanomaterials. Several platforms have been used for biomimetic production of metal oxides, including ferritin, viral capsids, and bacterial cells. These platforms provide exact conditions for NP formation and produce narrow distributions in shape and size. These methods' consideration of the environment and the possibility for expansion make them attractive. Due to its benefits, plant-based biomimetic synthesis of metal NPs has attracted interest recently, and researchers are investigating the mechanisms of nanomaterial synthesis and metal ion biological reduction in plants. Metallic BINMs have uses in the detection and eradication of contaminants. To develop sensors for biological substances and conduct research on nanotoxicology, biomimetic techniques can be used to make gold, silver, and bimetallic Ag-Au NPs [81]. For targeted drug delivery, biomimetic mineralization methods utilizing cubic nanostructures built on lipid membranes known as "cubosomes" are being investigated [82]. For the biomimetic synthesis of materials, metal-organic frameworks (MOFs) are another topic of study. They are useful for drug delivery, catalysis, and stabilizing biomacromolecules because they may be constructed with biomimetic active centers and restricted pockets [83].

3.3. Organic, Ceramic, and Hybrid BINMs

Organic, ceramic, and hybrid BINMs constitute a distinct category of NPs that exhibit various applications within biomedicine. Nanomaterials derive inspiration from natural structures and processes, showing distinctive properties that render them well suited for diverse biomedical applications. Organic BINMs are synthesized using naturally occurring molecules such as proteins, lipids, and carbohydrates. Biocompatibility is a characteristic exhibited by these entities, rendering them suitable for integration with biological systems. Moreover, they can be modified through engineering processes to acquire precise functionalities, such as targeted administration of pharmaceutical agents and facilitation of tissue regeneration. One illustrative instance involves the utilization of liposomes, which are lipid vesicles at the nanoscale level, to encapsulate drugs and facilitate their targeted delivery to precise locations within the human body [84]. This approach serves to mitigate adverse effects and enhance the efficacy of therapeutic interventions. Ceramic nanomaterials with bio-inspired characteristics, such as hydroxyapatite and silica NPs, exhibit a mineral composition resembling bones and teeth [85]. They are extensively utilized in bone tissue engineering, wherein they facilitate bone regeneration and augment the assimilation of implants into native bone tissue. Moreover, ceramic NPs exhibit considerable potential in drug delivery [86] and imaging [87] due to their inherent stability and biocompatibility. Hybrid BINMs amalgamate distinct material characteristics to attain heightened functionalities. Nanomaterials possess the capability to incorporate both organic and inorganic constituents, as well as to integrate magnetic attributes with organic coatings. An illustration can be found in the advancement of hybrid magnetic NPs designed for targeted drug delivery and hyperthermia-based cancer therapy [88]. A magnetic component facilitates convenient manipulation and precise localization within the human body. At the same time, implementing an organic coating ensures compatibility with biological systems and controlled release of therapeutic agents. In biomedicine, organic, ceramic, and hybrid BINMs exhibit significant promise. The valuable attributes of NPs, including their versatility, biocompatibility, and capacity for customization, render them highly advantageous in drug delivery, imaging, tissue engineering, and various other biomedical technologies. Nevertheless, it is imperative to thoroughly assess the safety and effectiveness of these novel nanomaterials before their extensive application in clinical settings.

Extensive research efforts have investigated BINMs comprising organic and ceramic constituents, demonstrating considerable potential in diverse biomedical domains. Proteins and peptides serve as templates for regulating the synthesis and self-assembly of organic

BINMs, facilitating the development of multifunctional materials possessing distinct structures and functionalities. Peptoids, a biomimetic polymer category, are widely recognized as versatile constituents for constructing hierarchical BINMs [89]. Research in the ceramic biomaterial and nanomaterial domain is primarily centered around developing scaffolds that emulate the structural and functional characteristics of natural tissues, particularly bone. Ceramic structures with a high degree of porosity, similar in structure to cancellous bone, have been successfully manufactured to facilitate the ingrowth of cells and the formation of new tissue. Biomimetic ceramic scaffolds are infused with therapeutic molecules to enhance their biological efficacy. Hybrid BINMs are being investigated for their potential applications in tissue engineering in orthopedics and dentistry. There is ongoing research and development in drug delivery carriers, specifically on membrane-camouflaged NPs.

4. Advantages of Bio-Inspired Nanomaterials in Micro/Nanodevices

The utilization of BINMs has emerged as a novel approach to advancing micro/nanodevices. Nanomaterials that draw inspiration from biological systems present a unique strategy for addressing the difficulties associated with device miniaturization while simultaneously improving performance, versatility, and biocompatibility. The utilization of BINMs in micro/nanodevices encompasses a wide range of disciplines, such as electronics, optics, environmental science, and biomedicine. The electronics field is investigating the potential of bio-inspired materials, such as protein-based nanowires and biogenic semiconductors, to develop novel electronic devices with distinctive electronic characteristics. BINMs have significantly transformed micro/nanodevices within the biomedical field. For example, nanomaterials have been utilized by drug delivery systems to augment the precision of drug administration and regulate the release of therapeutic agents, leading to notable advancements in treatment efficacy. BINMs exhibited enhanced sensitivity and specificity in detecting biomarkers, thus facilitating the potential for early disease diagnosis and disease monitoring. They possess numerous advantageous characteristics when integrated into micro/nanodevices, such as improved performance, biocompatibility, self-assembly capabilities, sustainability, and versatility (Figure 4).

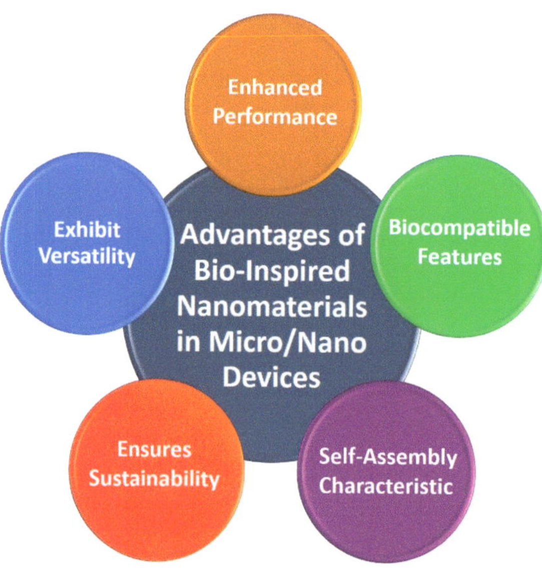

Figure 4. Advantages of BINMs in micro/nanodevices in biomedical applications.

4.1. Enhanced Performance

The utilization of BINMs has the potential to greatly enhance the performance of micro/nanodevices by capitalizing on their distinct properties. The observed enhancement in performance can be attributed to the remarkable characteristics that nature has developed over an extensive period of time. One potential application of nanomaterials is their ability to replicate the adhesive properties observed in gecko feet [90]. This unique characteristic has garnered significant attention due to its exceptional adhesive strength. Consequently, nanomaterials with gecko-inspired adhesion can potentially revolutionize the development of micro-robots and wearable devices, enabling them to adhere to diverse surfaces [91]. Likewise, the utilization of nanomaterials inspired by shark skin, which possesses characteristics that reduce drag, holds potential for enhancing the energy efficiency of microfluidic devices [92].

4.2. Biocompatibility

One notable advantage of numerous nanomaterials inspired by biological systems is their biocompatibility. Nanomaterials that draw inspiration from or are derived from biological entities possess an inherent compatibility with biological systems. The compatibility between these materials and living tissues or cells mitigates potential adverse reactions. In drug delivery, biocompatible nanomaterials enable the transportation of therapeutic agents within the human body while mitigating the risk of eliciting detrimental immune responses [93]. Likewise, in biosensing, NPs can be utilized for extended monitoring periods without inducing any adverse tissue irritation or rejection [94].

4.3. Self-Assembly

Self-assembly is a captivating characteristic observed in numerous biological systems, wherein complex structures are formed spontaneously [95]. BINMs frequently inherit this intriguing property. A range of factors, including pH, temperature, and ionic strength, can control the process of self-assembly [96]. This ability to direct self-assembly can be utilized to construct complex structures with minimal external intervention. This streamlined manufacturing process offers the potential to create intricate designs for devices that would present significant challenges or even be unattainable through conventional fabrication methods [97].

4.4. Sustainability

The design and synthesis of BINMs frequently incorporate the principles of green chemistry and biomimicry, contributing to the promotion of sustainability [98]. This methodology has the potential to facilitate the advancement of fabrication processes and devices that are environmentally sustainable. Numerous BINMs can be synthesized using gentle conditions, devoid of toxic solvents or by-products. Moreover, certain nanomaterials can undergo biodegradation, thereby mitigating their environmental impact upon completing their functional lifespan.

4.5. Versatility

The extensive array of biological systems that serve as sources of inspiration for the design of nanomaterials provides a wide spectrum of potential applications. The potential of BINMs is vast, as demonstrated by their ability to replicate the light-harvesting capabilities of photosynthetic organisms to enhance solar cells [99] or imitate the structural color found in butterfly wings to improve optical devices [100]. These capabilities hold significant promise in developing materials with customized properties that can effectively fulfill the distinct demands of diverse applications, thereby expanding the possibilities for advanced micro/nanodevices.

5. Bio-Inspired Nanomaterials in Micro/Nanodevices

The utilization of BINMs in micro/nanodevices represents the integration of intricate designs found in the natural world with the capabilities of modern nanotechnology. This convergence establishes a mutually beneficial relationship with potential remarkable advancements in various domains. These materials draw inspiration from the distinctive characteristics displayed by biological organisms. By leveraging these properties, they enable the development and production of micro/nanodevices that offer improved performance, biocompatibility, self-assembly capabilities, sustainability, and versatility. The effective integration of these materials in various devices is evident in multiple instances, such as using self-cleaning solar panels, dry adhesives for micro-robots, photonic sensors, drug delivery systems, and other notable applications [101–104]. The advent of BINMs in micro/nanodevices signifies a significant advancement in technology, medicine, and environmental sustainability, showcasing the extensive possibilities of biomimicry on the nanoscale. The applications of BINMs in micro/nanodevices other than biomedicine have been summarized in Figure 5, while Table 1 summarizes the subsections under this section and lists the applications of BINMs in various types of micro/nanodevices belonging to diverse nanotechnology domains.

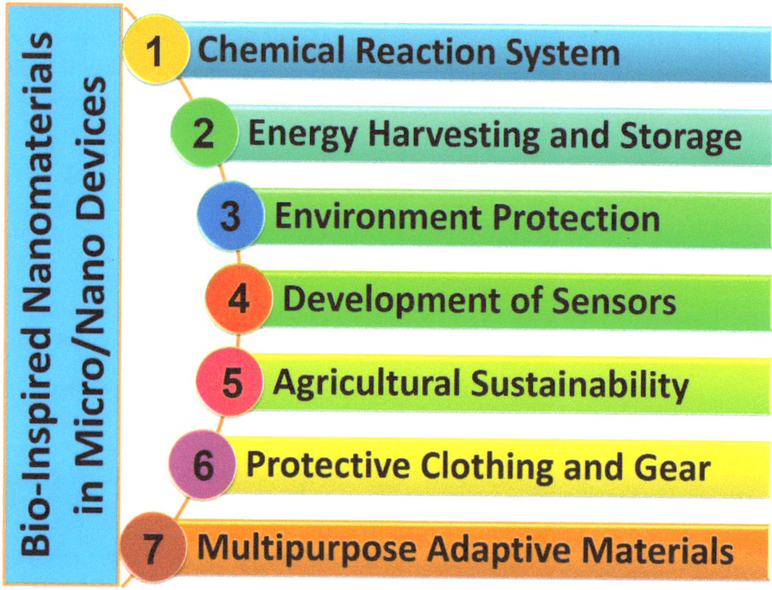

Figure 5. The applications of BINMs in micro/nanodevices other than biomedicine.

5.1. Chemical Reaction Systems

The utilization of BINMs is of significant importance in chemical synthesis and reactions, as it draws inspiration from nature's intricate designs to enhance various aspects such as efficiency, selectivity, and sustainability. Chemical processes are frequently improved by emulating the hierarchical structure and multifunctionality observed in biological systems, such as enzyme-catalyzed reactions. For example, the development of bio-inspired catalysts, which draw inspiration from the highly efficient and specific catalytic mechanisms observed in biological systems, facilitates the execution of reactions with enhanced selectivity and reduced environmental impact [105,106]. This approach effectively minimizes both waste generation and energy consumption. The phenomenon of self-assembly, which is widely observed in nature, is also utilized as a reliable technique in producing nanomaterials. This technique enables the creation of intricate structures using mild conditions. In addition, the utilization of BINMs that replicate the biomineralization mechanism observed in corals has enabled the synthesis of nanocrystalline materials under ambient conditions [107]. This advancement holds significant potential for their application in various chemical reactions. In general, the utilization of BINMs in the context of chemical synthesis and reactions is a compelling demonstration of the efficacy of leveraging natural principles to propel the field of synthetic chemistry forward [108,109].

The synthesis of ammonia through the electrocatalytic reduction of nitrate/nitrite can be conducted using bio-inspired metalloenzymes [110]. Drawing inspiration from the natural NO_3^- reductase, utilizing bio-inspired metalloenzymes presents a promising alternative to metal nanomaterials. This substitution can significantly enhance the electrocatalytic performance of NO_3^-/NO_2^- reduction to NH_3 (NRA) and improve NH_3 selectivity in a neutral environment. Ford and his colleagues conducted a research study to create an iron catalyst influenced by the active sites found in NO_3^- reductase enzymes. This catalyst was intended to treat industrial wastewater in challenging environmental conditions [111]. The catalyst possesses a secondary coordination sphere that assists in oxyanion deoxygenation. The reduction of oxyanions forms a Fe(III)-oxo species, which subsequently acts as a catalyst, facilitating regeneration while concurrently releasing water in the presence of protons and electrons. Using the bio-inspired iron catalyst in NRA has offered a sustainable approach for effectively utilizing the nitrogen resource. Through electrostatic interactions, a biomimetic nickel bis-diphosphine complex fixed on altered carbon nanotubes (CNTs) contained the amino acid arginine in the outer coordination sphere [112]. With a catalytic selectivity for H_2 oxidation at all pH levels, the functionalized redox nanomaterial demonstrates reversible electrocatalytic activity for the $H_2/2H^+$ interconversion from pH 0 to 9. The high activity of the complex over a broad pH range enables us to integrate this BINM either in a proton exchange membrane fuel cell (PEMFC) employing Pt/C at the cathode or in an enzymatic fuel cell in conjunction with a multicopper oxidase at the cathode. Comparing the Ni-based PEMFC to a full-Pt traditional PEMFC, its maximum output is just six times lower at 14 mW cm^2. A new efficiency record for a hydrogen biofuel cell using base metal catalysts is set by the Pt-free enzyme-based fuel cell, which produces 2 mW cm^2.

5.2. Energy Harvesting and Storage

Using BINMs in the energy sector has sparked the development of ground-breaking strategies for effective energy generation, storage, and conservation. These materials make creating more effective and sustainable energy systems easier by taking their design cues from nature's perfectly regulated energy processes. For instance, the distinctive light-harvesting systems seen in photosynthetic organisms have been imitated to increase the effectiveness of photovoltaic cells, allowing them to more efficiently gather and transform sunlight into electricity [113–115]. BINMs have also influenced the creation of durable and light-weight materials for wind turbine blades, improving both the performance and longevity of these devices [116,117]. Furthermore, progress has been made in designing BINMs for energy storage devices, including batteries and supercapacitors. For instance, bee-hive-inspired honeycomb shapes have been employed to make electrodes with a high surface-to-volume ratio, increasing their energy storage ability [118–120]. The cooling systems found in some animals have influenced the design of materials for effective heat dissipation in energy devices.

5.3. Environmental Protection and Sustainability

Using nature's design principles to address urgent environmental concerns, BINMs have opened new vistas for environmental preservation and sustainability. For example, the distinctive capacity of lotus leaves to self-clean has sparked the creation of coatings based on nanomaterials that lessen the need for harsh cleaning agents, thereby reducing water pollution. Similar to photocatalytic materials, which help break down pollutants when exposed to sunlight, photocatalytic materials are inspired by photosynthesis in plants and offer a promising alternative for air purification and wastewater treatment. Indirectly reducing the carbon footprint, BINMs have a considerable impact on the energy sector, helping to create better energy storage devices and increasing solar cell efficiency. Additionally, sustainable BINMs that are recyclable or biodegradable have been created using the concepts of biomimicry, supporting a circular economy.

5.4. Development of Sensors

BINMs are transforming sensor technology by boosting sensitivity, specificity, and dependability. For instance, photonic sensors that can sense minute changes in light, temperature, and pressure have been developed using structures that mirror the delicate architecture of butterfly wings [121]. Developing tactile sensors based on nanomaterials is highly sensitive to various stresses [122]. Similarly, acoustic sensors for detecting light sounds or vibrations have been created using the structure of spider silk, which is well recognized for its sensitivity to air movements [123]. The invention of microphones that simulate the highly directed hearing of the parasitic Ormiaochracea fly was made possible by the development of acoustic sensors, inspired by spider silk's sensitivity to air movements. Chemical sensors that can detect trace amounts of explosives, narcotics, or other compounds, similar to dogs' noses, have been developed. These sensors are based on the exceptional sense of smell that animals possess. Identical to the infrared-sensing organs of the pit viper, thermal sensors, modeled after the heat-sensing abilities of some snakes, have led to devices that can detect temperature changes without being impacted by ambient temperature. The biocompatibility and high surface-area-to-volume ratio of several of these materials make them excellent for detecting biomolecules at very low concentrations, which has important implications for biosensors. Utilizing the special qualities of BINMs, sensors can be developed that outperform conventional designs in terms of performance, flexibility, and versatility, with uses in various fields, including security, healthcare, and environmental monitoring.

5.5. Agricultural Sustainability

BINMs can potentially bring about significant transformations in the agriculture and food sectors, presenting viable solutions to key challenges these industries face. The applications of these technologies encompass precision agriculture, wherein bio-inspired nanosensors are employed to monitor soil conditions and crop health, drawing inspiration from the moisture detection mechanism found in plant roots [102,124,125]. This enables the optimization of resource utilization. In pest and disease management, NPs that draw inspiration from naturally occurring plant or microbial compounds, such as those that imitate the pyrethrins found in chrysanthemums, can selectively target particular pests or pathogens [126]. Within the realm of the food industry, nanomaterials play a significant role in the development of intelligent packaging [127]. One notable application involves utilizing nanosensors capable of detecting ethylene levels, a naturally occurring compound that indicates fruit ripening [128]. By employing such nanosensors, monitoring changes in food quality and minimizing wastage are possible. In addition, biosensors enhance food safety by emulating the immune response, enabling the prompt identification of foodborne pathogens such as *E. coli* or *Salmonella* [129]. The enhancement of nutrient delivery is achieved by employing nano-encapsulation techniques that draw inspiration from inherent cellular mechanisms, thereby augmenting the assimilation of nutrients or probiotics. The field of waste management stands to gain advantages from utilizing BINMs, specifically those that imitate the natural catalysts found in the gut of termites. These nanomaterials can expedite the decomposition process of agricultural waste, resulting in the production of valuable resources such as biofuel or compost [7,130]. Furthermore, implementing water purification techniques inspired by the physiological processes observed in xylem tissues of plant species such as pine trees plays a crucial role in enhancing the safety of water used for irrigation purposes [131]. Using BINM applications offers a collective approach toward achieving enhanced global food security and environmental sustainability, resulting in more sustainable, efficient, and effective solutions.

Table 1. A thorough summary of the subsections under this section and a list of the applications of BINMs in various types of micro/nanodevices belonging to diverse nanotechnology domains.

Sector	Devices	Bio-Inspiration	Mechanism	Applications	Refs.
Synthesis	Catalytic converter	Enzymes/natural catalyst Peptide sequence Luffa sponge	Electrocatalysis	Water splitting Oxygen reduction CO_2 reduction Metal nanoparticle (MNP) fabrication	[132–134]
	Photovoltaic device Fog harvester	Photosynthesis Butterfly wings	Electrochemical Photocatalysis	Water splitting Fog harvesting	[135–137]
Energy	Microbial biofuel cells	Photosynthesis Hydrogenases in microorganisms	Electrochemical Photocatalysis	Hydrogen production Energy production	[138–141]
	Electrodes	Bee honeycomb	Proton conduction as an electrode Electrochemical	Conversion of fuel energies into electricity Long-term energy conversion, transfer, and storage	[142–145]
	Solar cells	Photosynthesis Peptide nanomaterials Virus Cobweb	Photovoltaic	Electricity generation	[146–150]
	Battery and supercapacitors	Benzoquinone (BQ) in photosystem Adenine in DNA Human tissues Nanocluster arrays on a lotus leaf M13 virus protein shells Tobacco mosaic virus DNA	Electrochemical	Supercapacitor Li-ion battery electrode Li-sulfide batteries Rechargeable batteries	[136,151–158]
Environment	Hybrid photocatalysts Adsorbent Magnetic polymer nanocomposites Ultrafiltration membrane	Enzymes Peptides Biomolecules	Photocatalysis Adsorption Magnetism Ultrafiltration	Environmental detoxification Metal removal Dye removal Saline water separation Oil separation	[108,109,130,159–162]
	Electrochemical biosensor	RNA Nicking enzymes	Electrochemical sensing	Mercury identification Detection of *Salmonella enteritidis*	[163,164]
	Hybrid membranes	Aggregated amyloid protein fibrils	Filtration	Heavy metal removal	[165,166]
	Biosorbent	Bacteria	Adsorption	Elimination of Cd	[167]
	High-performance nanofilter	Tau protein *Moringaoleifera* pods	Nanofiltration	Air purification	[168,169]
Sensors	Potentiometric e-tongue	Tongue	Field-effect-transistor-based	Detecting the bitterness	[170,171]
	Organoid-based biosensor	Taste bud	Electrophysiological signals	Taste sensation	[172]
	Enzyme biosensor	Horseradish peroxidase	Electrochemical	Detection of H_2O_2	[173]
	Olfactory biosensor	Cardiomyocytes	Electrochemical	Odor detection	[174,175]
	Potentiometric sweetness sensor	Sweetness sensor GL1	Potentiometry	Detecting the sweetness	[176]
	Chemiresistive sensor	Sensor organ	Electrochemical	Detection of N_2	[177]
	Photonic sensor	Morpho butterfly scales	Optoelectrochemical	Detection of H_2, CO, and CO_2	[178]
	Photonic nose	Turkey skin M-13 bacteriophage	Colorimetric	Detection of molecules	[179–181]
	Acoustic sensors	Spider slit organ Lotus leaf	Electrical	Voice recognition	[182]
		OrmiaOchracea fly	Electro-mechanical	Direction finding sensor	[183]
	Infrared sensor	Snake skin	Photomechanical	IR sensing systems	[184]
	Hydrodynamic sensors	Fish and some amphibians	Electro-mechanical	Hydrodynamic artificial velocity sensor	[185]
	Humidity sensors	Spider silk	Electrochemical Biological structures Electrical conductivity Transduction mechanisms	Humidity and strain detection	[186]
	Motion sensors	Snake movement	Electro-mechanical	Robot	[187]

Table 1. *Cont.*

Sector	Devices	Bio-Inspiration	Mechanism	Applications	Refs.
	Magnetic sensors	Pigeons' magnetoreception ability	Electromagnetic Magnetoreception Signal transduction Miniaturization and integration	Wastewater treatment	[188]
Protective Clothing and Gear	Smart fabric	Lotus leaf Algae eyespot-stigmata design Touch-me-not (Mimosa sps.) pulvinus Pine cone Chameleon skin Fish scale Mammalian tissue Spider silk Firefly glow Shark skin Mammal skin Spider silk Chameleon Cactus spines	Microfabrication Mechanochromic Photonic elastomer Biomimetic structures Sensing and actuating system Self-regulation Energy harvesting and storage	Anti-dust cloth Water-repellent fabrics Light- and touch-sensitive apparel Smart breathing fabrics Camouflage apparel Self-healing fabric Anti-tear fabric design E-circuited luminescent fabrics Antibacterial clothing Thermo- and pressure-sensitive fabric Self-adapting textiles Self-hydrated clothing	[189–196]
Multipurpose Adaptive Materials	Self-healing materials	Wound healing Bone remodeling Plant healing and regeneration Blood clotting and vascular repair Microbial biofilms Skin healing	Molecular mobility Triggered response Microcapsulation Dynamic covalent bonding Autonomic healing	Field-effect transistors Pressure sensors Strain sensors Chemical sensors Triboelectric nanogenerators Soft actuators Smart coating	[197–200]
	Shape-memory materials	Creatures reducing impact damage Muscle contractions Insect wing folding Plant movements Tendons and ligaments Caterpillar and snake movements Soft tissues	Two-way shape memory effect Phase transitions Molecular reconfiguration Energy storage and release Microstructure design	Biomedical devices Aerospace engineering Robotics Textiles	[201–205]
	Responsive surfaces	Lotus leaf Gecko adhesion Butterfly wing Shark skin Mussel adhesion Pinecone closing Venus flytrap	Hierarchical structures Stimulus-responsive materials Self-assembly Wetting and capillary forces Surface gradients Biomolecular interactions	Self-cleaning coatings Anti-fouling surfaces Stimuli-responsive materials Environmental monitoring	[206–211]
	Multifunctional composites	Bone structure Plant fiber Nacre Spider silk Biomineralization processes	Synergy of materials Hierarchical structure Synergistic interfaces Functionalization and integration	Aerospace and automotive industries Energy harvesting and storage Protective coatings Robotics	[212–215]

5.6. Protective Clothing and Gear

The utilization of BINMs exhibits significant promise in augmenting the safety, performance, and comfort attributes of protective clothing and gear. For example, the distinct denticles found on the skin of sharks, which can impede the growth of bacteria, serve as a source of inspiration for developing materials used in protective clothing. These materials exhibit similar properties by effectively resisting microbial presence, thereby mitigating the risks of infection in environments where individuals are highly exposed to such hazards. The remarkable mechanical properties of spider silk have been replicated in nanoscale architectures, resulting in materials that possess a unique combination of strength, flexibility, and low weight. These materials have applications in various protective gear, such as bulletproof vests and helmets, where their exceptional properties are highly advantageous. The phenomenon of structural coloration observed in peacock feathers, wherein color changes occur in response to environmental stimuli, provides a valuable model for developing materials capable of indicating hazardous conditions via color alterations. The water retention abilities exhibited by cactus spines serve as a source of inspiration for developing survival suits that can extract moisture from the atmosphere in arid environments. Moreover, the phenomenon known as the self-cleaning "lotus effect," observed in the leaves of lotus plants, has prompted the advancement of self-cleaning materials that are particularly suitable for use in protective garments within environments characterized by dirt or contamination.

5.7. Multipurpose Adaptive Materials

The utilization of BINMs shows great potential in multipurpose adaptive materials (MAMs) for their ability to modify their properties according to environmental variations. Researchers have developed self-healing materials that exhibit automatic repair mechanisms, drawing inspiration from the regenerative capabilities observed in starfish and salamanders. These materials mimic the natural healing processes found in living organisms. Shape-memory materials have been engineered to imitate the adaptive response of the Venus flytrap to external stimuli. These materials have found applications in various fields, including smart textiles and precise administration of pharmaceuticals. The remarkable hydrophobic characteristics exhibited by lotus leaves have served as a source of inspiration for developing responsive surfaces that can repel water and dirt. These surfaces have practical utility in various domains, such as self-cleaning windows and anti-fouling coatings for maritime vessels. Scientists have utilized the heat-resistant adaptations of the Saharan silver ant as inspiration to create energy-efficient materials that can passively cool, thereby decreasing the need for energy-consuming air conditioning systems. Furthermore, the hierarchical arrangement of nacre, also known as the mother of pearl, has served as a source of inspiration for the development of multifunctional composites that possess a combination of strength and toughness. These composites have proven valuable in applications such as protective coatings and producing durable construction materials.

6. Synthesis of Bio-Inspired Biomedical Nanomaterials

Living things have developed the ideal arrangements of their structures, parts, and functions, giving rise to special qualities like the great strength of bones, the hardness of enamel, and the capacity of shark skin to lessen fluid drag. In recent decades, an attempt has been made to comprehend the connection between these elements in high-performance natural materials. Some processes, such as the layered structure of nacre for improved mechanical characteristics, the nanostructure of lotus leaves for hydrophobic properties, and the proteins in the feet of Mytilusedulis for adhesion, have been postulated. High-performance artificial materials have been created using these concepts in industries like energy, architecture, aircraft, and biomedicine. New synthesis techniques have also been put forth to better replicate the hierarchical components or structures of natural materials. Natural bioprocesses, including biomineralization, cell metabolism, and photosynthesis, that are essential to developing and operating natural biological systems have also received attention. These biological processes have advantages over artificial synthetic ones since they frequently occur in calm environments. A new study area called "bioprocess-inspired fabrication," which merges biology, life science, and material science, aims to create novel synthesis methods inspired by biological processes in nature. High-performance biomaterials are being developed using bio-inspired techniques for tissue regeneration, medication delivery, biosensing, and monitoring applications. To build biomimetic scaffolds for bone tissue engineering or to use mussel-inspired bioadhesives for skin wound healing, these methodologies imitate natural structures, such as the hierarchical structure of bone. The synthesis of BINMs might be divided into three categories (Figure 6): synthesis inspired by natural structure/components, synthesis inspired by biomolecules, and synthesis inspired by bioprocesses [21].

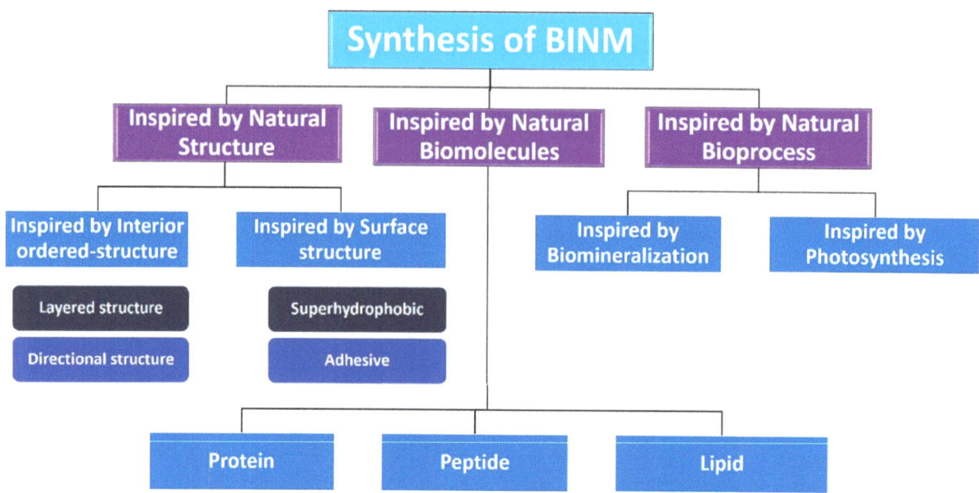

Figure 6. Synthesis of BINM suitable for biomedical applications.

6.1. Inspired by Natural Structures

The structural characteristics of a material are closely associated with its physical and chemical properties. Numerous natural substances exhibit intricate and well-organized patterns across various levels, such as the "brick and mortar" arrangement observed in nacre, the collagen fibers in a bone that are mineralized directionally, and the nanostructures resembling branches found on the surface of a lotus leaf. The exceptional performance of these distinct structures can be attributed to their specialized functions. At present, replicating the composition and arrangement found in natural materials is a prominent approach in biomedical materials to improve their mechanical strength, adhesion, and antibacterial characteristics.

6.1.1. Inspired by Interior Ordered Structure

Numerous natural materials exhibit remarkable mechanical properties, including strength, toughness, and lightness, despite comprising constituent components of lower inherent strength. The primary reason for this phenomenon can be attributed to the multi-scale hierarchical structures and diverse failure mechanisms observed in various materials. In tissue engineering, biomedical materials must have exceptional biocompatibility and sufficient mechanical properties. Within this framework, we shall present several noteworthy internal ordered structures observed in the natural world. These structures encompass the "brick and mortar" layered structure, the Bouligand structure, and the directional arrangement structures evident in enamel and bone. These structures are prototypes for developing biomedical materials that exhibit enhanced mechanical strength. Nanomaterials, with potential for biomedical applications, can be synthesized by the inspiration of the layered structure or multi-directional ordered structure of nature.

The distinctive layered structure of nacre, resembling bricks, has served as a significant source of inspiration in advancing biomaterials with exceptional performance capabilities. The fantastic mechanical qualities of this structure, renowned for its great strength and toughness, have motivated researchers to explore the intricacies of reproducing similar layered nanostructures in order to attain exceptional mechanical characteristics in synthetic biomaterials. The development of nacre-like films by Yoo et al. is a notable achievement in this field [216]. The aforementioned nanocomposite films, composed of structured boron nitride nanosheets (BNNSs) and gelatin, were developed by leveraging the electrostatic attractions between the charged groups of gelatin and BNNSs. To improve self-assembly and the connection at the interface of these components, BNNSs were enhanced with hyperbranched polyglycerol. The alignment of the BNNSs on a 2D plane could be adjusted by increasing the BNNS quantity or through a specific functionalization technique, resulting in a shift from a chaotic orientation to a structured brick-and-mortar arrangement. By varying the BNNS and gelatin mixture in the composite and adjusting the BNNS arrangement, one can modulate the nanocomposite's mechanical attributes, such as its strength and rigidity. This adjustment produces a substance with mechanical qualities mirroring human cortical bone. Preliminary in vitro tests showed that this BNNS/gelatin blend could promote attachment, sustenance, and growth of adipose-derived stem cells, marking its potential in the biomedical sector. The combined mechanical and biological results hint at the material's potential applications in medical fields, especially tissue restoration. Similarly, Zhang et al. took inspiration from nacre to address the longstanding issue of inadequate strength in conventional guided bone regeneration (GBR) membranes. By organizing graphene oxide nanosheets in a nacre-like fashion, they successfully amplified the mechanical strength of the GBR membrane [217]. Nevertheless, a constraint arose regarding the dimensions and structure of these planar membranes. The researchers were unable to manage substantial bone deficiencies effectively. One potential approach that has been suggested is the conversion of these membranes into cylindrical scaffolds with three-dimensional structures [218,219]. This particular strategy shows potential for future developments in bone restoration. The application of bi-directional freezing technology achieved the synthesis of a silicate-based bioceramic composite. This innovative approach resulted in the formation of an ordered lamellar microstructure, which bears resemblance to naturally occurring layered structures. The material exhibits increased strength and facilitates a regulated discharge of bioactive ions, hence expanding its possible uses [220]. Moving beyond nacre, the Bouligand structure, commonly found in entities like fish scales and crab shells, has also captured the attention of biomaterial developers. Li et al.'s development of a chitosan film derived from crab shells beautifully replicates this dense Bouligand structure, offering increased tensile strength and inherent antibacterial properties [221]. Furthermore, Han et al. innovatively transformed fish scales, using in situ mineralization of calcium silicate, into scaffolds for tendon repair, highlighting the diverse potential of these natural structures [222]. These investigations highlight the possibility of combining natural structural influences with contemporary technologies. They demonstrate the progress in synthesizing biomaterials inspired by biological systems and indicate the numerous avenues for further investigation. The incorporation of natural structures alongside bioactive molecules holds the potential to facilitate the development of a novel cohort of multifunctional biomaterials that are customized for distinct biological purposes. As we progress, a collaborative endeavor to extract and incorporate knowledge from these studies might establish the trajectory for future investigations in the field of multifunctional biomaterials.

Multi-directional ordered-structural arrangements can be seen in addition to layered structures in various natural materials, including tooth enamel, bone, muscle, and tendon. Natural materials have outstanding strength, toughness, and impact resistance thanks to these structures, which span the nanoscale to the macroscale. Bones, teeth, tendons, and ligaments are frequently replaced or repaired using biomedical materials that mimic these components. With a 96 wt% inorganic component, tooth enamel is a highly mineralized structure renowned for its extreme durability. It has been utilized as a bionic template to create materials for dental restoration. Notably, a sort of ceramic that resembles enamel was made by directing the growth of TiO_2 nanorods on a tin oxide substrate that had been doped [223]. Using layers of HA and ZrO_2, Zhao et al. developed a multiscale assembly process to produce bulk dental enamel [224]. Artificial tooth enamel (ATE) was developed, and it demonstrated exceptional levels of toughness, strength, and hardness, making it a suitable material for dental restoration. Type I collagen and hydroxyapatite nanocrystals combine to form mineralized collagen fibers in bone, another highly inorganic tissue. The basic network of the bone is made up of these fibers, which are grouped periodically. In the realm of biomaterials, materials that resemble bone are a popular issue. Through protein-induced intrafibrillar mineralization of cell-laden collagen,

Thrivikraman et al. created a scaffold that resembles bone, offering a model system for studying bone physiology and disease [225]. Furthermore, utilizing a multiscale cascade regulation technique, Zhao et al. designed a collagen/HA artificial lamellar bone (ALB) that is centimeter-sized and has mechanical characteristics close to real bone [226]. The successful development of a centimeter-sized artificial lamellar bone has been achieved for the first time by implementing a meticulously coordinated strategy known as "multiscale cascade regulation." This approach utilizes molecular self-assembly, electrospinning, and pressure-induced fusion methodologies implemented across multiple scales ranging from the molecular to the macroscopic (Figure 7). The artificial lamellar bone, predominantly composed of mineralized collagen fibrils organized hierarchically, successfully replicates the chemical composition, multiscale structural arrangement, and rotated plywood-like structure observed in natural lamellae. Notably, this achievement is accomplished without the incorporation of any synthetic polymer. Consequently, it presents a distinctive amalgamation of possessing a low weight and exhibiting a high degree of stiffness ($E_y \approx 15.2\,GPa$), strength ($\sigma_f \approx 118.4\,MPa$), and toughness ($K_{JC} \approx 9.3\,MPa\,m^{1/2}$). Implementing a multiscale cascade regulation strategy effectively addresses the limitations associated with individual techniques, enabling the development of advanced composite materials. This approach facilitates the precise control of hierarchical structural organizations at multiple scales, enhancing mechanical properties. The fabricated artificial lamellar bone (ALB) brilliantly mirrors the structural intricacies found in natural lamellar bone. It emulates the intricate hierarchical organization of mineralized collagen (MC) microfibrils, ranging from the nanoscale to the macroscale (Figure 8A). At the smallest scale, mineralized collagen microfibrils, each approximately 8 nm in diameter, form progressively through a biomimetic mineralization process. This process starts with nucleating amorphous calcium phosphate (ACP) precursors on the pre-constructed collagen microfibrils. The ACP transitions to crystalline apatite, simultaneously elevating the order of MC microfibrils. Transmission electron microscopy (TEM) images and fast Fourier transform (FFT) patterns give credence to the in situ co-assembly of nHAp and collagen microfibrils. Furthermore, the SAED patterns display diffraction rings characteristic of nHAp, mapped to the (002) and (211) crystallographic planes (Figure 8B). The electrospun MC fibrils, mimicking the spontaneous assembly seen in nature, have a consistent diameter of about a hundred nanometers, setting the stage for organization at more intricate levels (Figure 8C). Distributed uniformly within the electrospun MC fibrils, these MC microfibrils orientate along the fibril's length, directed by the strong electric field and shear force during electrospinning. The SAED pattern further emphasizes the polycrystalline nature of HAp, showcasing a (002) preferred orientation parallel to the collagen fibrils (Figure 8C insert). These electrospun MC fibrils accumulate in layers, either in aligned or random formations, resembling the pervasive structures in natural bones. Layer upon layer of aligned MC fibrils, adjusting by 30° each time, replicate a lamellar unit, eventually compacting into a bulk bone that retains the initial orientations and structures post pressure fusion (Figure 8E,F). Supplementing this, the focused ion beam (FIB) produced thin foils, which, when observed through scanning transmission electron microscopy (STEM), confirmed the maintained orientation of MC microfibrils and fibrils (Figure 8D). Lastly, the resulting ALB strikingly resembles the natural cortical bone, with its fracture surface showcasing a tightly packed lamellar-like microstructure (Figure 8F).

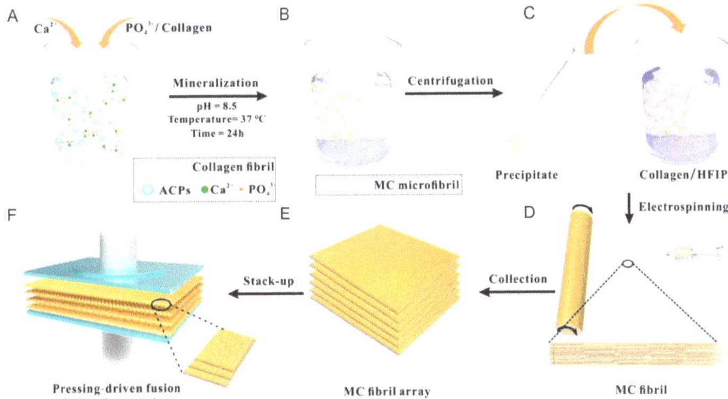

Figure 7. Schematic representation of the "multiscale cascade regulation" approach employed in the fabrication of artificial lamellar bone (ALB). (**A**) The process of calcium phosphate mineralization mediated by collagen. (**B**) The mineralized collagen (MC) microfibril precipitate obtained through centrifugation. (**C**) The electrospinning sols prepared by incorporating MC microfibril into a solution of collagen and HFIP. (**D**) Fabrication of MC fibrils through electrospinning. (**E**) An aligned array of MC (microcrystalline) fibrils obtained using a roller collector. (**F**) Bulk aluminum boride (ALB) forms through pressure-driven fusion at ambient temperature. Reprinted with permission from Ref. [226], Copyright 2023, Authors. CC-BY.

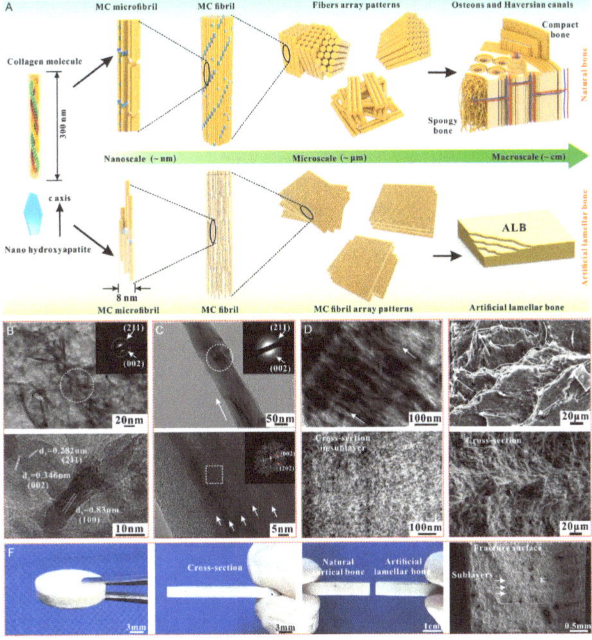

Figure 8. The illustration of the multiscale analysis and morphological features of the Asian longhorned beetle (ALB) and its comparison to natural bone. (**A**) Hierarchical organization from nanoscale to macroscale in natural bone and synthetic ALB is examined. (**B**) Transmission electron

microscopy (TEM) images reveal MC microfibrils, with the inset showing the electron diffraction pattern identified as HA. High-magnification TEM images at the bottom visualize the interplanar spacings of the HA crystalline lattice planes ((211), (002), and (100)). (**C**) TEM images display uniform MC fibril distribution. The accompanying selected area electron diffraction (SAED) pattern presents HA crystals' concentric rings, indicating a dominant alignment along the (002) plane. High-magnification TEM and fast Fourier transform (FFT) analysis evidence crystalline and amorphous calcium phosphate coexistence. (**D**) Scanning transmission electron microscopy (STEM) images of thin foils from a single sublayer, obtained through the focused ion beam (FIB) technique, show parallel (upper) and perpendicular (lower) fiber orientations. (**E**) Scanning electron microscopy (SEM) images reveal the ALB's fracture surfaces displaying a rotating layer pattern akin to plywood. (**F**) The synthetic ALB sample demonstrates morphological similarities to natural cortical bone, as seen in the various shapes and sizes, with the fracture surface exhibiting a lamellar structure, as shown on the right. Reprinted with permission from Ref. [226], Copyright 2023, Authors. CC-BY.

6.1.2. Inspired by Surface Structure

In conjunction with their inherent internal composition, the distinctive surface architectures of natural substances frequently play a substantial role in endowing them with exceptional characteristics, including enhanced wetting behavior, adhesive properties, and tactile perception capabilities. The external structures exhibit novel approaches in the development of functional biomedical materials. The superhydrophobic and adhesive surface structures could inspire synthetic nanomaterial routes.

Numerous natural creatures, including lotus leaves, red rose petals, mosquito complex eyes, butterfly wings, and water strider legs, exhibit superhydrophobicity, a unique wetting feature with a contact angle higher than 150 degrees. It has been demonstrated to possess antimicrobial qualities. The nanoarray surface structure of cicada wings, which kills bacteria in three minutes, was used by Ivanova et al. [227]. Using a reactive-ion etching process, the scientists also created black silicon with nano-protrusions modeled after dragonfly wings. This bio-inspired surface structure demonstrated strong antibacterial efficacy against Gram-positive and Gram-negative bacteria. To reduce the contact area and minimize biofouling, the lotus leaf's superhydrophobic feature, which results from hierarchical architectures of microsized protrusions and nanosized wax tubules, induces self-cleaning [228]. With over 98% bactericidal effectiveness against *Escherichia coli*, a team led by Jiang successfully developed a hydrophobic surface inspired by lotus leaves [229]. Li and colleagues used straightforward dip-coating techniques to design functional gauzes by coating regular gauze with a PDA hydrophobic layer and depositing Ag NPs to obtain a shape resembling a lotus leaf. In vivo experiments revealed that this has effective anti-adhesion properties, reducing wound adhesion and harming skin less when removed [230].

Through billions of years of evolution, several creatures, including the gecko, tree frog, and octopus, have acquired incredible sticky skills. Due to their hierarchical microstructures, which allow their toe pads to adhere to various surfaces, geckos have an exceptional capacity to attach to surfaces [231]. The suction cups with protuberances on an octopus' tentacles give them their adhesive properties, and tree frogs' toe pads are made of closely packed nanopillars and tightly arranged epithelial cells. To provide good adhesion to dry and wet surfaces, Huang et al. developed a wound patch inspired by octopi's adhesive structure [232]. The team has developed a biocompatible wound patch with a targeted design and selective stickiness by combining template replication and mask-guided lithography. Figure 9 depicts a biomimetic, skin-adhesive patch with customizable wound coverage. The distinctive GelMA-VEGF dressing, tailored to a specific wound shape, is created on the Ecoflex patch's surface by incorporating a UV mask. For adhesion to healthy skin, these patches use an Ecoflex film with microstructures that resemble the suction-cup effect, and for contact with the wound, they use a biocompatible gelatin methacryloyl (GelMA) hydrogel. Using a flexible ultraviolet mask, the GelMA hydrogel is customized to the geometry of each unique wound location, combining adhesion and non-adhesion properties into a single patch. Vascular endothelial growth factor is also included in the patch to hasten recovery. As the authors demonstrate, these characteristics enable the patches to adhere to different skin types and improve the healing of a rat skin wound model. The limitations of conventional patches are thus anticipated to be solved by this adaptable patch, making it an attractive option for wound healing and associated biomedical uses. A gravity-driven self-assembly technique allowed monodispersed steel microspheres to settle into the microcolumns. A wiper blade ensured each microsphere was properly placed and secured. The microspheres were designed to have a radius slightly bigger than the cavity's, enabling the creation of a negative mold with a convex bottom (Figure 10a). The yet-to-be-cured Ecoflex liquid precursor was poured over the mold's surface, filling the microcolumns that had convex bottoms. The micro-suction-cup array was finally produced after vacuum degassing, heat curing, and removal from the mold (Figure 10b–d). It is important to highlight that by altering the size of the negative mold, different-sized Ecoflex patches could be produced, showcasing the method's adaptability. When a vacuum is generated within the cavity, molecules at the boundary will restrict external liquids from entering, aiding in preserving the vacuum. This phenomenon becomes more pronounced on textured surfaces like pig skin. The research group tested four commonly used clinical liquids (water, ethanol, glycerin, and gelatin) for interface wetting. The outcomes revealed that the Ecoflex patch displayed exceptional adhesion on damp surfaces, holding up to a 0.2 kg weight in a tangential manner when stuck vertically to moist pig skin (Figure 10e). The standard adhesion and peel strength data further supported this observation (Figure 10f,g).

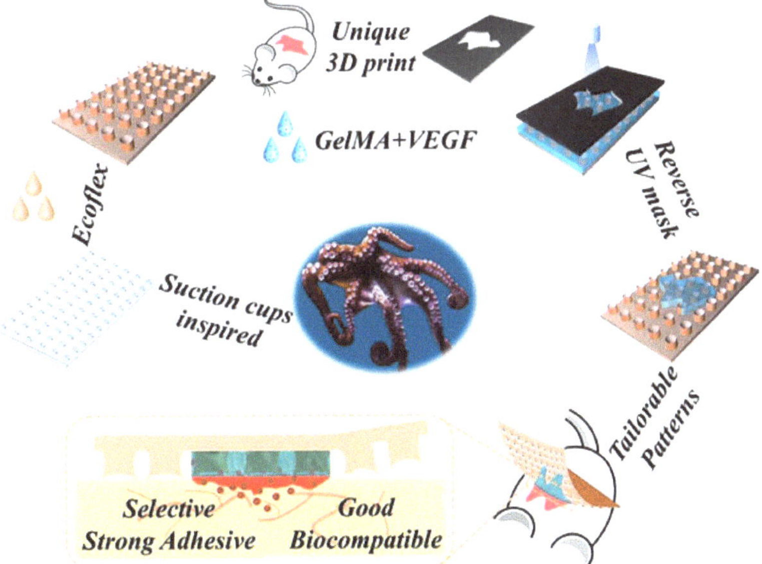

Figure 9. Schematic illustration of a biomimetic, skin-adhesive patch with customizable wound coverage. The distinctive GelMA-VEGF dressing, tailored to a specific wound shape, is developed on the surface of the Ecoflex patch by incorporating a UV mask. Reprinted with permission from Ref. [232], Copyright 2021, Authors.

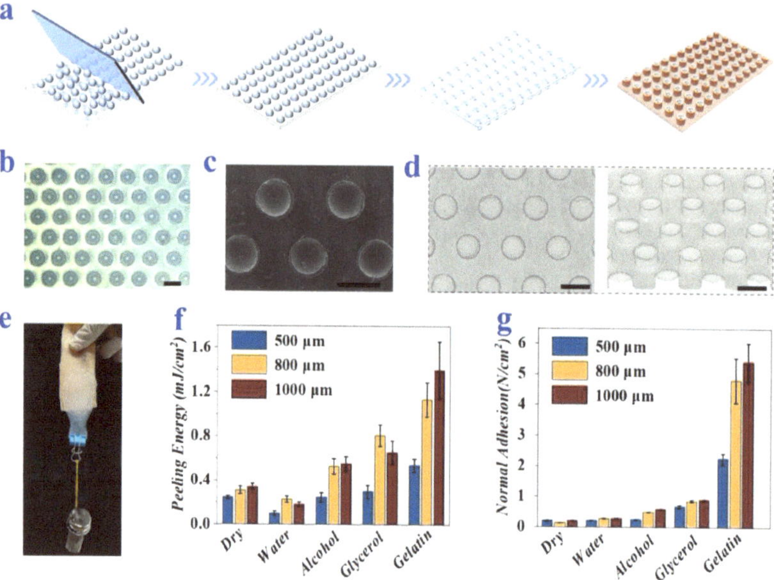

Figure 10. Creating and evaluating the adhesive capacity of the biomimetic patches. (**a**) An illustrative representation of the production procedure of the biomimetic patch. (**b,c**) Visual (**b**) and scanning electron microscopy (**c**) images showcasing the suction cups (measuring 800 μm in diameter). (**d**) Micro-CT scans of the suction cups (800 μm diameter) viewed from above and from the front. All accompanying scale bars represent 1000 μm. (**e**) An image capturing the adhesive patch adhered to pig skin, effectively bearing a weight of 0.2 kg. (**f,g**) Graphs indicating the resistance to peeling (**f**) and the vertical detachment and (**g**) forces of suction cups of three varied sizes when tested on both dry and damp surfaces. Reprinted with permission from Ref. [232], Copyright 2021, Authors.

GelMA hydrogels with unique shapes were produced using mask lithography on Ecoflex patches, as depicted in Figure 11a. In particular, a UV mask was fabricated initially using 3D printing technology to conform to the distinct contours of individual wounds. The purpose of this mask was to partially impede the transmission of UV light while allowing other portions to pass through. Following that, a solution of GelMA was administered onto the surface of the Ecoflex patch and subsequently concealed with the mask. Consequently, when subjected to ultraviolet (UV) light, solely the pregel solution located in the specific regions exposed to UV radiation would undergo polymerization, forming a solid hydrogel.

In contrast, the portions of the solution that were not exposed to UV light would remain in a liquid state and could subsequently be removed by wiping. The methodology resulted in developing customized hydrogel dressings to fit the diverse contours of UV masks, as depicted in Figure 11b. In addition, GelMA hydrogels of different concentrations were prepared, and their efficacy in generating distinct geometries was assessed. The results demonstrated that GelMA hydrogels with varying pregel concentrations (10, 15, and 20 wt%) exhibited high efficacy and reliability in achieving distinct geometries.

The hierarchical topologies of tree frog toe pads and the octopus suction-cup structure inspired Kim et al.'s development of a skin patch, which demonstrated increased peeling resistance and enhanced adhesion [233]. These biomimetic methods influence how sticky biomaterials are created for dry and moist environments. The preparation of materials with unique performances for various healthcare needs has been made easier by biomimicry. For example, high-strength NPs inspired by nacre are used for orthopedic implant materials, and hydrophobic surfaces inspired by lotus leaves are used for antibacterial purposes. Biomimetic structures have been created in artificial materials using a variety of production processes, including freeze casting, plasma etching, and self-assembly. To fully realize the potential of biomimetic nanomaterials in the biomedical field, additional research is required to address issues like scale-up fabrication, accurately reproducing the structures or properties of natural materials, and understanding the connection between bio-inspired structures and biological properties.

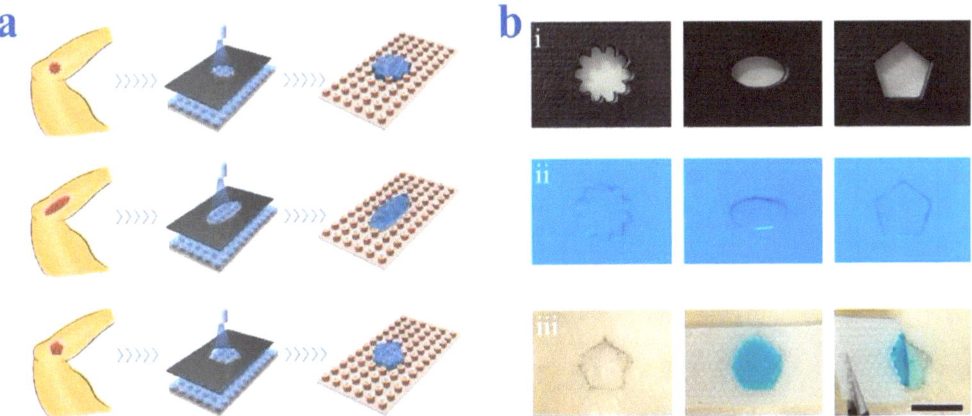

Figure 11. Creating GelMA dressings with customizable shapes. (**a**) Diagrams illustrating the process of creating GelMA dressings with adjustable shapes. (**b**) (**i**) Various shapes of UV masks produced through 3D printing. (**ii**) Creating GelMA hydrogels with shapes tailored to fit the UV masks. (**iii**) The GelMA hydrogel can accurately cover each pig skin wound area, and the contrasting properties of adhesion and anti-adhesion are incorporated within the same patch film. Blue dye was included in the GelMA solution for enhanced imaging. The scale bar indicates 1000 μm. Reprinted with permission from Ref. [232], Copyright 2021, Authors.

6.2. Inspired by Natural Biomolecule

Biomolecules serve as the essential building blocks of living organisms, fulfilling crucial functions in various physiological processes, including metabolism, transmission of genetic information, and immune responses, among other vital activities. Given the progress made in medical standards and the increasing demand for treatments, there is an urgent requirement to develop safe and efficient biomaterials. These biomaterials are crucial for various applications, such as targeted drug delivery and regenerative medicine. The utilization of biomimicry principles in the replication of biomolecules presents a viable method for the development and production of biomaterials with exceptional performance characteristics. This methodology enables the transformation of chemical compounds into artificial materials that imitate the composition and behavior of naturally occurring biomolecules, thereby meeting the rigorous requirements of biomedical uses. This section provides an overview of various biomedical nanomaterials that draw inspiration from biomolecules, encompassing their synthesis techniques and applications in biomedicine.

6.2.1. Protein-Inspired Nanomaterials

Proteins are essential to all living things and are involved in many biological processes, structure, and function preservation. Scientists have attempted to mimic the sticky surface structure of natural creatures to overcome the difficulty of obtaining strong adhesion in wet interfaces. For instance, mussels use the amino acid 3,4-dihydroxyphenylalanine (DOPA) to secrete their adhesive protein to cling to submerged surfaces. DOPA's catechol group promotes robust adherence in a wet environment. In response, Gan et al. created polydopamine (PDA), which has a molecular structure comparable to DOPA and is rich in catechol groups, to create antibacterial and sticky hydrogels for wound healing. This method increased the bacterial hydrogel's ability to adhere to surfaces, strengthening the sterilizing effect [234]. The hydrogels also showed a strong affinity with cells and tissues, which sped up the healing process. Barnacle cement proteins can also create potent hydrophobic and cation interactions that strengthen cohesion. Ni et al. created modified polyphosphazene/polyvinyl alcohol tissue adhesive hydrogels using this model. These demonstrated a potent adhesive force to tissue surfaces in aqueous conditions while avoiding the oxidation frequently associated with catechol-based hydrogels [235]. Additionally, the hydrogels favorably affected hemostasis and skin wound recovery, indicating a possible application in treating urgent wounds.

Bone and tooth proteins are essential to the biomineralization process. Notably, the tooth protein amelogenin directs the synthesis of HA, which is necessary for tooth enamel. Based on these findings, Chang et al. created a bioactive peptide to restore tooth enamel through biomimetic mineralization using low-complexity protein segments (LCPSs) found in the "fused in sarcoma" protein [236]. The LCPSs allowed amorphous calcium phosphate to convert more slowly into hydroxyapatite, producing a more integrated mineral structure and stronger enamel layers. This study offers a useful strategy for tooth enamel remineralization. Enzymes, primarily proteins, also catalyze biochemical events in the body and are crucial for detecting and managing diseases. The development of sophisticated artificial enzymes holds great promise for biomedicine. For instance, by incorporating MoO_3 into a metal-organic framework, Yang et al. created a bio-inspired spiky peroxidase-mimic enzyme to catch and kill germs. It has the potential for use in tissue engineering [237]. Additionally, Zhang et al. created a method to quickly gelate injectable hydrogels to fill tissue defects [238]. Glucose oxidase and ferrous glycine ($Fe[Gly]_2$), an enzyme complex inspired by biological systems, were used in this procedure. The ensuing reactions produced carbon free radicals, which facilitated quick polymerization. These hydrogels demonstrated adequate mechanical stability and an excellent capability for cartilage regeneration, making them viable fillers for cartilage repair.

6.2.2. Peptide-Inspired Nanomaterials

Short sequences of amino acids known as peptides are essential for controlling hormone release and metabolism. The design and manufacturing of peptide-inspired nanomaterials, which have several uses in the biomedical industry, have advanced significantly in recent years. Due to their specificities, peptide-inspired nanomaterials can be used as targeting agents. For instance, Yang et al. created the biomimetic peptide R4F with Apolipoprotein A-I as inspiration to target the SR-B1 receptors on M1 macrophages in rheumatoid arthritis cases [239]. This peptide was subsequently altered to coat an anti-inflammatory medication on a neutrophil membrane-wrapped F127 polymer, drastically reducing M1 macrophage polarization and increasing M2 macrophage polarization, suggesting successful rheumatoid arthritis treatment. Nanomaterials inspired by peptides are also widely used in cancer treatment. They enable the precise delivery of nanomedicines or photosensitizers to tumor locations. Under particular light wavelengths, these photosensitizers produce heat or reactive oxygen, which causes the death of tumor cells. This tactic works well and has fewer negative consequences.

In the antibacterial field, peptide-inspired nanomaterials are also promising. For instance, to combat methicillin-resistant Staphylococcus aureus (MRSA), Xie et al. prepared an antibacterial peptoid polymer based on host-defense peptides [240]. After prolonged use, this polymer showed outstanding anti-infection performance without producing drug resistance and reduced the growth of MRSA biofilms. In addition, tissue restoration has been performed with nanomaterials inspired by peptides. Self-assembled antimicrobial-antioxidative peptides enhancing infected wound healing in vivo have been the subject of certain studies. Another study described the usage of a bone-healing scaffold as a method for repairing bone defects in the presence of infection. This scaffold was made by altering an antibacterial peptide and an osteogenic growth peptide on a polyetheretherketone surface.

6.2.3. Lipid-Inspired Nanomaterials

Lipids, organic substances found in the body in structures like vesicles and cell membranes, are essential for energy storage and chemical communication among cells, tissues, and organs. As a result, the biomedical industry makes substantial use of lipid-inspired nanomaterials. Cell membranes, mostly made of lipids and proteins based on fatty acids, can isolate cells from their environment and regulate the flow of nutrients and waste products. They can transport NPs or medicines directly. For instance, Ying et al. improved tumor targeting effectiveness by encapsulating nanocamptothecin into macrophage membranes [241]. These macrophage-mimicking nanocamptothecin particles accumulated more at tumor sites than uncoated NPs in a mouse model of breast cancer. Additionally, nanocarriers coated with erythrocyte membranes were employed for in vivo biological imaging and medication delivery since they could resist macrophage clearance. Similarly, functional NPs were encapsulated in platelet membranes, reducing macrophage-like cell absorption and demonstrating preferential adherence to human blood vessel injury.

Vesicles, composed of lipid bilayers, store, digest, or transport materials. Recently, hybrid bio-inspired nanovesicles were developed for lung tissue transport and immunomodulatory activity by fusing lung-targeting liposomes and macrophage-derived nanovesicles. Extracellular vesicle-based collaborative anti-infective therapy was the subject of a straightforward biomimetic technique by Qiao et al. [242]. Pd-Pt nanosheets and ginger-derived extracellular vesicles gave EVs-Pd-Pt NPs good biocompatibility and sustained blood circulation, producing a potent antibacterial effect. Extracellular vesicles (EVs) have also been employed to encourage the differentiation of stem cells into several adult somatic cell types. In a recent method, EVs stimulate fibroblasts to transdifferentiate into cells that resemble induced cardiomyocytes. Direct conversion of embryonic fibroblasts toward mature induced cardiomyocytes may be accelerated using EVs generated from embryonic stem cells during cardiomyocyte differentiation. The electrical and typical cardiac calcium transient properties of these generated cardiomyocytes were similar to those of cardiomyocytes.

6.2.4. Other Biomolecule-Inspired Nanomaterials

Biomolecules outside proteins and lipids, such as viruses and saccharides, influence biomedical nanomaterial design. Complexes of saccharides and proteins are essential for biological processes like immunological control and drug transport. Taking this as a foundation, Duan et al. created a branched glycosyl polymer-pyropheophorbide-a conjugate that may be used as a drug delivery system [243]. Anticoagulants like heparin, a highly sulfated glycosaminoglycan produced by mast cells, are frequently utilized. Sodium alginate was used as a biological macromolecule model by Ma et al. to build a heparin-like anticoagulant biomolecule to reproduce its function. This biomolecule demonstrated good anticoagulant efficacy and biocompatibility in vitro and in vivo [244]. To develop a new design methodology for anticoagulant surfaces, Wang et al. explored the endothelialization of a novel heparin-like polymer [245]. Viruses, contagious organisms that multiply inside live cells, can be bionic models for artificial materials. Chen et al. developed a reversible and activatable near-infrared II nanoprobe by imitating a virus. This might help with future viral encephalitis interventions by tracking the progression of viral infection in real time [246]. In a different investigation, Li et al. created a nanodrug inspired by a virus that might avoid lysosomal hydrolysis while being delivered, potentially providing a unique method for treating tumors [247].

Nanomaterials with biomolecular inspiration have demonstrated promise in tumor therapy, medication transport, and tissue engineering. They increase the effectiveness of treatment, lessen adverse effects, and are frequently artificially produced, ensuring a plentiful supply of raw ingredients with strict quality control. Making multifunctional BINMs with intricate architectures resembling natural proteins is still difficult. These NPs also need to have their long-term safety, immunogenicity, and in vivo stability investigated. Emerging technologies like artificial intelligence may help the development of these nanomaterials in the future by providing fresh perspectives for individualized and targeted design [248].

6.3. Bioprocess-Inspired Nanomaterials

The bioprocess can efficiently and accurately construct hierarchical structures or synthesize essential substances in a natural environment while operating under mild conditions. The remarkable bioprocesses offer environmentally sustainable methods for the production of synthetic materials. The acquisition of knowledge regarding natural bioprocesses has facilitated the development and utilization of advanced synthesis technologies in creating high-performance biomaterials.

6.3.1. Inspired by Biomineralization Process

The production of organic–inorganic composites in structures like bone and teeth, known as biomineralization, inspires the creation of biomedical materials [1,249–251]. Researchers have used this procedure to produce materials for cancer therapy, medication creation, and hard tissue repair. By simulating the mineralization process, Li et al. created a dental replacement with enamel's mechanical qualities [252]. Tang et al. suggested a method for restoring enamel using a solution of calcium phosphate ion oligomers [253]. Zhou et al. produced extremely rigid DNA-HA bulk composites for dental applications using an engineering mineralization method [254]. Enhancing bone healing materials has also relied heavily on the biomineralization process. Bioactive nanominerals can be added to bone-repairing scaffolds using methods inspired by biomineralization to promote osteogenic activity. To better understand the process of bone formation, Ping et al. produced strontium carbonate crystals inside collagen fibers [255]. New cancer treatment approaches are also inspired by biomineralization. Zhao et al. suggested exploiting the buildup of calcium salts, which causes cell calcification and cancer cell death, as a drug-free tumor treatment method [256]. A macromolecular medication that causes cancer cells to calcify extracellularly was created through other studies. Last, a biomineralization-inspired technique was applied to safeguard tumor-targeted delivery carriers, highlighting its applicability in numerous biomedical research fields.

6.3.2. Inspired by Photosynthesis

Advanced artificial photosynthetic systems for biomedical materials have been developed due to photosynthesis, a crucial bioprocess found in plants, algae, and cyanobacteria. Microalgae, particularly cyanobacteria, have recently been used as effective oxygen sources to combat tumor hypoxia. For instance, Huo et al. produced oxygen using a hybrid of cyanobacteria and photosensitizers, which, when exposed to laser irradiation, transformed into singlet oxygen and killed cancer cells [257]. In addition to supplying oxygen for wound healing, microalgae also promote angiogenesis and collagen synthesis [258]. Artificial photosynthesis systems, like the one created by Chen et al. using spinach nano-thylakoid units (NTUs), show promise for degenerative disorders like osteoarthritis. The development of CM-NTUs, which combine NTUs with the chondrocyte membrane (CM), improved intracellular ATP and NADPH levels, enhancing cell metabolism and delaying the onset of osteoarthritis [259]. Nanomaterials with inspiration from biomineralization processes are widely used in biomedical disciplines and can safeguard the environment and promote the sustainable development of biomaterials. The inhomogeneous conveyance of mineralized media, slow reaction speeds, and difficulty in creating substantial, structurally sound structures continue to be problems. Like photosynthesis-inspired solutions that offer novel disease treatments, they are still in the early phases. They must fully address issues with light penetration into tissue, immunological responses, and biological safety.

6.4. Challenges for Bio-Inspired Synthesis

To improve performance in biomedical applications, bio-inspired design efficiently transfers special features and functionalities from natural materials to synthetic materials. Natural materials' hierarchical systems and individual components can be studied to provide solutions to various biomedical problems. The idea of manufacturing inspired by biological processes has evolved, enabling the creation of biomedical materials under benign circumstances. Designing biomaterials for various medical applications is made possible by certain biomimetic techniques. For instance, biomimetic nanovesicles are used for medicine administration, whereas bio-inspired layered nanostructures improve the mechanical qualities of bone implants. Biomedical nanomaterials inspired by biomineralization and photosynthesis also show great potential for treating cancer and tissue engineering.

Although notable advancements have been achieved in bio-inspired biomedical nanomaterials, unresolved obstacles still necessitate attention to facilitate their effective implementation. These materials frequently require optimal performance in fluidic environments characterized by dynamic flow, such as bodily fluids, where their properties may undergo compromise relative to their performance in dry or stabilized conditions. The occurrence of mechanical failure in moist environments is a prevalent concern. One of the primary obstacles encountered in the field is expanding the production of BINMs. This is due to the increasing complexity associated with achieving precise arrangement of structures and components at the nanoscale, which consequently restricts the size and yield of these materials. There is currently a deficiency in cost-effective and efficient techniques for producing goods on a large scale. Furthermore, it is imperative to conduct extensive research and analysis on the biological characteristics of BINMs. This is crucial because numerous materials with potential applications have not undergone thorough in vitro and in vivo assessments, which poses a significant obstacle to their successful implementation in clinical settings.

To enhance the efficacy of the development process for multifunctional biomedical nanomaterials, it is imperative to incorporate advanced methodologies such as integrating multiple bio-inspired strategies and utilizing computational simulations. Moreover, it is important to note that the study of biomedical materials encompasses various interdisciplinary domains, including material science, biology, and medicine. Consequently, the comprehensive assessment of BINMs necessitates the collaborative efforts of researchers specializing in materials, biologists, and medical professionals. Collaboration is pivotal in facilitating the effective translation of materials from the laboratory setting to clinical applications.

7. Design Principles of Bio-inspired Nanomaterials' Interfaces

Recent interest in research on bio-inspired interfaces has increased as a result of its expanding significance in the field of biomedical science. Enhancing medicine effectiveness while reducing adverse effects is one of the main goals in this sector. This includes increasing the effectiveness of distribution systems and giving targeting abilities more specificity. Another important factor is biomedical imaging, which provides essential information about molecular distributions in vitro and in vivo. The use of NPs in biomedical applications has been extensively explored due to their adaptability and accessibility [260]. When designing a delivery method, it is critical to consider the NPs' trajectory through the body and any potential roadblocks on the way to the illness site. NP-based delivery techniques typically include injection or oral consumption to penetrate the circulatory system.

Consequently, ensuring that the NPs reach the intended organs without being eliminated by the body is a crucial component of effective delivery. The reticuloendothelial system (RES), which comprises the liver and spleen, eliminates NPs from the circulatory system. When NPs are administered intravenously, they are frequently labeled as alien substances and go through hepatic Kupffer cell sequestration [261]. An additional issue emerges when NPs interact with various biomolecules, including plasma proteins, upon entering a complex biological milieu like blood, interstitial fluid, or the extracellular matrix, creating a protein corona. This corona development activates the mononuclear phagocyte system (MPS) for quick ejection as proteins build up on the NP surfaces [262]. To overcome these obstacles, NPs must have the proper mechanisms that will allow them to safely pass through the body and arrive at their target without being prematurely removed. One common technique to prevent RES clearance and early ejection is to cover NPs with the membranes of circulating blood cells [263]. Providing enough time in the circulatory system is crucial for the designed NPs to reach their target site. The capacity of NPs for selective targeting, or the capacity to interact just with the area of interest while disregarding other sites (cells, tissues, etc.), is another key aspect of NPs. NPs must also be securely cleared from the body when the task is completed without having any negative effects. They are essential for creating NPs since they ultimately dictate how the particles will behave when ingested by the body. Given its complexity as a biological environment, the human body can vary significantly from person to person. Designing NPs that can overcome these obstacles while keeping their intended functionality becomes even more difficult and complex. Figure 12 illustrates three distinct categories that can be drawn from the design guidelines for NP interfaces inspired by biological systems. First, membrane-coated NPs are produced using components like mammalian cells, cancer cells, bacteria, or viruses. The second category includes ligands that modify surfaces. This includes altering the NPs' surface using polyethylene glycol (PEG), zwitterions, positively or negatively charged ligands, or viral capsids. The third category of design concepts for these bio-inspired interfaces is the alteration of the geometric aspects of the NPs, such as their size or shape.

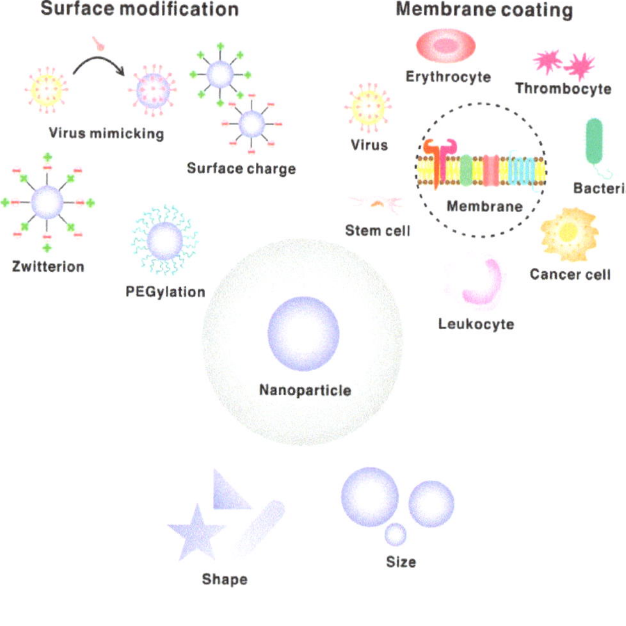

Figure 12. Three distinct categories can be drawn from the design guidelines for NP interfaces inspired by biological systems. First, membrane-coated NPs are produced using components like mammalian cells, cancer cells, bacteria, or viruses. The second category includes ligands that modify surfaces. The third category of design concepts for these bio-inspired interfaces is the alteration of the geometric aspects of the NPs, such as their size or shape. Reprinted with permission from Ref. [260]. Copyright 2022, Authors (CC-BY).

Researchers are interested in bio-inspired NPs because of the wide range of biomedical science applications they could have, including targeted drug administration, in vivo therapies, bioimaging, and cancer treatment. However, the human body's capacity for recognizing and getting rid of foreign compounds is one of the main obstacles to overcome. For example, the mononuclear phagocyte system (MPS) is quickly expelled from the body in response to opsonin interactions. The targeting abilities of ligand-functionalized NPs might also be lost due to excessive protein corona formation, which can also significantly affect the surface chemistry of NPs, negating or changing their desirable features. Without further surface alterations, NPs rely solely on the improved permeability retention effect, which is typically ineffective [264]. Numerous methods have been developed to address these problems by changing the way NPs function or their geometrical characteristics. These techniques are divided into various groups in the following sections based on the type of alterations. Applications of membrane-coated and surface-functionalized NPs span a broad spectrum of fields (Figure 13). One of these is bioimaging, which uses NPs to improve visual data at the cellular or molecular level. NPs can be used in targeted delivery to deliver medications to illness locations directly, boosting therapy effectiveness and lowering side effects. In multimodal theranostics, a discipline that integrates diagnostics and therapies, NPs are also employed to diagnose and treat diseases simultaneously. Finally, these NPs are used as drug carriers, acting as a means of controlled and precise drug delivery.

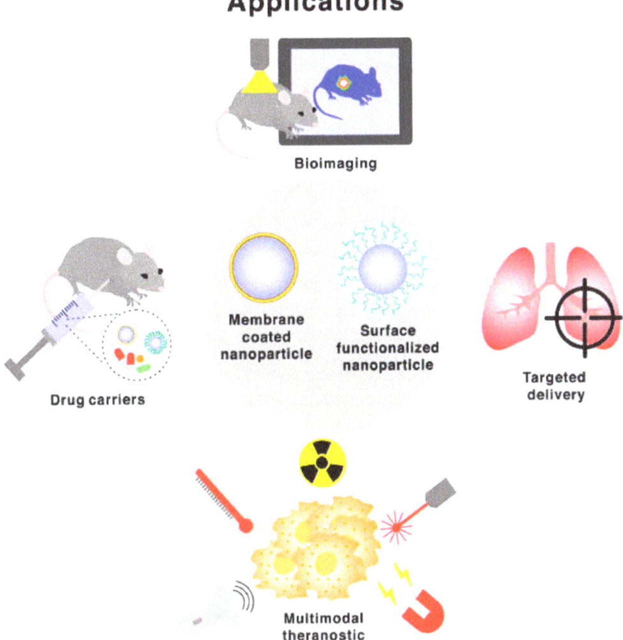

Figure 13. Membrane-coated and surface-functionalized NP applications include drug delivery, multimodal theranostic, and bioimaging. Reprinted with permission from Ref. [260]. Copyright 2022, Authors (CC-BY).

7.1. Biomimetic Functionalization of BINPs

Biomimetic substances can replicate or imitate the attributes, composition, chemical properties, and functionalities of their biological counterparts [265]. Both red and white blood cells can traverse the circulatory system without premature elimination by the mononuclear phagocyte system (MPS). Platelets can resist phagocytic activity and are equipped with surface receptors that enable them to selectively target specific sites, thereby aiding in tissue repair following an injury. Cancer and stem cells have also been investigated in this field of study (Figure 14). The emergence of membrane-coated nanoparticles (MCNPs) has resulted from the need to prolong circulation time upon entry into the bloodstream, enable precise drug delivery within the body, or function as contrast agents for bioimaging applications [266]. Moreover, scholarly investigations have been undertaken on materials derived from pathogens, including bacteria [267] and viruses [268], which exhibit distinctive capacities in the delivery of payloads. Consequently, it is crucial to comprehend the unique characteristics of different biological entities and exploit their therapeutic capabilities. This section examines the utilization of the intrinsic characteristics of biological materials by scientists to develop NPs for biomedical purposes while also addressing the potential obstacles they may face during this process.

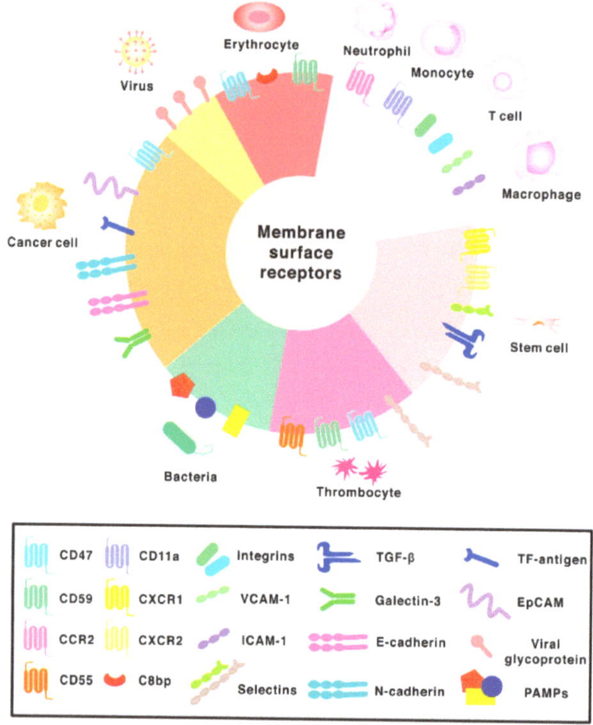

Figure 14. Various membranes host a diverse array of surface receptors. The abbreviations are listed below: CCR2 is an acronym for C-C chemokine receptor 2, CXCR1 represents C-X-C chemokine receptor 1, and CXCR2 denotes C-X-C chemokine receptor 2. C8bp is an abbreviation for C8 binding protein, VCAM-1 signifies vascular cell adhesion molecule-1, and ICAM-1 refers to intercellular adhesion molecule-1. TGF-β corresponds to transforming growth factor beta, TF-antigen refers to Thomsen–Friedenreich antigen, EpCAM stands for epithelial cell adhesion molecule, and PAMPs is the abbreviation for pathogen-associated molecular patterns. Reprinted with permission from Ref. [260]. Copyright 2022, Authors (CC-BY).

7.1.1. Erythrocyte Membrane

Red blood cells (RBCs) are the predominant cellular constituents within the human body. They possess notable attributes such as biocompatibility, prolonged circulation duration, and biodegradability, rendering them highly suitable for utilization as carriers of NPs. The CD47 proteins, which serve as markers of self, are found on the surfaces of red blood cells (RBCs) and play a role in preventing their phagocytosis by immune cells. As a result, the presence of these proteins leads to an extended circulation time for RBCs [269]. This characteristic has been employed to enhance drug delivery systems, wherein NPs are enveloped with red blood cell membranes (RBCMs), forming RBCM-NPs. The successful generation of RBCM-NPs has been achieved by encapsulating poly (lactic-co-glycolic acid) (PLGA) NPs within RBCMs. Applying this coating facilitates a 64% reduction in macrophage uptake of the NPs and prolongs their elimination half-life [270]. RBCM-NPs have demonstrated efficacy in selectively targeting particular conditions, such as atherosclerotic plaques, and in cancer therapy.

Nevertheless, the effectiveness of RBCMs is dependent on the enhanced permeability and retention (EPR) effect, given its inherent absence of natural tumor-targeting capabilities. To augment the targeting efficacy, ligands are introduced into red blood cell membranes (RBCMs) through chemical synthesis or lipid insertion techniques [271]. These supplementary components have substantially improved their capacity to target cancer cells selectively. Furthermore, RBCM-NPs have demonstrated bioimaging practicality, particularly in in vivo tumor imaging. Ongoing investigations and advancements in dual-functionalized RBCM-NPs show encouraging outcomes in enhancing bioimaging capabilities.

7.1.2. Leukocyte Membrane

Leukocytes, also known as white blood cells (WBCs), are important immune cells that protect the body from diseases and infections. WBC membrane-coated NPs able to self-recognize, penetrate biological barriers, and bind to receptors at disease locations have been developed due to their varied capabilities. For instance, WBC membrane-coated nanoporous silica particles that maintain WBC characteristics, such as avoiding the immune system and crossing the endothelium barrier, have been produced. Various WBCs, including neutrophils, macrophages, monocytes, and T cells, have been employed as NP carriers. By simulating the interaction of the cell with inflammatory tissues, neutrophil membrane-coated NPs, for example, can target and lower bacterial load at infection sites [272]. Due to their capacity to target inflammation and tumor endothelium specifically, macrophages have been incorporated into macrophage membrane-coated nanoparticles (MMCNPs) to slow the progression of atherosclerosis [273]. They also show promise in treating cancer and bioimaging, although they have drawbacks, such as the ability to target particular cancers [236,237] exclusively. The power of monocytes to infiltrate has been taken advantage of in a lipid NP-based drug-delivery platform, which enables monocytes to transport and deliver lipid NPs to sick areas [274]. Monocytes are normally recruited when physiological changes occur in the body. T cells have been employed to coat NPs' membranes, increasing their circulation duration and enhancing cancer targeting. T cells have a higher concentration of targeting proteins than other WBCs. For instance, coated NPs with azide-modified T-cell membranes exhibit strong fluorescence intensity and an improved photothermal response. However, the lack of tumor-specific indicators displayed by solid tumor cells limits their ability to treat solid tumors [275].

7.1.3. Thrombocyte Membrane

Blood components called platelets are in charge of starting blood clots when an artery is damaged. Their membranes' distinctive qualities make them advantageous for coating NPs, as they can support active and passive drug targeting. Platelet membranes include CD47 "marker-of-self" proteins, similar to those found in red blood cell membranes, which resist immune clearance, extending the circulation of platelet membrane-coated nanoparticles (PMCNPs) for passive drug targeting [276,277]. Additionally, platelet membranes include particular surface receptors such as glycoprotein Ib that can connect to exposed collagen in injured vascular tissues to stimulate tissue repair or directly bind to pathogenic bacteria to enable active drug targeting [278,279]. They have become quite popular in the creation of nanotherapeutics as a result. Platelet-like proteoliposomes that strongly interact with circulating monocytes have been created to enhance post-infarction therapy. Because PM-NPs naturally include platelet surface proteins that can interact with anti-platelet antibodies, they can be used as an antibody decoy in treating immune thrombocytopenia [280]. Additionally, they are utilized in tumor imaging, drug delivery, and the detection of cancer cells. In a mouse model of human lung cancer, for instance, a study revealed that docetaxel-loaded PLGA NPs coated with platelet membranes preserved the maximum drug concentration in tumors, successfully reducing tumor growth. Additionally, PM-NPs have been used in developing phototheranostic nanoprobes to target various tumors and deliver additional anti-cancer medications actively. The immune system can remove PM-NPs from patients with autoimmune disorders when platelet autoantibodies combine with them to form immune complexes [281].

7.1.4. Cancer Cell Membrane

Because they have surface receptors that permit adhesive contact, resulting in metastatic deposits, cancer cell adhesion molecules (CCAMs) play a significant role in cancer metastasis. Additionally, CD47 surface proteins are expressed by cancer cells, allowing them to avoid the immune system. Using these characteristics, researchers have coated NPs with cancer cell membranes (CCMs). CCM-coated NPs (CCM-NPs) can avoid immune detection and target malignant locations or tumors that are similar to them. They can be customized to satisfy the particular requirements of cancer therapy [282]. According to studies, compared to uncoated NPs and red blood cell membrane-coated NPs (RBCM-NPs), CCM-NPs have much higher cellular absorption and a strong affinity for source cancer cells [283]. CCM-NPs can achieve self-recognition, internalization by the source cancer cell lines, and highly selective targeting to the homologous tumor in vivo by adjusting the source of the cell membrane coating [284]. In tumor imaging, CCM-NPs have been widely employed and frequently functionalized with extra features [285,286]. For instance, recent research produced unique iridium compounds functionalized with black-titanium NPs coated with CCMs. These NPs have the potential to accumulate in malignant cells, accumulate in mitochondria, develop effective photothermal capability when exposed to NIR-II radiation, and form reactive oxygen species when exposed to ultrasonic radiation. This enables precise imaging of the tumor site and results in the elimination of tumor cells in mice models [287].

7.1.5. Stem Cell Membrane

Mesenchymal stem cells (MSCs), extensively researched in biomedical research, are renowned for their simplicity in separation and capacity to target tumors. MSCs have been effectively used in medication delivery systems based on NPs. According to research, the loading of NPs into MSCs preserved cell viability and differentiation. A human glioma model also shows great selectivity for MSC membrane-coated NPs (MSCM-NPs) [288]. Numerous cancer-related research studies have used MSCM-NPs. Due to various chemical recognition moieties on the MSC membrane, an MSC membrane-coated gelatin nanogel, for instance, displayed excellent stability and tumor selectivity both in vitro and in vivo [289]. In an orthotopic breast cancer model, PLGA NPs covered with an MSC membrane demonstrated strong anti-tumor effectiveness [290]. Taking advantage of MSCs' capacity for tumor homing, MSCM-NPs have been widely applied in bioimaging. A biocompatible MSCM-NP with multimodal imaging abilities for near-infrared fluorescence, magnetic resonance, and computed tomography has recently been developed [291].

7.1.6. Bacterial Membrane

Despite being frequently harmful, some aspects of bacteria can be utilized in therapeutic settings. Immunogenic antigens and adjuvants in bacterial membranes activate innate immunity and support adaptive immunological responses. These can be carefully coated onto NPs to simulate how the immune system reacts to antigens naturally when germs are present. Bacterial outer membrane vesicles (OMVs) are typically coated on NPs to create bacterial membrane-coated NPs (BM-NPs) [292]. BM-NPs are a relatively recent development in cell membrane-coated NP research, yet they have a number of special benefits. They showed bacterial-specific targeting; for example, S. aureus-infected macrophages and organs were selectively targeted by PLGA NPs coated with S. aureus OMVs, a feature not seen in other membrane-coated NPs. Given the rapid rise of bacterial drug resistance, BM-NPs may be used to create antibacterial vaccines as an alternative to antibiotics [293]. For instance, BM-NPs were created utilizing OMVs from CRKP, which improved the survival probability of immunized mouse models when exposed to a lethal dose of CRKP. Because some bacteria naturally target tumors, BM-NPs can be used for cancer treatment or tumor imaging. BM-NPs have been used to create a cancer vaccine that, when paired with radiotherapy, demonstrated increased tumor growth inhibition and made anti-cancer immunological memory.

However, research on using bacterial membranes as NP coating agents is ongoing. The impact of OMV size on host cell entrance and the cytotoxicity of BM-NPs for realistic biomedical applications are among the difficulties. To minimize the inflammatory response of BM-NPs, lipopolysaccharide neutralizing peptides have been suggested as a partial answer; nonetheless, these issues must be uniformly resolved before BM-NP-based vaccines and treatments can go further.

7.1.7. Virus-Derived Strategies

Viruses are frequently used in biomedical applications because they can protect and transfer nucleic acids into host cells while eluding the immune system. Adenoviruses and retroviruses have been utilized as viral gene vectors to introduce particular genes into host cells. Due to their pathogenicity, probable toxicity, mutagenesis, and constraints on size and cargo capacity, these vectors do have drawbacks. Alternatives include virosomes and virus-like particles (VLPs). While virosomes are liposome-like particles with integrated surface glycoproteins but lack capsid proteins, VLPs mimic the capsid architecture or envelope proteins of genuine viruses. Both can enclose a variety of payloads, yet neither contains viral genetic material. They hold potential for drug delivery, imaging, immunotherapy, and theranostics because they retain important virus traits, including cellular entrance, immune evasion, and precise targeting [294]. Viral proteins can be applied to NPs to give them additional functionality. For instance, magnetic NPs were enclosed in a hepatitis B core VLP, which improved cellular uptake and demonstrated potential for magnetic resonance imaging [295]. As an alternative, metallic NPs were joined to a specific adenoviral platform, producing an NP-labelled vector that could infect cells and target tumors. Surface alterations of NPs to resemble virus surfaces are a common component of other virus-derived techniques [296].

7.2. Surface Modification to Functionalize NPs

A crucial step in customizing NPs for particular biomedical applications is surface modification. The characteristics of NPs can be significantly altered using this rather simple procedure. For instance, coating NP surfaces with PEG might prevent the mononuclear phagocyte system (MPS) from clearing them away. It is possible to vary the surface electrical charges of NP surfaces by precisely adjusting the attachment of functional groups, which can affect how quickly cells absorb substances. Some strategies even try to mimic viruses to give NPs virus-like characteristics. This section lists the common ligands used in surface modifications and their benefits and drawbacks.

7.2.1. PEGylation

PEGylation, which involves bonding polyethylene glycol (PEG) molecules to NPs, was initially published in 1977 and is a widely used method to extend the period that NPs circulate in the bloodstream [297]. PEGylation creates an "anti-fouling" surface for the NPs by forming a hydrophilic brush coating. This layer inhibits aggregation, opsonization, and phagocytosis, preventing the NPs from being quickly eliminated from the body by the mononuclear phagocyte system (MPS). PEGylation has been crucial in developing therapeutic NP uses, from cellular pathways to sonodynamic therapy and tumor targeting for cancer. Additionally, PEGylated NPs have improved stability and biocompatibility in intricate biological contexts. mRNA-based COVID-19 vaccines have benefited greatly from PEG-decorated NPs [298].

PEGylation does have certain downsides, though. Unwanted immunological events, such as hypersensitivity reactions and the production of anti-PEG antibodies, have been documented, especially following the administration of numerous doses of PEGylated NPs [299]. Uncertainty surrounds the precise mechanisms causing these reactions. Despite these drawbacks, PEGylation's advantages—such as decreased immunogenicity, antigenicity, and toxicity—ensure its relevance in nanomedicine research.

7.2.2. Zwitterions

Researchers are looking toward substitutes like biodegradable poly(glutamic acid), non-biodegradable poly(glycerol), and zwitterionic compounds due to worries regarding PEG's drawbacks. These zwitterionic materials, which permit charge neutrality and super hydrophilicity by having equal amounts of cationic and anionic moieties, are of great interest. Similar to PEG, they can prolong the half-life of NPs in blood circulation without provoking an immunological reaction. Due to the robust hydration layer that zwitterionic materials create through electrostatic contact, they also resist nonspecific protein adsorption. This characteristic may, however, disrupt target cell communication and lessen the effectiveness of cellular absorption. This difficulty could be overcome by adding a variety of unique functional groups, enabling improved NP stability and biocompatibility, and offering a configurable interface for varied purposes [300]. Zwitterions have been used in some cutting-edge applications, including the development of ratiometric pH sensors based on quantum dots, the creation of reduction-responsive materials to boost intercellular drug-release rates in tumor cells, and the development of pH-sensitive materials to envelop NPs for effective tumor targeting. Additionally, zwitterionic materials can be functionalized to respond to additional stimuli, including temperature or light. For instance, multifunctional NPs coated with zwitterion have demonstrated great capabilities for imaging-guided cancer therapy [301].

7.2.3. Surface Electrical Charge

The absorption and subsequent behavior of NPs are greatly influenced by their surface charge [302]. Positively charged NPs are naturally drawn to cell membranes because they normally have a negative charge, but negatively charged NPs may be less readily taken up by cells. NPs can draw different proteins to them in a complex biological context, creating a "protein corona" on their surface. The qualities and functioning of the NPs could be altered by this process, possibly leading to them losing their intended role or acquiring undesirable traits. The protein corona may alter the surface charges or physical characteristics of NPs, which may enhance the likelihood of non-specific internalization. Surface functional group adsorption or environmental exposure are two ways to change the surface charge of NPs. As a result, when designing NPs, it should be decided whether to use or prevent protein adsorption.

7.2.4. Virus Mimicking

It has been demonstrated that surface alterations of NPs to imitate virus features can improve therapeutic effectiveness and efficiency of distribution. The surface topology of the virus highly influences viral interactions with host cells in question, such as enveloped viruses. Researchers have encouraged interactions between NPs and target cells, thereby improving delivery efficiency by mimicking these topological features, such as adding smaller silica NPs to larger ones to increase surface roughness [303]. Additionally, scientists are imitating viral design, particularly the viral capsid, which is essential for cellular entry and targeting. Viral capsids can be made with synthetic building pieces that give specialized targeting functions, unlike natural virus vectors like virus-like particles (VLPs) or virosomes. One illustration is a multifunctional viral mimic created from self-assembled amphiphilic dendritic lipopeptides that showed the ability to infect solid tumors and tumor cells like a virus and suppress tumor growth in both in vitro and in vivo experiments [304]. The ability of NPs to deconstruct and distribute their cargos—a benefit for intracellular interactions—can be achieved by integrating stimuli-responsive receptors to connect with specified places through viral capsid mimicry [305]. These virus-mimicking methods eliminate the necessity for viral components such as VLPs and virosomes that might cause infections. This field of study has the potential to aid in creating multifunctional artificial "viruses" that could get beyond the drawbacks of the NP drug delivery systems that are now in use [306].

7.2.5. Surface Modification by Bioactive Molecules

Specific biological functionalities can be bestowed on NP surfaces by adding bioactive chemicals [307]. With these alterations, bioactive compounds can selectively target particular cells or tissues, enabling targeted medication delivery. This guarantees effective delivery of medicinal substances to the intended site. NPs can also be made less hazardous through surface alterations, making them safer for biomedical applications. A further benefit is the improved biocompatibility, which enables NPs to operate in a biological context without inducing an unfavorable immune response. Surface modification can be achieved using a variety of methods. Physical adsorption is one technique in which bioactive chemicals are non-covalently attached to the surface of the NPs [308]. Covalent bonding is an alternative strategy that uses chemical reactions to permanently link the bioactive chemicals to the surface of the NPs. The encapsulation method traps the bioactive chemical inside the NP framework [309]. For these goals, several bioactive compounds are frequently used. Targeting certain cell receptors or enhancing the solubility and stability of NPs are possible with peptides [310] and proteins [311]. Chitosan [312] and hyaluronic acid [313] are natural polysaccharides that can facilitate targeted drug administration and improve biocompatibility. Antibodies can target particular cells or pathogens because of their high specificity, whilst some tiny compounds can increase the solubility of NPs or be used for targeting.

The utilization of bioactive compounds offers a number of benefits. Targeting ligands on the surface of NPs, for instance, can increase receptor-mediated endocytosis and increase the effectiveness of cellular uptake. Because some bioactive compounds can react to certain stimuli, such as pH changes [314], temperature changes [315], or the presence of specific enzymes [316], controlled medication release is possible. Additionally, NPs can reduce off-target and negative effects frequently associated with many therapeutic medicines by concentrating on particular tissues or cells. There are, however, issues to take into account [317]. Under physiological circumstances, the NP and the bioactive chemical bond must stay stable. It can be challenging to scale up from laboratory synthesis to larger-scale production without losing the effectiveness of bioactive molecule attachment. Furthermore, therapeutic NP change may be subject to regulatory review and rigorous testing. Using bioactive compounds to modify NPs holds great promise, particularly for biomedicine. NPs can work more precisely and effectively by utilizing these molecules, making them suitable for a range of tasks from imaging to medicine delivery. Although there are obstacles to overcome, current research in this field promises to produce ground-breaking and revolutionary answers.

7.3. Functionalization through Geometric Change

Even without any further surface alterations, the geometry of NPs, particularly their size and shape, significantly impacts their characteristics and interactions with cells. The size of the NPs taken up by cells directly affects how cytotoxic they are. Additionally, how NPs interact with cells is influenced by their form. For example, rod-shaped NPs interact with cells more effectively than spherical NPs because they have more accessible binding sites [318]. Designing more efficient NPs requires an understanding of the connection between the geometry of NPs and their functionality.

7.3.1. Size

The potential for endocytosis, a process that enables NPs to be internalized by cells, and the interactions of NPs with cell membranes are strongly influenced by their size. Increasing the number of ligands on NP surfaces to facilitate endocytosis is advantageous, but doing so requires bigger NP sizes. The size must fall within a specific range, though, as NPs less than 30 nm might not be able to drive the membrane-wrapping process as well as those larger than 60 nm. Regardless of the NP core and surface charge, in vitro studies indicate that the ideal range for cell uptake is between 10 and 60 nm [319]. There has been some variation in the effect of NP size on cellular internalization among studies. According to some studies, cellular internalization of functionalized Au NPs is inversely correlated with size [320], whereas, according to others, internalization of Au NPs larger than 50 nm is more common [321]. These differences could result from variances in the gold NPs' production processes or decorating ligands.

7.3.2. Shape

Beyond only size, NPs' form greatly impacts how well they interact with cells and deliver drugs. Spheres, rods, triangles, stars, and wires are examples of frequently produced NP shapes [322]. Due to their greater aspect ratio (AR), or length-to-width ratio, compared to spherical NPs, rod-shaped NPs have demonstrated increased effectiveness in cellular uptake [323]. This discovery has sparked additional in vivo research. For instance, after intravenous administration, NPs with bigger ARs were seen to collect in the spleen, while those with lower ARs were probably stuck in the liver of mice [324]. In a different study, mice given rod-shaped MSNs orally showed higher content in all organs than mice given spherical MSNs. This may be explained by the prolonged half-life of rod-shaped MSNs in blood circulation and their capacity to resist macrophage engulfment in RES (reticuloendothelial system) organs, including the liver and spleen [325]. The form of NPs also influences the cellular uptake mechanism. A comparison of gold NPs with star, rod, and triangle forms revealed a substantial relationship between the endocytosis pathway and NP shape, which calls for more research [326]. The decreased internalization rate seen for rod-shaped NPs relative to spherical NPs must be considered, even though larger AR NPs have been connected to higher cellular absorption efficiency [327]. Therefore, the benefits of modifying NP shape must be carefully assessed in the context of particular biomedical applications.

7.4. Challenges to Achieve Successful Designed Interfaces

Manufacturing challenges persist, and there is a lack of uniformity in the methods used to fuse cell membrane vesicles with NP cores. Extrusion is a technique that can create uniform particles, but it has complexity and manufacturing scale difficulties. Since cell membrane coating integrity affects internalization, current approaches frequently produce a mixture of fully, partially, and uncoated NPs. The classification of the therapeutic effects of various MCNPs will require further research to determine how to differentiate fully coated NPs. Future MCNP manufacturing efforts should improve procedures and create a successful, all-encompassing process for cell membrane extraction and fusing with NP cores. Moving toward industrial-level output requires automation. In the future, process development should take precedence over discovery more often.

NPs can be functionalized in a variety of ways employing surface modification techniques, sometimes with the addition of ligands or molecules for decoration. Despite its advantages, there are a few difficulties. Due to their effect on cellular absorption efficiency, the density and orientation of ligands on the NP surface must be considered. The complete evaluation of NPs with various surface changes also lacks defined approaches. The geometric features of NPs must also be considered because they can affect cellular absorption and possibly prevent planned functionalization. The next part will go into more detail regarding how NPs' geometric characteristics affect them.

8. Biomedical Applications of Bio-inspired Nanomaterials in Micro/Nanodevices

When the many uses of bio-inspired nanomaterials in micro/nanodevices are examined, a wide range of biological uses is found. From improving drug delivery methods to coming up with new ways to treat patients, these new materials are paving the way for big changes in healthcare. Drug delivery involves complicated systems, nanotheranostics for combined therapy and diagnostics, gene therapy for genetic disorders, and the creativity of self-propelled active nanovehicles and biohybrid micro/nanomotors that can move through complex biological environments. Bio-inspired nanobiosensors that can identify molecules in a complex way add to these achievements. Also, bio-inspired organ-on-a-chip technology gives us new ways to test drugs, and cancer-on-a-chip models change how we study cancer. Bio-inspired wound healing dressing mats and antimicrobial surfaces, such as those made from structure-oriented peptides, metal/metal oxide NPs, and chitosan, show how bio-inspired nanomaterials have a wide range of uses in medicinal applications. Bacteriophage-based antimicrobial surfaces use the power of nature to fight bacterial diseases. This all-around look at bio-inspired nanomaterials shows their importance and opens up a new era of opportunities for biomedical progress in micro/nanodevices. Figure 15 summarizes the biomedical applications of BINMs in micro/nanodevices.

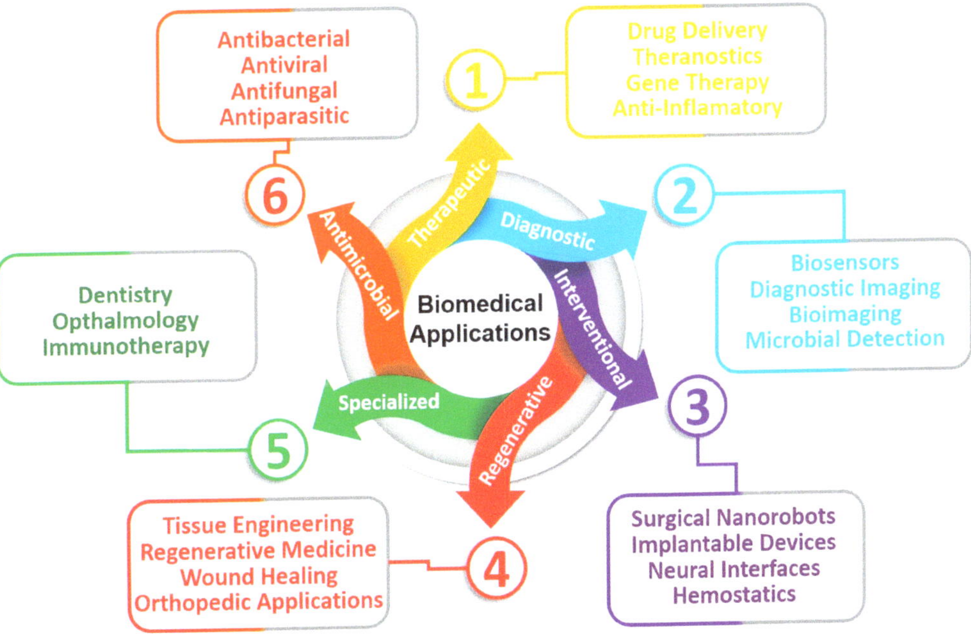

Figure 15. Biomedical applications of BINMs in micro/nanodevices.

8.1. Drug Delivery and Therapeutic Applications

8.1.1. Drug Delivery Systems

In the field of medicine, BINMs have made major advancements, particularly in the creation of sophisticated drug delivery systems. These nanomaterials can potentially address major difficulties in targeted medication delivery because they were created and constructed to resemble biological structures and processes. Micro- and nanodevices made from BINMs are at the core of these breakthroughs. In several ways, these devices can improve drug delivery. They can increase the specificity of drug delivery, ensuring that the medications reach the cells or tissues that require them most. This can reduce the risk of negative effects while significantly increasing the drug's effectiveness. Liposomes, for instance, nano-sized vesicles modeled after biological membranes, are frequently utilized as drug delivery systems. They can encapsulate both hydrophilic and hydrophobic medications, guard against deterioration, and release them gradually at the intended spot. Dendrimers, another category of BINMs, are spherical, highly-branching NPs. They are good candidates for targeted drug administration due to their well-defined structure, controllable size, and surface functional groups. Recently, scientists have started investigating the use of BINMs for targeted drug delivery that imitates the structure of cells, bacteria, and viruses. For instance, NPs with red blood cell (RBC) membrane coatings have shown promise in drug delivery while dodging the body's immunological response. But creating these biologically inspired micro- and nanodevices is difficult. Addressing concerns with stability, biocompatibility, scalability, and reproducibility is necessary. Extensive testing and clinical trials are also necessary to determine the safety and effectiveness of these technologies for human usage.

Liposomes are widely recognized as a promising and versatile means of drug delivery. Liposomes present several advantages in comparison to traditional drug delivery systems. These advantages include targeted delivery to specific sites, controlled and sustained release of drugs, protection against degradation and clearance, enhanced therapeutic outcomes, and reduced toxic side effects. These beneficial characteristics have contributed to the effective authorization and medical utilization of numerous liposomal pharmaceutical products within recent decades [328]. Liposomes can be divided into several categories depending on their lamellarity and compartment structure. These categories include multivesicular liposomes (MVLs), unilamellar vesicles (ULVs), oligolamellar vesicles (OLVs), and multilamellar vesicles (MLVs). OLVs and MLVs both have an onion-like structure; however, MLVs have more than five lipid bilayers, while OLVs only have two to five concentric lipid bilayers. On the other hand, MVLs have several non-concentric aqueous chambers that are each surrounded by a single bilayer lipid membrane, giving them the appearance of a honeycomb. Small unilamellar vesicles (SUVs, 30–100 nm), large unilamellar vesicles (LUVs, >100 nm), and giant unilamellar vesicles (GUVs, >1000 nm) are subcategories of ULVs based on particle size. According to several research studies, ULVs come in various sizes, including SUVs that are less than 200 nm and LUVs that are between 200 and 500 nm in size. Numerous techniques are applied for the preparation of liposomes. The manufacturing methods that are frequently utilized encompass thin-film hydration, ethanol injection, and double-emulsion techniques. The conventional procedures in these processes encompass several steps. Firstly, multilamellar vesicles (MLVs) or unilamellar vesicles (ULVs) are prepared, depending on the chosen method. Secondly, the size of the vesicles may be reduced if deemed necessary. Thirdly, the drug solution(s) are prepared and loaded into the liposomes. In the case of passive drug loading, this step is combined with step one. Fourthly, buffer exchange and concentration are performed if required. Fifthly, sterile filtration or aseptic processing is carried out. Lastly, if deemed necessary, lyophilization is conducted, followed by packaging. Figure 16 shows graphic representations of several polymersome- and liposome-related topics. In Panel A, the structural differences between polymersomes and liposomes are shown in the cross-section. The main physicochemical characteristics of nanocarriers are depicted in Panel B. The enhanced permeability and retention (EPR) effect, shown in Panel C, is a passive buildup of nanocarriers through fenestrated endothelial cells in tumor tissues. Due to their leaky vasculature and compromised lymphatic outflow, tumor tissues are more conducive to nanocarrier accumulation [329]. The liposome-inspired drug delivery system is very promising in multidimensional biomedical applications, including pulmonary nanotherapeutics [330], tumor-targeted therapy [331], anti-biofilm agents [332], and multiple diseases [333].

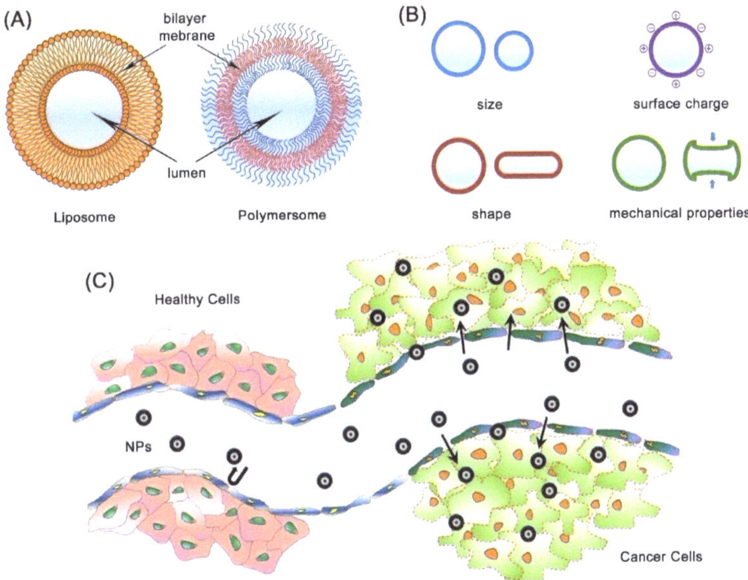

Figure 16. Graphical representations of several polymersome and liposome-related topics. (**A**) the structural differences between polymersomes and liposomes are shown in cross-section. (**B**) The main physicochemical characteristics of nanocarriers are depicted. (**C**) The enhanced permeability and retention (EPR) effect. It is a passive buildup of nanocarriers through fenestrated endothelial cells in tumor tissues. Due to their leaky vasculature and compromised lymphatic outflow, tumor tissues are more conducive to nanocarrier accumulation. Reprinted with permission from Ref. [329], Copyright 2023, Elsevier.

Dendrimers have emerged as crucial nanostructured carriers in nanomedicine for treating numerous diseases [334]. Thanks to their structural diversity, they can deliver medications and genes in various ways (Figure 17). For example, dendrimers with a hydrophobic center and a hydrophilic periphery can act like unimolecular micelles and successfully saturate hydrophobic medicines. The use of cationic dendrimers as non-viral gene carriers is widespread. Drugs and functional moieties can be attached to dendrimer surface groups to increase stability and solubility. Enhancing dendrimer compatibility and binding properties involves conjugating them with polymers like PEG or polysaccharides. Utilizing ligands like hyaluronic acid or mannose has improved tumor penetration and targeted distribution to specific cell types, such as macrophages. Compared to free medicines, dendrimer–drug conjugates have fewer systemic side effects and more localized efficacy. Dendrimer conjugation can lengthen the half-life of pharmaceuticals, improving medicinal efficacy and reducing administration frequency. Dendrimers increase the solubility of drugs, increasing their potency. When compared to timolol maleate, a study on a dendrimer–drug combination known as DenTimol demonstrated encouraging outcomes for the treatment of glaucoma. Various cleavable or stimuli-responsive linkages ensure that medications released from dendrimer–drug conjugates reach the intended area. For this aim, disulfide/thioketal linkers and pH-responsive linkers are frequently employed. Dendrimer–drug conjugates have promise as efficient drug delivery systems with controlled release mechanisms for better therapeutic results. Figure 18 illustrates the strategies for dendrimers in drug and gene delivery. Over free medicines, dendrimer–drug conjugates have several benefits, such as fewer systemic side effects and increased efficacy at the target site. They can lengthen a drug's half-life and make it more soluble, enhancing patient compliance and therapeutic results. For instance, PAMAM dendrimers have been utilized successfully to deliver antiglaucoma medications, exhibiting better effects on decreasing intraocular pressure. Examining drug release from such conjugates is crucial since regulatory agencies' classification of dendrimer–drug conjugates might be complicated. Drug release from dendrimers in tumor cells has been facilitated by cleavable linkers like disulfide and thioketal, increasing the effectiveness of cancer therapy. By serving as "unimolecular micelles" or "dendritic boxes," dendrimers also provide drug encapsulation through their hydrophobic cavities, enhancing the solubility of hydrophobic medicines in water. PAMAM dendrimers have also been extensively used as gene transfection vectors because of their great biocompatibility and capability for nucleic acid loading. They can improve endosomal escape and cellular uptake, increasing transfection effectiveness. Dendrimers can overcome intracellular gene delivery hurdles when decorated with functional moieties like peptides. This results in successful gene delivery and tumor growth inhibition. Overall, dendrimer-based medication and gene delivery systems provide considerable promise for treating various disorders using nanomedicine [335–337].

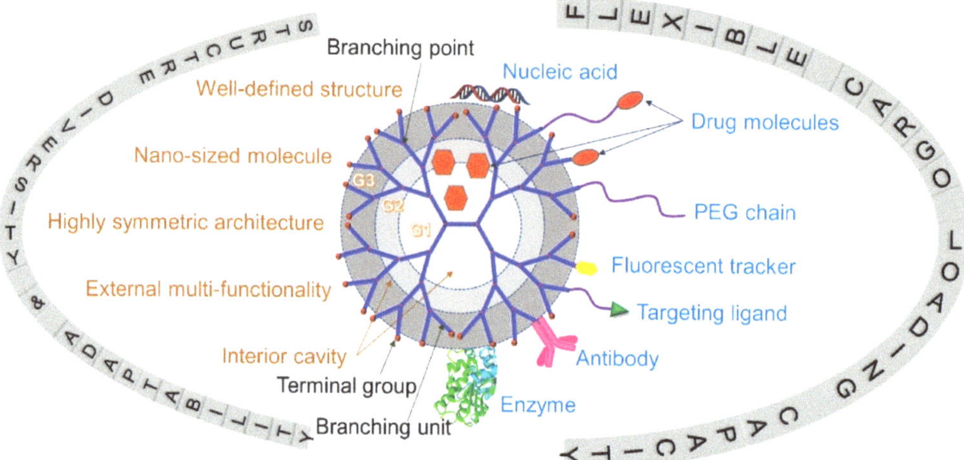

Figure 17. Dendrimers with a wide range of structural configurations and versatile cargo-loading capabilities for Generation 1, Generation 2, and Generation 3, correspondingly. Reprinted with permission from Ref. [334], Copyright 2022, Authors (CC BY 4.0).

Polymeric micelles (PMs) are nanostructures created through amphiphilic block copolymers (ABCs) self-assembled in an aqueous medium. These micelles possess a distinctive core-shell architecture. In conventional micelles, the hydrophobic segment of the polymer is oriented toward the interior, forming the core, whereas the hydrophilic segment is positioned on the outer surface. Reverse micelles exhibit an orientation that is opposite to that of regular micelles. Mixed micelles are generated by adding solubilizers to the existing surfactant micelle structure. The drug is expelled into the micelle core by the hydrophobic component of the copolymers, thereby enabling the solubilization of pharmaceuticals with low solubility. The intermolecular hydrophobic interactions between the drug and copolymers are important in modulating the drug release rate and enhancing its solubility. Numerous hydrophobic copolymers have undergone testing to solubilize drugs with low solubility efficiently. Polymeric micelles frequently contain a hydrophobic core enclosed by hydrophilic copolymers [338]. The hydrophilic portion of the polymer faces outward in normal micelles, while the lipophilic portion faces the core. The orientation of reverse micelles is the opposite. Solubilizers are included in the surfactant micelle to create mixed micelles. Pharmaceuticals that are difficult to dissolve are ejected into the micelle core by the copolymer's hydrophobic component. Copolymers' hydrophobic interactions with the medication are essential for reducing drug release and increasing solubility. It has been tested that different hydrophobic copolymers can successfully solubilize poorly soluble medicines. Drug leakage from polymeric micelles must be reduced during distribution, and drug release must be regulated to provide optimal therapeutic targeting. Either stable confinement of the drug payload within micellar cores or triggered release in response to internal or external stimuli are necessary for targeted drug delivery. While slowly released medications from capsules allow for pharmacological and toxicological effects, prolonged drug release from polymeric micelles in circulation assures congruent pharmacokinetics with the micelles. Stimuli-sensitive polymeric micelles use internal triggers, such as changes in pH, redox potential, temperature, enzyme profiles, and oxygen levels, to take advantage of disease-induced changes in target tissues. Outside stimuli, including heat, ultrasound, near-infrared light, or magnetic fields can also trigger drug release. These methods improve the adaptive drug carriers' ability for precise drug delivery, especially in sick tissues like malignant ones. Micelles are distinguished by their core-shell structure. The corona shell shields the drug from the mononuclear phagocyte system, allowing for longer blood circulation and less toxicity. They make it possible for hydrophobic medicines to become stable and water-soluble, facilitating effective medication delivery. The ideal micelles for dispensing hydrophobic medications feature a hydrophilic corona to protect and stabilize the medication. Medicines with a high water solubility can have their intravenous administration of hydrophobic medicines made possible by polymeric micelles. Although polymeric micellar systems have several drawbacks, different methods have been created to overcome these obstacles. With the right approaches to drug loading issues, scale-up options, and thorough research into their behavior in biological systems, polymeric micelles can successfully find their place in the market for various biomedical uses. Polymeric micelles present promising opportunities in biomedical applications [339,340].

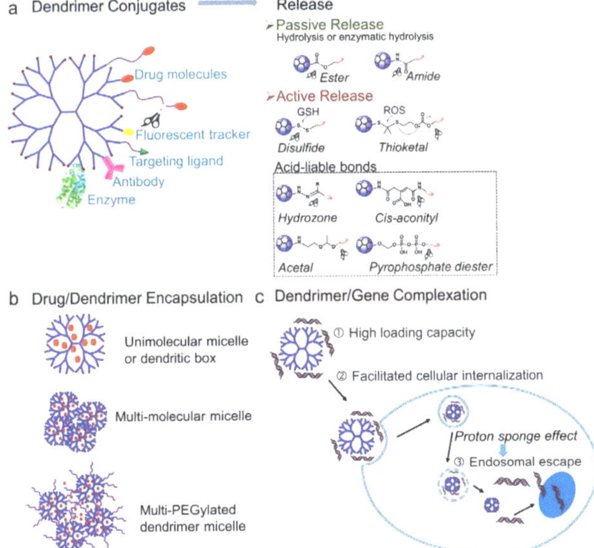

Figure 18. Strategies for dendrimers in drug and gene delivery: (**a**) dendrimer conjugation and cleavable linkers for enhanced delivery, (**b**) exploring three drug/dendrimer encapsulation models, (**c**) dendrimer-mediated gene complexation, and advancing transfection efficiency with cationic dendrimers. Reprinted with permission from Ref. [334], Copyright 2022, Authors (CC BY 4.0).

8.1.2. Nanotheranostic

For many illnesses, including AIDS, cancer, and microbial disorders, theranostic methods have been suggested. The medication is customized using this personalized treatment approach based on unique molecular profiles or the discovery of biomarkers. Nanotechnology advancements can now combine diagnostic and therapeutic approaches on a single platform. Nanomedicines increase the bioavailability of drugs, shield them from deterioration, and enable precise medication distribution within the body. Compared to conventional medicines, nanostages used in nanotheranostics provide simultaneous illness detection and treatment while improving medication penetration. This new field has the potential to significantly help the pharmaceutical and healthcare sectors by facilitating the creation of molecular sensors, imaging agents, and creative therapeutic agent carriers. Immunoassays and colorimetric tests, as well as gene therapy and targeted drug delivery, are nanotheranostic diagnostic and therapeutic tools that have the potential to transform the diagnosis and treatment of a wide range of illnesses, including cancer, AIDS, cardiovascular disease, infections, and burn wounds [341]. BINMs have significantly enhanced cancer diagnostics and treatments, largely due to their small size, ease of modification, high drug-loading capacity (thanks to their large surface-to-volume ratio), and efficient penetration and retention within targeted tissues. Furthermore, their superior biocompatibility, biodegradability, and multifaceted applications in bioimaging, bio-sensing, diagnostics, and therapeutics have escalated their potential in numerous biomedical fields [342]. Due to their potential to serve as alternative, biocompatible drug delivery systems in cancer theranostics, bio-inspired NPs that mimic natural body components have recently attracted a lot of attention. Unlike non-native drug delivery technologies, these NPs have the innate potential to change systemic bio-distribution, which is their main advantage. This review thoroughly explains numerous BINMs used in cancer theranostics, including liposomes, lipid NPs, bio-synthesized metal NPs, virus NPs, protein NPs, and others (Figure 19).

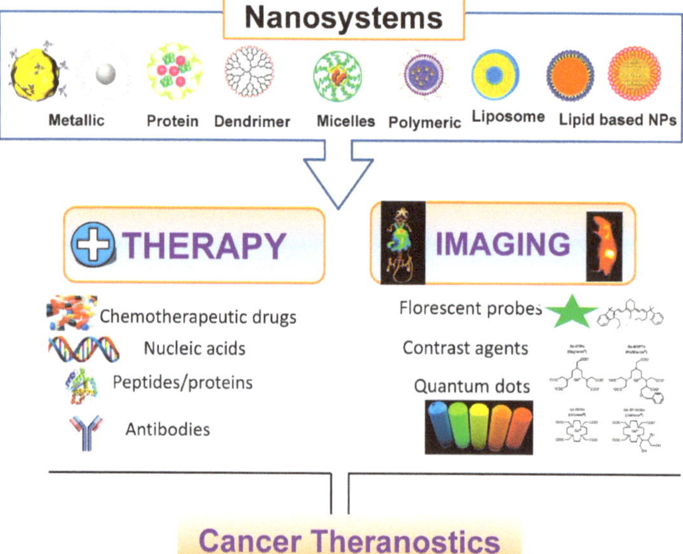

Figure 19. Nanotheranostics: metallic, protein, polymeric, lipid-based, micelle, and dendrimer nanomaterials for cancer theranostics. Reprinted with permission from Ref. [342], Copyright 2019, Authors (CC BY).

8.1.3. Gene Therapy

In several medical specialties, gene therapies are becoming more and more cutting-edge treatments. Gene therapy first proposed some 45 years ago as a viable treatment for hereditary monogenic illnesses, is currently being used to treat acquired conditions like cancer immunotherapy. The idea was that a single treatment could offer substantial, possibly curative advantages. For instance, gene-based therapies given to cells with a long lifespan may permit the continued production of crucial proteins. Hematopoietic stem cells (HSCs) that have undergone genetic engineering could provide long-lasting cell replacement, eliminating the requirement for ongoing enzyme administration or transfusion therapy. With initial clinical trials in the early 1990s producing poor findings, including little clinical benefit, unforeseen toxicities, and, in some cases, patient fatalities, converting gene therapy concepts into patient care has had difficulties. This caused a shift in attention to the fundamental science underlying gene therapy strategies. With a better understanding of viral vectors and target cells, a new wave of clinical trials in the late 1990s and early 2000s showed promise, but development was hampered by severe toxicities associated with high gene transfer efficiency. The discipline of gene therapy has made enormous strides over the past ten years, with improvements in safety, gene transfer effectiveness, and delivery spurring major clinical advancements. The FDA has approved several gene therapies, and other agencies worldwide have labeled others as "breakthrough therapies." The science of gene therapy is about to undergo another revolutionary change because of recent advancements in targeted genome editing [343].

The utilization of BINMs has played a crucial role in advancing the field of micro/nanodevices for gene therapy. These nanomaterials are derived from biological systems and can imitate the structures and functions of biological molecules. As a result, they exhibit enhanced biocompatibility and functionality. Nanomaterials, including lipid-based NPs, protein-based NPs, and DNA/RNA-based nanostructures, function as carriers for gene delivery, offering improved stability, specificity, and efficiency. One illustration of this concept involves the utilization of lipid-based NPs to encapsulate nucleic acids, thereby enabling their efficient transport into cells.

Additionally, DNA nanostructures can be purposefully engineered to serve as carriers for therapeutic genes, allowing for direct delivery. Significantly, BINMs possess the capability to undergo modification or functionalization to augment their targetability and mitigate potential toxicity. By capitalizing on these benefits, micro/nanodevices employing BINMs have demonstrated gene therapy potential. Gene therapy aims to address diseases at the fundamental genetic level by mending, activating, or eliminating specific genes. The ongoing investigation and advancement in this particular domain are anticipated to result in the emergence of gene therapy approaches that are both more efficient and secure.

There is a growing interest in the supramolecular self-assembly of dendrons and dendrimers as a powerful and challenging method for generating advanced nanostructures that exhibit exceptional properties. Xu et al. proposed a novel approach involving supramolecular hybridization to fabricate a dendritic system inspired by biological systems [344]. This system demonstrated remarkable versatility and can be utilized as an efficient nanoplatform for various delivery applications (Figure 20). Multifunctional supramolecular hybrid dendrimers (SHDs) were formed by integrating dual-functionalized low-generation peptide dendrons (PDs) onto inorganic NPs, facilitated by an intelligent design. The structural composition of these superhydrophobic surfaces (SHDs) exhibited a highly organized nanoarchitecture, accompanied by a substantial presence of arginine peptides, and demonstrated the ability to emit fluorescence signals. As predicted, the utilization of a bio-inspired supramolecular hybrid strategy dramatically enhances the gene transfection efficacy of self-assembled hydrogel NPs (SHDs) by approximately 50,000 times when compared to standalone polymeric NPs (PDs) at equivalent ratios of polymer to DNA. The bio-inspired self-assembled hydrogel NPs (SHDs) demonstrate several advantageous characteristics in gene delivery. Firstly, they possess low cytotoxicity and are resistant to serum, which enhances their safety and efficacy. Secondly, these SHDs have inherent fluorescence, monitoring various intracellular processes, including cellular uptake, escape from endosomes, and gene release. Lastly, they can serve as a valuable reference for tracking the expression of desired proteins, providing an alternative method for assessing gene delivery efficiency. Significantly, it is important to note that in vivo animal trials have shown that self-healing hydrogels (SHDs) exhibit considerable effectiveness in gene transfection, specifically in muscle tissue and HepG2 tumor xenografts. These trials have also demonstrated the ability of SHDs to perform real-time bioimaging. The anticipated outcome of these supramolecular hybrid dendritic (SHD) structures is the stimulation of research inquiries to utilize bio-inspired dendritic systems for biomedical purposes, encompassing laboratory-based and live organism studies.

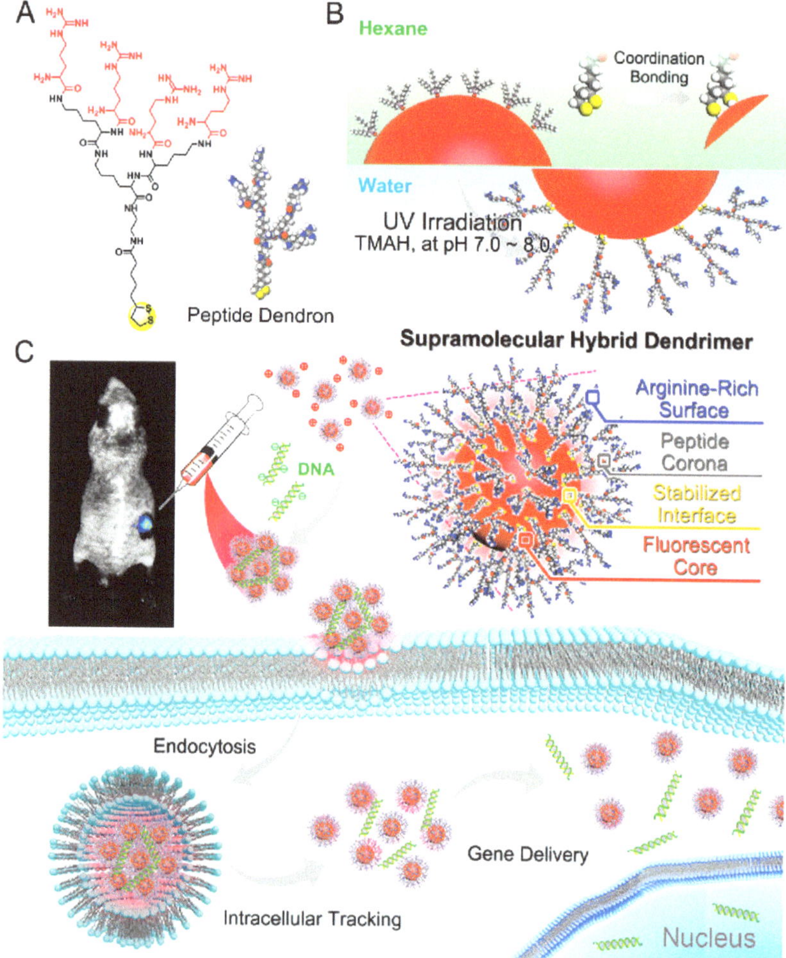

Figure 20. Diagrammatic representations of the self-assembly process and biomedical uses of SHDs. (**A**) Illustration of the chemically dual-functionalized PDs, (**B**) the process of PDs self-assembling onto quantum dots through coordination interactions, and (**C**) use of SHDs featuring complex nanostructures for delivering genes and for biological tracking both in vitro and in vivo. Reprinted with permission from Ref. [344], Copyright 2014, American Chemical Society.

With the progress of molecular biology, pharmacogenomics, and proteomics, there is an opportunity to customize the development of bio-inspired nanosystems to cater to individual patients' specific requirements. Bio-inspired nanomaterials provide numerous advantages, such as the ability to customize their surface, achieve targeted delivery, possess specific geometric properties, ensure biosafety, and facilitate proper disposition. Furthermore, the materials and procedures employed in fabricating these systems exhibit biocompatibility and environmental sustainability, as they necessitate limited processing compared to synthetic materials. Nevertheless, it is essential to acknowledge the potential apprehensions regarding residual solvents or reagents utilized during the synthesis process, which may harm biological systems. However, certain bio-inspired nanosystems exhibit enhanced pharmacological efficacy. There is expected to be a preference for gene carriers based on smart materials that are biocompatible, biodegradable, and safe in the future. These systems can detect the surrounding environment of the host following administration, enabling them to regulate the release of gene molecules that have been loaded within them at a particular target organ with accuracy and pre-determined control. This functionality serves to reduce the occurrence of undesired side effects. Therefore, utilizing bio-inspired gene delivery systems presents a distinct opportunity to develop anticipatory and individualized delivery systems for currently available medications, thereby holding great potential in shaping the future of the biomedical field [345,346].

8.1.4. Self-Propelling Active Nanovehicles

Chemical navigation is a crucial aspect of survival for a wide range of organisms, from bacteria to unicellular and multicellular organisms. The replication of these behaviors through artificial constructs is an emerging field of study, resulting in the development of active nanomaterials capable of converting external energy into mechanical work to achieve directed motion [347]. These nanomaterials can react to various stimuli, including chemical gradients, temperature changes, magnetic fields, and adhesion forces. Nevertheless, the development of self-propelling nano-constructs encounters various obstacles arising from physical limitations. For instance, water, which exhibits a high viscosity at the NP level, poses a significant challenge. Additionally, the randomizing effect of Brownian thermally driven fluctuations further complicates the control of the NPs' directionality. Two strategies can be employed to address these limitations. The first strategy involves inducing non-reciprocal movements by altering the body shape. The second strategy consists in taking advantage of gradients that modify the local environment of the nanomaterials. Illustrations of these strategies being implemented encompass the utilization of artificial bacterial flagella, which can be effectively manipulated by applying rotating magnetic fields to generate propulsion. The deployment of "spermbots" has been observed, wherein these microscale robots can facilitate the transportation of sperm cells toward the oocyte. Certain NPs can generate gradients autonomously, resulting in self-phoresis or self-propelled movement. Various innovative applications have been documented, including the utilization of silicon nanowires that exhibit a responsive behavior to externally manipulated electrical fields, as well as the development of "microbullets" capable of vaporizing biocompatible fuel and effectively penetrating and altering the structure of tissues. The utilization of bio-inspired methodologies in the design of NPs exhibits considerable promise for their application in drug delivery, targeted therapy, and various other domains within the biomedical field [348–350].

8.1.5. Biohybrid Micro/Nanomotors

Throughout history, human ingenuity has frequently drawn inspiration from diverse natural biological systems, exemplified by the development of radar technology, which was influenced by bats' utilization of ultrasonic waves. The advancement of autonomous artificial micro/nanomotors has been motivated by the existence of biological biomotors such as kinesins, dyneins, and sperm cells. Micro/nanomotors are devices capable of converting various types of energy into mechanical motion, enabling them to execute tasks that passive devices cannot accomplish [351]. Richard Feynman initially introduced the notion of these diminutive devices which has subsequently emerged as a prominent subject of scholarly investigation. Micro/nanomotors possess a diverse array of applications, particularly within biomedicine. These applications encompass drug delivery, biosensors, biological imaging, assisted fertilization, and microsurgery. Nevertheless, notable obstacles must be surmounted, with particular emphasis on the biocompatibility of said motors. To operate optimally, these entities must adapt effectively to the internal microenvironment of the organism, encompassing factors such as temperature, pH, and the immune system. Presently, the utilization of artificial motors is constrained by their inadequate biocompatibility, resulting in their susceptibility to immune system recognition upon introduction into the human body. To enhance biocompatibility, scholars are currently investigating the utilization of biocompatible and biodegradable substances such as polyethylene glycol (PEG) and magnesium. An emerging area of study pertains to biohybrid micro/nanomotors, wherein synthetic materials are integrated with biological constituents. Biohybrid motors exhibit enhanced biocompatibility, improved energy conversion efficiency, and the capacity to react to environmental stimuli intelligently. The cell is the building block of an organism and has receptors on its membrane that enable it to sense environmental inputs and modify its functions. Some cells have complex time-irreversible strokes and low-Reynolds-number autonomous motion processes. Researchers have used this autonomous mobility as motivation to build micro/nanomotors based on intact cells. High biocompatibility, adaptability to diverse internal conditions, and the ability to mix artificial micro/nanostructures with different cell properties like chemotaxis, magnetotaxis, and anaerobism to create multifunctional motors are all benefits of these biohybrid motors. A cell must be simple to grow and capable of large-scale, rapid multiplication to be a good candidate for biohybrid micro/nanomotors. There are several types of micro/nanomotors based on intact cells (sperm cells, bacteria, algae, blood cells, plant pollen, platelets, macrophages, etc.) and different biological components such as enzymes (catalase, urease, glucose oxidase, lipase, etc.) and cellular membrane (RBC, platelet, WBC, tumor cell, cancer cell, etc.) coating, operating in biomedical applications.

Sperm cells, also known as spermatozoa, are the male gametes that possess the ability to exhibit autonomous motility as a result of their flagellar structure. They possess the dual functionality of functioning as both a propulsion mechanism and a means of transporting goods, enabling autonomous movement and precise distribution capabilities. The micro-bio-robot design involved using sperm cells confined within a microtube, which demonstrated self-directed movement that was externally regulated through the implementation of a magnetic layer. Sperm cells possess considerable potential as vehicles for drug delivery, particularly in gynecologic ailments such as cervical cancer [352]. A micromotor propelled by sperm cells has been developed to facilitate the targeted release of drugs, demonstrating encouraging attributes for cancer treatment. Bacteria are abundant and come in various shapes, making them suitable candidates for biohybrid micro/nanomotors. Several bacteria, such as *Magnetococcusmarinus*, *Escherichia coli*, and others, have been utilized to fabricate biohybrid motors for biomedical applications. Magnetotactic bacteria can achieve self-propulsion using external magnetic fields, making them attractive for drug delivery and tumor targeting. Escherichia coli are frequently used due to their swimming ability, and they have been incorporated into micromotors for drug delivery and anti-tumor efficacy. Challenges include addressing safety concerns regarding pathogenic bacteria and ensuring the activity and fitness of bacteria on certain surfaces. Despite these challenges, bacterial biohybrid micro/nanomotors hold promise for advancing the field of micro/nanomotors.

Algae exhibit remarkable biological features despite lacking roots, stalks, and leaves. Spirulinaplatensis (Sp) is a suitable bio-template for biohybrid magnetic micromotors due to its naturally intact three-dimensional helical structure. Researchers used Sp to construct porous hollow micromotors to deliver medicinal and imaging chemicals in vivo. Sp-based biohybrid magnetic robots feature intrinsic fluorescence, MR signals, and low cytotoxicity, making them intriguing for blocking abnormal cell function, particularly malignant tumors, while retaining normal cell function. Sp-based magnet-powered microswimmers use ultrasonics to stimulate neural stem-like cell development. Ultrasound intensity can influence brain stem cell development, enabling minimally invasive neurodegenerative disease treatments. Algae, particularly Sp, have distinctive structures and intriguing biological features, making them promising in modern biotechnology. For biohybrid micro/nanomotors, different algae must be studied.

8.2. Bio-Inspired Nanobiosensors

Sensors are pivotal in many products, systems, and manufacturing processes, offering valuable feedback, monitoring capabilities, safety enhancements, and other advantageous features. When conventional sensor technology reaches a state of limited progress, exploring insights from non-engineering disciplines, such as biology, can foster innovative advancements. The field of biomimetic sensor technology is currently in its nascent stage, and it takes inspiration from the intricate sensory systems found in nature. These highly refined systems enable organisms to perform tasks such as navigation, spatial orientation, and prey detection with excellent efficiency. Engineers can construct various types of sensors by comprehending the fundamental principles of sensory physiology in biological systems. Biomimetic sensor designs can replicate biological systems directly or employ analogous principles. Both methodologies have demonstrated efficacy and yielded notable progress in sensor technology [353]. Biomimetic sensor designs offer distinct advantages in comparison to conventional sensor designs.

Retrieving archived sensor design information is a prevalent methodology in developing novel products that detect commonly encountered parameters. Nevertheless, employing unconventional approaches or drawing inspiration from diverse fields of study may be imperative in the context of atypical parameters. The field of sensor design has been influenced by nature, as it presents a wide range of sophisticated sensing and communication techniques observed in diverse organisms such as bacteria, plants, insects, mammals, and reptiles. The design of biomimetic sensors entails replicating various aspects of biological systems, including functional design, morphological design, principles, strategies, behaviors, and manufacturing techniques. The motivation behind these biomimetic devices is derived from rigorous methodologies, careful examination of natural phenomena, and the application of databases that document biological functionalities. Emulating the functionality, principles, morphology, or strategies observed in biological systems represents a form of biomimicry, which can be likened to the reverse-engineering process. In an alternative perspective, abstracting biological systems through analogical reasoning can be seen as an approach that aligns biology with engineering design principles. This approach involves seeking solutions to biological challenges by drawing inspiration from and imitating existing designs. Exploring natural phenomena to derive design inspiration or gain insights into the mechanisms by which biological systems process sensory information has resulted in notable advancements. The biological sensors found in nature have evolved over an extensive time spanning billions of years. These sensors provide enduring and efficient solutions that are well adapted to specific ecological niches. Frequently, these sensors demonstrate characteristics such as minimal energy consumption, heightened sensitivity, and redundancy. The redundancy concept serves as a valuable lesson derived from nature, as numerous biological systems exhibit multiple instances of redundancy to augment reliability and mitigate errors.

Chirality, also known as mirror dissymmetry, is an inherent characteristic observed in geometric entities, and it holds significant importance in biomolecules such as proteins and DNA. Circular dichroism spectroscopy is a technique used to evaluate the impact of a substance on biological, chemical, and physical characteristics. This method quantifies the disparity in the absorption of left and right circularly polarized light. Chirality is regarded as a principle inspired by biology in engineering. Chiral nanomaterials exhibit potential for various applications, such as sensing and catalysis, owing to their distinctive selectivity and specificity.

Nevertheless, there remains a lack of comprehensive understanding regarding the mechanisms that govern the transfer of chirality during the synthesis of inorganic nanomaterials possessing inherent chirality. Examining biological instances of chirality transfer can provide valuable insights for developing chiral inorganic nanomaterials across diverse applications. Chirality is a prevalent characteristic observed in biological entities, significantly influencing their geometries, properties, and behavior. Chiral objects, characterized by the absence of mirror symmetry, are widely observed in the natural world, and they have the potential to confer survival benefits to organisms. The phenomenon of chirality significantly influences the preferential incorporation of amino acids during the process of protein synthesis, thereby exerting a profound impact on the growth and behavior of both plants and animals. Organisms can utilize chiral structures to perceive polarized light and augment contrast within their surroundings. Inorganic materials also observe chirality due to molecular interactions and biological templates. Gaining insight into the processes by which chirality is transferred across various length scales is of utmost importance to effectively replicate and harness chiral nanostructures to design nanomaterials. The phenomenon of hierarchical chirality transfer, which occurs across multiple scales ranging from the molecular to the macroscopic level, has been documented in diverse biological systems. This observation has sparked interest and served as a source of inspiration for developing biomimetic materials and nanotechnologies [354]. Near-infrared (NIR) wavelengths are commonly favored in biomedical applications owing to their superior tissue penetration capabilities. The utilization of CdTe helices has been observed to effectively manipulate light within the near-infrared (NIR) wavelengths, rendering them valuable for various biomedicine and optical computing applications. Chiral molybdenum oxide NPs have the potential to be utilized in photothermal therapy, wherein they can selectively heat tumor tissue when exposed to circularly polarized light while minimizing damage to healthy tissue [355]. The bactericidal effects of gold nanobipyramids conjugated with D-Glu are enhanced, disrupting bacterial cell walls and facilitating the healing process in infected wounds when exposed to near-infrared (NIR) radiation [356]. The inherent structural chirality exhibited by gold nanomaterials can influence the immune system. Diverse immune responses are observed with left-handed and right-handed Au NPs, owing to their distinct interactions with specific receptors and subsequent activation of inflammasomes. The utilization of left-handed NPs as adjuvants in the influenza vaccine has been investigated, revealing a notable increase in antibody production and immune-related cell proliferation compared to right-handed NPs. These discoveries underscore the significance of nanoscale chirality within biological systems, alongside the molecular-scale chirality exhibited by L/D optical centers.

8.3. Bio-Inspired Organ-on-Chip (OOC)

Creating new medications is time-consuming and expensive, especially in the pre-clinical stage. Pre-clinical research has a history of using unethical animal experiments that do not always precisely anticipate how people will react to medications. Although they provide an alternative, two-dimensional cell culture models cannot match the intricacy of real tissues and organs. Three-dimensional cell culture models have been created to overcome these restrictions; however, they still lack some physiological elements. The development of organ-on-chip (OOC) systems, which are little devices that replicate the microenvironment of organs and tissues, has recently been made possible by microtechnology. OOCs can build human-based tissue-like structures, operate with minuscule drug concentrations for high-throughput screening, and add biosensors for real-time monitoring of cell survival and functionality, among other benefits. To replicate the response of different tissues to drug exposure methodically, several OOCs can be coupled. OOCs are a potential strategy for drug discovery because they combine the benefits of 2D and 3D cell culture models while offering a platform for drug testing and screening that is more physiologically appropriate [357].

8.3.1. Organ-on-Chip (OOC) Technology

The primary objective of OOC technology is to develop in vitro models that closely mimic the physiological conditions of human organs. This is achieved by integrating cell cultures within microfluidic channels and structures. These systems provide the benefits of a microenvironment that closely resembles physiological conditions and utilize human cell lines that have been extensively studied and characterized. Out-of-cell culture systems possess the inherent capability of parallelization and enhanced throughput, rendering them highly advantageous in drug screening. Nevertheless, constructing organic optoelectronic devices necessitates utilizing advanced manufacturing techniques and selecting meticulously chosen materials. Cell cultures are immobilized on substrates and structures in OOC devices. These chip systems' cells can be divided into three major categories: primary, immortalized, and stem cells. Primary cells are taken straight from tissues or organs without being altered. They closely mirror their in vivo counterparts' appearance and metabolism. Primary cells must be obtained, kept alive, and only cultivated temporarily.

Additionally, standardization might be challenging due to differences in cell populations and traits between extractions. Immortalized cells are standardized, easily accessible, and well-characterized cells. They can be made from clinical malignancies or by modifying original cells chemically or virologically so that continuous cell division lasts for a long time. However, relative to their initial in vivo state, the cells' phenotype may change during immortalization. Because of their physiological properties and regulated differentiation potential, stem cells hold great promise. Induced pluripotent stem cells, produced by reprogramming adult tissues to produce pluripotent stem cells, are becoming increasingly popular due to ethical considerations and the restricted availability of embryonic stem cells. These cells can be differentiated into multiple cell types, facilitating research like personalized drug testing and autologous tissue engineering. In 2012, induced pluripotent stem cells' discovery was given the Nobel Prize.

8.3.2. Organ Systems on Chips

Due to demographic shifts and the rising demand for new pharmaceuticals in pharmaceutical research, society is exposed to an increasing number of novel chemicals in today's modern, globalized world. Reliable testing methods are required to ensure these substances are safe and effective. Animals were used mostly in the early toxicological, pharmacological, and environmental testing stages. The 3R approach (reduction, refinement, and replacement of animal experiments) has prompted a move toward alternative techniques. The development of alternative techniques has been encouraged by regulatory bodies like the European Parliament and the Council of the European Union, which has led to the EU's prohibition on cosmetics containing chemicals that have undergone animal testing. According to industrial firms, alternative approaches offer the potential to advance fundamental research, medicine development, toxicity testing, and environmental studies. Excellent throughput screening, parallelization, excellent data quality, predictability in clinical trials, and cost savings are among the alternatives they are looking for to eliminate the usage of animals. Common in vitro systems, however, cannot fully satisfy all of these demands, which has increased demand for enhanced in vitro models and cutting-edge OOC technologies.

8.3.3. Two-Dimensional Cell Culture to OOC

Early in vitro cell culture models were two-dimensional (2D), but it soon became clear how important three dimensions were. Cell morphology and metabolism were improved by 3D cell culture employing extracellular matrix (ECM) components [358]. Predictability was improved by creating cell–cell interfaces by integrating various cell types onto the semiconductor. Microsystems, inspired by developments in the semiconductor industry, permitted controlled trials with smaller drug doses. They made it possible to imitate in vivo circumstances by precisely controlling the biological milieu and inducing physiological pressures or gradients. Surface alterations made possible by microtechnology also encourage cell self-organization. These developments boosted predictability and complexity without raising variability. For widespread usage in pre-clinical studies, OOC systems must accomplish simplicity, dependability, reproducibility, and ease of use.

8.3.4. Single OOC (SOOC)

Researchers have pursued the development of integrated body systems on a chip through a systematic approach, wherein the initial focus has been on creating individual organ chips that can subsequently be interconnected. The development of these SOOCs was facilitated through the utilization of microchip technology and the progress made in the semiconductor industry. The lung-on-chip was among the initial organ chips documented in Science magazine in 2010, garnering considerable interest [359]. Subsequently, many biomimetic organ systems on chips have been successfully developed. There are several SOOC systems, including liver-on-chip [360], kidney-on-chip [361,362], lung-on-chip [362], gut-on-chip [363], heart-on-chip [363], muscle-on-chip [364], blood–brain barrier-on-chip [365], splenon-on-chip [366], bone marrow-on-chip [367], etc.

8.3.5. Multi-OOC (MOOC)

The utilization of MOOCs represents an interim measure in investigating inter-organ interactions until a fully functional human organ system is realized on a chip. The integration of multiple organ chips enables the examination of intercellular communication and the assessment of different stages of drug metabolism. Two primary approaches exist to construct multi-organ chips: the linkage of pre-existing single-organ chips and the integration of multiple organs into a singular chip device. The latter methodology has been proposed by the TechnischeUniversität Berlin and TissUse GmbH, commencing with a biotechnological device that integrates two distinct compartments, namely the liver and skin [368]. The utilization of a chip system, comparable in size to a conventional microscope slide, facilitated enhanced spatial efficiency and ensured appropriate ratios of physiological fluid to tissue. Researchers have made progress in the step-by-step integration of supplementary organs, such as the small intestine, liver, renal secretion, and skin biopsies, thereby advancing the development of a comprehensive human-on-chip system [369,370].

8.3.6. Human-on-Chip (HOC)

The primary objective of OOC technology is to develop a human-on-chip (HOC) model that replicates the functionalities of several vital organs within a singular microfluidic platform. Numerous governmental initiatives, including those sponsored by the American Defense Advanced Research Projects Agency (DARPA) and the National Institutes of Health (NIH), provide financial backing for research endeavors in this particular domain. There exist two primary approaches in designing an HOC system: the first involves the interconnection of individual single-organ chips, while the second entails the integration of distinct organ compartments onto a single chip. It is imperative to surmount the obstacles encountered in engineering and implementation to attain precise emulation of physiological conditions and dependable predictions of drug effects [371]. One of the primary challenges in this context involves selecting appropriate cell types and mediums while also considering immune responses and the inherent variability in blood composition. Streamlining the culture conditions and chip construction is advisable to enhance the results' clarity. Additionally, implementing a modular plug-and-play system could facilitate the interconnection of compatible chips in subsequent endeavors [372]. The active participation of the pharmaceutical industry and regulatory agencies is imperative to achieve successful development and validation of organ systems on chips.

8.3.7. Patient-on-Chip (POC)

Stem cells are a type of cellular entity characterized by their undifferentiated state and their capacity to differentiate into diverse specialized cell lineages. There are two primary classifications of stem cells: embryonic stem cells, which are obtained from embryos, and adult stem cells, which are sourced from adult tissues. The utilization of embryonic stem cells in research is hindered by ethical considerations, thus leading to the prevalent use of human-induced pluripotent stem cells (HiPSCs) as a viable substitute. HiPSCs are derived through reprogramming mature cells, acquiring characteristics akin to embryonic stem cells. This reprogramming enables HiPSCs to undergo differentiation into diverse cell lineages upon exposure to specific molecular cues. HiPSCs present a multitude of benefits in the realm of scientific investigation and medical interventions. One potential application of these technologies is the generation of patient-specific tissues for tissue engineering purposes and facilitating patient-specific drug testing. These advancements have the potential to enhance the field of personalized medicine. Incorporation of patient-derived HiPSCs into OOC systems presents a promising approach that synergistically harnesses the advantages of microfluidic technology and genetically compatible human cells. This methodology enables the replication of pathological conditions and genetic variations, enhancing the applicability and precision of drug testing and research endeavors. In addition, utilizing HiPSC technology holds promise in facilitating the advancement of a comprehensive HOC platform. Utilizing HiPSCs derived from a single donor to generate diverse organ tissues can potentially mitigate the occurrence of immune reactions. Furthermore, examining patient-specific variables, including genetic factors, age, gender, and ethnicity, can be readily conducted, thereby facilitating the development of more individualized therapeutic strategies in subsequent endeavors [373]. In general, HiPSCs exhibit considerable potential in facilitating the progression of scientific inquiry and pharmaceutical innovation, culminating in enhanced medical interventions that are more efficacious and tailored to individual patients [374–376].

8.3.8. Applications of OOC

Using OOCs exhibits significant potential as in vitro testing platforms for diverse applications. They possess utility in toxicity assessment for cosmetics and chemicals, rendering them indispensable in novel and generic drug advancement alongside specialized domains such as radiation examination. A substantial transformation in pre-clinical test practices toward enhanced efficiency and accuracy can only be achieved through collaborative endeavors. Once successfully designed and validated, OOC systems will substantially impact pharmaceutical research and development. These methods can lessen the need for animal testing and produce more trustworthy outcomes for biowaiver studies for generic formulations and medication development. This enhancement will result in more accurate clinical study forecasts and fewer late-stage failures. Additionally, OOC systems will create new possibilities for pharmacological R&D. They will make it possible to simulate sick creatures, giving researchers a controlled environment to investigate the causes of disease and potential cures. Additionally, custom chips can be produced to customize drug testing for specific patients, resulting in more efficient and individualized therapy. Organ systems on chips can potentially advance pharmaceutical research and significantly transform pre-clinical testing procedures.

OOC presents exciting possibilities for modeling diseases and developing new medications. Researchers have created disease chips to simulate specific disease states and analyze treatment reactions in a controlled setting. A lung-on-chip system was developed by Huh et al. to simulate pulmonary edema, a potentially fatal condition brought on by inflammation and fluid buildup in the lungs [377]. The chip enabled the testing of possible therapeutic substances, including angiopoietin-1 and GSK2193874, and faithfully replicated the effects of interleukin-2 (IL-2) therapy. Nesmith et al. created a bronchial smooth muscle tissue chip to research allergic asthma [378]. The IL-13 and acetylcholine exposure successfully caused the chip to mimic the hypercontraction observed in asthmatic patients. The RhoA inhibitor HA-1077 was put to the test by the researchers, and it showed promise as a possible therapeutic candidate for the treatment of allergic asthma. Tumor spheroids, hydrogels, and ECM proteins have all been used to create in vitro cancer models. Microfluidic systems are being investigated to model tumor formation, tumor–tissue interactions, and metastasis to increase physiological relevance. Researchers can lessen their reliance on animal models by using OOCs to test prospective medications and properly analyze cancer causes. These disease-specific OOCs have demonstrated encouraging outcomes when simulating illness states and assessing medication responses. They have the potential to transform pre-clinical testing, lessen the need for animal testing, and enhance therapy approaches for a range of disorders.

8.4. Cancer-on-Chip (COC)

The utilization of specialized multichannel systems in cancer-on-chip (COC) models has emerged as a potent approach for studying the tumor microenvironment (TME) and its involvement in metastasis. The utilization of microfluidic channels enables the replication of tumors' biochemistry, geometry, and fluidic transport characteristics by these models, thereby facilitating the examination of intricate interactions associated with metastasis [379]. The functional COC platforms exhibit superior accuracy and capabilities to traditional models, enabling them to provide significant insights into the TME and cell interactions during metastasis. Invasion, intravasation, extravasation, and angiogenesis are all components of the intricate process known as metastasis (Figure 21). Micrometastases are formed when tumor cells extravasate to colonize other organs after invading the extracellular matrix or vascular endothelium and entering circulation. The endothelial blood vessel wall must be broken during the key phases of intravasation and extravasation. For early diagnosis, prognosis prediction, and treatment planning, microfluidic technology holds promise for isolating and counting circulating tumor cells (CTCs). Microfluidic-based COC models are crucial to examine the complex interactions between tumor cells and the TME during invasion, intravasation, and re-growth in secondary organs. These models offer insights that conventional approaches cannot. The primary TME, circulatory microenvironment, and secondary TME are three different tumor microenvironments that interact during metastasis. For cancer cells to pass the endothelium and enter the bloodstream, the extracellular matrix (ECM) must be broken down. Surviving circulating tumor cells (CTCs) multiply and create secondary cancers in distant organs by adjusting to the local microenvironment (Figure 22). The TME contains various stromal elements, including fibroblasts, immune cells, vessels, and ECM. Microfluidic models that continuously expose tumor cell development to biological fluids in a biologically appropriate microenvironment are used to research cancer. Cancer invasion, intravasation, extravasation, and the evaluation of anti-cancer medications can all be studied with the aid of these models.

COC and tumor-on-chip (TOC) are intricately interconnected and serve as mutually reinforcing methodologies within cancer investigation. The primary objective of TOC models is to gain insights into the behavior and characteristics of tumor cells within a precisely regulated microenvironment. These models frequently employ microfluidic channels and compartments to replicate the biochemical and biophysical stimuli that impact the development and advancement of tumors. TOC models have the potential to enhance their fidelity to the TME by including non-tumor cells, such as stromal components, immune cells, and endothelial cells, thereby increasing their complexity. The COC and TOC models employ microfluidic technology to facilitate the uninterrupted provision of essential nutrients, oxygen, and other factors crucial for cellular proliferation and intercellular communication. By establishing physiologically relevant conditions, these models offer more precise depictions of tumor behavior than traditional in vitro cell culture systems. Furthermore, both methodologies can be employed to screen anti-cancer pharmaceuticals and examine the effectiveness of prospective therapeutic interventions. Figure 23 depicts the basic components of a common TOC. The utilization of a tumor-on-chip system fabricated through 3D printing techniques facilitated the cultivation of cells for an extended duration, thereby replicating the intricate process of nutrient and anti-cancer drug transportation within authentic TMEs [380]. The analysis of convective and diffusive transport within the culture chamber was conducted by employing a fluorescent tracer. The GelMA/alginate microbeads were the most efficient in facilitating transport. The microbeads were utilized to cultivate Caco2 cells, and subsequent drug assays mimicking chemotherapy exhibited a rise in cell death and a decline in cellular metabolism. Hypoxic conditions were artificially created within the microspheres, emulating the oxygen-deprived environment typically found in avascular tumors observed in patients. The study showcased the capacity of TOC platforms created through 3D printing for drug testing and examining cancer biology. The increased dimensions of the chip facilitated a higher quantity of biological material that could be used for analysis. The transport characterization demonstrated efficient convective and thoroughly mixed conditions, rendering it a valuable instrument for replicating tumor scenarios and other tissue environments. Additional investigation is warranted to delve into the potential benefits of employing 3D-printed tumor-on-chip systems, specifically regarding their design flexibility and the feasibility of their fabrication and utilization.

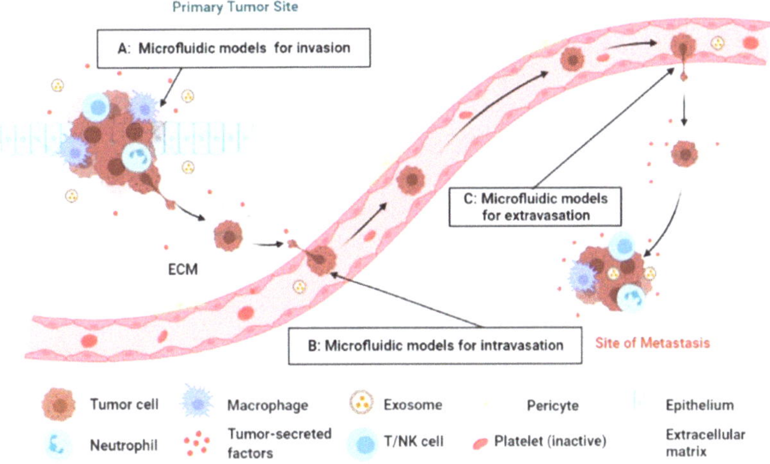

Figure 21. Microfluidic models for tnvasion, intravasation, and extravasation. Reprint with permission of [379], Copyright 2022, Authors (CC BY 4.0).

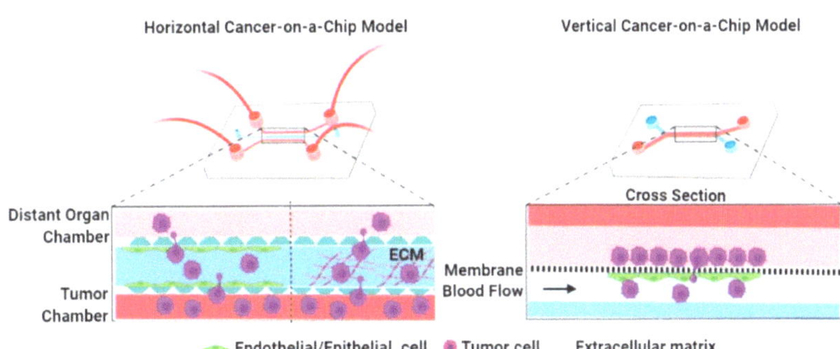

Figure 22. Simple COC models using microfluidics show how cancer cells can spread from their initial site to distant tissues. These designs can be broadly divided into chips that are horizontal and vertical. Different cell types can grow in their zones while still being able to interact with one another thanks to diverse signaling processes in horizontal models where chambers are divided by pillars that are only a few microns in diameter. Channels in vertical chips that simulate both cancer intravasation and extravasation processes are divided by membranes. To replicate more sophisticated tumor behavior, some models may incorporate vertical and horizontal layers, as with the ovarian TME OOC model. Reprinted with permission from Ref. [379], Copyright 2022, Authors (CC BY 4.0).

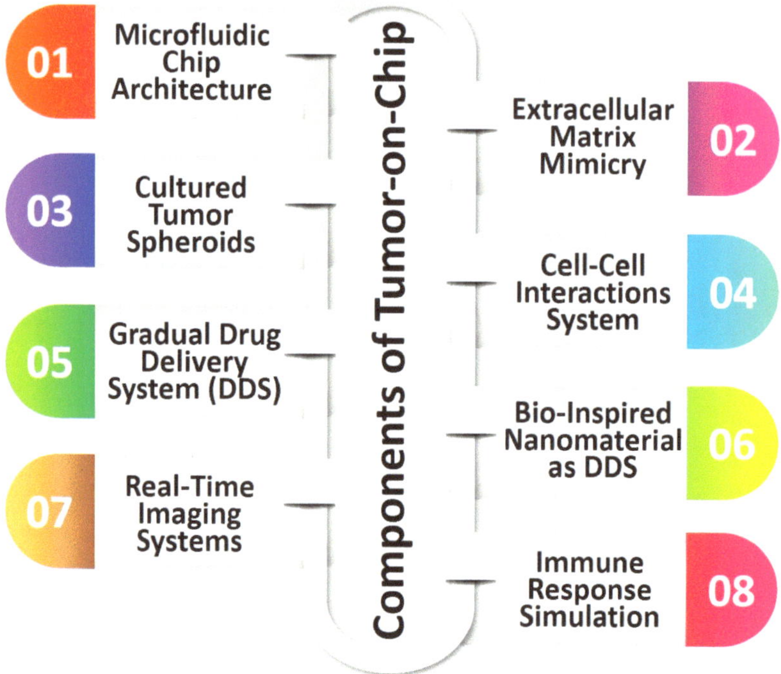

Figure 23. The basic components of TOC.

8.5. Bio-Inspired Wound Healing Dressing Mat

Bio-inspired wound healing dressing mats are a novel class of biomaterials specifically engineered to enhance the process of wound healing by leveraging inspiration from biological systems. These mats are designed to imitate the natural extracellular matrix, creating an environment that promotes cell growth, angiogenesis, and tissue regeneration. These environments provide advantageous conditions for wound healing, diminish the presence of microorganisms, and facilitate the regulated discharge of therapeutic substances such as growth factors and cytokines. Biomaterials derived from natural sources, such as silk proteins (fibroin and sericin), have demonstrated significant promise in wound dressings owing to their biocompatible nature and capacity to stimulate skin tissue regeneration. Integrating regenerative medicine and nanotechnologies offers a potentially effective strategy for tackling the complexities of wound management and promoting improving the healing process.

The integumentary system serves a vital role in numerous physiological processes. However, when the skin becomes compromised due to injuries, it can give rise to significant medical complications, such as heightened morbidity and mortality rates. Non-healing chronic wounds pose a significant challenge, particularly for individuals diagnosed with diabetes, as they may experience limb ulcers that can lead to severe consequences. The principal objective of wound management is to achieve expeditious healing while ensuring both functional and aesthetically satisfactory results. The wound healing process is intricate and encompasses various cellular interactions, secretion of factors, and interactions with the ECM. Comprehending these processes is imperative to formulate efficacious wound management strategies. The impact of diabetes on wound healing is detrimental, highlighting the need for a comprehensive comprehension of the wound environment and pathophysiology to develop more effective strategies for promoting wound healing. Utilizing biomaterials that can release signaling molecules, such as growth factors and cytokines, in a controlled manner has been shown to facilitate the process of angiogenesis and tissue regeneration. The development of effective biomaterials for tissue repair, including three-dimensional living tissues, has been facilitated by advancements in regenerative medicine, nanotechnologies, and bioengineering. Biomaterials derived from biological sources have demonstrated significant potential in treating tissue injuries and enhancing wound healing, owing to their biocompatible nature and capacity to stimulate skin tissue repair [381]. The investigation of biomaterials possessing wound-healing properties has been undertaken for diverse purposes in wound management. These biomaterials create advantageous microenvironments that promote cellular proliferation, inhibit microbial colonization, and facilitate the controlled release of therapeutic agents. The recent progress in wound healing approaches has created novel opportunities within the realm of regenerative medicine and tissue engineering.

As a promising biomaterial for tissue repair and regeneration, Bombyxmori's silk fibroin has attracted much interest [382]. Numerous research teams have investigated the potential of silk fibroin to create cutting-edge methodologies for tissue engineering and wound healing applications, either alone or in combination with other materials and processed in various ways. As biomaterials for wound dressing, many forms of silk fibroin, such as hydrogels, sponges, films, and nanofibers, have been suggested. During various stages of wound healing, these materials maintain moist conditions, permit gas permeability, and improve cell responsiveness. Silkworm cocoon goods include fibroin hydrogel, electrospun fibroin, sponge, film, solution, and powder. These products have a variety of uses in bioengineering, particularly in the treatment of wounds (Figure 24).

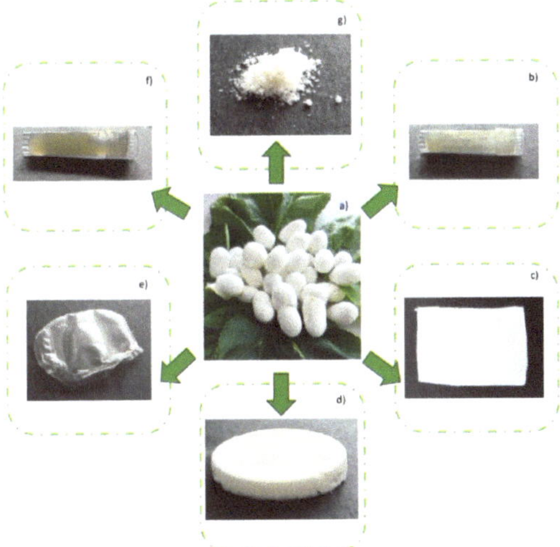

Figure 24. Representation of the instances of BINMs in devices fabricated by processing silkworm cocoons (**a**) for wound healing application: fibroin hydrogel (**b**), electrospun fibroin (**c**), sponge (**d**), film (**e**), solution (**f**), and powder (**g**). Reprinted with permission from Ref. [381].

8.6. Antimicrobial Surface

Using bio-inspired antimicrobial surfaces has generated considerable attention in diverse biomedical contexts owing to their efficacy in combating microbial hazards. These surfaces are influenced by natural defense mechanisms found in plants and microorganisms. They employ light-activated compounds or other biomimetic strategies to generate antimicrobial effects. Within medicine, these surfaces are utilized in various capacities, such as wound dressings, medical implants, and surgical instruments. Their primary function is to mitigate the risk of infections and expedite the healing process. In dentistry, dental implants and orthodontic devices are utilized to minimize bacterial colonization and the formation of biofilms, thereby improving oral health. In addition, implementing bio-inspired antimicrobial coatings on medical equipment and surfaces within hospitals enhances infection control measures and mitigates the potential for healthcare-associated infections. The potential of these bio-inspired antimicrobial surfaces, with their adaptability and inspiration drawn from biological systems, holds significant promise for enhancing healthcare and improving patient outcomes. This is achieved by offering robust protection against microbial pathogens in various biomedical environments.

A range of bio-inspired antimicrobial surfaces have been developed, each exhibiting unique mechanisms of action. Certain surfaces are designed to mimic the micro/nanostructures observed in natural entities such as cicada wings or lotus leaves. These surfaces possess rough and hydrophobic topographies, which effectively hinder the adhesion and colonization of bacteria. Consequently, these surfaces exhibit self-cleaning properties. Some researchers utilize synthetic antimicrobial peptides (AMPs), short chains of amino acids with a wide range of antimicrobial properties. These synthetic AMPs are employed to either disrupt the cell membranes of bacteria or hinder crucial cellular processes. Cationic polymers, which draw inspiration from the positive charge exhibited by antimicrobial peptides (AMPs), interact with bacterial cell membranes with a negative charge. These interactions ultimately disrupt the membrane structure, leading to the demise of the bacterial cells. Moreover, incorporating silver and other metal NPs into surfaces enables the gradual release of antimicrobial ions upon interaction with bacteria. This process disrupts bacterial metabolism and hinders DNA replication. Chitosan-based surfaces, derived from the exoskeletons of crustaceans, offer a biopolymer barrier of natural origin that effectively inhibits bacterial colonization. Mussel-inspired coatings, which utilize polymers functionalized with catechol groups, serve as effective platforms for integrating antimicrobial agents onto diverse surfaces. In addition, photodynamic antimicrobial therapy (PDT) involves the immobilization of light-sensitive agents on various surfaces, which, upon exposure to light, generates reactive oxygen species that effectively eliminate bacteria. In conclusion, using bio-inspired surfaces designed to release bacteriophages, viruses that selectively target and infect bacteria, results in the targeted eradication of bacterial pathogens. Utilizing a wide range of bio-inspired antimicrobial surfaces presents a potential avenue for improving healthcare outcomes and addressing the challenges posed by microbial infections.

8.6.1. Structure-Oriented Surface

Many plants and animals have evolved distinctive surface structures throughout millions of years of evolution, enabling them to endure external threats in difficult environmental circumstances. These organic and synthetic antimicrobial nanostructures, which are crucial to bioengineering, have piqued the curiosity of researchers. These surfaces' superhydrophobicity and micro/nanotopographies are thought to be responsible for their antibiofouling qualities. Taro leaves, for example, have a unique uneven structural distribution that prevents Gram-negative bacteria from adhering even in humid situations. Staphylococcus aureus has been discovered to resist shark skin's antibacterial properties. Studies using naturally occurring bactericidal surfaces, such as cicada wings with nanoneedle arrays, have demonstrated the ability to kill bacteria instantly upon direct contact in just 5 min. Dragonfly wings and gecko skin are examples of other species with mechanobactericidal surfaces. Although animal surfaces may have a lower water contact angle than plant surfaces, the antibacterial impact is similar, indicating that hydrophobicity is not the only factor affecting bactericidal effectiveness [383]. Creating next-generation bactericidal surfaces with physico-antimicrobial characteristics has drawn heavily on inspiration from nature. Research into naturally occurring nanostructures has sparked a number of ground-breaking innovations. Researchers fabricate artificial nanostructures on various substrates using bottom-up chemical synthesis and top-down multiway etching techniques. These nanostructures, like carbon nanotubes and ZnO nanorods, have physical and mechanical antibacterial properties that can damage bacterial cell membranes and prevent bacterial adherence. Hydrothermal synthesis and chemical deposition are two surface topography coatings and changes that improve the bactericidal effects. It is possible to replicate nanostructures seen in nature, such as those on cicada wings and pitcher plant surfaces, by combining several processes.

One or both probable manifestations of the antibacterial activity of naturally occurring nanostructured surfaces are the biocidal effect (total destruction of the cellular envelope) or the antibiofouling effect (inhibition of bacterial proliferation). The internalization or insertion of NPs that disrupt membrane function (nanotoxicological effects), physical puncturing, physical tearing, and chemical destructive extraction through oxidative stress are just a few of the factors contributing to the killing mechanism of physico-mechanical antibacterial materials. Mechanical antibacterial materials like carbon nanotubes and graphene impact bacterial adherence, internal cell architecture, and cell migration. These substances repeatedly rupture bacterial cells as part of a cumulative process. According to other studies, bacterial cells are prevented from approaching nanocolumn arrays by the height and spacing of the arrays, leading to a contacting physical puncturing mechanism. When the cells try to migrate on the nanocolumns' surface, they are broken apart. Some scientists suggest that bacteria produce an extracellular polymeric substance (EPS) under external mechanical stress rather than being directly pierced, which causes bacterial membrane damage through strong attachment to nanocolumns. Since diverse materials and structural configurations may have unique antibacterial effects on different microbes, the precise mechanism is still unclear and up for debate. However, the success of physico-antimicrobial surfaces depends on their specific structures, which are required for their antibacterial capabilities. Compared to chemical mechanisms, the physical antibacterial mechanism is typically faster, and the material's structure is key to generating an efficient antibacterial effect.

8.6.2. Peptide-Based Surface

Antimicrobial peptides (AMPs) exhibit considerable potential as viable therapeutic options for various diseases, particularly in combating multidrug-resistant bacteria. The global emergence of antibiotic resistance has garnered significant attention, prompting the exploration of alternative solutions, such as AMPs that possess wide-ranging antimicrobial properties. AMPs are synthesized by diverse organisms, identifying more than 5000 distinct AMPs to date [384]. These substances exhibit a specific mode of action by selectively interacting with microbial membranes, resulting in the formation of pores and ultimately leading to the demise of bacteria. Moreover, AMPs exhibit anti-inflammatory, regenerative, and anti-cancer characteristics.

Nevertheless, despite their considerable potential, AMPs encounter certain obstacles in their application. These challenges encompass the toxicity exhibited toward mammalian cells, vulnerability to proteases, and the high costs associated with their production methods. To tackle these concerns, there have been suggestions for using nanotechnology-based delivery methods to augment the stability and biological efficacy of AMPs. The development of bio-inspired NPs has been undertaken to preserve the activity of AMPs while mitigating any potential adverse effects. In addition, AMPs have been employed as surface coatings on implants to mitigate the risk of implant-related infections and promote bone regeneration. AMPs have demonstrated potential in cancer therapy due to their ability to specifically target malignant cells and facilitate the delivery of cancer medications or nucleic acids. AMP-based materials have demonstrated high efficacy in condensing and delivering nucleic acids, thereby protecting against degradation.

8.6.3. Metal/Metal Oxide NP-Based Surface

The wide range of features that metal and metal oxide NPs possess, such as non-toxicity, antibacterial activity, and anti-insecticidal activity, make them useful in the biomedical industry for identifying and treating serious illnesses [385]. Different bio-inspired metals and metal oxide NPs are essential for maintaining life processes, and deficiencies in these substances can cause diseases. For instance, Co NPs exhibit good magnetic, optical, and mechanical properties, making them useful for biomedical applications like magnetic resonance imaging (MRI) and drug delivery, while nanoceria, despite its lack of stability in living systems, shows promising applications in battling cancer and Alzheimer's disease. Other NPs, such as those made of Au, Ag, Fe, MgO, Ni, Se, and ZnO, also exhibit distinctive properties that can be used in energy storage, biosensing, imaging, and therapies. Chemoresistive nanosensors are created using nanomaterials such as nanorods, nanotubes, and nanobelts, broadening the range of biomedical applications. Overall, diverse features of metal and metal oxide NPs hold considerable promise for increasing biomedical research and healthcare.

8.6.4. Chitosan-Based Surfaces

The antibacterial properties of chitosan have been thoroughly investigated for various uses in the biomedical, cosmetic, food, and agricultural industries. Researchers have studied its usage in self-preserving materials, which have created various goods with antibacterial qualities, including beads, films, fibers, membranes, and hydrogels. Studies on the antimicrobial effects of chitosan have changed over the past 20 years, moving from studies on foodborne and soilborne pathogenic fungi to studies on bacteria, with varied assays and methodologies revealing the underlying mechanisms and factors determining its efficiency. Chitosan's potential as an antibacterial agent has been further boosted by the development of nanotechnology, which has made it possible to create materials with nanostructures that are more effective at the atomic level [386]. Although the precise mechanism underlying chitosan's antibacterial activity is not entirely understood, many independent factors have an impact. According to the major hypothesized mechanism, Chitosan adhering to the bacterial cell wall causes cell disruption, changes in membrane permeability, and inhibition of DNA replication, which results in cell death. Chitosan's polycationic structure, which interacts electrostatically with the anionic components of microbial surfaces, is essential to the substance's antibacterial effect. Additionally, the shape and size of chitosan particles can affect how they interact with bacterial cell surfaces, with larger NPs behaving differently from smaller ones. Chitosan also has antifungal properties that limit spore germination and radial growth in fungi. Studies have demonstrated its effectiveness against several fungi linked to food and plant rotting. Additionally, chitosan can activate enzymes called chitinases in plant tissues, which act on various fungus species.

8.6.5. Mussel-Inspired Antimicrobial Coatings

Mussel-inspired coatings are developed from the adhesive properties of mussel foot proteins in marine mussels [387]. The proteins in question facilitate the strong attachment of mussels to diverse surfaces within moist and turbulent environments, such as coastal areas in the ocean. Researchers have successfully replicated the bioadhesive chemistry found in mussel foot proteins, resulting in coatings exhibiting strong surface adhesion and antimicrobial characteristics. The fundamental operational principle underlying mussel-inspired coatings centers on integrating catechol molecules. Catechol is a prevalent chemical functional group abundantly present in mussel foot proteins. This collective facilitates robust and enduring adherence to various surfaces, encompassing metals, polymers, ceramics, and even biological tissues. The coatings under consideration utilize catechol groups to establish a durable attachment mechanism by forming covalent and non-covalent bonds with the desired substrate. The antimicrobial properties of these coatings are derived from the distinctive amalgamation of adhesive chemistry and the intrinsic characteristics of catechol. The application of these coatings demonstrates a high level of efficacy in inhibiting bacterial colonization and the formation of biofilms. The presence of adhesive catechol groups results in the disruption of microbial cell membranes, causing destabilization of the membranes and subsequent leakage of vital cellular constituents. Furthermore, the surface roughness and hydrophobicity of the coatings also impede the attachment and proliferation of bacteria. Mussel-inspired antimicrobial coatings significantly advance healthcare outcomes and biomedicine by improving biocompatibility, preventing infections, and supporting tissue regeneration.

8.6.6. Bacteriophage-Based Antimicrobial Surface

Bacteriophages are widespread and are viruses that attack bacterial cells. When Ernest Hankin examined the waters of the Ganges and Jamuna Rivers in India in 1896, he discovered the initial signs of bacterial parasites in the environment. Hankin showed an unidentified material in the river water that has antibacterial capabilities against *Vibrio cholerae*, without specifically identifying phages. While working with Bacillus subtilis two years later, Russian bacteriologist Nikolay Gamaleya noticed a comparable incident [388]. The immobilization of bacteriophages plays a crucial role in advancing biotechnologies, presenting new prospects for detecting pathogenic microorganisms at minimal concentrations, developing materials possessing distinctive antimicrobial characteristics, and facilitating fundamental research on bacteriophages. Bacteriophages of indigenous origin exhibit a notable propensity for a particular bacterial species, and in some cases, a discernible subspecies, primarily due to the recognition of epitopes on the capsid proteins [389]. By means of chemical or genetic modifications, the binding specificity can be modified, thereby enabling redirection toward a diverse range of substrates and analytes beyond the scope of bacteria. Therefore, the attachment of bacteriophages to flat and particulate surfaces is a rapidly growing area of significant scientific fascination [390]. Table 2 thoroughly summarizes all the subsections under this section and lists the most recent biomedical applications of micro/nanodevices fabricated from BINMs.

Table 2. A summary of all the subsections under this section and list of the most recent biomedical applications of micro/nanodevices fabricated from BINMs.

Sector	Devices	Bio-Inspiration	Mechanism	Applications	Refs.
Drug Delivery and Therapeutic Applications	Liposome-based drug delivery system	Liposomes	Loaded with chemotherapy drugs	Laryngeal cancer cells	[391]
	Liposome-based nanoarchitectonics	Liposomes	Loaded Ag NP	Cancer management	[392]
	Nano-liposome-based transdermal hydrogel		Targeted delivery of dexamethasone	Rheumatoid arthritis therapy	[393]
	Liposome-based nanocomposite drug delivery system		Loaded with Ag NPs, hyaluronic acid, lipid NPs	Cancer treatment	[394]
	Dendrimer-based nanocomposites	Dendrimers	RNA delivery	Cancer vaccination	[395]
	Dendrimernanosystems	Dendrimer nanomicelles	Adaptive tumor-assisted drug delivery via extracellular vesicle hijacking	Tumor treatment	[396]
	Lipid-coated ruthenium dendrimer conjugation	Dendrimers	Hydrophobic locking protocol	Cancer treatment	[397]
	Dendrimer-gel-derived drug delivery systems		Encapsulation or chemical coupling	Glaucoma medications	[398]
	Erlotinib-loaded dendrimer nanocomposites		Entrapment or encapsulation of drug	Targeted lung cancer chemotherapy	[399]
	Antiglycolytic cancer treatment	Micelles	Considering the unique metabolism of cancer cells	Antiglycolytic cancer treatment	[400]
	Core-(shell-cross-linking)-corona micelles (CSCCMs)		Shell protection strategy	Photo- and pH dual-sensitive drug delivery	[401]
	Polymeric nanocomposite		Polymeric micelles using citraconic amide bonds	Cancer treatment	[402]
	Hyaluronic acid-coated polymeric micelles		Through specific cellular uptake	Liver fibrosis therapy	[403]

Table 2. *Cont.*

Sector	Devices	Bio-Inspiration	Mechanism	Applications	Refs.
	Membrane-coated nanosystems	Blood cells Cancer cells Stem cells Extracellular vesicles Viral capsids Bacteria	Targeting Diagnosis Drug delivery	Theranostic applications	[404]
	Surface plasmon resonance (SPR)-based sensors	Cancer cells	Detected via a change in SPR angle	Detection of cancer	[405]
	Metal–organic framework–azidosugar complex	Cancer cell membrane	Metabolic glycan labeling (MGL)	Breast cancer treatment	[406]
	Gene delivery system	Viral vectors	Nucleic acid molecule in a protein coat	Gene therapy and imaging	[407,408]
		Virus-like particles	Self-assembled capsules composed of viral capsid or envelope proteins that preserve antigenicity	Vaccine, drug, and gene delivery	[409]
		Virosomes	Virion-like phospholipid bilayer vesicle containing an incorporated glycoprotein within an empty compartment	Vaccine, gene, and drug delivery	[410–412]
	Enzyme-powered nanomotors	Natural molecular motor	Self-propulsion	Protein delivery and imaging	[413]
	Micro/nanomotors	Natural mobile microorganisms	Self-propulsion	Diagnostics, therapeutics, and theranostics	[414]
		Sperm cells	Propulsion of its own flagellum Directional guidance by chemotaxis, thermotaxis, and rheotaxis	Diagnostics, therapeutics, and theranostics	[415–418]
		Bacteria	Self-propulsion and driven by stimuli	Diagnostics, therapeutics, and theranostics	[419–421]
		Algae	Electro-magnetic field propulsion and driven by stimuli	Diagnostics, therapeutics, and theranostics	[422–424]
Biomimetic Nanobiosensors	Biomimetic nanophotonic biosensors	Nature's photonic crystal	Natural structural color and stimuli-responsive photochemical reaction	Enzyme detection, detection of spiked human serum	[425–427]
		Structural colors in chitin-constituted insect shells	Hierarchical structures of carbohydrate nanofibrils such as chitin and cellulose	Nanobiosensor label-free detection of urinary venous thromboembolism biomarker	[428–430]
	Metallic nanobiosensors	Bio-inspired metallic NPs	Fluorescence signal variations that depend on the size, color, and surroundings	Electrochemical, colorimetric, and fluorescence nanobiosensors	[431–437]
	Polymer-composite-based nanobiosensors	Nature's biorecognition components such as antibody, enzyme, antigen, protein, DNA, etc.	Detection of biological reactions and conversion to signals	Detection of interleukin-8 (IL-8), TNF-α, cancer biomarkers, and neuron-specific enolase (NSE)	[438–442]
	Hydrogel-based piezoelectric sensors	Biological tissues and organisms that exhibit piezoelectric properties	Due to the asymmetric arrangement of atoms or molecules in the crystal structure of the material	Wound healing, ultrasound simulation, and imaging	[443–445]
Organ-on-Chip	Single organ-on-chip	Different human organs individually such as liver, kidney, lung, gut, heart, muscle, blood–brain barrier, seplon, bone marrow, etc.	Through the replication and simulation of the physiological functions and interactions of human organs in a microfluidic platform	Drug development and testing, disease modeling, personalized medicine, toxicity screening, and reducing the reliance on animal testing in pharmaceutical research	[363,446–451]
	Multiple organs-on-chip	Multiple organs are interconnected similar to the human body	Replicating the physiological and biochemical characteristics of organs in a controlled and interconnected manner	Enabling the study of organ–organ crosstalk, drug responses, and disease progression with higher accuracy and relevance compared to traditional in vitro models	[369,370,452]

Table 2. *Cont.*

Sector	Devices	Bio-Inspiration	Mechanism	Applications	Refs.
	Human-on-chip	Whole human body	Replicating the complexity and functionality of the entire human body in a miniature, interconnected platform	Applicable for accurate and predictive pre-clinical studies, drug testing, and disease modeling	[371,453,454]
	Patients-on-chip	Patient-specific cells, tissues, or induced pluripotent stem cells (iPSCs)	Replicating the unique characteristics of an individual's biology, including their genetic background, disease conditions, and drug response, in a microscale platform	Particularly valuable for precision medicine, where tailored therapies are developed based on a patient's specific needs and response	[455,456]
Cancer-on-Chip	3D breast COC	Breast tumors	Replicating key aspects of the breast tumor microenvironment, such as the extracellular matrix composition, stiffness, and architecture	Therapeutic evaluation of drug delivery systems	[457–459]
	Pancreatic COC	Pancreatic tumors	The use of microfluidic technology to recreate the microenvironment of pancreatic tumors	Platform that recapitulates the tumor microenvironment and enables detailed investigations into pancreatic cancer biology, drug responses, and potential therapeutic strategies	[456,460,461]
	Lung COC	Lung tumors	Platform that replicates the key features of lung tumors and their microenvironment	Provides a potent tool for studying the biology of lung cancer and creating tailored treatments for this terrible illness by simulating the lung tumor microenvironment, including mechanical forces, oxygen levels, and cellular interactions	[462–464]
Wound Healing Dressing Mat	Silk-fibroin-based wound dressings	Silkworm cocoons	Creating a favorable microenvironment for wound healing and tissue regeneration	Wound healing, tissue regeneration, bioactive molecule delivery, cosmetic uses, drug delivery, hemostatic dressings, and tissue engineering scaffolds	[465–467]
	Biopolymer nanocomposite thin film	Natural ECM	Creating a flexible and biocompatible film that can adhere to the wound, maintain a moist environment, and release bioactive molecules to promote wound healing and tissue regeneration	Wound healing and tissue regeneration	[468–470]
	Bio-inspired adhesive formulations	Gecko feet, mussel adhesive proteins, or insect adhesives	Mimicking the chemical and physical properties of natural adhesives	Medical adhesives for wound closure and healing	[471–473]
Bio-inspired Antimicrobial Surface	Structure-oriented antimicrobial surface	Natural surface morphology of plants, animals, and insects	Affecting microbial adhesion, internal cell structures, and cell migration	Antibacterial, antiviral, and antifungal applications	[474–477]
	Peptide-based surface	Naturally occurring peptide molecules found in various organisms	By specifically interacting with the negatively charged membranes of bacteria, fungi, and viruses, AMPs produce their antimicrobial actions by inducing holes to develop and ultimately cell death	Antimicrobial coating, wound dressing, detection of pathogens, nanomedicine, and medical implants	[478–481]
	Metal/metal oxide NP-based antimicrobial surface	Natural metallic NPs	Antimicrobial properties of metal/metal oxide NPs	Antimicrobial coating, wound dressing, detection of pathogens, nanomedicine, and medical implants	[385,482,483]
	Chitosan-based antimicrobial surfaces	Chitosan	Interacting with bacterial cell membranes, disrupting their structure, and leading to cell death	Antimicrobial coating, wound dressing, detection of pathogens, nanomedicine, and medical implants	[484–486]

Table 2. *Cont.*

Sector	Devices	Bio-Inspiration	Mechanism	Applications	Refs.
	Mussel-inspired antimicrobial coatings	Adhesive properties of mussel foot proteins	The catechol groups facilitate strong and stable interactions with surfaces, providing long-lasting antimicrobial properties	Coating medical devices, implants, and wound dressings	[487–489]
	Bacteriophage-based antimicrobial surface	Naturally occurring viruses that specifically target and infect bacteria	These surfaces provide a promising substitute for conventional antibiotics by using bacteriophage selectivity to eradicate particular bacterial infections	Wound dressings, medical implants, and catheters	[490–492]

9. Challenges for Bio-Inspired Nanomaterials in Biomedical Applications

The application of BINMs in biomedicine holds great promise and potential. However, several challenges must be addressed to incorporate these materials and ensure widespread adoption. The biomedical applications of BINMs are still facing some challenges related to biocompatibility and safety concerns, biological complexity, synthesis scalability, targeting and delivery precision, long-term stability, regulatory and ethical considerations, interdisciplinary collaboration, cost and accessibility, standardization, and quality control (Figure 25).

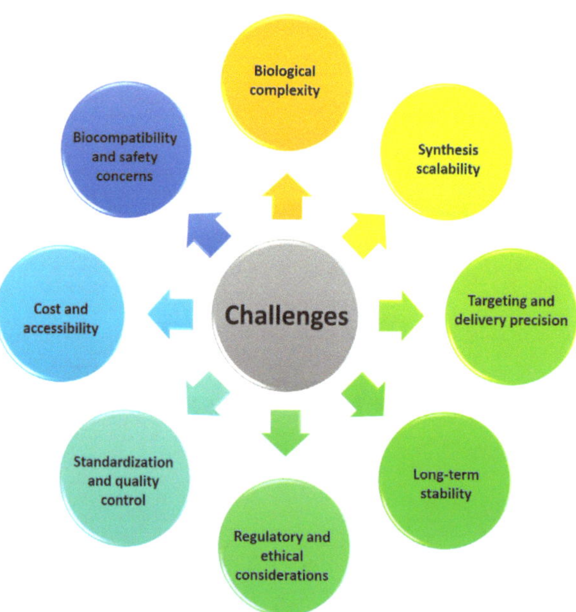

Figure 25. The challenges faced by the BINM-based micro/nanodevices in biomedical applications.

9.1. Biocompatibility and Safety

One of the foremost challenges lies in the assurance of biocompatibility and safety of BINMs within intricate biological systems. The cytotoxicity (effects on cell activities and survival) and biocompatibility of BINMs intended for biomedical applications must be assessed [493]. Both in vitro and in vivo tests are included in this biocompatibility assessment. They must pass tests for cytotoxicity, carcinogenicity, reproductive toxicity, immunotoxicity, irritation, sensitization, hemocompatibility, systemic toxicity, and pyrogenicity on BINMs. These assessments are essential for the security of manufacturing employees and patients receiving BINM therapies [494]. Furthermore, a fundamental understanding of the connection between the BINMs' physicochemical characteristics and their particular biological effects is required to improve their application. For instance, studies show that BINMs can limit cancer formation and manage the scarring process [495]. However, because of the wide variety of BINMs, the abundance of testing model systems, the absence of standardized testing techniques, and the difficulties involved with in vivo tests, difficulties occur when measuring bio-interactions.

9.1.1. Cytotoxicity and Genotoxicity of BINMs

The cytotoxicity of BINMs might be caused via direct necrosis, induced apoptosis, or immunological clearance, depending on their composition, molecular structure, and size. Cytotoxicity studies frequently evaluate metabolic impairment, cell-death marker production, and damaged cell membranes. Several particular tests such as the LDH test (tracks the release of LDH, which is generally present in healthy cells), Caspase-3/7 (measures the amount of caspases produced, which are responsible for apoptosis), and MTT assay (this assay particularly measures the reduction of the tetrazolium salt utilizing redox indicators to gauge metabolic activity changes to evaluate cell viability) are suitable for assessing BINMs. Cell viability may be impacted by BINMs, depending on their composition or geometry [496]. Even though some substances may not result in cell death, they can nonetheless have sub-lethal effects on the genome and epigenome, particularly at lower dosages. The genotoxicity of BINMs has been extensively studied, and common procedures include the Ames test, comet test, micronuclei test, DNA laddering test, and chromosome aberration test. Recently, chemical mutagenicity assessments have also used next-generation sequencing [497]. The ability of BINMs to directly penetrate cells or to catalyze intracellular OH radical generation, which can enhance ROS and DNA damage, is crucial to comprehend [498].

9.1.2. Immunomodulation of BINMs

Both direct and indirect immunomodulation, including immunosuppression and immunostimulation, can be caused by BINMs [499]. Although BINMs can be used to deliver drugs or vaccines utilizing NPs, this article focuses on the immunological reactions that the BINMs cause. Adaptive immunity involves T cells and B cells creating antigen-specific reactions, and innate immunity, which deals with non-specific interactions with immune cells like macrophages, is involved in this. For instance, zinc oxide and silver BINMs have been found to increase the production of IL-6 and IL-8 in kidney cells, indicating improved innate and adaptive responses [500]. Size, surface chemistry, molecule structure, and chemical content of BINMs all impact their immunomodulatory activities. Particularly important in immunomodulation is particle size. For example, NPs (193 nm) induced a stronger immunological response than their microparticle equivalents (1530 nm) [501]. Larger particles may stimulate a stronger serum immunoglobulin response [499]. Identifying sensitivities to specific BINM features is difficult due to the diversity of immunological responses from different BINMs and the varied testing procedures. The intricacy of interpretations can also be increased by the synergistic influence of several components on these processes [502].

9.1.3. Fibrosis Induced by BINMs

BINMs can cause fibrosis, a condition characterized by the excessive accumulation and modification of the extracellular matrix (ECM). Since fibrosis can develop without causing an immediate reaction and because there are no established prognostic tests, our understanding of it is limited. Cells are examined for the expression of fibrosis-related proteins, such as α-smooth muscle actin, transforming growth factor beta (TGF-β), and other ECM components like fibronectin, laminin, and COL I, during in vitro tests for fibrotic reactions. Cells are routinely grown on rigid substrates (in 2D cultures), which might alter gene expression due to the nucleus pore opening from mechanical stresses. This is a known problem. This implies the necessity of in vivo research [503]. Due to their tiny size, NPs can easily enter lung alveoli. They can trigger a variety of reactions, including fibrosis when they come into touch with lung cells. For instance, NPs can increase the production of TGF-β and reactive oxygen species (ROS) [504]. The relationship between NPs and these negative reactions may be complicated because these consequences may not cause immediate discomfort. There are several different ways that NMs cause fibrosis. Unintended immunogenic responses, cytotoxicity, and potential long-term impacts on human health are significant issues that necessitate comprehensive examination and resolution. The nanomaterials produced through biosynthesis exhibited negligible toxicity concerning hematological, biochemical, histological, and DNA damage assessments [505].

9.2. Biological Complexity

Nanomaterials must navigate a complicated biological environment with its many cellular networks, interlaced routes, and multiscale mechanisms. These tiny structures meet a continually changing physiological and metabolic milieu when introduced into such systems. Therefore, understanding and forecasting nanomaterial behavior and, more significantly, biological system response is a huge problem. Proteins in the living environment can build a "protein corona" on NPs. Unintended adsorption can drastically change the nanomaterial's biodistribution, half-life, and therapeutic efficacy. The protein corona can also elicit immunological responses, which may remove nanomaterials quickly or cause unexpected adverse effects. Nanomaterials interact dynamically with cells, life's building components. Depending on size, charge, and surface properties, nanomaterials may be endocytosed or diffused by cells. On the other hand, cellular absorption can be harmful. It may be useful for delivering treatments directly into cells but may also cause cytotoxicity or interfere with biological functioning. Tissues and cell clumps increase this intricacy. Some tissues are nanomaterial-permeable, while others are impenetrable. The blood–brain barrier blocks most chemicals, including nanomaterials, from entering the brain. Overcoming such constraints without harming the nanomaterial's functionality demands a delicate balance of design and innovation.

9.3. Synthesis Scalability

The synthesis of BINMs frequently encompasses intricate procedures, elaborate molecular architectures, and meticulous functionalization, necessitating a high level of complexity and precision. Scaling up these processes to achieve mass production while preserving their inherent properties and quality poses a substantial technical challenge. A comprehensive approach is employed to address the intricate biological complexities related to using BINMs in biomedical applications. The process entails comprehensive biocompatibility evaluations, including in vitro and in vivo investigations to assess potential cytotoxicity, immunogenicity, and long-term consequences. Furthermore, customized surface modifications and functionalizations are implemented to augment the biocompatibility and targeting specificity of nanomaterials, thereby minimizing any detrimental interactions with biological systems. Again, there has been significant progress in developing sophisticated computational models capable of simulating and predicting the interactions of nanomaterials within intricate biological environments. This advancement has greatly facilitated the process of designing and optimizing the performance of these nanomaterials. Implementing these collaborative approaches plays a significant role in effectively navigating the complex biological environment and facilitating the secure and efficient incorporation of BINMs across various biomedical contexts.

9.4. Targeting Delivery

BINMs have expanded biomedical applications, including medication delivery and diagnostics. These NPs can selectively reach desired tissues or cells without disrupting healthy ones, maximizing therapeutic efficacy and minimizing negative effects. Molecular or cellular interaction is a major benefit of nanomaterials in medicine. Their small size lets them negotiate the complex blood vessel network and reach even the most inaccessible body areas. Simply being small is not enough. Surface features, including charge, hydrophobicity, and functional groups determine how these NPs interact with biological substances and cells. Complex systems like the human body have several defenses to identify and destroy alien things. Avoiding the immune system is difficult when using nanomaterials for therapy. Researchers can make these NPs "invisible" to immune cells or use specific biological processes to improve their targeting by altering their surfaces. Active targeting is another trending method. NPs are functionalized with ligands or antibodies that bind to target cell receptors. NPs can bind selectively to cancer cells that overexpress certain receptors, delivering therapeutic chemicals precisely where they're required without harming healthy cells. The regulated release of medicinal compounds from NPs is crucial. A delayed or inadequate release may not be helpful, whereas an abrupt or excessive release may be harmful. Scientists can sustain and control medication release by modifying nanomaterial composition and structure, keeping drug concentration within the therapeutic window. External stimuli like pH, temperature, or light can also regulate release. Certain nanomaterials release their therapeutic payload in reaction to tumor cells' acidic environment or inflamed tissues' high temperatures.

9.5. Stability of BINMs

The long-term stability of BINMs within the biological environment is paramount in ensuring sustained therapeutic efficacy. Managing factors such as degradation, aggregation, or alteration of properties over time is necessary. Several techniques can be used to increase the BINMs' long-term stability for biomedical applications. These include meticulously choosing biocompatible and stable materials, surface modification to prevent degradation, controlled release systems to control therapeutic agent release, encapsulation within protective matrices, use of crosslinking agents to increase stability, thorough biocompatibility testing, in vivo research for in-the-moment insights, and computational modeling to forecast behavior. Researchers can address stability issues and guarantee the long-term effectiveness of BINMs in challenging biological settings by employing these techniques.

9.6. Regulatory and Ethical Considerations

The field of nanomaterials in biological applications is characterized by both technological advancements and complex regulatory and ethical considerations. As scientific advancements approach the limits of feasibility, it becomes increasingly imperative to exercise prudence, guaranteeing the responsible introduction of innovations, taking into account considerations of personal well-being and broader societal ramifications. From a regulatory standpoint, the journey from a laboratory notion to a commercially viable product is intricate and complex. Regulatory agencies, such as the Food and Drug Administration (FDA) or the European Medicines Agency (EMA), require a thorough compilation of supporting documentation prior to granting permission. The data presented encompass more than simply the clinical effectiveness of the treatment and explore the possible long-term adverse effects, environmental consequences, and wider societal implications associated with introducing a novel therapeutic approach. This frequently entails conducting extended clinical trials, comprehensive toxicity assessments, and implementing rigorous production protocols to guarantee uniformity and excellence.

The ethical considerations are similarly, if not more, complex. As the manipulation of materials at the nanoscale and their subsequent introduction into the human body are undertaken, inquiries emerge: Has full informed consent been acquired from the patients? Do individuals possess an understanding of the potential long-term ramifications, particularly in cases when these ramifications remain uncertain? How can we effectively promote equitable access to these potentially transformative therapies? Does the potential exist for the exacerbation of pre-existing inequality in healthcare? The issue of privacy arises, especially when these nanomaterials interact with digital technologies. Which individuals or entities are granted access to the data, and what measures are being implemented to ensure their protection?

The environment constitutes an additional dimension inside the ethical framework. The environmental implications of the production, utilization, and disposal of NPs warrant investigation. Do these entities undergo decomposition or persist, resulting in unanticipated ecological disturbances? The multifaceted nature of BINMs also gives rise to many prospects and complexities. Integrating knowledge from other fields can contribute to developing comprehensive solutions, underscoring the importance of proficient interdisciplinary communication. Every academic field is characterized by its own unique vocabulary, methodology, and priorities. It is of utmost importance to prioritize establishing effective communication and a shared vision to prevent the fragmentation of efforts, which may result in overlooking crucial aspects. Fundamentally, the field of bio-inspired nanomaterials for biomedical purposes exhibits significant potential. However, this pursuit is accompanied by many factors surpassing mere scientific obstacles. Each progression is a nuanced choreography of originality and accountability, necessitating attentiveness, anticipation, and a dedication to the collective welfare.

9.7. Cost and Accessibility

BINMs are expensive because they require high-end facilities, specialized researchers, and rare raw ingredients. Between basic research and commercial production are many stages, each with its own costs. Fundamental research might take years, followed by synthesis refining, safety studies, and rigorous clinical trials before medicinal use. These trials are necessary to ensure nanomaterial safety and efficacy but are time- and resource-intensive. Other economic aspects must be considered besides production costs. Medical material safety and efficacy regulations can be lengthy and costly. Many nanomaterials are unique, and therefore, regulatory authorities may require extra testing, increasing expenses. Accessibility is another issue, especially for global healthcare. Developed nations may have the facilities and resources to invest in these modern therapies, whereas emerging or undeveloped places may not have the funds. Even if these components are obtained, a lack of skilled staff or facilities to conduct treatments may render the technology worthless. The issue goes beyond procurement. Specialized equipment or conditions may be needed to store and preserve these fragile nanomaterials, straining limited resources. Distributing these items worldwide, especially to rural areas, might be a logistical nightmare without compromising their efficacy. While nanomaterials are expensive, their potential benefits are great. Their capacity to target specific cells or tissues, limit side effects, and even offer new treatments puts them at the forefront of modern medicine. This makes cost and accessibility issues even more important. The effectiveness and accessibility of BINMs must transcend geographical and economic boundaries to improve healthcare. Researchers, industry leaders, and politicians must work together to achieve this balance.

9.8. Standardization and Quality Control

Establishing standardization protocols and implementing quality control measures play a crucial role in guaranteeing uniform quality, reproducibility, and adherence to standardized testing methodologies for BINMs. These factors are of utmost importance in facilitating the effective translation of such materials into clinical applications. Overcoming these challenges necessitates the integration of scientific advancements, meticulous experimentation, adherence to regulatory standards, and cooperative endeavors involving researchers, medical professionals, policymakers, and industry stakeholders. The successful resolution of these obstacles will facilitate the efficient implementation of BINMs in various biomedical domains, encompassing diagnostics, drug administration, tissue engineering, and regenerative medicine.

9.9. Recent Developments and Commercial Viability of BINM-based Micro/Nanodevices

9.9.1. Commercially Available Nanobiosensors

When a biosensor demonstrates superior performance during real-sample testing, it can proceed to commercial production. The analytical capability of the sensor in practical scenarios will determine its market feasibility [506]. Several biosensors have achieved considerable commercial success, tracking metrics such as blood glucose, cholesterol, malaria, HIV, and uric acid [507]. Furthermore, sensors monitoring cancer and cardiac diseases have garnered significant commercial interest [508]. The affordability of production and scalability are essential for these devices to thrive in the market. To make them more economical, innovations like paper-based and chip-based microfluidic technologies have been introduced [509]. These tools ensure accurate sample management and precise analyte detection. Due to the low cost of paper, paper-based biosensors are attractive for both manufacturers and consumers. A variety of clinical biosensors such as a blood profiler from abbott (iSTAT) (https://www.pointofcare.abbott, accessed on 15 August 2023), glucose monitoring system from allmedicus (GlucoDr) (https://www.lelong.com, accessed on 15 August 2023), blood hemoglobin analyzer (AimStrip) (https://www.1cascade.com, accessed on 15 August 2023), uric acid detector from ApexBio (UASure) (https://redmed.pl, accessed on 15 August 2023), integrated printed circuit biosensor from Acreo (https://www.acreo.se, accessed on 15 August 2023), and pregnancy dipstick from alere (hCG combo) (https://www.alere.com, accessed on 15 August 2023) have been adopted commercially that offer both excellent analytical results and speedy detections. There is also a surge in emerging technologies, including cost-effective prototypes built on paper, elastomers, and combinations of the two [510]. Traditional diagnostics, which can be slow and require bulky equipment, are inadequate for urgent or remote medical situations. In contrast, advanced biosensors address these challenges with swift, on-the-spot testing capabilities. Yet, the high diagnostic needs often outpace traditional methods and current commercial biosensing devices in areas hit hard by epidemics or pandemics. Therefore, there is an immediate demand for scalable, cost-effective prototypes to handle future disease outbreaks better.

9.9.2. Commercially Available Drug Delivery System

Drug delivery systems (DDSs) composed of tiny molecules, peptides, nucleic acids, proteins, and cells commonly encounter obstacles in the delivery process, impeding commercial product development progress. The investigation has uncovered three primary solutions for addressing delivery challenges, which encompass the modification of the medication itself, manipulating the drug's surrounding environment, and developing a delivery system that can effectively regulate drug interactions within its microenvironment [511]. Small-molecule-based DDSs face challenges related to biodistribution, half-life, exposure, concentration, solubility, permeability, target development, and off-target toxicity. Commercial manufacturers addressed these issues by developing an osmotically controlled release oral-delivery system for methylphenidate HCl (Concerta) to address the drug tolerance issue by regulating its pharmacokinetics [512]. Ritonavir, commercially known as Norvir, is a protease inhibitor often used to treat HIV. It has been chemically modified with thiazole to enhance its metabolic stability and solubility in aqueous environments [513]. Benazepril, commercially known as Lotensin, is an alkyl ester prodrug that is designed to conceal ionizable groups and enhance its lipophilicity [514]. Ezetimibe (Zetia) is a pharmacological agent that is a selective inhibitor of cholesterol absorption. This compound was initially identified through a process known as library screening [515]. Naloxegol, commercially known as Movantik, is a derivative of naloxone that has been PEGylated to inhibit its ability to pass the blood–brain barrier [516].

DDSs composed of protein and peptide molecules confront challenges like enhancing physical stability, managing pharmacokinetic (PK) attributes like half-life and biodistribution, non-invasive application, overcoming biological barriers, minimizing immune reactions, and refining target precision. To address these challenges, commercial producers have adopted a range of strategies. For example, Desmopressin (DDAVP) was developed as a vasopressin analog, incorporating a non-standard amino acid to boost its stability and half-life [517]. Leuprolide acetate's depot suspension, Lupron Depot, offers a prolonged-release microsphere formula of the luteinizing-hormone-releasing hormone, enhancing its half-life [518]. Insulin human inhalation powder, known as Afrezza, is an inhaled insulin variant comprising microparticles mixed with fumaryl diketopiperazine, ensuring it is apt for inhalation [519]. Semaglutide, or Rybelsus, is an orally administered GLP-1 agonist, blended with SNAC to enhance stomach absorption [520]. Pegademase bovine, or Adagen, is a PEGylated protein treatment designed to prolong half-life while decreasing immunogenic responses [521]. Belatacept, or Nulojix, is an innovatively designed fusion protein that showcases amino acid modifications to heighten its selectivity for CD86 and CD80 [522].

Antibody-based DDSs suffer from similar kinds of obstacles faced by protein- and peptide-based DDSs. Some examples of issues resolved by commercial producers are cited here. Certolizumab pegol, also known as Cimza, is the initial PEGylated antibody fragment to receive approval from the FDA [523]. This modification enhances the half-life of the antibody fragment and improves its solubility. Blinatumomab, commercially known as Blincyto, is a lyophilized antibody formulation that incorporates trehalose to enhance the stability of the antibody structure [524]. Trastuzumab and hyaluronidase-oysk (commercially known as Herceptin Hylecta) are a novel subcutaneous depot formulation that incorporates hyaluronidases to enable Herceptin's controlled and prolonged release [525]. Panitumumab, commercially known as Vectibix, is the initial entirely human antibody to receive approval from the U.S. FDA [526]. This therapeutic agent effectively mitigates immunogenicity and inhibits the production of anti-antibodies.

Nucleic acid-based DDSs encounter various obstacles in their implementation, including the regulation of pharmacokinetic parameters, maintenance of stability, facilitation of efficient cell membrane penetration, attainment of access to the cytosol or nucleus after uptake, mitigation of immunogenic responses, and prevention of unwanted gene modifications. There are various commercial solutions available to address these concerns. Patisiran, also known as Onpattro, is a therapeutic agent that has received approval from the United States FDA [527]. This therapy utilizes small interfering RNA (siRNA) and lipid NPs to facilitate effective transportation to the liver and uptake by target cells. The therapy also utilizes ionizable cationic lipids to facilitate the drug's release from endosomes following endocytosis. Fomivirsen, also known as Vitravene, represents the inaugural antisense oligonucleotide to receive approval from the FDA. It incorporates a modification in the form of a phosphorothioate backbone, which enhances its resilience against nucleases [528]. Givosiran, also known as Givlaari, is a GalNAc-siRNA compound that facilitates enhanced cellular absorption in liver hepatocytes [529]. Nusinersen, commercially known as Spinraza, has been granted approval for treating spinal muscular atrophy. This therapeutic intervention incorporates a modification known as 2'-O-methoxyethyl phosphorothioate, which serves the dual purpose of diminishing immunogenicity and enhancing stability [530]. CRISPR technology is utilized to modify CD34$^+$ cells with CCR5 to combat HIV-1. The CD34$^+$ cells, which are subject to clinical trials with the identifier NCT03164135, undergo ex vivo editing procedures. In order to detect any unintentional modifications, the process of whole-genome sequencing is employed subsequent to post-editing and engraftment [531].

Nucleic acid-based DDSs encounter challenges pertaining to the unpredictability of pharmacokinetic parameters, the need to sustain persistence and viability within the body, the imperative to minimize immune reactions, the preservation of therapeutic properties of cells, the assurance of precise delivery to the intended disease site, and the scalability of manufacturing processes. In light of these challenges, a number of novel ways have been devised. Preclinical alginate implants have been developed to regulate the release of chimeric antigen receptor (CAR) T cells to the specific illness site [532]. The SIG-001 treatment utilizes genetically engineered cells embedded in an antifibrotic matrix to achieve a prolonged therapeutic effect [533]. The administration of fludarabine conditioning chemotherapy is employed to mitigate the immunological rejection response toward infused chimeric antigen receptor (CAR) T cells [534]. Sipuleucel-T, an innovative immunotherapy utilizing dendritic cells that has received approval from the FDA, employs ex vivo antigen presentation to initiate and maintain a therapeutic cellular phenotype [535]. Matrix-induced autologous chondrocyte implantation (MACI) is a novel technique that effectively retains chondrocytes at the intended site [536], making it the initial cell-embedded scaffold product to receive approval from the FDA. Furthermore, Tisagenlecleucel, the first CAR T-cell therapy to receive approval from the FDA, established the standard for manufacturing autologous cell therapies [537].

10. Future Perspectives and Concluding Remarks

The ongoing progress in the domain of BINMs for micro/nanodevices in biomedical applications presents significant potential for transforming the healthcare sector and other related fields. The progression from the initial idea to the actualization of these materials has demonstrated their significant capacity to improve the performance of devices, achieve compatibility with biological systems, facilitate self-assembly processes, promote sustainability, and provide a wide range of applications. The utilization of the bio-inspired approach, which draws inspiration from nature's efficiency and elegance, has not only facilitated the creation of innovative materials but has also provided a new lens through which to tackle intricate problems. Looking toward the future, the field of BINMs is anticipated to explore novel frontiers. The advancement of innovation can be propelled by incorporating various biological inspirations and elucidating complex structure–function relationships inherent in organisms.

Furthermore, the progress made in utilizing biomimetic materials and implementing energy-minimizing designs will facilitate the development of micro/nanodevices that are both highly efficient and environmentally sustainable. The potential applications within the micro/nanodevices domain are extensive and encompass various fields beyond biomedicine. The application of BINMs in various domains such as chemical reaction systems, energy harvesting and storage, environmental protection, sensors, agricultural sustainability, protective clothing, and adaptive materials demonstrates this approach's wide-ranging capabilities and significant influence. These applications possess the capacity to transform industries and effectively tackle urgent global challenges fundamentally.

Although substantial advancements have been made thus far, there are still obstacles to fully overcome to harness the capabilities of BINMs for biomedical purposes. The successful resolution of challenges related to synthesis intricacies, attainment of accurate interfaces, mitigation of biocompatibility issues, assurance of long-term stability, and negotiation of regulatory and ethical considerations necessitates the collaborative endeavors of interdisciplinary groups. Incorporating knowledge from various disciplines such as biology, chemistry, material science, medicine, engineering, and other related fields will play a crucial role in determining the trajectory of BINMs in the future. BINMs signify a novel biomedical application era characterized by inventive designs, improved functionality, and diverse possibilities. The remarkable trajectory from inspiration to realization underscores the significance of biomimicry in propelling scientific progress. As scholars persist in investigating and enhancing these materials' design principles, synthesis techniques, and applications, the range of potential outcomes will broaden, leading to a forthcoming era in which BINMs assume a crucial position in influencing our approach to healthcare, technology, and the global landscape.

Author Contributions: Conceptualization, A.B.I. and M.H.-U.-R.; methodology, A.B.I. and M.H.-U.-R.; resources, A.B.I.; writing—original draft preparation, M.H.-U.-R., I.J., T.F. and A.B.I.; writing—review and editing, M.H.-U.-R., I.J., T.F. and A.B.I.; visualization, M.H.-U.-R., I.J., T.F. and A.B.I.; supervision, A.B.I.; project administration, A.B.I.; funding acquisition, A.B.I. All authors have read and agreed to the published version of the manuscript.

Funding: This research received no external funding.

Data Availability Statement: Data is contained within the article.

Acknowledgments: A.B.I. gratefully recognizes the support provided by the Committee for Advanced Studies and Research (CASR) at the Bangladesh University of Engineering and Technology.

Conflicts of Interest: The authors declare no conflict of interest.

References

1. Chen, Y.; Feng, Y.; Deveaux, J.G.; Masoud, M.A.; Chandra, F.S.; Chen, H.; Zhang, D.; Feng, L. Biomineralization Forming Process and Bio-inspired Nanomaterials for Biomedical Application: A Review. *Minerals* **2019**, *9*, 68. [CrossRef]
2. Li, W.; Chen, Y.; Jiao, Z. Efficient Anti-Fog and Anti-Reflection Functions of the Bio-Inspired, Hierarchically-Architectured Surfaces of Multiscale Columnar Structures. *Nanomaterials* **2023**, *13*, 1570. [CrossRef]
3. Ullah, R.; Bibi, S.; Khan, M.N.; Al Mohaimeed, A.M.; Naz, Q.; Kamal, A. Application of Bio-Inspired Gold Nanoparticles as Advanced Nanomaterial in Halt Nociceptive Pathway and Hepatotoxicity via Triggering Antioxidation System. *Catalysts* **2023**, *13*, 786. [CrossRef]
4. Arshad, R.; Razlansari, M.; Maryam Hosseinikhah, S.; Tiwari Pandey, A.; Ajalli, N.; Ezra Manicum, A.-L.; Thorat, N.; Rahdar, A.; Zhu, Y.; Tabish, T.A. Antimicrobial and anti-biofilm activities of bio-inspired nanomaterials for wound healing applications. *Drug Discov. Today* **2023**, *28*, 103673. [CrossRef] [PubMed]
5. Azarnoush, A.; Dambri, O.A.; Karatop, E.Ü.; Makrakis, D.; Cherkaoui, S. Simulation and Performance Evaluation of a Bio-inspired Nanogenerator for Medical Applications. *IEEE Trans. Biomed. Eng.* **2023**, *70*, 2616–2623. [CrossRef] [PubMed]
6. Ding, M.; Chen, G.; Xu, W.; Jia, C.; Luo, H. Bio-inspired synthesis of nanomaterials and smart structures for electrochemical energy storage and conversion. *Nano Mater. Sci.* **2020**, *2*, 264–280. [CrossRef]
7. Dutta, V.; Verma, R.; Gopalkrishnan, C.; Yuan, M.-H.; Batoo, K.M.; Jayavel, R.; Chauhan, A.; Lin, K.-Y.A.; Balasubramani, R.; Ghotekar, S. Bio-Inspired Synthesis of Carbon-Based Nanomaterials and Their Potential Environmental Applications: A State-of-the-Art Review. *Inorganics* **2022**, *10*, 169. [CrossRef]
8. Li, H.-Y.; Feng, J.-K.; Xiang, L.; Huang, J.; Xie, B. Facile synthesis of bio-inspired anemone-like VS4 nanomaterials for long-life supercapacitors with high energy density. *J. Power Sources* **2020**, *457*, 228031. [CrossRef]
9. Xiang, H.; Li, Z.; Wang, W.; Wu, H.; Zhou, H.; Ni, Y.; Liu, H. Enabling a Paper-Based Flexible Sensor to Work under Water with Exceptional Long-Term Durability through Biomimetic Reassembling of Nanomaterials from Natural Wood. *ACS Sustain. Chem. Eng.* **2023**, *11*, 8667–8674. [CrossRef]
10. Duan, S.; Shi, Q.; Hong, J.; Zhu, D.; Lin, Y.; Li, Y.; Lei, W.; Lee, C.; Wu, J. Water-Modulated Biomimetic Hyper-Attribute-Gel Electronic Skin for Robotics and Skin-Attachable Wearables. *ACS Nano* **2023**, *17*, 1355–1371. [CrossRef]
11. Chi, L.; Zhang, C.; Wu, X.; Qian, X.; Sun, H.; He, M.; Guo, C. Research Progress on Biomimetic Nanomaterials for Electrochemical Glucose Sensors. *Biomimetics* **2023**, *8*, 167. [CrossRef] [PubMed]
12. Li, Q.; Wang, Y.; Zhang, G.; Su, R.; Qi, W. Biomimetic mineralization based on self-assembling peptides. *Chem. Soc. Rev.* **2023**, *52*, 1549–1590. [CrossRef] [PubMed]
13. Chen, M.; Sun, Y.; Liu, H. Cell membrane biomimetic nanomedicines for cancer phototherapy. *Interdiscip. Med.* **2023**, *1*, e20220012. [CrossRef]
14. Jiang, L.; Sun, Y.; Chen, Y.; Nan, P. From DNA to Nerve Agents—The Biomimetic Catalysts for the Hydrolysis of Phosphate Esters. *ChemistrySelect* **2020**, *5*, 9492–9516. [CrossRef]
15. Lee, J.; Sands, I.; Zhang, W.; Zhou, L.; Chen, Y. DNA-inspired nanomaterials for enhanced endosomal escape. *Proc. Natl. Acad. Sci. USA* **2021**, *118*, e2104511118. [CrossRef] [PubMed]
16. Wu, K.C.-W.; Yang, C.-Y.; Cheng, C.-M. Using cell structures to develop functional nanomaterials and nanostructures—case studies of actin filaments and microtubules. *Chem. Commun.* **2014**, *50*, 4148–4157. [CrossRef]
17. Griffin, S.; Masood, M.I.; Nasim, M.J.; Sarfraz, M.; Ebokaiwe, A.P.; Schäfer, K.-H.; Keck, C.M.; Jacob, C. Natural Nanoparticles: A Particular Matter Inspired by Nature. *Antioxidants* **2018**, *7*, 3. [CrossRef]
18. Zhang, Y.-P.; Reimer, D.L.; Zhang, G.; Lee, P.H.; Bally, M.B. Self-Assembling DNA-Lipid Particles for Gene Transfer. *Pharm. Res.* **1997**, *14*, 190–196. [CrossRef]
19. Liu, K.; Du, J.; Wu, J.; Jiang, L. Superhydrophobic gecko feet with high adhesive forces towards water and their bio-inspired materials. *Nanoscale* **2012**, *4*, 768–772. [CrossRef]
20. Xia, X.; Liu, J.; Liu, Y.; Lei, Z.; Han, Y.; Zheng, Z.; Yin, J. Preparation and Characterization of Biomimetic SiO2-TiO2-PDMS Composite Hydrophobic Coating with Self-Cleaning Properties for Wall Protection Applications. *Coatings* **2023**, *13*, 224. [CrossRef]
21. Tang, K.; Xue, J.; Zhu, Y.; Wu, C. Design and synthesis of bioinspired nanomaterials for biomedical application. *WIREs Nanomed. Nanobiotechnology* **2023**, e1914. [CrossRef]
22. Malakar, A.; Kanel, S.R.; Ray, C.; Snow, D.D.; Nadagouda, M.N. Nanomaterials in the environment, human exposure pathway, and health effects: A review. *Sci. Total Environ.* **2021**, *759*, 143470. [CrossRef]
23. Harun-Ur-Rashid, M.; Seki, T.; Takeoka, Y. Structural colored gels for tunable soft photonic crystals. *Chem. Rec.* **2009**, *9*, 87–105. [CrossRef]
24. Karim, M.R.; Harun-Ur-Rashid, M.; Imran, A.B. Effect of sizes of vinyl modified narrow-dispersed silica cross-linker on the mechanical properties of acrylamide based hydrogel. *Sci. Rep.* **2023**, *13*, 5089. [CrossRef] [PubMed]
25. Harun-Ur-Rashid, M.; Imran, A.B. Superabsorbent Hydrogels from Carboxymethyl Cellulose. In *Carboxymethyl Cellulose*; Mondal, M.I.H., Ed.; Nova Science Publisher, Inc.: New York, NY, USA, 2019; p. 159.
26. Karim, M.R.; Harun-Ur-Rashid, M.; Imran, A.B. Highly Stretchable Hydrogel Using Vinyl Modified Narrow Dispersed Silica Particles as Cross-Linker. *ChemistrySelect* **2020**, *5*, 10556–10561. [CrossRef]
27. Huda, N.; Rashid, M.; Harun-Ur-Rashid, M. A Review on Stimuli-responsive grafted membranes Based on Facile Synthesis Process and Extensive Applications. *Int. J. Innov. Appl. Stud.* **2014**, *8*, 1296–1312.

28. Harun-Ur-Rashid, M.; Foyez, T.; Imran, A.B. Fabrication of Stretchable Composite Thin Film for Superconductor Applications. In *Sensors for Stretchable Electronics in Nanotechnology*; Pal, K., Ed.; CRC Press: Boca Raton, FL, USA, 2021; pp. 63–78. [CrossRef]
29. Huda, M.N.; Seki, T.; Suzuki, H.; ANM, H.K.; Harunur-Rashid, M.; Takeoka, Y. Characteristics of High-Density Poly (N-isopropylacrylamide)(PNIPA) Brushes on Silicon Surface by Atom Transfer Radical Polymerization. *Trans. Mater. Res. Soc. Jpn.* **2010**, *35*, 845–848. [CrossRef]
30. Harun-Ur-Rashid, M.; Imran, A.B.; Seki, T.; Takeoka, Y.; Ishii, M.; Nakamura, H. Template synthesis for stimuli-responsive angle independent structural colored smart materials. *Trans. Mater. Res. Soc. Jpn.* **2009**, *34*, 333–337. [CrossRef]
31. Takeoka, Y.; Yoshioka, S.; Teshima, M.; Takano, A.; Harun-Ur-Rashid, M.; Seki, T. Structurally coloured secondary particles composed of black and white colloidal particles. *Sci. Rep.* **2013**, *3*, 2371. [CrossRef] [PubMed]
32. Harun-Ur-Rashid, M.; Bin Imran, A.; Seki, T.; Ishii, M.; Nakamura, H.; Takeoka, Y. Angle-independent structural color in colloidal amorphous arrays. *Chemphyschem* **2010**, *11*, 579–583. [CrossRef]
33. Harun-Ur-Rashid, M.; Imran, A.B. Organometallic Nanomaterials Synthesis and Sustainable Green Nanotechnology Applications. In *Green Nanoarchitectonics*; Pal, K., Ed.; Jenny Stanford Publishing: New York, NY, USA, 2022; pp. 249–270. [CrossRef]
34. Imran, A.B.; Harun-Ur-Rashid, M.; Takeoka, Y. Polyrotaxane Actuators. In *Soft Actuators: Materials, Modeling, Applications, and Future Perspectives*; Asaka, K., Okuzaki, H., Eds.; Springer Nature: Singapore, 2019; pp. 81–147. [CrossRef]
35. Harun-Ur-Rashid, M.; Foyez, T.; Jahan, I.; Pal, K.; Imran, A.B. Rapid diagnosis of COVID-19 via nano-biosensor-implemented biomedical utilization: A systematic review. *RSC Adv.* **2022**, *12*, 9445–9465. [CrossRef] [PubMed]
36. Harun-Ur-Rashid, M.; Imran, A.B. Nanomaterials in the Automobile Sector. In *Emerging Applications of Nanomaterials*; Singh, N.B., Susan, M.A.B.H., Chaudhary, R.G., Eds.; Materials Research Foundations: Millersville, PA, USA, 2023; pp. 124–150.
37. Harun-Ur-Rashid, M.; Imran, A.B.; Susan, M. Prospective Nanomaterials for Food Packaging and Safety. In *Emerging Applications of Nanomaterials, Materials Research Forum LLC*; Singh, N.B., Susan, M.A.B.H., Chaudhary, R.G., Eds.; Materials Research Foundations: Millersville, PA, USA, 2023; pp. 327–352.
38. Harun-Ur-Rashid, M.; Pal, K.; Imran, A.B. Hybrid Nanocomposite Fabrication of Nanocatalyst with Enhanced and Stable Photocatalytic Activity. *Top. Catal.* **2023**. [CrossRef]
39. Harun-Ur-Rashid, M.; Imran, A.B.; Susan, M.; Hasan, A.B. Green Polymer Nanocomposites in Automotive and Packaging Industries. *Curr. Pharm. Biotechnol.* **2023**, *24*, 145–163. [CrossRef] [PubMed]
40. Harun-Ur-Rashid, M.; Imran, A.B. Engineered Nanomaterials for Energy Conversion Cells. In *Applications of Emerging Nanomaterials and Nanotechnology*; Singh, N.B., Susan, M.A.B.H., Chaudhary, R.G., Eds.; Materials Research Forum LLC: Millersville, PA, USA, 2023; pp. 103–126.
41. Harun-Ur-Rashid, M.; Imran, A.B.; Susan, M.A.B.H. Fire-Resistant Polymeric Foams and Their Applications. In *Polymeric Foams: Applications of Polymeric Foams*; Gupta, R.K., Ed.; ACS Publications: Washington, CO, USA, 2023; pp. 97–121.
42. Khair, A.; Dey, N.K.; Harun-Ur-Rashid, M.; Alim, M.A.; Bahadur, N.M.; Mahamud, S.; Ahmed, S. Diffusimetry renounces graham's law, achieves diffusive convection, concentration gradient induced diffusion, heat and mass transfer. *Defect Diffus. Forum* **2021**, *407*, 173–184.
43. Gao, J.; Yang, P.; Li, H.; Li, N.; Liu, X.; Cai, K.; Li, J. Advances in anti-tumor research based on bionic micro-nano technology. *J. Drug Deliv. Sci. Technol.* **2023**, *86*, 104674. [CrossRef]
44. Miao, S.; Cao, X.; Lu, M.; Liu, X. Tailoring micro/nano-materials with special wettability for biomedical devices. *Biomed. Technol.* **2023**, *2*, 15–30. [CrossRef]
45. Mahfuz, M.U.; Ahmed, K. A review of micro-nano-scale wireless sensor networks for environmental protection: Prospects and challenges. *Sci. Technol. Adv. Mater.* **2005**, *6*, 302–306. [CrossRef]
46. Iannacci, J. A perspective vision of micro/nano systems and technologies as enablers of 6g, super-iot, and tactile internet [point of view]. *Proc. IEEE* **2023**, *111*, 5–18. [CrossRef]
47. Pan, X.; Hong, X.; Xu, L.; Li, Y.; Yan, M.; Mai, L. On-chip micro/nano devices for energy conversion and storage. *Nano Today* **2019**, *28*, 100764. [CrossRef]
48. Zhang, W.; Yu, X.; Li, Y.; Su, Z.; Jandt, K.D.; Wei, G. Protein-mimetic peptide nanofibers: Motif design, self-assembly synthesis, and sequence-specific biomedical applications. *Prog. Polym. Sci.* **2018**, *80*, 94–124. [CrossRef]
49. Prianka, T.R.; Subhan, N.; Reza, H.M.; Hosain, M.K.; Rahman, M.A.; Lee, H.; Sharker, S.M. Recent exploration of bio-mimetic nanomaterial for potential biomedical applications. *Mater. Sci. Eng. C* **2018**, *93*, 1104–1115. [CrossRef] [PubMed]
50. Weidner, B.V.; Nagel, J.; Weber, H.-J. Facilitation method for the translation of biological systems to technical design solutions. *Int. J. Des. Creat. Innov.* **2018**, *6*, 211–234. [CrossRef]
51. Ibrahim, M.D.; Philip, S.; Lam, S.S.; Sunami, Y. Evaluation of an Antifouling Surface Inspired by Malaysian Sharks Negaprion Brevirostris and Carcharhinus Leucas Riblets. *Tribol. Online* **2021**, *16*, 70–80. [CrossRef]
52. Hayes, S.; Desha, C.; Gibbs, M. Findings of Case-Study Analysis: System-Level Biomimicry in Built-Environment Design. *Biomimetics* **2019**, *4*, 73. [CrossRef] [PubMed]
53. Jamei, E.; Vrcelj, Z. Biomimicry and the Built Environment, Learning from Nature's Solutions. *Appl. Sci.* **2021**, *11*, 7514. [CrossRef]
54. Anjum, S.; Rahman, F.; Pandey, P.; Arya, D.K.; Alam, M.; Rajinikanth, P.S.; Ao, Q. Electrospun Biomimetic Nanofibrous Scaffolds: A Promising Prospect for Bone Tissue Engineering and Regenerative Medicine. *Int. J. Mol. Sci.* **2022**, *23*, 9206. [CrossRef]
55. Saylan, Y.; Erdem, Ö.; Inci, F.; Denizli, A. Advances in Biomimetic Systems for Molecular Recognition and Biosensing. *Biomimetics* **2020**, *5*, 20. [CrossRef]

56. Raheem, A.A.; Hameed, P.; Whenish, R.; Elsen, R.S.; Jaiswal, A.K.; Prashanth, K.G.; Manivasagam, G. A Review on Development of Bio-Inspired Implants Using 3D Printing. *Biomimetics* **2021**, *6*, 65. [CrossRef]

57. Choi, J.S.; An, J.H.; Lee, J.-K.; Lee, J.Y.; Kang, S.M. Optimization of Shapes and Sizes of Moth-Eye-Inspired Structures for the Enhancement of Their Antireflective Properties. *Polymers* **2020**, *12*, 296. [CrossRef]

58. Sikdar, S.; Rahman, M.H.; Siddaiah, A.; Menezes, P.L. Gecko-Inspired Adhesive Mechanisms and Adhesives for Robots—A Review. *Robotics* **2022**, *11*, 143.

59. Wang, R.; Yang, H.; Fu, R.; Su, Y.; Lin, X.; Jin, X.; Du, W.; Shan, X.; Huang, G. Biomimetic Upconversion Nanoparticles and Gold Nanoparticles for Novel Simultaneous Dual-Modal Imaging-Guided Photothermal Therapy of Cancer. *Cancers* **2020**, *12*, 3136. [CrossRef] [PubMed]

60. Alfei, S.; Schito, A.M. From Nanobiotechnology, Positively Charged Biomimetic Dendrimers as Novel Antibacterial Agents: A Review. *Nanomaterials* **2020**, *10*, 2022. [CrossRef] [PubMed]

61. Gebeshuber, I.C. Biomimetics—Prospects and Developments. *Biomimetics* **2022**, *7*, 29. [CrossRef] [PubMed]

62. Basak, S. Walking through the biomimetic bandages inspired by Gecko's feet. *Bio. Des. Manuf.* **2020**, *3*, 148–154. [CrossRef]

63. Yoon, J.-C.; Yoon, C.-S.; Lee, J.-S.; Jang, J.-H. Lotus leaf-inspired CVD grown graphene for a water repellant flexible transparent electrode. *Chem. Commun.* **2013**, *49*, 10626–10628. [CrossRef]

64. Huda, M.N.; Nafiujjaman, M.; Deaguero, I.G.; Okonkwo, J.; Hill, M.L.; Kim, T.; Nurunnabi, M. Potential Use of Exosomes as Diagnostic Biomarkers and in Targeted Drug Delivery: Progress in Clinical and Preclinical Applications. *ACS Biomater. Sci. Eng.* **2021**, *7*, 2106–2149. [CrossRef]

65. Gareev, K.G.; Grouzdev, D.S.; Koziaeva, V.V.; Sitkov, N.O.; Gao, H.; Zimina, T.M.; Shevtsov, M. Biomimetic Nanomaterials: Diversity, Technology, and Biomedical Applications. *Nanomaterials* **2022**, *12*, 2485. [CrossRef]

66. Filippi, M.; Garello, F.; Yasa, O.; Kasamkattil, J.; Scherberich, A.; Katzschmann, R.K. Engineered Magnetic Nanocomposites to Modulate Cellular Function. *Small* **2022**, *18*, 2104079. [CrossRef]

67. Klem, M.T.; Young, M.; Douglas, T. Biomimetic magnetic nanoparticles. *Mater. Today* **2005**, *8*, 28–37. [CrossRef]

68. Kralj, S.; Marchesan, S. Bioinspired Magnetic Nanochains for Medicine. *Pharmaceutics* **2021**, *13*, 1262. [CrossRef]

69. Peigneux, A.; Jabalera, Y.; Vivas, M.A.F.; Casares, S.; Azuaga, A.I.; Jimenez-Lopez, C. Tuning properties of biomimetic magnetic nanoparticles by combining magnetosome associated proteins. *Sci. Rep.* **2019**, *9*, 8804. [CrossRef] [PubMed]

70. Peigneux, A.; Glitscher, E.A.; Charbaji, R.; Weise, C.; Wedepohl, S.; Calderón, M.; Jimenez-Lopez, C.; Hedtrich, S. Protein corona formation and its influence on biomimetic magnetite nanoparticles. *J. Mater. Chem. B* **2020**, *8*, 4870–4882. [CrossRef] [PubMed]

71. Vurro, F.; Jabalera, Y.; Mannucci, S.; Glorani, G.; Sola-Leyva, A.; Gerosa, M.; Romeo, A.; Romanelli, M.G.; Malatesta, M.; Calderan, L.; et al. Improving the Cellular Uptake of Biomimetic Magnetic Nanoparticles. *Nanomaterials* **2021**, *11*, 766. [CrossRef] [PubMed]

72. Taher, Z.; Legge, C.; Winder, N.; Lysyganicz, P.; Rawlings, A.; Bryant, H.; Muthana, M.; Staniland, S. Magnetosomes and Magnetosome Mimics: Preparation, Cancer Cell Uptake and Functionalization for Future Cancer Therapies. *Pharmaceutics* **2021**, *13*, 367. [CrossRef]

73. Mondal, S.; Manivasagan, P.; Bharathiraja, S.; Santha Moorthy, M.; Kim, H.H.; Seo, H.; Lee, K.D.; Oh, J. Magnetic hydroxyapatite: A promising multifunctional platform for nanomedicine application. *Int. J. Nanomed.* **2017**, *12*, 8389–8410. [CrossRef]

74. Bianco, L.D.; Lesci, I.G.; Fracasso, G.; Barucca, G.; Spizzo, F.; Tamisari, M.; Scotti, R.; Ciocca, L. Synthesis of nanogranular Fe3O4/biomimetic hydroxyapatite for potential applications in nanomedicine: Structural and magnetic characterization. *Mater. Res. Express* **2015**, *2*, 065002. [CrossRef]

75. Correa, S.; Puertas, S.; Gutiérrez, L.; Asín, L.; De La Fuente, J.M.; Grazú, V.; Betancor, L. Design of stable magnetic hybrid nanoparticles of Si-entrapped HRP. *PLoS ONE* **2019**, *14*, e0214004. [CrossRef]

76. Gareev, K.G.; Grouzdev, D.S.; Kharitonskii, P.V.; Kosterov, A.; Koziaeva, V.V.; Sergienko, E.S.; Shevtsov, M.A. Magnetotactic Bacteria and Magnetosomes: Basic Properties and Applications. *Magnetochemistry* **2021**, *7*, 86. [CrossRef]

77. Yang, C.; Cao, C.; Cai, Y.; Xu, H.; Zhang, T.; Pan, Y. Effects of PEGylation on biomimetic synthesis of magnetoferritin nanoparticles. *J. Nanopart. Res.* **2017**, *19*, 101. [CrossRef]

78. Cruz, E.; Kayser, V. Synthesis and Enhanced Cellular Uptake In Vitro of Anti-HER2 Multifunctional Gold Nanoparticles. *Cancers* **2019**, *11*, 870. [CrossRef]

79. Paladini, F.; Pollini, M. Antimicrobial Silver Nanoparticles for Wound Healing Application: Progress and Future Trends. *Materials* **2019**, *12*, 2540. [CrossRef]

80. Dadfar, S.M.; Camozzi, D.; Darguzyte, M.; Roemhild, K.; Varvarà, P.; Metselaar, J.; Banala, S.; Straub, M.; Güvener, N.; Engelmann, U.; et al. Size-isolation of superparamagnetic iron oxide nanoparticles improves MRI, MPI and hyperthermia performance. *J. Nanobiotechnology* **2020**, *18*, 22. [CrossRef] [PubMed]

81. Liu, Y.; Liu, X.; Wang, X. Biomimetic Synthesis of Gelatin Polypeptide-Assisted Noble-Metal Nanoparticles and Their Interaction Study. *Nanoscale Res. Lett.* **2010**, *6*, 22. [CrossRef]

82. Cardellini, J.; Montis, C.; Barbero, F.; De Santis, I.; Caselli, L.; Berti, D. Interaction of Metallic Nanoparticles With Biomimetic Lipid Liquid Crystalline Cubic Interfaces. *Front. Bioeng. Biotechnol.* **2022**, *10*, 848687. [CrossRef] [PubMed]

83. Li, S.-F.; Chen, Y.; Wang, Y.-S.; Mo, H.-L.; Zang, S.-Q. Integration of enzyme immobilization and biomimetic catalysis in hierarchically porous metal-organic frameworks for multi-enzymatic cascade reactions. *Sci. China Chem.* **2022**, *65*, 1122–1128. [CrossRef]

84. Guimarães, D.; Cavaco-Paulo, A.; Nogueira, E. Design of liposomes as drug delivery system for therapeutic applications. *Int. J. Pharm.* **2021**, *601*, 120571. [CrossRef] [PubMed]

85. Engelmann, T.; Desante, G.; Labude, N.; Rütten, S.; Telle, R.; Neuss, S.; Schickle, K. Coatings Based on Organic/Non-Organic Composites on Bioinert Ceramics by Using Biomimetic Co-Precipitation. *Ceramics* **2019**, *2*, 260–270. [CrossRef]

86. Chan, M.-H.; Li, C.-H.; Chang, Y.-C.; Hsiao, M. Iron-Based Ceramic Composite Nanomaterials for Magnetic Fluid Hyperthermia and Drug Delivery. *Pharmaceutics* **2022**, *14*, 2584. [CrossRef]

87. Lempicki, A.; Brecher, C.; Szupryczynski, P.; Lingertat, H.; Nagarkar, V.V.; Tipnis, S.V.; Miller, S.R. A new lutetia-based ceramic scintillator for X-ray imaging. *Nucl. Instrum. Methods Phys. Res. Sect. A Accel. Spectrometers Detect. Assoc. Equip.* **2002**, *488*, 579–590. [CrossRef]

88. Liu, L.; Bai, X.; Martikainen, M.-V.; Kårlund, A.; Roponen, M.; Xu, W.; Hu, G.; Tasciotti, E.; Lehto, V.-P. Cell membrane coating integrity affects the internalization mechanism of biomimetic nanoparticles. *Nat. Commun.* **2021**, *12*, 5726. [CrossRef]

89. Okuno, Y.; Iwasaki, Y. Well-Defined Anisotropic Self-Assembly from Peptoids and Their Biomedical Applications. *ChemMedChem* **2023**, *18*, e202300217. [CrossRef] [PubMed]

90. Xiao, Z.; Zhao, Q.; Niu, Y.; Zhao, D. Adhesion advances: From nanomaterials to biomimetic adhesion and applications. *Soft Matter* **2022**, *18*, 3447–3464. [CrossRef] [PubMed]

91. Shahsavan, H.; Yu, L.; Jákli, A.; Zhao, B. Smart biomimetic micro/nanostructures based on liquid crystal elastomers and networks. *Soft Matter* **2017**, *13*, 8006–8022. [CrossRef] [PubMed]

92. Qin, L.; Huang, X.; Sun, Z.; Ma, Z.; Mawignon, F.J.; Lv, B.; Shan, L.; Dong, G. Synergistic effect of sharkskin-inspired morphologies and surface chemistry on regulating stick-slip friction. *Tribol. Int.* **2023**, *187*, 108765. [CrossRef]

93. Kim, M.W.; Kwon, S.-H.; Choi, J.H.; Lee, A. A Promising Biocompatible Platform: Lipid-Based and Bio-Inspired Smart Drug Delivery Systems for Cancer Therapy. *Int. J. Mol. Sci.* **2018**, *19*, 3859. [CrossRef] [PubMed]

94. Bisht, N.; Patel, M.; Dwivedi, N.; Kumar, P.; Mondal, D.P.; Srivastava, A.K.; Dhand, C. Bio-inspired polynorepinephrine based nanocoatings for reduced graphene oxide/gold nanoparticles composite for high-performance biosensing of Mycobacterium tuberculosis. *Environ. Res.* **2023**, *227*, 115684. [CrossRef] [PubMed]

95. Whitesides, G.M.; Grzybowski, B. Self-Assembly at All Scales. *Science* **2002**, *295*, 2418–2421. [CrossRef]

96. Mohanty, A.; Mithra, K.; Jena, S.S.; Behera, R.K. Kinetics of Ferritin Self-Assembly by Laser Light Scattering: Impact of Subunit Concentration, pH, and Ionic Strength. *Biomacromolecules* **2021**, *22*, 1389–1398. [CrossRef]

97. Mandal, D.; Shirazi, A.N.; Parang, K. Self-assembly of peptides to nanostructures. *Org. Biomol. Chem.* **2014**, *12*, 3544–3561. [CrossRef] [PubMed]

98. Naikoo, G.A.; Mustaqeem, M.; Hassan, I.U.; Awan, T.; Arshad, F.; Salim, H.; Qurashi, A. Bioinspired and green synthesis of nanoparticles from plant extracts with antiviral and antimicrobial properties: A critical review. *J. Saudi Chem. Soc.* **2021**, *25*, 101304. [CrossRef]

99. Al-Faouri, T.; Buguis, F.L.; Azizi Soldouz, S.; Sarycheva, O.V.; Hussein, B.A.; Mahmood, R.; Koivisto, B.D. Exploring Structure-Property Relationships in a Bio-Inspired Family of Bipodal and Electronically-Coupled Bistriphenylamine Dyes for Dye-Sensitized Solar Cell Applications. *Molecules* **2020**, *25*, 2260. [CrossRef] [PubMed]

100. Zhao, Y.; Zhao, Y.; Hu, S.; Lv, J.; Ying, Y.; Gervinskas, G.; Si, G. Artificial Structural Color Pixels: A Review. *Materials* **2017**, *10*, 944. [CrossRef] [PubMed]

101. Himel, M.H.; Sikder, B.; Ahmed, T.; Choudhury, S.M. Biomimicry in nanotechnology: A comprehensive review. *Nanoscale Adv.* **2023**, *5*, 596–614. [CrossRef] [PubMed]

102. Chaudhry, N.; Dwivedi, S.; Chaudhry, V.; Singh, A.; Saquib, Q.; Azam, A.; Musarrat, J. Bio-inspired nanomaterials in agriculture and food: Current status, foreseen applications and challenges. *Microb. Pathog.* **2018**, *123*, 196–200. [CrossRef]

103. Chowdhury, A.N.; Shapter, J.; Imran, A.B. *Innovations in Nanomaterials*; Nova Publishers: Hauppauge, NY, USA, 2015.

104. Zhang, Z.; Vogelbacher, F.; Song, Y.; Tian, Y.; Li, M. Bio-inspired optical structures for enhancing luminescence. *Exploration* **2023**, *3*, 20220052. [CrossRef]

105. Huan, T.N.; Jane, R.T.; Benayad, A.; Guetaz, L.; Tran, P.D.; Artero, V. Bio-inspired noble metal-free nanomaterials approaching platinum performances for H2 evolution and uptake. *Energy Environ. Sci.* **2016**, *9*, 940–947. [CrossRef]

106. Niu, J.; Zhao, C.; Liu, C.; Ren, J.; Qu, X. Bio-Inspired Bimetallic Enzyme Mimics as Bio-Orthogonal Catalysts for Enhanced Bacterial Capture and Inhibition. *Chem. Mater.* **2021**, *33*, 8052–8058. [CrossRef]

107. Levy, S.; Mass, T. The Skeleton and Biomineralization Mechanism as Part of the Innate Immune System of Stony Corals. *Front. Immunol.* **2022**, *13*, 850338. [CrossRef]

108. Kumar, A.; Sharma, G.; Naushad, M.; Al-Muhtaseb, A.a.H.; García-Peñas, A.; Mola, G.T.; Si, C.; Stadler, F.J. Bio-inspired and biomaterials-based hybrid photocatalysts for environmental detoxification: A review. *Chem. Eng. J.* **2020**, *382*, 122937. [CrossRef]

109. Prasad, C.; Sreenivasulu, K.; Gangadhara, S.; Venkateswarlu, P. Bio inspired green synthesis of Ni/Fe3O4 magnetic nanoparticles using Moringa oleifera leaves extract: A magnetically recoverable catalyst for organic dye degradation in aqueous solution. *J. Alloys Compd.* **2017**, *700*, 252–258. [CrossRef]

110. Wang, J.; Feng, T.; Chen, J.; Ramalingam, V.; Li, Z.; Kabtamu, D.M.; He, J.-H.; Fang, X. Electrocatalytic nitrate/nitrite reduction to ammonia synthesis using metal nanocatalysts and bio-inspired metalloenzymes. *Nano Energy* **2021**, *86*, 106088. [CrossRef]

111. Ford, C.L.; Park, Y.J.; Matson, E.M.; Gordon, Z.; Fout, A.R. A bioinspired iron catalyst for nitrate and perchlorate reduction. *Science* **2016**, *354*, 741–743. [CrossRef] [PubMed]

112. Gentil, S.; Lalaoui, N.; Dutta, A.; Nedellec, Y.; Cosnier, S.; Shaw, W.J.; Artero, V.; Le Goff, A. Carbon-Nanotube-Supported Bio-Inspired Nickel Catalyst and Its Integration in Hybrid Hydrogen/Air Fuel Cells. *Angew. Chem.* **2017**, *129*, 1871–1875. [CrossRef]
113. Kathpalia, R.; Verma, A.K. Bio-inspired nanoparticles for artificial photosynthesis. *Mater. Today Proc.* **2021**, *45*, 3825–3832. [CrossRef]
114. Ocakoglu, K.; Joya, K.S.; Harputlu, E.; Tarnowska, A.; Gryko, D.T. A nanoscale bio-inspired light-harvesting system developed from self-assembled alkyl-functionalized metallochlorin nano-aggregates. *Nanoscale* **2014**, *6*, 9625–9631. [CrossRef]
115. Liu, Z.; Leow, W.R.; Chen, X. Bio-Inspired Plasmonic Photocatalysts. *Small Methods* **2019**, *3*, 1800295. [CrossRef]
116. Mishra, S.; Tripathi, B.; Garg, S.; Kumar, A.; Kumar, P. Design and Development of a Bio-Inspired Flapping Wing Type Micro Air Vehicle. *Procedia Mater. Sci.* **2015**, *10*, 519–526. [CrossRef]
117. Yao, H.-B.; Fang, H.-Y.; Wang, X.-H.; Yu, S.-H. Hierarchical assembly of micro-/nano-building blocks: Bio-inspired rigid structural functional materials. *Chem. Soc. Rev.* **2011**, *40*, 3764–3785. [CrossRef]
118. Tian, W.; Gao, Q.; Tan, Y.; Yang, K.; Zhu, L.; Yang, C.; Zhang, H. Bio-inspired beehive-like hierarchical nanoporous carbon derived from bamboo-based industrial by-product as a high performance supercapacitor electrode material. *J. Mater. Chem. A* **2015**, *3*, 5656–5664. [CrossRef]
119. Hussain, I.; Lamiel, C.; Sahoo, S.; Javed, M.S.; Ahmad, M.; Chen, X.; Gu, S.; Qin, N.; Assiri, M.A.; Zhang, K. Animal- and Human-Inspired Nanostructures as Supercapacitor Electrode Materials: A Review. *Nano Micro. Lett.* **2022**, *14*, 199. [CrossRef]
120. Zhang, Q.; Yang, X.; Li, P.; Huang, G.; Feng, S.; Shen, C.; Han, B.; Zhang, X.; Jin, F.; Xu, F.; et al. Bioinspired engineering of honeycomb structure—Using nature to inspire human innovation. *Prog. Mater Sci.* **2015**, *74*, 332–400. [CrossRef]
121. Atanasijevic, P.; Grujic, D.; Krajinic, F.; Mihailovic, P.; Pantelic, D. Characterization of a bioderived imaging sensor based on a Morpho butterfly's wing. *Opt. Laser Technol.* **2023**, *159*, 108919. [CrossRef]
122. Amoli, V.; Kim, S.Y.; Kim, J.S.; Choi, H.; Koo, J.; Kim, D.H. Biomimetics for high-performance flexible tactile sensors and advanced artificial sensory systems. *J. Mater. Chem. C* **2019**, *7*, 14816–14844. [CrossRef]
123. Zhou, J.; Miles, R.N. Sensing fluctuating airflow with spider silk. *Proc. Natl. Acad. Sci. USA* **2017**, *114*, 12120–12125. [CrossRef] [PubMed]
124. He, X.; Deng, H.; Hwang, H.-M. The current application of nanotechnology in food and agriculture. *J. Food Drug Anal.* **2019**, *27*, 1–21. [CrossRef]
125. Malini, S.; Raj, K.; Madhumathy, S.; El-Hady, K.M.; Islam, S.; Dutta, M. Bioinspired Advances in Nanomaterials for Sustainable Agriculture. *J. Nanomater.* **2022**, *2022*, 8926133. [CrossRef]
126. Liang, J.; Yu, M.; Guo, L.; Cui, B.; Zhao, X.; Sun, C.; Wang, Y.; Liu, G.; Cui, H.; Zeng, Z. Bioinspired Development of P(St–MAA)–Avermectin Nanoparticles with High Affinity for Foliage To Enhance Folia Retention. *J. Agric. Food Chem.* **2018**, *66*, 6578–6584. [CrossRef] [PubMed]
127. Pandey, S.; Sharma, K.; Gundabala, V. Antimicrobial bio-inspired active packaging materials for shelf life and safety development: A review. *Food Biosci.* **2022**, *48*, 101730. [CrossRef]
128. Chakraborty, S.; Bera, D.; Roy, L.; Ghosh, C.K. Biomimetic and Bioinspired Nanostructures. In *Bioinspired and Green Synthesis of Nanostructures*; John Wiley & Sons: Hoboken, NJ, USA, 2023; pp. 353–404. [CrossRef]
129. Huang, L.; Sun, D.-W.; Pu, H.; Zhang, C.; Zhang, D. Nanocellulose-based polymeric nanozyme as bioinspired spray coating for fruit preservation. *Food Hydrocoll.* **2023**, *135*, 108138. [CrossRef]
130. Maarisetty, D.; Sow, P.K.; Baral, S.S. Chapter 2—Bioinspired nanomaterials for remediation of toxic metal ions from wastewater. In *Advances in Nano and Biochemistry*; Morajkar, P., Naik, M., Eds.; Academic Press: Cambridge, MA, USA, 2023; pp. 39–55. [CrossRef]
131. Gonzalez-Perez, A.; Persson, K.M. Bioinspired Materials for Water Purification. *Materials* **2016**, *9*, 447. [CrossRef]
132. Barrio, J.; Pedersen, A.; Favero, S.; Luo, H.; Wang, M.; Sarma, S.C.; Feng, J.; Ngoc, L.T.T.; Kellner, S.; Li, A.Y.; et al. Bioinspired and Bioderived Aqueous Electrocatalysis. *Chem. Rev.* **2023**, *123*, 2311–2348. [CrossRef]
133. Li, Y.; Zhang, W.; Zhang, L.; Li, J.; Su, Z.; Wei, G. Sequence-Designed Peptide Nanofibers Bridged Conjugation of Graphene Quantum Dots with Graphene Oxide for High Performance Electrochemical Hydrogen Peroxide Biosensor. *Adv. Mater. Interfaces* **2017**, *4*, 1600895. [CrossRef]
134. Zampieri, A.; Mabande, G.T.P.; Selvam, T.; Schwieger, W.; Rudolph, A.; Hermann, R.; Sieber, H.; Greil, P. Biotemplating of Luffa cylindrica sponges to self-supporting hierarchical zeolite macrostructures for bio-inspired structured catalytic reactors. *Mater. Sci. Eng. C* **2006**, *26*, 130–135. [CrossRef]
135. Zhang, X.; Shi, X.-R.; Wang, P.; Bao, Z.; Huang, M.; Xu, Y.; Xu, S. Bio-inspired design of NiFeP nanoparticles embedded in (N,P) co-doped carbon for boosting overall water splitting. *Dalton Trans.* **2023**, *52*, 6860–6869. [CrossRef] [PubMed]
136. Harimohan, E.; Nanaji, K.; Rao, B.V.A.; Rao, T.N. A facile one-step synthesis of bio-inspired porous graphitic carbon sheets for improved lithium-sulfur battery performance. *Int. J. Energy Res.* **2022**, *46*, 4339–4351. [CrossRef]
137. Zheng, Y.; Bai, H.; Huang, Z.; Tian, X.; Nie, F.-Q.; Zhao, Y.; Zhai, J.; Jiang, L. Directional water collection on wetted spider silk. *Nature* **2010**, *463*, 640–643. [CrossRef] [PubMed]
138. Kapoor, R.T.; Rafatullah, M.; Qamar, M.; Qutob, M.; Alosaimi, A.M.; Alorfi, H.S.; Hussein, M.A. Review on Recent Developments in Bioinspired-Materials for Sustainable Energy and Environmental Applications. *Sustainability* **2022**, *14*, 16931. [CrossRef]

139. Leone, L.; Sgueglia, G.; La Gatta, S.; Chino, M.; Nastri, F.; Lombardi, A. Enzymatic and Bioinspired Systems for Hydrogen Production. *Int. J. Mol. Sci.* **2023**, *24*, 8605. [CrossRef]

140. Ma, J.; Zhang, J.; Zhang, Y.; Guo, Q.; Hu, T.; Xiao, H.; Lu, W.; Jia, J. Progress on anodic modification materials and future development directions in microbial fuel cells. *J. Power Sources* **2023**, *556*, 232486. [CrossRef]

141. Idris, M.O.; Noh, N.A.M.; Ibrahim, M.N.M.; Yaqoob, A.A. Sustainable microbial fuel cell functionalized with a bio-waste: A feasible route to formaldehyde bioremediation along with bioelectricity generation. *Chem. Eng. J.* **2023**, *455*, 140781. [CrossRef]

142. Tang, R.; Men, X.; Zhang, L.; Bi, L.; Liu, Z. Bio-inspired honeycomb-shaped La0·5Sr0·5Fe0·9P0·1O3-δ as a high-performing cathode for proton-conducting SOFCs. *Int. J. Hydrogen Energy* **2023**, *48*, 15248–15257. [CrossRef]

143. Sajjad, M.; Lu, W. Honeycomb-based heterostructures: An emerging platform for advanced energy applications: A review on energy systems. *Electrochem. Sci. Adv.* **2022**, *2*, e202100075. [CrossRef]

144. Fan, H.; Zhou, S.; Wei, Q.; Hu, X. Honeycomb-like carbon for electrochemical energy storage and conversion. *Renew. Sustain. Energy Rev.* **2022**, *165*, 112585. [CrossRef]

145. Bera, A.; Maitra, A.; Das, A.K.; Halder, L.; Paria, S.; Si, S.K.; De, A.; Ojha, S.; Khatua, B.B. A Quasi-Solid-State Asymmetric Supercapacitor Device Based on Honeycomb-like Nickel–Copper–Carbonate–Hydroxide as a Positive and Iron Oxide as a Negative Electrode with Superior Electrochemical Performances. *ACS Appl. Electron. Mater.* **2020**, *2*, 177–185. [CrossRef]

146. Dang, X.; Yi, H.; Ham, M.-H.; Qi, J.; Yun, D.S.; Ladewski, R.; Strano, M.S.; Hammond, P.T.; Belcher, A.M. Virus-templated self-assembled single-walled carbon nanotubes for highly efficient electron collection in photovoltaic devices. *Nat. Nanotechnol.* **2011**, *6*, 377–384. [CrossRef] [PubMed]

147. Weng, M.; Ding, M.; Zhou, P.; Ye, Y.; Luo, Z.; Ye, X.; Guo, Q.; Chen, L. Multi-functional and integrated actuators made with bio-inspired cobweb carbon nanotube–Polymer composites. *Chem. Eng. J.* **2023**, *452*, 139146. [CrossRef]

148. Gatto, E.; Venanzi, M. The Impervious Route to Peptide-Based Dye-Sensitized Solar Cells. *Isr. J. Chem.* **2015**, *55*, 671–681. [CrossRef]

149. Gardères, J.; Elkhooly, T.A.; Link, T.; Markl, J.S.; Müller, W.E.G.; Renkel, J.; Korzhev, M.; Wiens, M. Self-assembly and photocatalytic activity of branched silicatein/silintaphin filaments decorated with silicatein-synthesized TiO$_2$ nanoparticles. *Bioprocess Biosyst. Eng.* **2016**, *39*, 1477–1486. [CrossRef]

150. Wang, J.; Zhu, T.; Ho, G.W. Nature-Inspired Design of Artificial Solar-to-Fuel Conversion Systems based on Copper Phosphate Microflowers. *ChemSusChem* **2016**, *9*, 1575–1578. [CrossRef]

151. Biradar, M.R.; Rao, C.R.K.; Bhosale, S.V.; Bhosale, S.V. Bio-inspired adenine-benzoquinone-adenine pillar grafted graphene oxide materials with excellent cycle stability for high energy and power density supercapacitor applications. *J. Energy Storage* **2023**, *58*, 106399. [CrossRef]

152. Shimada, K. Elucidation of Response and Electrochemical Mechanisms of Bio-Inspired Rubber Sensors with Supercapacitor Paradigm. *Electronics* **2023**, *12*, 2304. [CrossRef]

153. Zhu, M.; Han, T.; Zhu, L.; Yang, S.; Lin, X.; Hu, C.; Liu, J. Engineering nanocluster arrays on lotus leaf as free-standing high areal capacity Li-ion battery anodes: A cost-effective and general bio-inspired approach. *J. Alloys Compd.* **2022**, *892*, 162136. [CrossRef]

154. Nam, K.T.; Kim, D.-W.; Yoo, P.J.; Chiang, C.-Y.; Meethong, N.; Hammond, P.T.; Chiang, Y.-M.; Belcher, A.M. Virus-Enabled Synthesis and Assembly of Nanowires for Lithium Ion Battery Electrodes. *Science* **2006**, *312*, 885–888. [CrossRef] [PubMed]

155. Chen, X.; Gerasopoulos, K.; Guo, J.; Brown, A.; Wang, C.; Ghodssi, R.; Culver, J.N. Virus-Enabled Silicon Anode for Lithium-Ion Batteries. *ACS Nano* **2010**, *4*, 5366–5372. [CrossRef] [PubMed]

156. Lu, Y.; Zhou, Y.P.; Yan, Q.Y.; Fong, E. Bio-inspired synthesis of N,F co-doped 3D graphitized carbon foams containing manganese fluoride nanocrystals for lithium ion batteries. *J. Mater. Chem. A* **2016**, *4*, 2691–2698. [CrossRef]

157. Fu, X.; Li, C.; Wang, Y.; Scudiero, L.; Liu, J.; Zhong, W.-H. Self-Assembled Protein Nanofilter for Trapping Polysulfides and Promoting Li+ Transport in Lithium–Sulfur Batteries. *J. Phys. Chem. Lett.* **2018**, *9*, 2450–2459. [CrossRef] [PubMed]

158. Shin, S.R.; Lee, C.K.; So, I.S.; Jeon, J.H.; Kang, T.M.; Kee, C.W.; Kim, S.I.; Spinks, G.M.; Wallace, G.G.; Kim, S.J. DNA-Wrapped Single-Walled Carbon Nanotube Hybrid Fibers for supercapacitors and Artificial Muscles. *Adv. Mater.* **2008**, *20*, 466–470. [CrossRef]

159. Huang, J.; Lin, L.; Sun, D.; Chen, H.; Yang, D.; Li, Q. Bio-inspired synthesis of metal nanomaterials and applications. *Chem. Soc. Rev.* **2015**, *44*, 6330–6374. [CrossRef]

160. Tang, J.; Song, Y.; Zhao, F.; Spinney, S.; da Silva Bernardes, J.; Tam, K.C. Compressible cellulose nanofibril (CNF) based aerogels produced via a bio-inspired strategy for heavy metal ion and dye removal. *Carbohydr. Polym.* **2019**, *208*, 404–412. [CrossRef]

161. Ajibade, T.F.; Tian, H.; Lasisi, K.H.; Zhang, K. Bio-inspired PDA@WS2 polyacrylonitrile ultrafiltration membrane for the effective separation of saline oily wastewater and the removal of soluble dye. *Sep. Purif. Technol.* **2022**, *299*, 121711. [CrossRef]

162. Nandi, N.; Baral, A.; Basu, K.; Roy, S.; Banerjee, A. A dipeptide-based superhydrogel: Removal of toxic dyes and heavy metal ions from waste water. *Pept. Sci.* **2017**, *108*, e22915. [CrossRef]

163. Li, D.; Zhang, W.; Yu, X.; Wang, Z.; Su, Z.; Wei, G. When biomolecules meet graphene: From molecular level interactions to material design and applications. *Nanoscale* **2016**, *8*, 19491–19509. [CrossRef] [PubMed]

164. Song, Y.; Li, W.; Duan, Y.; Li, Z.; Deng, L. Nicking enzyme-assisted biosensor for Salmonella enteritidis detection based on fluorescence resonance energy transfer. *Biosens. Bioelectron.* **2014**, *55*, 400–404. [CrossRef]

165. Bolisetty, S.; Mezzenga, R. Amyloid–carbon hybrid membranes for universal water purification. *Nat. Nanotechnol.* **2016**, *11*, 365–371. [CrossRef] [PubMed]

166. Zhang, Y.; Mei, J.; Yan, C.; Liao, T.; Bell, J.; Sun, Z. Bioinspired 2D Nanomaterials for Sustainable Applications. *Adv. Mater.* **2020**, *32*, 1902806. [CrossRef] [PubMed]
167. Xu, L.; Zhu, Z.; Sun, D.-W. Bioinspired Nanomodification Strategies: Moving from Chemical-Based Agrosystems to Sustainable Agriculture. *ACS Nano* **2021**, *15*, 12655–12686. [CrossRef]
168. Li, X.; Wang, R.; Wicaksana, F.; Tang, C.; Torres, J.; Fane, A.G. Preparation of high performance nanofiltration (NF) membranes incorporated with aquaporin Z. *J. Membr. Sci.* **2014**, *450*, 181–188. [CrossRef]
169. Radhakrishnan, K.; Kalyanasundharam, S.; Ravichandran, N.; Thiyagarajan, S.; Richard Thilagaraj, W. A novel method of unburned hydrocarbons and NOx gases capture from vehicular exhaust using natural biosorbent. *Sep. Sci. Technol.* **2018**, *53*, 13–21. [CrossRef]
170. Nath, A.; Eren, B.A.; Zinia Zaukuu, J.-L.; Koris, A.; Pásztorné-Huszár, K.; Szerdahelyi, E.; Kovacs, Z. Detecting the Bitterness of Milk-Protein-Derived Peptides Using an Electronic Tongue. *Chemosensors* **2022**, *10*, 215. [CrossRef]
171. Qin, C.; Zhang, S.; Yuan, Q.; Liu, M.; Jiang, N.; Zhuang, L.; Huang, L.; Wang, P. A Cell Co-Culture Taste Sensor Using Different Proportions of Caco-2 and SH-SY5Y Cells for Bitterness Detection. *Chemosensors* **2022**, *10*, 173. [CrossRef]
172. Liu, S.; Zhu, P.; Tian, Y.; Chen, Y.; Liu, Y.; Wang, M.; Chen, W.; Du, L.; Wu, C. A Taste Bud Organoid-Based Microelectrode Array Biosensor for Taste Sensing. *Chemosensors* **2022**, *10*, 208. [CrossRef]
173. Vahidpour, F.; Alghazali, Y.; Akca, S.; Hommes, G.; Schöning, M.J. An Enzyme-Based Interdigitated Electrode-Type Biosensor for Detecting Low Concentrations of H_2O_2 Vapor/Aerosol. *Chemosensors* **2022**, *10*, 202. [CrossRef]
174. Yuan, Q.; Qin, C.; Zhang, S.; Wu, J.; Qiu, Y.; Chen, C.; Huang, L.; Wang, P.; Jiang, D.; Zhuang, L. An In Vitro HL-1 Cardiomyocyte-Based Olfactory Biosensor for Olfr558-Inhibited Efficiency Detection. *Chemosensors* **2022**, *10*, 200. [CrossRef]
175. Liu, Y.; Furuno, S.; Akagawa, S.; Yatabe, R.; Onodera, T.; Fujiwara, N.; Takeda, H.; Uchida, S.; Toko, K. Odor Recognition of Thermal Decomposition Products of Electric Cables Using Odor Sensing Arrays. *Chemosensors* **2021**, *9*, 261. [CrossRef]
176. Ye, Z.; Ai, T.; Wu, X.; Onodera, T.; Ikezaki, H.; Toko, K. Elucidation of Response Mechanism of a Potentiometric Sweetness Sensor with a Lipid/Polymer Membrane for Uncharged Sweeteners. *Chemosensors* **2022**, *10*, 166. [CrossRef]
177. Gu, D.; Liu, W.; Wang, J.; Yu, J.; Zhang, J.; Huang, B.; Rumyantseva, M.N.; Li, X. Au Functionalized SnS2 Nanosheets Based Chemiresistive NO_2 Sensors. *Chemosensors* **2022**, *10*, 165. [CrossRef]
178. Potyrailo, R.A.; Karker, N.; Carpenter, M.A.; Minnick, A. Multivariable bio-inspired photonic sensors for non-condensable gases. *J. Opt.* **2018**, *20*, 024006. [CrossRef]
179. Yildirim, A.; Vural, M.; Yaman, M.; Bayindir, M. Bioinspired Optoelectronic Nose with Nanostructured Wavelength-Scalable Hollow-Core Infrared Fibers. *Adv. Mater.* **2011**, *23*, 1263–1267. [CrossRef]
180. Li, P.-H.; Madhaiyan, G.; Shin, Y.-Y.; Tsai, H.-Y.; Meng, H.-F.; Horng, S.-F.; Zan, H.-W. Facile Fabrication of a Bio-Inspired Leaf Vein-Based Ultra-Sensitive Humidity Sensor with a Hygroscopic Polymer. *Polymers* **2022**, *14*, 5030. [CrossRef]
181. Hurot, C.; Scaramozzino, N.; Buhot, A.; Hou, Y. Bio-Inspired Strategies for Improving the Selectivity and Sensitivity of Artificial Noses: A Review. *Sensors* **2020**, *20*, 1803. [CrossRef]
182. Dinh Le, T.-S.; An, J.; Huang, Y.; Vo, Q.; Boonruangkan, J.; Tran, T.; Kim, S.-W.; Sun, G.; Kim, Y.-J. Ultrasensitive Anti-Interference Voice Recognition by Bio-Inspired Skin-Attachable Self-Cleaning Acoustic Sensors. *ACS Nano* **2019**, *13*, 13293–13303. [CrossRef] [PubMed]
183. Wilmott, D.; Alves, F.; Karunasiri, G. Bio-Inspired Miniature Direction Finding Acoustic Sensor. *Sci. Rep.* **2016**, *6*, 29957. [CrossRef]
184. Shen, Q.; Luo, Z.; Ma, S.; Tao, P.; Song, C.; Wu, J.; Shang, W.; Deng, T. Bioinspired Infrared Sensing Materials and Systems. *Adv. Mater.* **2018**, *30*, 1707632. [CrossRef]
185. Hu, X.; Jiang, Y.; Ma, Z.; Xu, Y.; Zhang, D. Bio-inspired Flexible Lateral Line Sensor Based on P(VDF-TrFE)/BTO Nanofiber Mat for Hydrodynamic Perception. *Sensors* **2019**, *19*, 5384. [CrossRef] [PubMed]
186. Liu, L.; Xu, X.; Zhu, M.; Cui, X.; Feng, J.; Rad, Z.F.; Wang, H.; Song, P. Bioinspired Strong, Tough, and Biodegradable Poly(Vinyl Alcohol) and its Applications as Substrates for Humidity Sensors. *Adv. Mater. Technol.* **2023**, *8*, 2201414. [CrossRef]
187. Seetohul, J.; Shafiee, M. Snake Robots for Surgical Applications: A Review. *Robotics* **2022**, *11*, 57. [CrossRef]
188. Chang, P.; Bao, X.; Meng, F.; Lu, R. Multi-objective Pigeon-inspired Optimized feature enhancement soft-sensing model of Wastewater Treatment Process. *Expert Syst. Appl.* **2023**, *215*, 119193. [CrossRef]
189. Singh, A.V.; Rahman, A.; Sudhir Kumar, N.V.G.; Aditi, A.S.; Galluzzi, M.; Bovio, S.; Barozzi, S.; Montani, E.; Parazzoli, D. Bio-inspired approaches to design smart fabrics. *Mater. Des.* **2012**, *36*, 829–839. [CrossRef]
190. Wen, J.; Prawel, D.; Li, Y.V. A study on the mechanical and antimicrobial properties of biomimetic shark skin fabrics with different denticle size via 3D printing technology. *Phys. Scr.* **2023**, *98*, 035031. [CrossRef]
191. Wood, J. Bioinspiration in Fashion—A Review. *Biomimetics* **2019**, *4*, 16. [CrossRef]
192. Ahmad, I.; Kan, C.-W. A Review on Development and Applications of Bio-Inspired Superhydrophobic Textiles. *Materials* **2016**, *9*, 892. [CrossRef]
193. Richards, C.; Slaimi, A.; O'Connor, N.E.; Barrett, A.; Kwiatkowska, S.; Regan, F. Bio-inspired Surface Texture Modification as a Viable Feature of Future Aquatic Antifouling Strategies: A Review. *Int. J. Mol. Sci.* **2020**, *21*, 5063. [CrossRef] [PubMed]
194. Khan, A.Q.; Yu, K.; Li, J.; Leng, X.; Wang, M.; Zhang, X.; An, B.; Fei, B.; Wei, W.; Zhuang, H.; et al. Spider Silk Supercontraction-Inspired Cotton-Hydrogel Self-Adapting Textiles. *Adv. Fiber Mater.* **2022**, *4*, 1572–1583. [CrossRef]
195. Huang, H.; Cao, E.; Zhao, M.; Alamri, S.; Li, B. Spider Web-Inspired Lightweight Membrane-Type Acoustic Metamaterials for Broadband Low-Frequency Sound Isolation. *Polymers* **2021**, *13*, 1146. [CrossRef] [PubMed]

196. Miao, D.; Cheng, N.; Wang, X.; Yu, J. Bioinspired electrospun fibrous materials for directional water transport. *J. Text. Inst.* **2022**, 1–15. [CrossRef]

197. Qi, M.; Yang, R.; Wang, Z.; Liu, Y.; Zhang, Q.; He, B.; Li, K.; Yang, Q.; Wei, L.; Pan, C.; et al. Bioinspired Self-healing Soft Electronics. *Adv. Funct. Mater.* **2023**, *33*, 2214479. [CrossRef]

198. Li, X.; Xue, Z.; Sun, W.; Chu, J.; Wang, Q.; Tong, L.; Wang, K. Bio-inspired self-healing MXene/polyurethane coating with superior active/passive anticorrosion performance for Mg alloy. *Chem. Eng. J.* **2023**, *454*, 140187. [CrossRef]

199. Zhang, K.; Chen, J.; Shi, X.; Qian, H.; Wu, G.; Jiang, B.; Qi, D.; Huang, Y. Bioinspired self-healing and robust elastomer via tailored slipping semi-crystalline arrays for multifunctional electronics. *Chem. Eng. J.* **2023**, *454*, 139982. [CrossRef]

200. Yang, J.; Zhang, X.; Zhang, X.; Wang, L.; Feng, W.; Li, Q. Beyond the Visible: Bioinspired Infrared Adaptive Materials. *Adv. Mater.* **2021**, *33*, 2004754. [CrossRef]

201. Wang, W.; Wang, S.; Zhou, J.; Deng, H.; Sun, S.; Xue, T.; Ma, Y.; Gong, X. Bio-Inspired Semi-Active Safeguarding Design with Enhanced Impact Resistance via Shape Memory Effect. *Adv. Funct. Mater.* **2023**, *33*, 2212093. [CrossRef]

202. Muralidharan, M.; Saini, P.; Ameta, P.; Palani, I.A. Bio-inspired soft jellyfish robot: A novel polyimide-based structure actuated by shape memory alloy. *Int. J. Intell. Robot. Appl.* **2023**. [CrossRef]

203. Wang, Y.; Zhang, Z.; Chen, H.; Zhang, H.; Zhang, H.; Zhao, Y. Bio-inspired shape-memory structural color hydrogel film. *Sci. Bull.* **2022**, *67*, 512–519. [CrossRef]

204. Cera, L.; Gonzalez, G.M.; Liu, Q.; Choi, S.; Chantre, C.O.; Lee, J.; Gabardi, R.; Choi, M.C.; Shin, K.; Parker, K.K. A bioinspired and hierarchically structured shape-memory material. *Nat. Mater.* **2021**, *20*, 242–249. [CrossRef] [PubMed]

205. Cruz Ulloa, C.; Terrile, S.; Barrientos, A. Soft Underwater Robot Actuated by Shape-Memory Alloys "JellyRobcib" for Path Tracking through Fuzzy Visual Control. *Appl. Sci.* **2020**, *10*, 7160. [CrossRef]

206. Zhang, R.; Lei, J.; Xu, J.; Fu, H.; Jing, Y.; Chen, B.; Hou, X. Bioinspired Photo-Responsive Liquid Gating Membrane. *Biomimetics* **2022**, *7*, 47. [CrossRef]

207. Zhu, H.; Huang, Y.; Lou, X.; Xia, F. Bioinspired superwetting surfaces for biosensing. *View* **2021**, *2*, 20200053. [CrossRef]

208. Sun, T.; Feng, L.; Gao, X.; Jiang, L. Bioinspired Surfaces with Special Wettability. *Acc. Chem. Res.* **2005**, *38*, 644–652. [CrossRef] [PubMed]

209. Liu, Y.; Wang, X.; Fei, B.; Hu, H.; Lai, C.; Xin, J.H. Bioinspired, Stimuli-Responsive, Multifunctional Superhydrophobic Surface with Directional Wetting, Adhesion, and Transport of Water. *Adv. Funct. Mater.* **2015**, *25*, 5047–5056. [CrossRef]

210. Zhu, H.; Cai, S.; Liao, G.; Gao, Z.F.; Min, X.; Huang, Y.; Jin, S.; Xia, F. Recent Advances in Photocatalysis Based on Bioinspired Superwettabilities. *ACS Catal.* **2021**, *11*, 14751–14771. [CrossRef]

211. Reddy, S.; Arzt, E.; del Campo, A. Bioinspired Surfaces with Switchable Adhesion. *Adv. Mater.* **2007**, *19*, 3833–3837. [CrossRef]

212. Nepal, D.; Kang, S.; Adstedt, K.M.; Kanhaiya, K.; Bockstaller, M.R.; Brinson, L.C.; Buehler, M.J.; Coveney, P.V.; Dayal, K.; El-Awady, J.A.; et al. Hierarchically structured bioinspired nanocomposites. *Nat. Mater.* **2023**, *22*, 18–35. [CrossRef]

213. Wei, J.; Pan, F.; Ping, H.; Yang, K.; Wang, Y.; Wang, Q.; Fu, Z. Bioinspired Additive Manufacturing of Hierarchical Materials: From Biostructures to Functions. *Research* **2023**, *6*, 0164. [CrossRef] [PubMed]

214. Zhao, X.; Bhagia, S.; Gomez-Maldonado, D.; Tang, X.; Wasti, S.; Lu, S.; Zhang, S.; Parit, M.; Rencheck, M.L.; Korey, M.; et al. Bioinspired design toward nanocellulose-based materials. *Mater. Today* **2023**, *66*, 409–430. [CrossRef]

215. Saleemi, S.; Aouraghe, M.A.; Wei, X.; Liu, W.; Liu, L.; Siyal, M.I.; Bae, J.; Xu, F. Bio-Inspired Hierarchical Carbon Nanotube Yarn with Ester Bond Cross-Linkages towards High Conductivity for Multifunctional Applications. *Nanomaterials* **2022**, *12*, 208. [CrossRef] [PubMed]

216. Yoo, S.C.; Park, Y.K.; Park, C.; Ryu, H.; Hong, S.H. Biomimetic Artificial Nacre: Boron Nitride Nanosheets/Gelatin Nanocomposites for Biomedical Applications. *Adv. Funct. Mater.* **2018**, *28*, 1805948. [CrossRef]

217. Zhang, K.-R.; Gao, H.-L.; Pan, X.-F.; Zhou, P.; Xing, X.; Xu, R.; Pan, Z.; Wang, S.; Zhu, Y.; Hu, B.; et al. Multifunctional Bilayer Nanocomposite Guided Bone Regeneration Membrane. *Matter* **2019**, *1*, 770–781. [CrossRef]

218. Feng, C.; Xue, J.; Yu, X.; Zhai, D.; Lin, R.; Zhang, M.; Xia, L.; Wang, X.; Yao, Q.; Chang, J.; et al. Co-inspired hydroxyapatite-based scaffolds for vascularized bone regeneration. *Acta Biomater.* **2021**, *119*, 419–431. [CrossRef]

219. Xue, J.; Feng, C.; Xia, L.; Zhai, D.; Ma, B.; Wang, X.; Fang, B.; Chang, J.; Wu, C. Assembly Preparation of Multilayered Biomaterials with High Mechanical Strength and Bone-Forming Bioactivity. *Chem. Mater.* **2018**, *30*, 4646–4657. [CrossRef]

220. Li, T.; Ma, B.; Xue, J.; Zhai, D.; Zhao, P.; Chang, J.; Wu, C. Bioinspired Biomaterials with a Brick-and-Mortar Microstructure Combining Mechanical and Biological Performance. *Adv. Healthc. Mater.* **2020**, *9*, 1901211. [CrossRef] [PubMed]

221. Li, S.; Wang, H.; Huang, D.; Liu, J.; Chen, C.; Li, D.; Zhu, M.; Chen, Y. An Ultra-Strong, Water Stable and Antimicrobial Chitosan Film with Interdigitated Bouligand Structure. *Adv. Sustain. Syst.* **2022**, *6*, 2200033. [CrossRef]

222. Han, F.; Li, T.; Li, M.; Zhang, B.; Wang, Y.; Zhu, Y.; Wu, C. Nano-calcium silicate mineralized fish scale scaffolds for enhancing tendon-bone healing. *Bioact. Mater.* **2023**, *20*, 29–40. [CrossRef]

223. Wei, J.; Ping, H.; Xie, J.; Zou, Z.; Wang, K.; Xie, H.; Wang, W.; Lei, L.; Fu, Z. Bioprocess-Inspired Microscale Additive Manufacturing of Multilayered TiO_2/Polymer Composites with Enamel-Like Structures and High Mechanical Properties. *Adv. Funct. Mater.* **2020**, *30*, 1904880. [CrossRef]

224. Zhao, H.; Liu, S.; Wei, Y.; Yue, Y.; Gao, M.; Li, Y.; Zeng, X.; Deng, X.; Kotov, N.A.; Guo, L.; et al. Multiscale engineered artificial tooth enamel. *Science* **2022**, *375*, 551–556. [CrossRef] [PubMed]

225. Thrivikraman, G.; Athirasala, A.; Gordon, R.; Zhang, L.; Bergan, R.; Keene, D.R.; Jones, J.M.; Xie, H.; Chen, Z.; Tao, J.; et al. Rapid fabrication of vascularized and innervated cell-laden bone models with biomimetic intrafibrillar collagen mineralization. *Nat. Commun.* **2019**, *10*, 3520. [CrossRef]

226. Zhao, Y.; Zheng, J.; Xiong, Y.; Wang, H.; Yang, S.; Sun, X.; Zhao, L.; Mikos, A.G.; Wang, X. Hierarchically Engineered Artificial Lamellar Bone with High Strength and Toughness. *Small Struct.* **2023**, *4*, 2200256. [CrossRef]

227. Ivanova, E.P.; Hasan, J.; Webb, H.K.; Truong, V.K.; Watson, G.S.; Watson, J.A.; Baulin, V.A.; Pogodin, S.; Wang, J.Y.; Tobin, M.J.; et al. Natural Bactericidal Surfaces: Mechanical Rupture of Pseudomonas aeruginosa Cells by Cicada Wings. *Small* **2012**, *8*, 2489–2494. [CrossRef]

228. Zhang, P.; Lin, L.; Zang, D.; Guo, X.; Liu, M. Designing Bioinspired Anti-Biofouling Surfaces based on a Superwettability Strategy. *Small* **2017**, *13*, 1503334. [CrossRef]

229. Jiang, R.; Hao, L.; Song, L.; Tian, L.; Fan, Y.; Zhao, J.; Liu, C.; Ming, W.; Ren, L. Lotus-leaf-inspired hierarchical structured surface with non-fouling and mechanical bactericidal performances. *Chem. Eng. J.* **2020**, *398*, 125609. [CrossRef]

230. Li, S.; Chen, A.; Chen, Y.; Yang, Y.; Zhang, Q.; Luo, S.; Ye, M.; Zhou, Y.; An, Y.; Huang, W.; et al. Lotus leaf inspired antiadhesive and antibacterial gauze for enhanced infected dermal wound regeneration. *Chem. Eng. J.* **2020**, *402*, 126202. [CrossRef]

231. Shahsavan, H.; Salili, S.M.; Jákli, A.; Zhao, B. Thermally Active Liquid Crystal Network Gripper Mimicking the Self-Peeling of Gecko Toe Pads. *Adv. Mater.* **2017**, *29*, 1604021. [CrossRef]

232. Huang, R.; Zhang, X.; Li, W.; Shang, L.; Wang, H.; Zhao, Y. Suction Cups-Inspired Adhesive Patch with Tailorable Patterns for Versatile Wound Healing. *Adv. Sci.* **2021**, *8*, 2100201. [CrossRef] [PubMed]

233. Kim, D.W.; Baik, S.; Min, H.; Chun, S.; Lee, H.J.; Kim, K.H.; Lee, J.Y.; Pang, C. Highly Permeable Skin Patch with Conductive Hierarchical Architectures Inspired by Amphibians and Octopi for Omnidirectionally Enhanced Wet Adhesion. *Adv. Funct. Mater.* **2019**, *29*, 1807614. [CrossRef]

234. Gan, D.; Xu, T.; Xing, W.; Ge, X.; Fang, L.; Wang, K.; Ren, F.; Lu, X. Mussel-Inspired Contact-Active Antibacterial Hydrogel with High Cell Affinity, Toughness, and Recoverability. *Adv. Funct. Mater.* **2019**, *29*, 1805964. [CrossRef]

235. Ni, Z.; Yu, H.; Wang, L.; Liu, X.; Shen, D.; Chen, X.; Liu, J.; Wang, N.; Huang, Y.; Sheng, Y. Polyphosphazene and Non-Catechol-Based Antibacterial Injectable Hydrogel for Adhesion of Wet Tissues as Wound Dressing. *Adv. Healthc. Mater.* **2022**, *11*, 2101421. [CrossRef]

236. Chang, R.; Liu, Y.-J.; Zhang, Y.-L.; Zhang, S.-Y.; Han, B.-B.; Chen, F.; Chen, Y.-X. Phosphorylated and Phosphonated Low-Complexity Protein Segments for Biomimetic Mineralization and Repair of Tooth Enamel. *Adv. Sci.* **2022**, *9*, 2103829. [CrossRef] [PubMed]

237. Yang, Y.; Wu, X.; Ma, L.; He, C.; Cao, S.; Long, Y.; Huang, J.; Rodriguez, R.D.; Cheng, C.; Zhao, C.; et al. Bioinspired Spiky Peroxidase-Mimics for Localized Bacterial Capture and Synergistic Catalytic Sterilization. *Adv. Mater.* **2021**, *33*, 2005477. [CrossRef] [PubMed]

238. Zhang, Q.; Xu, H.; Wu, C.; Shang, Y.; Wu, Q.; Wei, Q.; Zhang, Q.; Sun, Y.; Wang, Q. Tissue Fluid Triggered Enzyme Polymerization for Ultrafast Gelation and Cartilage Repair. *Angew. Chem. Int. Ed.* **2021**, *60*, 19982–19987. [CrossRef]

239. Yang, N.; Li, M.; Wu, L.; Song, Y.; Yu, S.; Wan, Y.; Cheng, W.; Yang, B.; Mou, X.; Yu, H.; et al. Peptide-anchored neutrophil membrane-coated biomimetic nanodrug for targeted treatment of rheumatoid arthritis. *J. Nanobiotechnology* **2023**, *21*, 13. [CrossRef] [PubMed]

240. Xie, J.; Zhou, M.; Qian, Y.; Cong, Z.; Chen, S.; Zhang, W.; Jiang, W.; Dai, C.; Shao, N.; Ji, Z.; et al. Addressing MRSA infection and antibacterial resistance with peptoid polymers. *Nat. Commun.* **2021**, *12*, 5898. [CrossRef]

241. Ying, K.; Zhu, Y.; Wan, J.; Zhan, C.; Wang, Y.; Xie, B.; Xu, P.; Pan, H.; Wang, H. Macrophage membrane-biomimetic adhesive polycaprolactone nanocamptothecin for improving cancer-targeting efficiency and impairing metastasis. *Bioact. Mater.* **2023**, *20*, 449–462. [CrossRef]

242. Qiao, Z.; Zhang, K.; Liu, J.; Cheng, D.; Yu, B.; Zhao, N.; Xu, F.-J. Biomimetic electrodynamic nanoparticles comprising ginger-derived extracellular vesicles for synergistic anti-infective therapy. *Nat. Commun.* **2022**, *13*, 7164. [CrossRef] [PubMed]

243. Duan, Z.; Luo, Q.; Dai, X.; Li, X.; Gu, L.; Zhu, H.; Tian, X.; Zhang, H.; Gong, Q.; Gu, Z.; et al. Synergistic Therapy of a Naturally Inspired Glycopolymer-Based Biomimetic Nanomedicine Harnessing Tumor Genomic Instability. *Adv. Mater.* **2021**, *33*, 2104594. [CrossRef]

244. Ma, L.; Huang, J.; Zhu, X.; Zhu, B.; Wang, L.; Zhao, W.; Qiu, L.; Song, B.; Zhao, C.; Yan, F. In vitro and in vivo anticoagulant activity of heparin-like biomacromolecules and the mechanism analysis for heparin-mimicking activity. *Int. J. Biol. Macromol.* **2019**, *122*, 784–792. [CrossRef]

245. Wang, H.; Wang, J.; Feng, J.; Rao, Y.; Xu, Z.; Zu, J.; Wang, H.; Zhang, Z.; Chen, H. Artificial Extracellular Matrix Composed of Heparin-Mimicking Polymers for Efficient Anticoagulation and Promotion of Endothelial Cell Proliferation. *ACS Appl. Mater. Interfaces* **2022**, *14*, 50142–50151. [CrossRef] [PubMed]

246. Chen, H.-J.; Qin, Y.; Wang, Z.-G.; Wang, L.; Pang, D.-W.; Zhao, D.; Liu, S.-L. An Activatable and Reversible Virus-Mimicking NIR-II Nanoprobe for Monitoring the Progression of Viral Encephalitis. *Angew. Chem. Int. Ed.* **2022**, *61*, e202210285. [CrossRef] [PubMed]

247. Li, X.; Wang, Z.-G.; Zhu, H.; Wen, H.-P.; Ning, D.; Liu, H.-Y.; Pang, D.-W.; Liu, S.-L. Inducing Autophagy and Blocking Autophagic Flux via a Virus-Mimicking Nanodrug for Cancer Therapy. *Nano Lett.* **2022**, *22*, 9163–9173. [CrossRef] [PubMed]

248. Tan, P.; Chen, X.; Zhang, H.; Wei, Q.; Luo, K. Artificial intelligence aids in development of nanomedicines for cancer management. *Semin. Cancer Biol.* **2023**, *89*, 61–75. [CrossRef] [PubMed]

249. Campodoni, E.; Montanari, M.; Artusi, C.; Bassi, G.; Furlani, F.; Montesi, M.; Panseri, S.; Sandri, M.; Tampieri, A. Calcium-Based Biomineralization: A Smart Approach for the Design of Novel Multifunctional Hybrid Materials. *J. Compos. Sci.* **2021**, *5*, 278. [CrossRef]

250. Zambonino, M.C.; Quizhpe, E.M.; Jaramillo, F.E.; Rahman, A.; Santiago Vispo, N.; Jeffryes, C.; Dahoumane, S.A. Green Synthesis of Selenium and Tellurium Nanoparticles: Current Trends, Biological Properties and Biomedical Applications. *Int. J. Mol. Sci.* **2021**, *22*, 989. [CrossRef] [PubMed]

251. Liu, J.; Gu, T.; Li, L.; Li, L. Synthesis of MnO/C/NiO-Doped Porous Multiphasic Composites for Lithium-Ion Batteries by Biomineralized Mn Oxides from Engineered Pseudomonas putida Cells. *Nanomaterials* **2021**, *11*, 361. [CrossRef]

252. Li, Y.; Kong, Y.; Xue, B.; Dai, J.; Sha, G.; Ping, H.; Lei, L.; Wang, W.; Wang, K.; Fu, Z. Mechanically Reinforced Artificial Enamel by Mg2+-Induced Amorphous Intergranular Phases. *ACS Nano* **2022**, *16*, 10422–10430. [CrossRef]

253. Shao, C.; Jin, B.; Mu, Z.; Lu, H.; Zhao, Y.; Wu, Z.; Yan, L.; Zhang, Z.; Zhou, Y.; Pan, H.; et al. Repair of tooth enamel by a biomimetic mineralization frontier ensuring epitaxial growth. *Sci. Adv.* **2019**, *5*, eaaw9569. [CrossRef] [PubMed]

254. Zhou, Y.; Deng, J.; Zhang, Y.; Li, C.; Wei, Z.; Shen, J.; Li, J.; Wang, F.; Han, B.; Chen, D.; et al. Engineering DNA-Guided Hydroxyapatite Bulk Materials with High Stiffness and Outstanding Antimicrobial Ability for Dental Inlay Applications. *Adv. Mater.* **2022**, *34*, 2202180. [CrossRef] [PubMed]

255. Ping, H.; Wagermaier, W.; Horbelt, N.; Scoppola, E.; Li, C.; Werner, P.; Fu, Z.; Fratzl, P. Mineralization generates megapascal contractile stresses in collagen fibrils. *Science* **2022**, *376*, 188–192. [CrossRef]

256. Zhao, R.; Wang, B.; Yang, X.; Xiao, Y.; Wang, X.; Shao, C.; Tang, R. A Drug-Free Tumor Therapy Strategy: Cancer-Cell-Targeting Calcification. *Angew. Chem. Int. Ed.* **2016**, *55*, 5225–5229. [CrossRef] [PubMed]

257. Huo, M.; Wang, L.; Zhang, L.; Wei, C.; Chen, Y.; Shi, J. Photosynthetic Tumor Oxygenation by Photosensitizer-Containing Cyanobacteria for Enhanced Photodynamic Therapy. *Angew. Chem. Int. Ed.* **2020**, *59*, 1906–1913. [CrossRef] [PubMed]

258. Wang, X.; Yang, C.; Yu, Y.; Zhao, Y. In Situ 3D Bioprinting Living Photosynthetic Scaffolds for Autotrophic Wound Healing. *Research* **2022**, *2022*, 9794745. [CrossRef] [PubMed]

259. Chen, P.; Liu, X.; Gu, C.; Zhong, P.; Song, N.; Li, M.; Dai, Z.; Fang, X.; Liu, Z.; Zhang, J.; et al. A plant-derived natural photosynthetic system for improving cell anabolism. *Nature* **2022**, *612*, 546–554. [CrossRef] [PubMed]

260. Shih, C.-P.; Tang, X.; Kuo, C.W.; Chueh, D.-Y.; Chen, P. Design principles of bioinspired interfaces for biomedical applications in therapeutics and imaging. *Front. Chem.* **2022**, *10*, 990171. [CrossRef]

261. Tang, Y.; Wang, X.; Li, J.; Nie, Y.; Liao, G.; Yu, Y.; Li, C. Overcoming the Reticuloendothelial System Barrier to Drug Delivery with a "Don't-Eat-Us" Strategy. *ACS Nano* **2019**, *13*, 13015–13026. [CrossRef]

262. Mirkasymov, A.B.; Zelepukin, I.V.; Nikitin, P.I.; Nikitin, M.P.; Deyev, S.M. In vivo blockade of mononuclear phagocyte system with solid nanoparticles: Efficiency and affecting factors. *J. Control. Release* **2021**, *330*, 111–118. [CrossRef] [PubMed]

263. Xie, J.; Shen, Q.; Huang, K.; Zheng, T.; Cheng, L.; Zhang, Z.; Yu, Y.; Liao, G.; Wang, X.; Li, C. Oriented Assembly of Cell-Mimicking Nanoparticles via a Molecular Affinity Strategy for Targeted Drug Delivery. *ACS Nano* **2019**, *13*, 5268–5277. [CrossRef] [PubMed]

264. Subhan, M.A.; Yalamarty, S.S.; Filipczak, N.; Parveen, F.; Torchilin, V.P. Recent Advances in Tumor Targeting via EPR Effect for Cancer Treatment. *J. Pers. Med.* **2021**, *11*, 571. [CrossRef]

265. Perera, A.S.; Coppens, M.-O. Re-designing materials for biomedical applications: From biomimicry to nature-inspired chemical engineering. *Philos. Trans. R. Soc. A Math. Phys. Eng. Sci.* **2018**, *377*, 20180268. [CrossRef] [PubMed]

266. Zou, S.; Wang, B.; Wang, C.; Wang, Q.; Zhang, L. Cell membrane-coated nanoparticles: Research advances. *Nanomedicine* **2020**, *15*, 625–641. [CrossRef]

267. Jiménez-Jiménez, C.; Moreno, V.M.; Vallet-Regí, M. Bacteria-Assisted Transport of Nanomaterials to Improve Drug Delivery in Cancer Therapy. *Nanomaterials* **2022**, *12*, 288. [CrossRef] [PubMed]

268. Park, J.H.; Mohapatra, A.; Zhou, J.; Holay, M.; Krishnan, N.; Gao, W.; Fang, R.H.; Zhang, L. Virus-Mimicking Cell Membrane-Coated Nanoparticles for Cytosolic Delivery of mRNA. *Angew. Chem. Int. Ed.* **2022**, *61*, e202113671. [CrossRef]

269. Guo, M.; Xia, C.; Wu, Y.; Zhou, N.; Chen, Z.; Li, W. Research Progress on Cell Membrane-Coated Biomimetic Delivery Systems. *Front. Bioeng. Biotechnol.* **2021**, *9*, 772522. [CrossRef] [PubMed]

270. Hu, C.-M.J.; Fang, R.H.; Luk, B.T.; Chen, K.N.H.; Carpenter, C.; Gao, W.; Zhang, K.; Zhang, L. 'Marker-of-self' functionalization of nanoscale particles through a top-down cellular membrane coating approach. *Nanoscale* **2013**, *5*, 2664–2668. [CrossRef]

271. Wu, M.; Le, W.; Mei, T.; Wang, Y.; Chen, B.; Liu, Z.; Xue, C. Cell membrane camouflaged nanoparticles: A new biomimetic platform for cancer photothermal therapy. *Int. J. Nanomed.* **2019**, *14*, 4431–4448. [CrossRef]

272. Zhang, C.; Zhang, L.; Wu, W.; Gao, F.; Li, R.-Q.; Song, W.; Zhuang, Z.-N.; Liu, C.-J.; Zhang, X.-Z. Artificial Super Neutrophils for Inflammation Targeting and HClO Generation against Tumors and Infections. *Adv. Mater.* **2019**, *31*, 1901179. [CrossRef] [PubMed]

273. Chen, L.-J.; Zhao, X.; Liu, Y.-Y.; Yan, X.-P. Macrophage membrane coated persistent luminescence nanoparticle@MOF-derived mesoporous carbon core–shell nanocomposites for autofluorescence-free imaging-guided chemotherapy. *J. Mater. Chem. B* **2020**, *8*, 8071–8083. [CrossRef]

274. Huang, S.-S.; Lee, K.-J.; Chen, H.-C.; Prajnamitra, R.P.; Hsu, C.-H.; Jian, C.-B.; Yu, X.-E.; Chueh, D.-Y.; Kuo, C.W.; Chiang, T.-C.; et al. Immune cell shuttle for precise delivery of nanotherapeutics for heart disease and cancer. *Sci. Adv.* **2021**, *7*, eabf2400. [CrossRef] [PubMed]

275. He, Q.; Jiang, X.; Zhou, X.; Weng, J. Targeting cancers through TCR-peptide/MHC interactions. *J. Hematol. Oncol.* **2019**, *12*, 139. [CrossRef]
276. Olsson, M.; Bruhns, P.; Frazier, W.A.; Ravetch, J.V.; Oldenborg, P.-A. Platelet homeostasis is regulated by platelet expression of CD47 under normal conditions and in passive immune thrombocytopenia. *Blood* **2005**, *105*, 3577–3582. [CrossRef]
277. Han, H.; Bártolo, R.; Li, J.; Shahbazi, M.-A.; Santos, H.A. Biomimetic platelet membrane-coated nanoparticles for targeted therapy. *Eur. J. Pharm. Biopharm.* **2022**, *172*, 1–15. [CrossRef]
278. Deppermann, C.; Kubes, P. Start a fire, kill the bug: The role of platelets in inflammation and infection. *Innate Immun.* **2018**, *24*, 335–348. [CrossRef] [PubMed]
279. Hu, C.-M.J.; Fang, R.H.; Wang, K.-C.; Luk, B.T.; Thamphiwatana, S.; Dehaini, D.; Nguyen, P.; Angsantikul, P.; Wen, C.H.; Kroll, A.V.; et al. Nanoparticle biointerfacing by platelet membrane cloaking. *Nature* **2015**, *526*, 118–121. [CrossRef]
280. Wei, X.; Gao, J.; Fang, R.H.; Luk, B.T.; Kroll, A.V.; Dehaini, D.; Zhou, J.; Kim, H.W.; Gao, W.; Lu, W.; et al. Nanoparticles camouflaged in platelet membrane coating as an antibody decoy for the treatment of immune thrombocytopenia. *Biomaterials* **2016**, *111*, 116–123. [CrossRef] [PubMed]
281. Nelson, V.S.; Jolink, A.-T.C.; Amini, S.N.; Zwaginga, J.J.; Netelenbos, T.; Semple, J.W.; Porcelijn, L.; de Haas, M.; Schipperus, M.R.; Kapur, R. Platelets in ITP: Victims in Charge of Their Own Fate? *Cells* **2021**, *10*, 3235. [CrossRef] [PubMed]
282. Li, S.-Y.; Cheng, H.; Qiu, W.-X.; Zhang, L.; Wan, S.-S.; Zeng, J.-Y.; Zhang, X.-Z. Cancer cell membrane-coated biomimetic platform for tumor targeted photodynamic therapy and hypoxia-amplified bioreductive therapy. *Biomaterials* **2017**, *142*, 149–161. [CrossRef] [PubMed]
283. Fang, R.H.; Hu, C.-M.J.; Luk, B.T.; Gao, W.; Copp, J.A.; Tai, Y.; O'Connor, D.E.; Zhang, L. Cancer Cell Membrane-Coated Nanoparticles for Anticancer Vaccination and Drug Delivery. *Nano Lett.* **2014**, *14*, 2181–2188. [CrossRef]
284. Gdowski, A.S.; Lampe, J.B.; Lin, V.J.T.; Joshi, R.; Wang, Y.-C.; Mukerjee, A.; Vishwanatha, J.K.; Ranjan, A.P. Bioinspired Nanoparticles Engineered for Enhanced Delivery to the Bone. *ACS Appl. Nano Mater.* **2019**, *2*, 6249–6257. [CrossRef] [PubMed]
285. Rao, L.; Bu, L.-L.; Cai, B.; Xu, J.-H.; Li, A.; Zhang, W.-F.; Sun, Z.-J.; Guo, S.-S.; Liu, W.; Wang, T.-H.; et al. Cancer Cell Membrane-Coated Upconversion Nanoprobes for Highly Specific Tumor Imaging. *Adv. Mater.* **2016**, *28*, 3460–3466. [CrossRef] [PubMed]
286. Li, J.; Wang, X.; Zheng, D.; Lin, X.; Wei, Z.; Zhang, D.; Li, Z.; Zhang, Y.; Wu, M.; Liu, X. Cancer cell membrane-coated magnetic nanoparticles for MR/NIR fluorescence dual-modal imaging and photodynamic therapy. *Biomater. Sci.* **2018**, *6*, 1834–1845. [CrossRef]
287. Shen, J.; Karges, J.; Xiong, K.; Chen, Y.; Ji, L.; Chao, H. Cancer cell membrane camouflaged iridium complexes functionalized black-titanium nanoparticles for hierarchical-targeted synergistic NIR-II photothermal and sonodynamic therapy. *Biomaterials* **2021**, *275*, 120979. [CrossRef]
288. Roger, M.; Clavreul, A.; Venier-Julienne, M.-C.; Passirani, C.; Sindji, L.; Schiller, P.; Montero-Menei, C.; Menei, P. Mesenchymal stem cells as cellular vehicles for delivery of nanoparticles to brain tumors. *Biomaterials* **2010**, *31*, 8393–8401. [CrossRef]
289. Gao, C.; Lin, Z.; Jurado-Sánchez, B.; Lin, X.; Wu, Z.; He, Q. Stem Cell Membrane-Coated Nanogels for Highly Efficient In Vivo Tumor Targeted Drug Delivery. *Small* **2016**, *12*, 4056–4062. [CrossRef]
290. Tian, W.; Lu, J.; Jiao, D. Stem cell membrane vesicle–coated nanoparticles for efficient tumor-targeted therapy of orthotopic breast cancer. *Polym. Adv. Technol.* **2019**, *30*, 1051–1060. [CrossRef]
291. Chetty, S.S.; Praneetha, S.; Vadivel Murugan, A.; Govarthanan, K.; Verma, R.S. Human Umbilical Cord Wharton's Jelly-Derived Mesenchymal Stem Cells Labeled with Mn^{2+} and Gd^{3+} Co-Doped CuInS2–ZnS Nanocrystals for Multimodality Imaging in a Tumor Mice Model. *ACS Appl. Mater. Interfaces* **2020**, *12*, 3415–3429. [CrossRef]
292. Naskar, A.; Cho, H.; Lee, S.; Kim, K.-S. Biomimetic Nanoparticles Coated with Bacterial Outer Membrane Vesicles as a New-Generation Platform for Biomedical Applications. *Pharmaceutics* **2021**, *13*, 1887. [CrossRef] [PubMed]
293. Naskar, A.; Kim, K.-s. Nanomaterials as Delivery Vehicles and Components of New Strategies to Combat Bacterial Infections: Advantages and Limitations. *Microorganisms* **2019**, *7*, 356. [CrossRef] [PubMed]
294. Chung, Y.H.; Cai, H.; Steinmetz, N.F. Viral nanoparticles for drug delivery, imaging, immunotherapy, and theranostic applications. *Adv. Drug Del. Rev.* **2020**, *156*, 214–235. [CrossRef] [PubMed]
295. Shen, L.; Zhou, J.; Wang, Y.; Kang, N.; Ke, X.; Bi, S.; Ren, L. Efficient Encapsulation of Fe3O4 Nanoparticles into Genetically Engineered Hepatitis B Core Virus-Like Particles Through a Specific Interaction for Potential Bioapplications. *Small* **2015**, *11*, 1190–1196. [CrossRef]
296. Saini, V.; Martyshkin, D.V.; Mirov, S.B.; Perez, A.; Perkins, G.; Ellisman, M.H.; Towner, V.D.; Wu, H.; Pereboeva, L.; Borovjagin, A.; et al. An Adenoviral Platform for Selective Self-Assembly and Targeted Delivery of Nanoparticles. *Small* **2008**, *4*, 262–269. [CrossRef] [PubMed]
297. Gref, R.; Minamitake, Y.; Peracchia, M.T.; Trubetskoy, V.; Torchilin, V.; Langer, R. Biodegradable Long-Circulating Polymeric Nanospheres. *Science* **1994**, *263*, 1600–1603. [CrossRef]
298. Corbett, K.S.; Edwards, D.K.; Leist, S.R.; Abiona, O.M.; Boyoglu-Barnum, S.; Gillespie, R.A.; Himansu, S.; Schäfer, A.; Ziwawo, C.T.; DiPiazza, A.T.; et al. SARS-CoV-2 mRNA vaccine design enabled by prototype pathogen preparedness. *Nature* **2020**, *586*, 567–571. [CrossRef]
299. Kozma, G.T.; Shimizu, T.; Ishida, T.; Szebeni, J. Anti-PEG antibodies: Properties, formation, testing and role in adverse immune reactions to PEGylated nano-biopharmaceuticals. *Adv. Drug Del. Rev.* **2020**, *154–155*, 163–175. [CrossRef]

300. Debayle, M.; Balloul, E.; Dembele, F.; Xu, X.; Hanafi, M.; Ribot, F.; Monzel, C.; Coppey, M.; Fragola, A.; Dahan, M.; et al. Zwitterionic polymer ligands: An ideal surface coating to totally suppress protein-nanoparticle corona formation? *Biomaterials* **2019**, *219*, 119357. [CrossRef]

301. Zheng, T.; Wang, W.; Wu, F.; Zhang, M.; Shen, J.; Sun, Y. Zwitterionic Polymer-Gated Au@TiO$_2$ Core-Shell Nanoparticles for Imaging-Guided Combined Cancer Therapy. *Theranostics* **2019**, *9*, 5035–5048. [CrossRef]

302. Sabourian, P.; Yazdani, G.; Ashraf, S.S.; Frounchi, M.; Mashayekhan, S.; Kiani, S.; Kakkar, A. Effect of Physico-Chemical Properties of Nanoparticles on Their Intracellular Uptake. *Int. J. Mol. Sci.* **2020**, *21*, 8019. [CrossRef]

303. Niu, Y.; Yu, M.; Hartono, S.B.; Yang, J.; Xu, H.; Zhang, H.; Zhang, J.; Zou, J.; Dexter, A.; Gu, W.; et al. Nanoparticles Mimicking Viral Surface Topography for Enhanced Cellular Delivery. *Adv. Mater.* **2013**, *25*, 6233–6237. [CrossRef]

304. Zhang, Z.; Zhang, X.; Xu, X.; Li, Y.; Li, Y.; Zhong, D.; He, Y.; Gu, Z. Virus-Inspired Mimics Based on Dendritic Lipopeptides for Efficient Tumor-Specific Infection and Systemic Drug Delivery. *Adv. Funct. Mater.* **2015**, *25*, 5250–5260. [CrossRef]

305. Chen, M.Y.; Butler, S.S.; Chen, W.; Suh, J. Physical, chemical, and synthetic virology: Reprogramming viruses as controllable nanodevices. *WIREs Nanomed. Nanobiotechnology* **2019**, *11*, e1545. [CrossRef] [PubMed]

306. Figueroa, S.M.; Fleischmann, D.; Goepferich, A. Biomedical nanoparticle design: What we can learn from viruses. *J. Control. Release* **2021**, *329*, 552–569. [CrossRef] [PubMed]

307. Chen, Y.; Xianyu, Y.; Jiang, X. Surface Modification of Gold Nanoparticles with Small Molecules for Biochemical Analysis. *Acc. Chem. Res.* **2017**, *50*, 310–319. [CrossRef] [PubMed]

308. Luo, M.; Shen, C.; Feltis, B.N.; Martin, L.L.; Hughes, A.E.; Wright, P.F.A.; Turney, T.W. Reducing ZnO nanoparticle cytotoxicity by surface modification. *Nanoscale* **2014**, *6*, 5791–5798. [CrossRef]

309. Xu, C.; Cao, L.; Zhao, P.; Zhou, Z.; Cao, C.; Li, F.; Huang, Q. Emulsion-based synchronous pesticide encapsulation and surface modification of mesoporous silica nanoparticles with carboxymethyl chitosan for controlled azoxystrobin release. *Chem. Eng. J.* **2018**, *348*, 244–254. [CrossRef]

310. Sun, Y.; Ye, X.; Cai, M.; Liu, X.; Xiao, J.; Zhang, C.; Wang, Y.; Yang, L.; Liu, J.; Li, S.; et al. Osteoblast-Targeting-Peptide Modified Nanoparticle for siRNA/microRNA Delivery. *ACS Nano* **2016**, *10*, 5759–5768. [CrossRef]

311. Treuel, L.; Brandholt, S.; Maffre, P.; Wiegele, S.; Shang, L.; Nienhaus, G.U. Impact of Protein Modification on the Protein Corona on Nanoparticles and Nanoparticle–Cell Interactions. *ACS Nano* **2014**, *8*, 503–513. [CrossRef]

312. Lu, B.; Lv, X.; Le, Y. Chitosan-Modified PLGA Nanoparticles for Control-Released Drug Delivery. *Polymers* **2019**, *11*, 304. [CrossRef] [PubMed]

313. Fu, C.-P.; Cai, X.-Y.; Chen, S.-L.; Yu, H.-W.; Fang, Y.; Feng, X.-C.; Zhang, L.-M.; Li, C.-Y. Hyaluronic Acid-Based Nanocarriers for Anticancer Drug Delivery. *Polymers* **2023**, *15*, 2317. [CrossRef] [PubMed]

314. Demin, A.M.; Pershina, A.G.; Minin, A.S.; Brikunova, O.Y.; Murzakaev, A.M.; Perekucha, N.A.; Romashchenko, A.V.; Shevelev, O.B.; Uimin, M.A.; Byzov, I.V.; et al. Smart Design of a pH-Responsive System Based on pHLIP-Modified Magnetite Nanoparticles for Tumor MRI. *ACS Appl. Mater. Interfaces* **2021**, *13*, 36800–36815. [CrossRef] [PubMed]

315. Tian, Z.; Yu, X.; Ruan, Z.; Zhu, M.; Zhu, Y.; Hanagata, N. Magnetic mesoporous silica nanoparticles coated with thermo-responsive copolymer for potential chemo- and magnetic hyperthermia therapy. *Microporous Mesoporous Mater.* **2018**, *256*, 1–9. [CrossRef]

316. Adam Smith, R.; Sewell, S.L.; Giorgio, T.D. Proximity-activated nanoparticles: In vitro performance of specific structural modification by enzymatic cleavage. *Int. J. Nanomed.* **2008**, *3*, 95–103. [CrossRef]

317. Granot, Y.; Peer, D. Delivering the right message: Challenges and opportunities in lipid nanoparticles-mediated modified mRNA therapeutics—An innate immune system standpoint. *Semin. Immunol.* **2017**, *34*, 68–77. [CrossRef]

318. Choo, P.; Liu, T.; Odom, T.W. Nanoparticle Shape Determines Dynamics of Targeting Nanoconstructs on Cell Membranes. *J. Am. Chem. Soc.* **2021**, *143*, 4550–4555. [CrossRef] [PubMed]

319. Hoshyar, N.; Gray, S.; Han, H.; Bao, G. The effect of nanoparticle size on in vivo pharmacokinetics and cellular interaction. *Nanomedicine* **2016**, *11*, 673–692. [CrossRef]

320. Wong, A.C.; Wright, D.W. Size-Dependent Cellular Uptake of DNA Functionalized Gold Nanoparticles. *Small* **2016**, *12*, 5592–5600. [CrossRef]

321. Liu, X.; Huang, N.; Li, H.; Jin, Q.; Ji, J. Surface and Size Effects on Cell Interaction of Gold Nanoparticles with Both Phagocytic and Nonphagocytic Cells. *Langmuir* **2013**, *29*, 9138–9148. [CrossRef]

322. Shiohara, A.; Novikov, S.M.; Solís, D.M.; Taboada, J.M.; Obelleiro, F.; Liz-Marzán, L.M. Plasmon Modes and Hot Spots in Gold Nanostar–Satellite Clusters. *J. Phys. Chem. C* **2015**, *119*, 10836–10843. [CrossRef]

323. Huang, X.; Teng, X.; Chen, D.; Tang, F.; He, J. The effect of the shape of mesoporous silica nanoparticles on cellular uptake and cell function. *Biomaterials* **2010**, *31*, 438–448. [CrossRef] [PubMed]

324. Huang, X.; Li, L.; Liu, T.; Hao, N.; Liu, H.; Chen, D.; Tang, F. The Shape Effect of Mesoporous Silica Nanoparticles on Biodistribution, Clearance, and Biocompatibility in Vivo. *ACS Nano* **2011**, *5*, 5390–5399. [CrossRef]

325. Zhao, Y.; Wang, Y.; Ran, F.; Cui, Y.; Liu, C.; Zhao, Q.; Gao, Y.; Wang, D.; Wang, S. A comparison between sphere and rod nanoparticles regarding their in vivo biological behavior and pharmacokinetics. *Sci. Rep.* **2017**, *7*, 4131. [CrossRef]

326. Xie, X.; Liao, J.; Shao, X.; Li, Q.; Lin, Y. The Effect of shape on Cellular Uptake of Gold Nanoparticles in the forms of Stars, Rods, and Triangles. *Sci. Rep.* **2017**, *7*, 3827. [CrossRef] [PubMed]

327. Verma, A.; Stellacci, F. Effect of Surface Properties on Nanoparticle–Cell Interactions. *Small* **2010**, *6*, 12–21. [CrossRef]

328. Liu, P.; Chen, G.; Zhang, J. A Review of Liposomes as a Drug Delivery System: Current Status of Approved Products, Regulatory Environments, and Future Perspectives. *Molecules* **2022**, *27*, 1372. [CrossRef]

329. Kansız, S.; Elçin, Y.M. Advanced liposome and polymersome-based drug delivery systems: Considerations for physicochemical properties, targeting strategies and stimuli-sensitive approaches. *Adv. Colloid Interface Sci.* **2023**, *317*, 102930. [CrossRef]

330. Al Badri, Y.N.; Chaw, C.S.; Elkordy, A.A. Insights into Asymmetric Liposomes as a Potential Intervention for Drug Delivery Including Pulmonary Nanotherapeutics. *Pharmaceutics* **2023**, *15*, 294. [CrossRef] [PubMed]

331. Wang, S.; Chen, Y.; Guo, J.; Huang, Q. Liposomes for Tumor Targeted Therapy: A Review. *Int. J. Mol. Sci.* **2023**, *24*, 2643. [CrossRef]

332. Makhlouf, Z.; Ali, A.A.; Al-Sayah, M.H. Liposomes-Based Drug Delivery Systems of Anti-Biofilm Agents to Combat Bacterial Biofilm Formation. *Antibiotics* **2023**, *12*, 875. [CrossRef] [PubMed]

333. Gbian, D.L.; Omri, A. Lipid-Based Drug Delivery Systems for Diseases Managements. *Biomedicines* **2022**, *10*, 2137. [CrossRef]

334. Wang, J.; Li, B.; Qiu, L.; Qiao, X.; Yang, H. Dendrimer-based drug delivery systems: History, challenges, and latest developments. *J. Biol. Eng.* **2022**, *16*, 18. [CrossRef] [PubMed]

335. Zhu, Y.; Liu, C.; Pang, Z. Dendrimer-Based Drug Delivery Systems for Brain Targeting. *Biomolecules* **2019**, *9*, 790. [CrossRef]

336. Chauhan, A.S. Dendrimers for Drug Delivery. *Molecules* **2018**, *23*, 938. [CrossRef] [PubMed]

337. Palmerston Mendes, L.; Pan, J.; Torchilin, V.P. Dendrimers as Nanocarriers for Nucleic Acid and Drug Delivery in Cancer Therapy. *Molecules* **2017**, *22*, 1401. [CrossRef] [PubMed]

338. Negut, I.; Bita, B. Polymeric Micellar Systems—A Special Emphasis on "Smart" Drug Delivery. *Pharmaceutics* **2023**, *15*, 976.

339. Jain, A.; Bhardwaj, K.; Bansal, M. Polymeric Micelles as Drug Delivery System: Recent Advances, Approaches, Applications and Patents. *Curr. Drug Saf.* **2023**. [CrossRef]

340. Wang, Q.; Atluri, K.; Tiwari, A.K.; Babu, R.J. Exploring the Application of Micellar Drug Delivery Systems in Cancer Nanomedicine. *Pharmaceuticals* **2023**, *16*, 433. [CrossRef]

341. Jamil, B.; Rai, M. Nanotheranostics: An Emerging Nanoscience. In *Nanotheranostics: Applications and Limitations*; Rai, M., Jamil, B., Eds.; Springer International Publishing: Cham, Switzerland, 2019; pp. 1–18. [CrossRef]

342. Madamsetty, V.S.; Mukherjee, A.; Mukherjee, S. Recent Trends of the Bio-Inspired Nanoparticles in Cancer Theranostics. *Front. Pharmacol.* **2019**, *10*, 1264. [CrossRef]

343. Dunbar, C.E.; High, K.A.; Joung, J.K.; Kohn, D.B.; Ozawa, K.; Sadelain, M. Gene therapy comes of age. *Science* **2018**, *359*, eaan4672. [CrossRef] [PubMed]

344. Xu, X.; Jian, Y.; Li, Y.; Zhang, X.; Tu, Z.; Gu, Z. Bio-Inspired Supramolecular Hybrid Dendrimers Self-Assembled from Low-Generation Peptide Dendrons for Highly Efficient Gene Delivery and Biological Tracking. *ACS Nano* **2014**, *8*, 9255–9264. [CrossRef]

345. Sharma, S.; Javed, M.N.; Pottoo, F.H.; Rabbani, S.A.; Barkat, M.A.; Sarafroz, M.; Amir, M. Bioresponse inspired nanomaterials for targeted drug and gene delivery. *Pharm. Nanotechnol.* **2019**, *7*, 220–233. [CrossRef] [PubMed]

346. Sabu, C.; Rejo, C.; Kotta, S.; Pramod, K. Bioinspired and biomimetic systems for advanced drug and gene delivery. *J. Control. Release* **2018**, *287*, 142–155. [CrossRef] [PubMed]

347. De Matteis, V.; Rizzello, L.; Cascione, M.; Liatsi-Douvitsa, E.; Apriceno, A.; Rinaldi, R. Green Plasmonic Nanoparticles and Bio-Inspired Stimuli-Responsive Vesicles in Cancer Therapy Application. *Nanomaterials* **2020**, *10*, 1083. [CrossRef]

348. Rühle, B.; Saint-Cricq, P.; Zink, J.I. Externally Controlled Nanomachines on Mesoporous Silica Nanoparticles for Biomedical Applications. *Chemphyschem* **2016**, *17*, 1769–1779. [CrossRef]

349. Xu, D.; Wang, Y.; Liang, C.; You, Y.; Sanchez, S.; Ma, X. Self-Propelled Micro/Nanomotors for On-Demand Biomedical Cargo Transportation. *Small* **2020**, *16*, 1902464. [CrossRef]

350. Guix, M.; Weiz, S.M.; Schmidt, O.G.; Medina-Sánchez, M. Self-Propelled Micro/Nanoparticle Motors. *Part. Part. Syst. Charact.* **2018**, *35*, 1700382. [CrossRef]

351. Xu, W.; Qin, H.; Tian, H.; Liu, L.; Gao, J.; Peng, F.; Tu, Y. Biohybrid micro/nanomotors for biomedical applications. *Appl. Mater. Today* **2022**, *27*, 101482. [CrossRef]

352. Magdanz, V.; Schmidt, O.G. Spermbots: Potential impact for drug delivery and assisted reproductive technologies. *Expert Opin. Drug Deliv.* **2014**, *11*, 1125–1129. [CrossRef]

353. Stroble, J.K.; Stone, R.B.; Watkins, S.E. An overview of biomimetic sensor technology. *Sens. Rev.* **2009**, *29*, 112–119. [CrossRef]

354. Cho, N.H.; Guerrero-Martínez, A.; Ma, J.; Bals, S.; Kotov, N.A.; Liz-Marzán, L.M.; Nam, K.T. Bioinspired chiral inorganic nanomaterials. *Nat. Rev. Bioeng.* **2023**, *1*, 88–106. [CrossRef]

355. Li, Y.; Miao, Z.; Shang, Z.; Cai, Y.; Cheng, J.; Xu, X. A Visible- and NIR-Light Responsive Photothermal Therapy Agent by Chirality-Dependent MoO3−x Nanoparticles. *Adv. Funct. Mater.* **2020**, *30*, 1906311. [CrossRef]

356. Xu, L.; Wang, X.; Wang, W.; Sun, M.; Choi, W.J.; Kim, J.-Y.; Hao, C.; Li, S.; Qu, A.; Lu, M.; et al. Enantiomer-dependent immunological response to chiral nanoparticles. *Nature* **2022**, *601*, 366–373. [CrossRef] [PubMed]

357. Beißner, N.; Lorenz, T.; Reichl, S. Organ on Chip. In *Microsystems for Pharmatechnology: Manipulation of Fluids, Particles, Droplets, and Cells*; Dietzel, A., Ed.; Springer International Publishing: Cham, Switzerland, 2016; pp. 299–339. [CrossRef]

358. Kimlin, L.; Kassis, J.; Virador, V. 3D in vitro tissue models and their potential for drug screening. *Expert Opin. Drug Discov.* **2013**, *8*, 1455–1466. [CrossRef]

359. Huh, D.; Matthews, B.D.; Mammoto, A.; Montoya-Zavala, M.; Hsin, H.Y.; Ingber, D.E. Reconstituting Organ-Level Lung Functions on a Chip. *Science* **2010**, *328*, 1662–1668. [CrossRef] [PubMed]

360. Deng, J.; Wei, W.; Chen, Z.; Lin, B.; Zhao, W.; Luo, Y.; Zhang, X. Engineered Liver-On-A-Chip Platform to Mimic Liver Functions and Its Biomedical Applications: A Review. *Micromachines* **2019**, *10*, 676. [CrossRef]

361. Wang, D.; Gust, M.; Ferrell, N. Kidney-on-a-Chip: Mechanical Stimulation and Sensor Integration. *Sensors* **2022**, *22*, 6889. [CrossRef]

362. Tan, J.; Sun, X.; Zhang, J.; Li, H.; Kuang, J.; Xu, L.; Gao, X.; Zhou, C. Exploratory Evaluation of EGFR-Targeted Anti-Tumor Drugs for Lung Cancer Based on Lung-on-a-Chip. *Biosensors* **2022**, *12*, 618. [CrossRef]

363. Guo, Y.; Chen, X.; Gong, P.; Li, G.; Yao, W.; Yang, W. The Gut–Organ-Axis Concept: Advances the Application of Gut-on-Chip Technology. *Int. J. Mol. Sci.* **2023**, *24*, 4089. [CrossRef] [PubMed]

364. Wan, L.; Flegle, J.; Ozdoganlar, B.; LeDuc, P.R. Toward Vasculature in Skeletal Muscle-on-a-Chip through Thermo-Responsive Sacrificial Templates. *Micromachines* **2020**, *11*, 907. [CrossRef]

365. Staicu, C.E.; Jipa, F.; Axente, E.; Radu, M.; Radu, B.M.; Sima, F. Lab-on-a-Chip Platforms as Tools for Drug Screening in Neuropathologies Associated with Blood–Brain Barrier Alterations. *Biomolecules* **2021**, *11*, 916. [CrossRef] [PubMed]

366. Rigat-Brugarolas, L.G.; Bernabeu, M.; Elizalde, A.; de Niz, M.; Martin-Jaular, L.; Fernandez-Becerra, C.; Homs-Corbera, A.; del Portillo, H.A.; Samitier, J. Human Splenon-on-a-chip: Design and Validation of a Microfluidic Model Resembling the Interstitial Slits and the Close/Fast and Open/Slow Microcirculations. In *XIII Mediterranean Conference on Medical and Biological Engineering and Computing 2013*; Springer International Publishing: Cham, Switzerland, 2014.

367. Sarkar, R.; Pampaloni, F. In Vitro Models of Bone Marrow Remodelling and Immune Dysfunction in Space: Present State and Future Directions. *Biomedicines* **2022**, *10*, 766. [CrossRef]

368. Wagner, I.; Materne, E.-M.; Brincker, S.; Süßbier, U.; Frädrich, C.; Busek, M.; Sonntag, F.; Sakharov, D.A.; Trushkin, E.V.; Tonevitsky, A.G.; et al. A dynamic multi-organ-chip for long-term cultivation and substance testing proven by 3D human liver and skin tissue co-culture. *Lab A Chip* **2013**, *13*, 3538–3547. [CrossRef]

369. Zhao, Y.; Kankala, R.K.; Wang, S.-B.; Chen, A.-Z. Multi-Organs-on-Chips: Towards Long-Term Biomedical Investigations. *Molecules* **2019**, *24*, 675. [CrossRef]

370. Cecen, B.; Karavasili, C.; Nazir, M.; Bhusal, A.; Dogan, E.; Shahriyari, F.; Tamburaci, S.; Buyukoz, M.; Kozaci, L.D.; Miri, A.K. Multi-Organs-on-Chips for Testing Small-Molecule Drugs: Challenges and Perspectives. *Pharmaceutics* **2021**, *13*, 1657. [CrossRef]

371. Luni, C.; Serena, E.; Elvassore, N. Human-on-chip for therapy development and fundamental science. *Curr. Opin. Biotechnol.* **2014**, *25*, 45–50. [CrossRef]

372. van der Meer, A.D.; van den Berg, A. Organs-on-chips: Breaking the in vitro impasse. *Integr. Biol.* **2012**, *4*, 461–470. [CrossRef]

373. Williamson, A.; Singh, S.; Fernekorn, U.; Schober, A. The future of the patient-specific Body-on-a-chip. *Lab A Chip* **2013**, *13*, 3471–3480. [CrossRef]

374. Caballero, D.; Kaushik, S.; Correlo, V.M.; Oliveira, J.M.; Reis, R.L.; Kundu, S.C. Organ-on-chip models of cancer metastasis for future personalized medicine: From chip to the patient. *Biomaterials* **2017**, *149*, 98–115. [CrossRef]

375. Chou, D.B.; Frismantas, V.; Milton, Y.; David, R.; Pop-Damkov, P.; Ferguson, D.; MacDonald, A.; Bölükbaşı, V.; Joyce, C.E.; Teixeira, L.S.M.; et al. On-chip recapitulation of clinical bone marrow toxicities and patient-specific pathophysiology. *Nat. Biomed. Eng.* **2020**, *4*, 394–406. [CrossRef] [PubMed]

376. Peck, R.W.; Hinojosa, C.D.; Hamilton, G.A. Organs-on-Chips in Clinical Pharmacology: Putting the Patient Into the Center of Treatment Selection and Drug Development. *Clin. Pharmacol. Ther.* **2020**, *107*, 181–185. [CrossRef] [PubMed]

377. Huh, D.; Leslie, D.C.; Matthews, B.D.; Fraser, J.P.; Jurek, S.; Hamilton, G.A.; Thorneloe, K.S.; McAlexander, M.A.; Ingber, D.E. A Human Disease Model of Drug Toxicity—Induced Pulmonary Edema in a Lung-on-a-Chip Microdevice. *Sci. Transl. Med.* **2012**, *4*, 159ra147. [CrossRef] [PubMed]

378. Nesmith, A.P.; Agarwal, A.; McCain, M.L.; Parker, K.K. Human airway musculature on a chip: An in vitro model of allergic asthmatic bronchoconstriction and bronchodilation. *Lab A Chip* **2014**, *14*, 3925–3936. [CrossRef] [PubMed]

379. Zhang, X.; Karim, M.; Hasan, M.M.; Hooper, J.; Wahab, R.; Roy, S.; Al-Hilal, T.A. Cancer-on-a-Chip: Models for Studying Metastasis. *Cancers* **2022**, *14*, 648. [CrossRef] [PubMed]

380. Sánchez-Salazar, M.G.; Crespo-López Oliver, R.; Ramos-Meizoso, S.; Jerezano-Flores, V.S.; Gallegos-Martínez, S.; Bolívar-Monsalve, E.J.; Ceballos-González, C.F.; Trujillo-de Santiago, G.; Álvarez, M.M. 3D-Printed Tumor-on-Chip for the Culture of Colorectal Cancer Microspheres: Mass Transport Characterization and Anti-Cancer Drug Assays. *Bioengineering* **2023**, *10*, 554. [CrossRef] [PubMed]

381. Pollini, M.; Paladini, F. Bioinspired Materials for Wound Healing Application: The Potential of Silk Fibroin. *Materials* **2020**, *13*, 3361. [CrossRef]

382. Umuhoza, D.; Yang, F.; Long, D.; Hao, Z.; Dai, J.; Zhao, A. Strategies for Tuning the Biodegradation of Silk Fibroin-Based Materials for Tissue Engineering Applications. *ACS Biomater. Sci. Eng.* **2020**, *6*, 1290–1310. [CrossRef] [PubMed]

383. Luo, L.; Zhou, Y.; Xu, X.; Shi, W.; Hu, J.; Li, G.; Qu, X.; Guo, Y.; Tian, X.; Zaman, A.; et al. Progress in construction of bio-inspired physico-antimicrobial surfaces. *Nanotechnol. Rev.* **2020**, *9*, 1562–1575. [CrossRef]

384. Rai, A.; Ferrão, R.; Palma, P.; Patricio, T.; Parreira, P.; Anes, E.; Tonda-Turo, C.; Martins, M.C.L.; Alves, N.; Ferreira, L. Antimicrobial peptide-based materials: Opportunities and challenges. *J. Mater. Chem. B* **2022**, *10*, 2384–2429. [CrossRef]

385. Singh, K.R.; Nayak, V.; Singh, J.; Singh, A.K.; Singh, R.P. Potentialities of bioinspired metal and metal oxide nanoparticles in biomedical sciences. *RSC Adv.* **2021**, *11*, 24722–24746. [CrossRef]

386. Yilmaz Atay, H. Antibacterial Activity of Chitosan-Based Systems. In *Functional Chitosan: Drug Delivery and Biomedical Applications*; Jana, S., Jana, S., Eds.; Springer: Singapore, 2019; pp. 457–489. [CrossRef]

387. Ilyas, K.; Akhtar, M.A.; Ammar, E.B.; Boccaccini, A.R. Surface Modification of 3D-Printed PCL/BG Composite Scaffolds via Mussel-Inspired Polydopamine and Effective Antibacterial Coatings for Biomedical Applications. *Materials* **2022**, *15*, 8289. [CrossRef] [PubMed]

388. Zalewska-Piątek, B.; Piątek, R. Bacteriophages as Potential Tools for Use in Antimicrobial Therapy and Vaccine Development. *Pharmaceuticals* **2021**, *14*, 331. [CrossRef] [PubMed]

389. O'Connell, L.; Marcoux, P.R.; Roupioz, Y. Strategies for Surface Immobilization of Whole Bacteriophages: A Review. *ACS Biomater. Sci. Eng.* **2021**, *7*, 1987–2014. [CrossRef] [PubMed]

390. Calfee, M.W.; Ryan, S.P.; Abdel-Hady, A.; Monge, M.; Aslett, D.; Touati, A.; Stewart, M.; Lawrence, S.; Willis, K. Virucidal efficacy of antimicrobial surface coatings against the enveloped bacteriophage Φ6. *J. Appl. Microbiol.* **2022**, *132*, 1813–1824. [CrossRef]

391. Moya-Garcia, C.R.; Li-Jessen, N.Y.K.; Tabrizian, M. Chitosomes Loaded with Docetaxel as a Promising Drug Delivery System to Laryngeal Cancer Cells: An In Vitro Cytotoxic Study. *Int. J. Mol. Sci.* **2023**, *24*, 9902. [CrossRef]

392. Jayachandran, P.; Ilango, S.; Suseela, V.; Nirmaladevi, R.; Shaik, M.R.; Khan, M.; Khan, M.; Shaik, B. Green Synthesized Silver Nanoparticle-Loaded Liposome-Based Nanoarchitectonics for Cancer Management: In Vitro Drug Release Analysis. *Biomedicines* **2023**, *11*, 217. [CrossRef]

393. Zhao, Y.-P.; Han, J.-F.; Zhang, F.-Y.; Liao, T.-T.; Na, R.; Yuan, X.-F.; He, G.-b.; Ye, W. Flexible nano-liposomes-based transdermal hydrogel for targeted delivery of dexamethasone for rheumatoid arthritis therapy. *Drug Deliv.* **2022**, *29*, 2269–2282. [CrossRef]

394. Hussein, H.A.; Abdullah, M.A. Novel drug delivery systems based on silver nanoparticles, hyaluronic acid, lipid nanoparticles and liposomes for cancer treatment. *Appl. Nanosci.* **2022**, *12*, 3071–3096. [CrossRef]

395. Lim, C.C.; Chia, L.Y.; Kumar, P.V. Dendrimer-based nanocomposites for the production of RNA delivery systems. *OpenNano* **2023**, *13*, 100173. [CrossRef]

396. Jiang, Y.; Lyu, Z.; Ralahy, B.; Liu, J.; Roussel, T.; Ding, L.; Tang, J.; Kosta, A.; Giorgio, S.; Tomasini, R.; et al. Dendrimer nanosystems for adaptive tumor-assisted drug delivery via extracellular vesicle hijacking. *Proc. Natl. Acad. Sci. USA* **2023**, *120*, e2215308120. [CrossRef] [PubMed]

397. Michlewska, S.; Garaiova, Z.; Šubjakova, V.; Hołota, M.; Kubczak, M.; Grodzicka, M.; Okła, E.; Naziris, N.; Balcerzak, Ł.; Ortega, P.; et al. Lipid-coated ruthenium dendrimer conjugated with doxorubicin in anti-cancer drug delivery: Introducing protocols. *Colloids Surf. B. Biointerfaces* **2023**, *227*, 113371. [CrossRef]

398. Wang, J.; Li, B.; Kompella, U.B.; Yang, H. Dendrimer and dendrimer gel-derived drug delivery systems: Breaking bottlenecks of topical administration of glaucoma medications. *MedComm Biomater. Appl.* **2023**, *2*, e30. [CrossRef]

399. Fatani, W.K.; Aleanizy, F.S.; Alqahtani, F.Y.; Alanazi, M.M.; Aldossari, A.A.; Shakeel, F.; Haq, N.; Abdelhady, H.; Alkahtani, H.M.; Alsarra, I.A. Erlotinib-Loaded Dendrimer Nanocomposites as a Targeted Lung Cancer Chemotherapy. *Molecules* **2023**, *28*, 3974. [CrossRef]

400. Lee, J.; Kim, K.; Kwon, I.C.; Lee, K.Y. Intracellular Glucose-Depriving Polymer Micelles for Antiglycolytic Cancer Treatment. *Adv. Mater.* **2023**, *35*, 2207342. [CrossRef] [PubMed]

401. Feng, Y.; Bai, J.; Du, X.; Zhao, X. Shell-cross-linking of polymeric micelles by Zn coordination for Photo- and pH dual-sensitive drug delivery. *Colloids Surf. Physicochem. Eng. Asp.* **2023**, *666*, 131369. [CrossRef]

402. Jo, M.J.; Shin, H.J.; Yoon, M.S.; Kim, S.Y.; Jin, C.E.; Park, C.-W.; Kim, J.-S.; Shin, D.H. Evaluation of pH-Sensitive Polymeric Micelles Using Citraconic Amide Bonds for the Co-Delivery of Paclitaxel, Etoposide, and Rapamycin. *Pharmaceutics* **2023**, *15*, 154. [CrossRef]

403. Yoshizaki, Y.; Yamasaki, M.; Nagata, T.; Suzuki, K.; Yamada, R.; Kato, T.; Murase, N.; Kuzuya, A.; Asai, A.; Higuchi, K.; et al. Drug Delivery with Hyaluronic Acid-Coated Polymeric Micelles in Liver Fibrosis Therapy. *ACS Biomater. Sci. Eng.* **2023**, *9*, 3414–3424. [CrossRef] [PubMed]

404. Desai, N.; Rana, D.; Pande, S.; Salave, S.; Giri, J.; Benival, D.; Kommineni, N. "Bioinspired" Membrane-Coated Nanosystems in Cancer Theranostics: A Comprehensive Review. *Pharmaceutics* **2023**, *15*, 1677. [CrossRef]

405. Huang, X.; Hu, X.; Song, S.; Mao, D.; Lee, J.; Koh, K.; Zhu, Z.; Chen, H. Triple-enhanced surface plasmon resonance spectroscopy based on cell membrane and folic acid functionalized gold nanoparticles for dual-selective circulating tumor cell sensing. *Sens. Actuators B Chem.* **2020**, *305*, 127543. [CrossRef]

406. Liu, Z.; Zhang, L.; Cui, T.; Ma, M.; Ren, J.; Qu, X. A Nature-Inspired Metal–Organic Framework Discriminator for Differential Diagnosis of Cancer Cell Subtypes. *Angew. Chem. Int. Ed.* **2021**, *60*, 15436–15444. [CrossRef]

407. Schnierle, B.S.; Stitz, J.; Bosch, V.; Nocken, F.; Merget-Millitzer, H.; Engelstädter, M.; Kurth, R.; Groner, B.; Cichutek, K. Pseudotyping of murine leukemia virus with the envelope glycoproteins of HIV generates a retroviral vector with specificity of infection for CD4-expressing cells. *Proc. Natl. Acad. Sci. USA* **1997**, *94*, 8640–8645. [CrossRef] [PubMed]

408. Fan, G.; Fan, M.; Wang, Q.; Jiang, J.; Wan, Y.; Gong, T.; Zhang, Z.; Sun, X. Bio-inspired polymer envelopes around adenoviral vectors to reduce immunogenicity and improve in vivo kinetics. *Acta Biomater.* **2016**, *30*, 94–105. [CrossRef] [PubMed]

409. Garland, S.M.; Hernandez-Avila, M.; Wheeler, C.M.; Perez, G.; Harper, D.M.; Leodolter, S.; Tang, G.W.K.; Ferris, D.G.; Steben, M.; Bryan, J.; et al. Quadrivalent Vaccine against Human Papillomavirus to Prevent Anogenital Diseases. *N. Engl. J. Med.* **2007**, *356*, 1928–1943. [CrossRef] [PubMed]

410. de Jonge, J.; Leenhouts, J.M.; Holtrop, M.; Schoen, P.; Scherrer, P.; Cullis, P.R.; Wilschut, J.; Huckriede, A. Cellular gene transfer mediated by influenza virosomes with encapsulated plasmid DNA. *Biochem. J.* **2007**, *405*, 41–49. [CrossRef]

411. Lakadamyali, M.; Rust, M.J.; Zhuang, X. Endocytosis of influenza viruses. *Microb. Infect.* **2004**, *6*, 929–936. [CrossRef]

412. Leroux-Roels, G. Unmet needs in modern vaccinology: Adjuvants to improve the immune response. *Vaccine* **2010**, *28*, C25–C36. [CrossRef]

413. Sun, J.; Mathesh, M.; Li, W.; Wilson, D.A. Enzyme-Powered Nanomotors with Controlled Size for Biomedical Applications. *ACS Nano* **2019**, *13*, 10191–10200. [CrossRef]

414. Liu, T.; Xie, L.; Price, C.-A.H.; Liu, J.; He, Q.; Kong, B. Controlled propulsion of micro/nanomotors: Operational mechanisms, motion manipulation and potential biomedical applications. *Chem. Soc. Rev.* **2022**, *51*, 10083–10119. [CrossRef]

415. Liang, H.; Peng, F.; Tu, Y. Active therapy based on the byproducts of micro/nanomotors. *Nanoscale* **2023**, *15*, 953–962. [CrossRef]

416. Xu, H.; Medina-Sánchez, M.; Magdanz, V.; Schwarz, L.; Hebenstreit, F.; Schmidt, O.G. Sperm-Hybrid Micromotor for Targeted Drug Delivery. *ACS Nano* **2018**, *12*, 327–337. [CrossRef]

417. Liu, X.; Wang, C.; Zhang, Z. Chapter 4—Biohybrid microrobots driven by sperm. In *Untethered Small-Scale Robots for Biomedical Applications*; Lu, H., Wang, X., Zhang, S., Eds.; Academic Press: Cambridge, MA, USA, 2023; pp. 63–75. [CrossRef]

418. Shah, P.; Shende, P. Biotherapy using Sperm Cell-oriented Transportation of Therapeutics in Female Reproductive Tract Cancer. *Curr. Pharm. Biotechnol.* **2022**, *23*, 1359–1366. [CrossRef] [PubMed]

419. Stanton, M.M.; Simmchen, J.; Ma, X.; Miguel-López, A.; Sánchez, S. Biohybrid Janus Motors Driven by Escherichia coli. *Adv. Mater. Interfaces* **2016**, *3*, 1500505. [CrossRef]

420. Martel, S.; Mohammadi, M.; Felfoul, O.; Zhao, L.; Pouponneau, P. Flagellated Magnetotactic Bacteria as Controlled MRI-trackable Propulsion and Steering Systems for Medical Nanorobots Operating in the Human Microvasculature. *Int. J. Robot. Res.* **2009**, *28*, 571–582. [CrossRef] [PubMed]

421. Felfoul, O.; Mohammadi, M.; Taherkhani, S.; de Lanauze, D.; Zhong Xu, Y.; Loghin, D.; Essa, S.; Jancik, S.; Houle, D.; Lafleur, M.; et al. Magneto-aerotactic bacteria deliver drug-containing nanoliposomes to tumour hypoxic regions. *Nat. Nanotechnol.* **2016**, *11*, 941–947. [CrossRef] [PubMed]

422. Liu, L.; Wu, J.; Wang, S.; Kun, L.; Gao, J.; Chen, B.; Ye, Y.; Wang, F.; Tong, F.; Jiang, J.; et al. Control the Neural Stem Cell Fate with Biohybrid Piezoelectrical Magnetite Micromotors. *Nano Lett.* **2021**, *21*, 3518–3526. [CrossRef] [PubMed]

423. Liu, L.; Chen, B.; Liu, K.; Gao, J.; Ye, Y.; Wang, Z.; Qin, N.; Wilson, D.A.; Tu, Y.; Peng, F. Wireless Manipulation of Magnetic/Piezoelectric Micromotors for Precise Neural Stem-Like Cell Stimulation. *Adv. Funct. Mater.* **2020**, *30*, 1910108. [CrossRef]

424. Yan, X.; Zhou, Q.; Yu, J.; Xu, T.; Deng, Y.; Tang, T.; Feng, Q.; Bian, L.; Zhang, Y.; Ferreira, A.; et al. Magnetite Nanostructured Porous Hollow Helical Microswimmers for Targeted Delivery. *Adv. Funct. Mater.* **2015**, *25*, 5333–5342. [CrossRef]

425. Qin, J.; Dong, B.; Wang, W.; Cao, L. Self-regulating bioinspired supramolecular photonic hydrogels based on chemical reaction networks for monitoring activities of enzymes and biofuels. *J. Colloid Interface Sci.* **2023**, *649*, 344–354. [CrossRef]

426. Suleimenova, A.; Frasco, M.F.; Soares da Silva, F.A.G.; Gama, M.; Fortunato, E.; Sales, M.G.F. Bacterial nanocellulose membrane as novel substrate for biomimetic structural color materials: Application to lysozyme sensing. *Biosens. Bioelectron. X* **2023**, *13*, 100310. [CrossRef]

427. Altug, H.; Oh, S.-H.; Maier, S.A.; Homola, J. Advances and applications of nanophotonic biosensors. *Nat. Nanotechnol.* **2022**, *17*, 5–16. [CrossRef] [PubMed]

428. Lizundia, E.; Nguyen, T.-D.; Winnick, R.J.; MacLachlan, M.J. Biomimetic photonic materials derived from chitin and chitosan. *J. Mater. Chem. C* **2021**, *9*, 796–817. [CrossRef]

429. Resende, S.; Frasco, M.F.; Sales, M.G.F. A biomimetic photonic crystal sensor for label-free detection of urinary venous thromboembolism biomarker. *Sens. Actuators B Chem.* **2020**, *312*, 127947. [CrossRef]

430. Yu, H.; Li, H.; Sun, X.; Pan, L. Biomimetic Flexible Sensors and Their Applications in Human Health Detection. *Biomimetics* **2023**, *8*, 293. [CrossRef] [PubMed]

431. Zhu, Q.; Yang, H.; Luo, J.; Huang, H.; Fang, L.; Deng, J.; Li, C.; Li, Y.; Zeng, T.; Zheng, J. 3D matrixed DNA self-nanocatalyzer as electrochemical sensitizers for ultrasensitive investigation of DNA 5-methylcytosine. *Anal. Chim. Acta* **2021**, *1142*, 127–134. [CrossRef] [PubMed]

432. Xue, C.; Han, Q.; Wang, Y.; Wu, J.; Wen, T.; Wang, R.; Hong, J.; Zhou, X.; Jiang, H. Amperometric detection of dopamine in human serumby electrochemical sensor based on gold nanoparticles doped molecularly imprinted polymers. *Biosens. Bioelectron.* **2013**, *49*, 199–203. [CrossRef] [PubMed]

433. Vural, T.; Yaman, Y.T.; Ozturk, S.; Abaci, S.; Denkbas, E.B. Electrochemical immunoassay for detection of prostate specific antigen based on peptide nanotube-gold nanoparticle-polyaniline immobilized pencil graphite electrode. *J. Colloid Interface Sci.* **2018**, *510*, 318–326. [CrossRef]

434. Li, H.; Rothberg, L. Colorimetric detection of DNA sequences based on electrostatic interactions with unmodified gold nanoparticles. *Proc. Natl. Acad. Sci. USA* **2004**, *101*, 14036–14039. [CrossRef]

435. Tan, Y.N.; Lee, K.H.; Su, X. Study of Single-Stranded DNA Binding Protein–Nucleic Acids Interactions using Unmodified Gold Nanoparticles and Its Application for Detection of Single Nucleotide Polymorphisms. *Anal. Chem.* **2011**, *83*, 4251–4257. [CrossRef]

436. Aparna, R.S.; Anjali Devi, J.S.; Anjana, R.R.; Nebu, J.; George, S. Reversible fluorescence modulation of BSA stabilised copper nanoclusters for the selective detection of protamine and heparin. *Analyst* **2019**, *144*, 1799–1808. [CrossRef]

437. Tang, Y.; Sun, Z.; Shen, J.; Yu, J.; Wang, S.; Hao, J.; Wang, B.; Huang, Y.; Liu, X.; Zhuang, H. Peptide modified gold nanoclusters as a novel fluorescence detector based on quenching system of detecting Allura red. *Anal. Methods* **2018**, *10*, 5672–5678. [CrossRef]

438. Azad, A.K.; Paul, P.; Abdul Majid, A.M.S.; Mozafari, M.R. Polymer composite sensors for biomedical applications. In *Polymeric Nanocomposite Materials for Sensor Applications*; Parameswaranpillai, J., Ganguly, S., Eds.; Woodhead Publishing: Sawston, UK, 2023; pp. 501–520. [CrossRef]

439. Carvalho, W.S.P.; Wei, M.; Ikpo, N.; Gao, Y.; Serpe, M.J. Polymer-based technologies for sensing applications. *Anal. Chem.* **2017**, *90*, 459–479. [CrossRef] [PubMed]

440. Aydın, E.B.; Aydın, M.; Sezgintürk, M.K. A highly sensitive immunosensor based on ITO thin films covered by a new semi-conductive conjugated polymer for the determination of TNFα in human saliva and serum samples. *Biosens. Bioelectron.* **2017**, *97*, 169–176. [CrossRef] [PubMed]

441. Wang, H.; Han, H.; Ma, Z. Conductive hydrogel composed of 1,3,5-benzenetricarboxylic acid and Fe3+ used as enhanced electrochemical immunosensing substrate for tumor biomarker. *Bioelectrochemistry* **2017**, *114*, 48–53. [CrossRef]

442. Miodek, A.; Mejri-Omrani, N.; Khoder, R.; Korri-Youssoufi, H. Electrochemical functionalization of polypyrrole through amine oxidation of poly(amidoamine) dendrimers: Application to DNA biosensor. *Talanta* **2016**, *154*, 446–454. [CrossRef]

443. Du, Y.; Du, W.; Lin, D.; Ai, M.; Li, S.; Zhang, L. Recent Progress on Hydrogel-Based Piezoelectric Devices for Biomedical Applications. *Micromachines* **2023**, *14*, 167. [CrossRef]

444. Zaidi, S.F.A.; Saeed, A.; Heo, J.H.; Lee, J.H. Multifunctional small biomolecules as key building blocks in the development of hydrogel-based strain sensors. *J. Mater. Chem. A* **2023**, *11*, 13844–13875. [CrossRef]

445. Wang, Y.; Zhu, P.; Tan, M.; Niu, M.; Liang, S.; Mao, Y. Recent Advances in Hydrogel-Based Self-Powered Artificial Skins for Human-Machine Interfaces. *Adv. Intell. Syst.* **2023**, 2300162. [CrossRef]

446. Messelmani, T.; Le Goff, A.; Souguir, Z.; Maes, V.; Roudaut, M.; Vandenhaute, E.; Maubon, N.; Legallais, C.; Leclerc, E.; Jellali, R. Development of Liver-on-Chip Integrating a Hydroscaffold Mimicking the Liver's Extracellular Matrix. *Bioengineering* **2022**, *9*, 443. [CrossRef]

447. Butkutė, A.; Jurkšas, T.; Baravykas, T.; Leber, B.; Merkininkaitė, G.; Žilėnaitė, R.; Čereška, D.; Gulla, A.; Kvietkauskas, M.; Marcinkevičiūtė, K.; et al. Combined Femtosecond Laser Glass Microprocessing for Liver-on-Chip Device Fabrication. *Materials* **2023**, *16*, 2174. [CrossRef]

448. Kanabekova, P.; Kadyrova, A.; Kulsharova, G. Microfluidic Organ-on-a-Chip Devices for Liver Disease Modeling In Vitro. *Micromachines* **2022**, *13*, 428. [CrossRef]

449. Shinde, A.; Illath, K.; Kasiviswanathan, U.; Nagabooshanam, S.; Gupta, P.; Dey, K.; Chakrabarty, P.; Nagai, M.; Rao, S.; Kar, S.; et al. Recent Advances of Biosensor-Integrated Organ-on-a-Chip Technologies for Diagnostics and Therapeutics. *Anal. Chem.* **2023**, *95*, 3121–3146. [CrossRef]

450. Li, Q.; Tong, Z.; Mao, H. Microfluidic Based Organ-on-Chips and Biomedical Application. *Biosensors* **2023**, *13*, 436. [CrossRef]

451. Dai, M.; Xiao, G.; Shao, M.; Zhang, Y.S. The Synergy between Deep Learning and Organs-on-Chips for High-Throughput Drug Screening: A Review. *Biosensors* **2023**, *13*, 389. [CrossRef]

452. Boeri, L.; Izzo, L.; Sardelli, L.; Tunesi, M.; Albani, D.; Giordano, C. Advanced Organ-on-a-Chip Devices to Investigate Liver Multi-Organ Communication: Focus on Gut, Microbiota and Brain. *Bioengineering* **2019**, *6*, 91. [CrossRef]

453. Ingber, D.E. Is it Time for Reviewer 3 to Request Human Organ Chip Experiments Instead of Animal Validation Studies? *Adv. Sci.* **2020**, *7*, 2002030. [CrossRef]

454. Zhang, C.; Zhao, Z.; Abdul Rahim, N.A.; van Noort, D.; Yu, H. Towards a human-on-chip: Culturing multiple cell types on a chip with compartmentalized microenvironments. *Lab A Chip* **2009**, *9*, 3185–3192. [CrossRef]

455. Yi, H.-G.; Jeong, Y.H.; Kim, Y.; Choi, Y.-J.; Moon, H.E.; Park, S.H.; Kang, K.S.; Bae, M.; Jang, J.; Youn, H.; et al. A bioprinted human-glioblastoma-on-a-chip for the identification of patient-specific responses to chemoradiotherapy. *Nat. Biomed. Eng.* **2019**, *3*, 509–519. [CrossRef]

456. Haque, M.R.; Wessel, C.R.; Leary, D.D.; Wang, C.; Bhushan, A.; Bishehsari, F. Patient-derived pancreatic cancer-on-a-chip recapitulates the tumor microenvironment. *Microsyst. Nanoeng.* **2022**, *8*, 36. [CrossRef]

457. Chen, Y.; Gao, D.; Wang, Y.; Lin, S.; Jiang, Y. A novel 3D breast-cancer-on-chip platform for therapeutic evaluation of drug delivery systems. *Anal. Chim. Acta* **2018**, *1036*, 97–106. [CrossRef]

458. Banerjee, M.; Devi Rajeswari, V. A novel cross-communication of HIF-1α and HIF-2α with Wnt signaling in TNBC and influence of hypoxic microenvironment in the formation of an organ-on-chip model of breast cancer. *Med. Oncol.* **2023**, *40*, 245. [CrossRef]

459. Guo, M.; Deng, Y.; Huang, J.; Huang, Y.; Deng, J.; Wu, H. Fabrication and Validation of a 3D Portable PEGDA Microfluidic Chip for Visual Colorimetric Detection of Captured Breast Cancer Cells. *Polymers* **2023**, *15*, 3183. [CrossRef]

460. Nguyen, D.-H.T.; Lee, E.; Alimperti, S.; Norgard, R.J.; Wong, A.; Lee, J.J.-K.; Eyckmans, J.; Stanger, B.Z.; Chen, C.S. A biomimetic pancreatic cancer on-chip reveals endothelial ablation via ALK7 signaling. *Sci. Adv.* **2019**, *5*, eaav6789. [CrossRef]

461. Haque, M.R.; Rempert, T.H.; Al-Hilal, T.A.; Wang, C.; Bhushan, A.; Bishehsari, F. Organ-Chip Models: Opportunities for Precision Medicine in Pancreatic Cancer. *Cancers* **2021**, *13*, 4487. [CrossRef]

462. Khalid, M.A.U.; Kim, Y.S.; Ali, M.; Lee, B.G.; Cho, Y.-J.; Choi, K.H. A lung cancer-on-chip platform with integrated biosensors for physiological monitoring and toxicity assessment. *Biochem. Eng. J.* **2020**, *155*, 107469. [CrossRef]

463. Park, S.; Kim, T.H.; Kim, S.H.; You, S.; Jung, Y. Three-Dimensional Vascularized Lung Cancer-on-a-Chip with Lung Extracellular Matrix Hydrogels for In Vitro Screening. *Cancers* **2021**, *13*, 3930. [CrossRef]

464. Carvalho, Â.; Ferreira, G.; Seixas, D.; Guimarães-Teixeira, C.; Henrique, R.; Monteiro, F.J.; Jerónimo, C. Emerging Lab-on-a-Chip Approaches for Liquid Biopsy in Lung Cancer: Status in CTCs and ctDNA Research and Clinical Validation. *Cancers* **2021**, *13*, 2101. [CrossRef]

465. Syromiatnikova, V.; Gupta, S.; Zhuravleva, M.; Masgutova, G.; Zakirova, E.; Aimaletdinov, A.; Rizvanov, A.; Salafutdinov, I.; Naumenko, E.; Bit, A. Engineered GO-Silk Fibroin-Based Hydrogel for the Promotion of Collagen Synthesis in Full-Thickness Skin Defect. *J. Compos. Sci.* **2023**, *7*, 186. [CrossRef]

466. Lyu, Y.; Liu, Y.; He, H.; Wang, H. Application of Silk-Fibroin-Based Hydrogels in Tissue Engineering. *Gels* **2023**, *9*, 431. [CrossRef]

467. Bayraktar, O.; Oder, G.; Erdem, C.; Kose, M.D.; Cheaburu-Yilmaz, C.N. Selective Encapsulation of the Polyphenols on Silk Fibroin Nanoparticles: Optimization Approaches. *Int. J. Mol. Sci.* **2023**, *24*, 9327. [CrossRef] [PubMed]

468. Cadinoiu, A.N.; Rata, D.M.; Daraba, O.M.; Ichim, D.L.; Popescu, I.; Solcan, C.; Solcan, G. Silver Nanoparticles Biocomposite Films with Antimicrobial Activity: In Vitro and In Vivo Tests. *Int. J. Mol. Sci.* **2022**, *23*, 10671. [CrossRef] [PubMed]

469. Hodel, K.V.; Fonseca, L.M.; Santos, I.M.; Cerqueira, J.C.; Santos-Júnior, R.E.; Nunes, S.B.; Barbosa, J.D.; Machado, B.A. Evaluation of Different Methods for Cultivating Gluconacetobacter hansenii for Bacterial Cellulose and Montmorillonite Biocomposite Production: Wound-Dressing Applications. *Polymers* **2020**, *12*, 267. [CrossRef] [PubMed]

470. Neacsu, I.A.; Leau, S.-A.; Marin, S.; Holban, A.M.; Vasile, B.-S.; Nicoara, A.-I.; Ene, V.L.; Bleotu, C.; Albu Kaya, M.G.; Ficai, A. Collagen-Carboxymethylcellulose Biocomposite Wound-Dressings with Antimicrobial Activity. *Materials* **2021**, *14*, 1153. [CrossRef]

471. Ma, X.; Bian, Q.; Hu, J.; Gao, J. Stem from nature: Bioinspired adhesive formulations for wound healing. *J. Control. Release* **2022**, *345*, 292–305. [CrossRef] [PubMed]

472. George, B.; Bhatia, N.; Kumar, A.; Gnanamani, A.; Thilagam, R.; Shanuja, S.K.; Meethal, K.V.; Shiji, T.M.; Suchithra, T.V. Bioinspired gelatin based sticky hydrogel for diverse surfaces in burn wound care. *Sci. Rep.* **2022**, *12*, 13735. [CrossRef]

473. Zheng, Y.; Zhang, K.; Yao, Y.; Li, X.; Yu, J.; Ding, B. Bioinspired sequentially crosslinked nanofibrous hydrogels with robust adhesive and stretchable capability for joint wound dressing. *Compos. Commun.* **2021**, *26*, 100785. [CrossRef]

474. Soni, A.; Brightwell, G. Nature-Inspired Antimicrobial Surfaces and Their Potential Applications in Food Industries. *Foods* **2022**, *11*, 844. [CrossRef]

475. Yang, X.; Zhang, W.; Qin, X.; Cui, M.; Guo, Y.; Wang, T.; Wang, K.; Shi, Z.; Zhang, C.; Li, W.; et al. Recent Progress on Bioinspired Antibacterial Surfaces for Biomedical Application. *Biomimetics* **2022**, *7*, 88. [CrossRef]

476. da Silva, D.J.; Duran, A.; Cabral, A.D.; Fonseca, F.L.A.; Wang, S.H.; Parra, D.F.; Bueno, R.F.; Pereyra, I.; Rosa, D.S. Bioinspired Antimicrobial PLA with Nanocones on the Surface for Rapid Deactivation of Omicron SARS-CoV-2. *ACS Biomater. Sci. Eng.* **2023**, *9*, 1891–1899. [CrossRef]

477. Segura, T.; Puga, A.M.; Burillo, G.; Llovo, J.; Brackman, G.; Coenye, T.; Concheiro, A.; Alvarez-Lorenzo, C. Materials with Fungi-Bioinspired Surface for Efficient Binding and Fungi-Sensitive Release of Antifungal Agents. *Biomacromolecules* **2014**, *15*, 1860–1870. [CrossRef]

478. Tetorya, M.; Li, H.; Djami-Tchatchou, A.T.; Buchko, G.W.; Czymmek, K.J.; Shah, D.M. Plant defensin MtDef4-derived antifungal peptide with multiple modes of action and potential as a bio-inspired fungicide. *Mol. Plant Pathol.* **2023**, *24*, 896–913. [CrossRef] [PubMed]

479. Qiao, Z.; Fu, Y.; Lei, C.; Li, Y. Advances in antimicrobial peptides-based biosensing methods for detection of foodborne pathogens: A review. *Food Control.* **2020**, *112*, 107116. [CrossRef]

480. Teixeira, M.C.; Carbone, C.; Sousa, M.C.; Espina, M.; Garcia, M.L.; Sanchez-Lopez, E.; Souto, E.B. Nanomedicines for the Delivery of Antimicrobial Peptides (AMPs). *Nanomaterials* **2020**, *10*, 560. [CrossRef]

481. Patra, A.; Das, J.; Agrawal, N.R.; Kushwaha, G.S.; Ghosh, M.; Son, Y.-O. Marine Antimicrobial Peptides-Based Strategies for Tackling Bacterial Biofilm and Biofouling Challenges. *Molecules* **2022**, *27*, 7546. [CrossRef] [PubMed]

482. Chouke, P.B.; Shrirame, T.; Potbhare, A.K.; Mondal, A.; Chaudhary, A.R.; Mondal, S.; Thakare, S.R.; Nepovimova, E.; Valis, M.; Kuca, K.; et al. Bioinspired metal/metal oxide nanoparticles: A road map to potential applications. *Mater. Today Adv.* **2022**, *16*, 100314. [CrossRef]

483. García, D.G.; Garzón-Romero, C.; Salazar, M.A.; Lagos, K.J.; Campaña, K.O.; Debut, A.; Vizuete, K.; Rivera, M.R.; Niebieskikwiat, D.; Benitez, M.J.; et al. Bioinspired Synthesis of Magnetic Nanoparticles Based on Iron Oxides Using Orange Waste and Their Application as Photo-Activated Antibacterial Agents. *Int. J. Mol. Sci.* **2023**, *24*, 4770. [CrossRef]

484. Khubiev, O.M.; Egorov, A.R.; Kirichuk, A.A.; Khrustalev, V.N.; Tskhovrebov, A.G.; Kritchenkov, A.S. Chitosan-Based Antibacterial Films for Biomedical and Food Applications. *Int. J. Mol. Sci.* **2023**, *24*, 10738. [CrossRef]

485. Khubiev, O.M.; Egorov, A.R.; Lobanov, N.N.; Fortalnova, E.A.; Kirichuk, A.A.; Tskhovrebov, A.G.; Kritchenkov, A.S. Novel Highly Efficient Antibacterial Chitosan-Based Films. *BioTech* **2023**, *12*, 50. [CrossRef] [PubMed]

486. Kumar, A.; Yadav, S.; Pramanik, J.; Sivamaruthi, B.S.; Jayeoye, T.J.; Prajapati, B.G.; Chaiyasut, C. Chitosan-Based Composites: Development and Perspective in Food Preservation and Biomedical Applications. *Polymers* **2023**, *15*, 3150. [CrossRef]

487. Li, L.; Smitthipong, W.; Zeng, H. Mussel-inspired hydrogels for biomedical and environmental applications. *Polym. Chem.* **2015**, *6*, 353–358. [CrossRef]

488. Mao, S.; Zhang, D.; He, X.; Yang, Y.; Protsak, I.; Li, Y.; Wang, J.; Ma, C.; Tan, J.; Yang, J. Mussel-Inspired Polymeric Coatings to Realize Functions from Single and Dual to Multiple Antimicrobial Mechanisms. *ACS Appl. Mater. Interfaces* **2021**, *13*, 3089–3097. [CrossRef] [PubMed]

489. Li, M.; Schlaich, C.; Willem Kulka, M.; Donskyi, I.S.; Schwerdtle, T.; Unger, W.E.S.; Haag, R. Mussel-inspired coatings with tunable wettability, for enhanced antibacterial efficiency and reduced bacterial adhesion. *J. Mater. Chem. B* **2019**, *7*, 3438–3445. [CrossRef]
490. Raza, S.; Matuła, K.; Karoń, S.; Paczesny, J. Resistance and Adaptation of Bacteria to Non-Antibiotic Antibacterial Agents: Physical Stressors, Nanoparticles, and Bacteriophages. *Antibiotics* **2021**, *10*, 435. [CrossRef]
491. Abedon, S.T. Phage-Antibiotic Combination Treatments: Antagonistic Impacts of Antibiotics on the Pharmacodynamics of Phage Therapy? *Antibiotics* **2019**, *8*, 182. [CrossRef]
492. Liu, C.; Hong, Q.; Chang, R.Y.; Kwok, P.C.; Chan, H.-K. Phage-Antibiotic Therapy as a Promising Strategy to Combat Multidrug-Resistant Infections and to Enhance Antimicrobial Efficiency. *Antibiotics* **2022**, *11*, 570. [CrossRef]
493. Kyriakides, T.R.; Raj, A.; Tseng, T.H.; Xiao, H.; Nguyen, R.; Mohammed, F.S.; Halder, S.; Xu, M.; Wu, M.J.; Bao, S.; et al. Biocompatibility of nanomaterials and their immunological properties. *Biomed. Mater.* **2021**, *16*, 042005. [CrossRef]
494. Rodríguez-Ibarra, C.; Déciga-Alcaraz, A.; Ispanixtlahuatl-Meráz, O.; Medina-Reyes, E.I.; Delgado-Buenrostro, N.L.; Chirino, Y.I. International landscape of limits and recommendations for occupational exposure to engineered nanomaterials. *Toxicol. Lett.* **2020**, *322*, 111–119. [CrossRef] [PubMed]
495. Soenen, S.J.; Parak, W.J.; Rejman, J.; Manshian, B. (Intra)Cellular Stability of Inorganic Nanoparticles: Effects on Cytotoxicity, Particle Functionality, and Biomedical Applications. *Chem. Rev.* **2015**, *115*, 2109–2135. [CrossRef] [PubMed]
496. Karlsson, H.L.; Gustafsson, J.; Cronholm, P.; Möller, L. Size-dependent toxicity of metal oxide particles—A comparison between nano- and micrometer size. *Toxicol. Lett.* **2009**, *188*, 112–118. [CrossRef] [PubMed]
497. Du, H.; Pan, B.; Chen, T. Evaluation of chemical mutagenicity using next generation sequencing: A review. *J. Environ. Sci. Health Part C* **2017**, *35*, 140–158. [CrossRef]
498. Singh, N.; Manshian, B.; Jenkins, G.J.S.; Griffiths, S.M.; Williams, P.M.; Maffeis, T.G.G.; Wright, C.J.; Doak, S.H. NanoGenotoxicology: The DNA damaging potential of engineered nanomaterials. *Biomaterials* **2009**, *30*, 3891–3914. [CrossRef]
499. Gutierro, I.; Hernández, R.M.; Igartua, M.; Gascón, A.R.; Pedraz, J.L. Size dependent immune response after subcutaneous, oral and intranasal administration of BSA loaded nanospheres. *Vaccine* **2002**, *21*, 67–77. [CrossRef]
500. Kermanizadeh, A.; Vranic, S.; Boland, S.; Moreau, K.; Baeza-Squiban, A.; Gaiser, B.K.; Andrzejczuk, L.A.; Stone, V. An in vitro assessment of panel of engineered nanomaterials using a human renal cell line: Cytotoxicity, pro-inflammatory response, oxidative stress and genotoxicity. *BMC Nephrol.* **2013**, *14*, 96. [CrossRef] [PubMed]
501. Kumar, S.; Anselmo, A.C.; Banerjee, A.; Zakrewsky, M.; Mitragotri, S. Shape and size-dependent immune response to antigen-carrying nanoparticles. *J. Control. Release* **2015**, *220*, 141–148. [CrossRef] [PubMed]
502. Rahmati, M.; Mozafari, M. Nano-immunoengineering: Opportunities and challenges. *Curr. Opin. Biomed. Eng.* **2019**, *10*, 51–59. [CrossRef]
503. Porras, A.M.; Hutson, H.N.; Berger, A.J.; Masters, K.S. Engineering approaches to study fibrosis in 3-D in vitro systems. *Curr. Opin. Biotechnol.* **2016**, *40*, 24–30. [CrossRef] [PubMed]
504. Gonzalez-Gonzalez, F.J.; Chandel, N.S.; Jain, M.; Budinger, G.R.S. Reactive oxygen species as signaling molecules in the development of lung fibrosis. *Transl. Res.* **2017**, *190*, 61–68. [CrossRef]
505. Mittal, A.K.; Banerjee, U.C. In vivo safety, toxicity, biocompatibility and anti-tumour efficacy of bioinspired silver and selenium nanoparticles. *Mater. Today Commun.* **2021**, *26*, 102001. [CrossRef]
506. Mahato, K.; Maurya, P.K.; Chandra, P. Fundamentals and commercial aspects of nanobiosensors in point-of-care clinical diagnostics. *3 Biotech* **2018**, *8*, 149. [CrossRef]
507. Bahadır, E.B.; Sezgintürk, M.K. Applications of commercial biosensors in clinical, food, environmental, and biothreat/biowarfare analyses. *Anal. Biochem.* **2015**, *478*, 107–120. [CrossRef]
508. Turner, A.P.F. Biosensors: Sense and sensibility. *Chem. Soc. Rev.* **2013**, *42*, 3184–3196. [CrossRef] [PubMed]
509. Mahato, K.; Srivastava, A.; Chandra, P. Paper based diagnostics for personalized health care: Emerging technologies and commercial aspects. *Biosens. Bioelectron.* **2017**, *96*, 246–259. [CrossRef]
510. Liu, S.; Su, W.; Ding, X. A Review on Microfluidic Paper-Based Analytical Devices for Glucose Detection. *Sensors* **2016**, *16*, 2086. [CrossRef] [PubMed]
511. Vargason, A.M.; Anselmo, A.C.; Mitragotri, S. The evolution of commercial drug delivery technologies. *Nat. Biomed. Eng.* **2021**, *5*, 951–967. [CrossRef] [PubMed]
512. Adler, L.A.; Zimmerman, B.; Starr, H.L.; Silber, S.; Palumbo, J.; Orman, C.; Spencer, T. Efficacy and Safety of OROS Methylphenidate in Adults With Attention-Deficit/Hyperactivity Disorder: A Randomized, Placebo-Controlled, Double-Blind, Parallel Group, Dose-Escalation Study. *J. Clin. Psychopharmacol.* **2009**, *29*, 239–247. [CrossRef] [PubMed]
513. Kempf, D.J.; Sham, H.L.; Marsh, K.C.; Flentge, C.A.; Betebenner, D.; Green, B.E.; McDonald, E.; Vasavanonda, S.; Saldivar, A.; Wideburg, N.E.; et al. Discovery of Ritonavir, a Potent Inhibitor of HIV Protease with High Oral Bioavailability and Clinical Efficacy. *J. Med. Chem.* **1998**, *41*, 602–617. [CrossRef]
514. Jana, S.; Mandlekar, S.; Marathe, P. Prodrug design to improve pharmacokinetic and drug delivery properties: Challenges to the discovery scientists. *Curr. Med. Chem.* **2010**, *17*, 3874–3908. [CrossRef]
515. Swinney, D.C.; Anthony, J. How were new medicines discovered? *Nat. Rev. Drug Discov.* **2011**, *10*, 507–519. [CrossRef]
516. Chey, W.D.; Webster, L.; Sostek, M.; Lappalainen, J.; Barker, P.N.; Tack, J. Naloxegol for Opioid-Induced Constipation in Patients with Noncancer Pain. *N. Engl. J. Med.* **2014**, *370*, 2387–2396. [CrossRef]

517. Agersø, H.; Seiding Larsen, L.; Riis, A.; Lövgren, U.; Karlsson, M.O.; Senderovitz, T. Pharmacokinetics and renal excretion of desmopressin after intravenous administration to healthy subjects and renally impaired patients. *Br. J. Clin. Pharmacol.* **2004**, *58*, 352–358. [CrossRef]

518. Dlugi, A.M.; Miller, J.D.; Knittle, J. Lupron TAP Pharmaceuticals, North Chicago, Illinois. depot (leuprolide acetate for depot suspension) in the treatment of endometriosis: A randomized, placebo-controlled, double-blind study. *Fertil. Steril.* **1990**, *54*, 419–427. [CrossRef]

519. Al-Tabakha, M.M. Future prospect of insulin inhalation for diabetic patients: The case of Afrezza versus Exubera. *J. Control. Release* **2015**, *215*, 25–38. [CrossRef] [PubMed]

520. Suzuki, R.; Brown, G.A.; Christopher, J.A.; Scully, C.C.G.; Congreve, M. Recent Developments in Therapeutic Peptides for the Glucagon-like Peptide 1 and 2 Receptors. *J. Med. Chem.* **2020**, *63*, 905–927. [CrossRef]

521. Booth, C.; Gaspar, H.B. Pegademase bovine (PEG-ADA) for the treatment of infants and children with severe combined immunodeficiency (SCID). *Biol. Targets Ther.* **2009**, *3*, 349–358.

522. Larsen, C.P.; Pearson, T.C.; Adams, A.B.; Tso, P.; Shirasugi, N.; Strobertb, E.; Anderson, D.; Cowan, S.; Price, K.; Naemura, J.; et al. Rational Development of LEA29Y (belatacept), a High-Affinity Variant of CTLA4-Ig with Potent Immunosuppressive Properties. *Am. J. Transplant.* **2005**, *5*, 443–453. [CrossRef] [PubMed]

523. Pasut, G. Pegylation of Biological Molecules and Potential Benefits: Pharmacological Properties of Certolizumab Pegol. *Biodrugs* **2014**, *28*, 15–23. [CrossRef] [PubMed]

524. Mensink, M.A.; Frijlink, H.W.; van der Voort Maarschalk, K.; Hinrichs, W.L.J. How sugars protect proteins in the solid state and during drying (review): Mechanisms of stabilization in relation to stress conditions. *Eur. J. Pharm. Biopharm.* **2017**, *114*, 288–295. [CrossRef]

525. Sanford, M. Subcutaneous trastuzumab: A review of its use in HER2-positive breast cancer. *Target. Oncol.* **2014**, *9*, 85–94. [CrossRef]

526. Cohenuram, M.; Saif, M.W. Panitumumab the first fully human monoclonal antibody: From the bench to the clinic. *Anti-Cancer Drugs* **2007**, *18*, 7–15. [CrossRef]

527. Akinc, A.; Maier, M.A.; Manoharan, M.; Fitzgerald, K.; Jayaraman, M.; Barros, S.; Ansell, S.; Du, X.; Hope, M.J.; Madden, T.D.; et al. The Onpattro story and the clinical translation of nanomedicines containing nucleic acid-based drugs. *Nat. Nanotechnol.* **2019**, *14*, 1084–1087. [CrossRef]

528. Eckstein, F. Phosphorothioates, Essential Components of Therapeutic Oligonucleotides. *Nucleic Acid Ther.* **2014**, *24*, 374–387. [CrossRef] [PubMed]

529. Springer, A.D.; Dowdy, S.F. GalNAc-siRNA Conjugates: Leading the Way for Delivery of RNAi Therapeutics. *Nucleic Acid Ther.* **2018**, *28*, 109–118. [CrossRef]

530. Corey, D.R. Nusinersen, an antisense oligonucleotide drug for spinal muscular atrophy. *Nat. Neurosci.* **2017**, *20*, 497–499. [CrossRef] [PubMed]

531. Xu, L.; Wang, J.; Liu, Y.; Xie, L.; Su, B.; Mou, D.; Wang, L.; Liu, T.; Wang, X.; Zhang, B.; et al. CRISPR-Edited Stem Cells in a Patient with HIV and Acute Lymphocytic Leukemia. *N. Engl. J. Med.* **2019**, *381*, 1240–1247. [CrossRef] [PubMed]

532. Stephan, S.B.; Taber, A.M.; Jileaeva, I.; Pegues, E.P.; Sentman, C.L.; Stephan, M.T. Biopolymer implants enhance the efficacy of adoptive T-cell therapy. *Nat. Biotechnol.* **2015**, *33*, 97–101. [CrossRef] [PubMed]

533. Carmona, G.; Barney, L.; Sewell, J.; Newman, R.; Carroll, C.; Beauregard, M.; Huang, J.; Heidebrecht, R.W.; Corzo, D.; Moller, D.; et al. Correcting Rare Blood Disorders Using Coagulation Factors Produced In Vivo By Shielded Living Therapeutics™ Products. *Blood* **2019**, *134*, 2065. [CrossRef]

534. Brudno, J.N.; Kochenderfer, J.N. Chimeric antigen receptor T-cell therapies for lymphoma. *Nat. Rev. Clin. Oncol.* **2018**, *15*, 31–46. [CrossRef]

535. Cheever, M.A.; Higano, C.S. PROVENGE (Sipuleucel-T) in Prostate Cancer: The First FDA-Approved Therapeutic Cancer Vaccine. *Clin. Cancer. Res.* **2011**, *17*, 3520–3526. [CrossRef]

536. Bartlett, W.; Skinner, J.A.; Gooding, C.R.; Carrington, R.W.J.; Flanagan, A.M.; Briggs, T.W.R.; Bentley, G. Autologous chondrocyte implantation versus matrix-induced autologous chondrocyte implantation for osteochondral defects of the knee. *J. Bone Jt. Surg. Br. Vol.* **2005**, *87-B*, 640–645. [CrossRef] [PubMed]

537. Maude, S.L.; Laetsch, T.W.; Buechner, J.; Rives, S.; Boyer, M.; Bittencourt, H.; Bader, P.; Verneris, M.R.; Stefanski, H.E.; Myers, G.D.; et al. Tisagenlecleucel in Children and Young Adults with B-Cell Lymphoblastic Leukemia. *N. Engl. J. Med.* **2018**, *378*, 439–448. [CrossRef]

 *micromachines*

Article

Fabrication and Characterization of Dielectric $ZnCr_2O_4$ Nanopowders and Thin Films for Parallel-Plate Capacitor Applications

Vasyl Mykhailovych [1,2,3], Gabriel Caruntu [1,4,5,*], Adrian Graur [1], Mariia Mykhailovych [1,2], Petro Fochuk [2], Igor Fodchuk [3], Gelu-Marius Rotaru [6] and Aurelian Rotaru [1,*]

1 Department of Electrical Engineering and Computer Science & Research Center MANSiD, Stefan cel Mare University of Suceava, 13, University St., No. 13, 720229 Suceava, Romania; vasyl.mykhailovych@usm.ro (V.M.); adriang@usm.ro (A.G.); m.mykhailovych@gmail.com (M.M.)

2 Department of General Chemistry and Material Science, Yuriy Fedkovych Chernivtsi National University, 2, Kotsjubynskyi St., 58012 Chernivtsi, Ukraine; fochukp@gmail.com

3 Physical, Technical and Computer Science Institute, Yuriy Fedkovych Chernivtsi National University, 2, Kotsjubynskyi St., 58012 Chernivtsi, Ukraine

4 Department of Chemistry and Biochemistry, Central Michigan University, 1200 S. Franklin St., Mount Pleasant, MI 48859, USA

5 Science of Advanced Materials Program, Central Michigan University, 1200 S. Franklin St., Mount Pleasant, MI 48859, USA

6 Faculty of Mechanical Engineering Mechatronics and Management & Research Center MANSiD, Stefan cel Mare University, 720229 Suceava, Romania; gelu.rotaru@usm.ro

* Correspondence: g.caruntu@cmich.edu (G.C.); aurelian.rotaru@usm.ro (A.R.)

Citation: Mykhailovych, V.; Caruntu, G.; Graur, A.; Mykhailovych, M.; Fochuk, P.; Fodchuk, I.; Rotaru, G.-M.; Rotaru, A. Fabrication and Characterization of Dielectric $ZnCr_2O_4$ Nanopowders and Thin Films for Parallel-Plate Capacitor Applications. *Micromachines* **2023**, *14*, 1759. https://doi.org/10.3390/mi14091759

Academic Editor: Amir Hussain Idrisi

Received: 3 July 2023
Revised: 4 September 2023
Accepted: 8 September 2023
Published: 12 September 2023

Abstract: We report here the successful shape-controlled synthesis of dielectric spinel-type $ZnCr_2O_4$ nanoparticles by using a simple sol-gel auto-combustion method followed by successive heat treatment steps of the resulting powders at temperatures from 500 to 900 °C and from 5 to 11 h, in air. A systematic study of the dependence of the morphology of the nanoparticles on the annealing time and temperature was performed by using field effect scanning electron microscopy (FE-SEM), powder X-ray diffraction (PXRD) and structure refinement by the Rietveld method, dynamic lattice analysis and broadband dielectric spectrometry, respectively. It was observed for the first time that when the aerobic post-synthesis heat treatment temperature increases progressively from 500 to 900 °C, the $ZnCr_2O_4$ nanoparticles: (i) increase in size from 10 to 350 nm and (ii) develop well-defined facets, changing their shape from shapeless to truncated octahedrons and eventually pseudo-octahedra. The samples were found to exhibit high dielectric constant values and low dielectric losses with the best dielectric performance characteristics displayed by the 350 nm pseudo-octahedral nanoparticles whose permittivity reaches a value of $\varepsilon = 1500$ and a dielectric loss $\tan \delta = 5 \times 10^{-4}$ at a frequency of 1 Hz. Nanoparticulate $ZnCr_2O_4$-based thin films with a thickness varying from 0.5 to 2 μm were fabricated by the drop-casting method and subsequently incorporated into planar capacitors whose dielectric performance was characterized. This study undoubtedly shows that the dielectric properties of nanostructured zinc chromite powders can be engineered by the rational control of their morphology upon the variation of the post-synthesis heat treatment process.

Keywords: high-k material; $ZnCr_2O_4$ nanoparticles; shape-controlled synthesis; dielectric properties; thin films; planar capacitor

1. Introduction

  The continuous advances in microelectronics and computing require the development and improvement of active elements of electronic circuitry. Spinel type materials have been the workhorse in electronics due to their high stability, unique tunable magnetic and electric properties, along with easy processability [1]. Unlike their ferrite counterparts, transition metal chromites adopt, in bulk, the normal spinel structure, due to the large crystal field stabilization energy of the Cr^{3+} ions crystallizing in the cubic system (space group $F\bar{d}3m$, No. 227). [2] Among chromites, zinc chromite ($ZnCr_2O_4$) has emerged as an interesting catalytic material in the oxidation of CO, the catalytic combustion of hydrocarbons and others [3–5], synthesis of methanol [6], photocatalysis [7,8], sensing [9–12], the design of near-infrared spectral emitters [13,14], and broadband photo-detectors [15]. Moreover, $ZnCr_2O_4$ is a wide bandgap semiconductor whose energy gap varies between 3.0 and 3.5 eV, depending on the size of the constituting particles and the existence of structural defects [11,16,17], which makes it a potential dielectric material for applications in micro- and nanoelectronics. Interestingly, bulk zinc chromite is a typical spin-frustrated material that undergoes an antiferromagnetic transition at the Néel temperature T_N = 12 K. Kagomyia and colleagues demonstrated that the relaxation of this magnetic frustration is lattice-mediated, leading to the onset of an anomaly in the variation of the dielectric permittivity at the Néel temperature along with a dispersion of the dielectric permittivity at temperatures below 70 K [18]. In recent years, extensive efforts have been devoted to the investigation of the dielectric properties of both bulk and nanostructured spinel $ZnCr_2O_4$ [18–21]. Javed et al. [19] investigated the dielectric properties of $ZnCr_2O_4$ nanoparticles with an average size of 144 nm, obtained by the sol-gel auto-combustion method, showing that the dielectric constant varied from ε = 44 to ε = 20 within a frequency range between 70 Hz and 1 MHz. Similarly, Shafqat et al. [22] investigated the structural, morphological and dielectric properties of nanoscale spinel transition metal chromites XCr_2O_4 (X = Zn, Mn, Cu and Fe), finding that the permittivity of pelletized powders is the highest for $ZnCr_2O_4$ nanoparticles, ranging from ε = 100,000 to ε = 100 for frequencies varying between 20 Hz and 20 MHz, respectively. In an impedance analysis study of nanostructured zinc chromite by Naz et al. [20], the dielectric permittivity of 50 nm nanoparticles synthesized through a hydrothermal route was found to decrease from ε = 40,000 to ε = 7 in the frequency range between 1 Hz and 10 MHz. The comprehensive assessment of the dielectric permittivity of nanoparticles generally relies on various factors, including the crystal structure, morphology and porosity. These three main parameters significantly impact the results of the dielectric spectroscopy measurements; hence a step-by-step and thorough investigation of each parameter is necessary for the reliable evaluation of the dielectric properties of these materials. The determination of the structure and the purity of spinel $ZnCr_2O_4$ nanopowders can be achieved through detailed X-ray analysis. Additionally, valuable structural information can be obtained using Raman and Energy Dispersive X-ray analyses. Once the spinel structure is determined, the next parameter to consider is morphology. The morphology of particles plays a significant role in dielectric investigation as it directly impacts the properties of the material and the porosity of the final product.

A noteworthy study is that of Binks et al., who simulated the crystal morphology of $ZnCr_2O_4$ ceramic-predicted four morphologies of cubic spinel lattice $ZnCr_2O_4$ crystallites, namely: octahedral, slightly trunked octahedral in vertices, cubic and heavily capped octahedron [23]. Furthermore, based on surface energy considerations, they also calculated that the main crystallite growth of $ZnCr_2O_4$ occurs into a regular octahedral geometry with the (111) crystallographic plane dominating. However, it is important to note that the morphology of spinel-type oxides can be influenced by several factors. For example, Xiao and co-authors demonstrated a high level of control over the morphology of spinel Co_3O_4 nanoparticles. By adjusting the ratio of cobalt nitrate hexahydrate and co-precipitant sodium hydroxide during hydrothermal synthesis, they were able to obtain cubic, truncated octahedral and octahedral shapes [24]. This clearly indicates that the surface or the attachment energies can be influenced by the precursor concentration and reaction condition, ultimately dictating the morphology of the nanoparticles. In a different study, Parhi et al. synthesized $ZnCr_2O_4$ nanoparticles with similar octahedral morphology as predicted by Binks and colleagues [23]. They synthesized the nanoparticles using the microwave metathetic approach, which predominantly led to the formation of an octahedral morphology. Similarly, Mancic et al. investigated $ZnCr_2O_4$ nanoparticles synthesized through the aerosol reaction of precursors, considering their stoichiometry and morphology [25]. Interestingly, the stoichiometry of the particles was found to be influenced by the aerosol residence time (3, 6 and 9 s at 700 °C), while an additional annealing at 1000 °C led to the transformation from pseudospherical shape to a mainly octahedral shape of the particles. These studies highlight the existence of several approaches to change or control the morphology of $ZnCr_2O_4$ nanoparticles, such as using different synthesis methods or varying the reaction conditions.

However, an investigation of $ZnCr_2O_4$ nanoparticle morphology changes until the complete formation of the spinel structure and its dependence on the thermal treatment history of the samples has not been previously conducted. Therefore, we investigated the structure, morphology progress and dielectric properties of $ZnCr_2O_4$ nanoparticles as a function of annealing temperature and time. The main goal was to find the optimal conditions that allow for a reliable evaluation of the dielectric constant for nanoparticles obtained at different temperatures. Additionally, we fabricated a series of thin films and a high-k capacitor using a thin film of $ZnCr_2O_4$ nanoparticles that exhibited the most promising dielectric properties.

2. Materials and Methods

2.1. Synthesis and Reagents

The analytical grade reagents, namely zinc acetate dehydrate ($Zn(CH_3COO)_2 \cdot 2H_2O$, 98%), Chromium (III) nitrate nonahydrate ($Cr(NO_3)_3 \cdot 9H_2O$, 99%) and tartaric acid, were purchased from Sigma-Aldrich (St. Louis, Missouri, United States) and used as received for $ZnCr_2O_4$ nanoparticle synthesis by the sol-gel auto-combustion method. Tartaric acid was used as a chelating-fuel agent. As in a typical synthesis, the reagents were mixed in their respective molar ratio followed by their dissolution in distilled water. The molar ratio for metal cations $Zn^{2+}:Cr^{3+}$ was 1:2, while for chromite: tartaric acid it was 1:3. The resulting dark violet solution obtained after mixing the precursors was stirred at 75 °C until the excess water evaporated. Subsequently, the resulting gel mixture was kept at ambient temperature for up to 24 h to allow for the complete gel formation. Afterwards, the gel mixture was thermally treated in a sand bath while the temperature was increased from 100 °C to 350 °C with a step of 50 °C per hour. The thermal treatment led to an auto-combustion reaction, which was followed by post-synthesis annealing at 500 °C, 700 °C, 800 °C and 900 °C for 5, 7, 9 and 11 h, respectively. The optimization of reaction conditions was achieved by analyzing the samples at each stage of the synthesis process. We systematically investigated the influence of annealing temperature and annealing time on the nanoparticle size and morphology. The samples were kept at the same temperature for 1, 3, and 5 h (see Figure S1 from ESI). In addition, on the $ZnCr_2O_4$ nanoparticles annealed at 500 °C, we compared the size evolution for annealing times of 5 h and 21 h. As can be seen in Figure S2 from ESI, by increasing the annealing temperature, only a slight increase in the particle's size was observed in the samples annealed at 500 °C. However, the distinctive octahedral morphology was exclusively observed only in the samples that were maintained at 900 °C for a duration of 11 h.

2.2. Preparation of $ZnCr_2O_4$ Nanoparticle Suspensions

The resulting $ZnCr_2O_4$ nanopowders were processed into nanoparticle-based thin film structures by suspending the nanopowders into toluene (99.5% purity), in which a small amount (0.45 mg/mL) of elastomer butadiene styrene (SBS) from Merck was dissolved. The suspension was sonicated to ensure the proper dispersion of the nanoparticles. After sonication, the suspension was left undisturbed for 15 min, allowing excess particles to precipitate. Finally, the supernatant was collected and used for the thin film deposition by the drop-casting method. A toluene solution of SBS was used for several reasons. Firstly, the evaporation of toluene is relatively slow to minimize the number of cracks in the thin film. In addition, SBS exerts a complementary effect in minimizing the number of cracks, as it is an elastomer, which will minimize the stress during the formation of the films, thereby improving their quality. The deposition of $ZnCr_2O_4$ nanoparticles from suspension was directed on silver-coated glass substrate. The prepared thin film was allowed to dry under ambient conditions and used for further device fabrication.

2.3. Characterization of $ZnCr_2O_4$ Nanoparticles, Pellets, Films and Fabricated Devices

The morphological and elemental analyses of both nanoparticles and nanoparticle-based films were performed using a Hitachi SU-70 Field Emission Scanning Electron Microscope (FE-SEM) equipped with an Oxford Instrument EDX-detector. The phase formation and purity of the crystal structure $ZnCr_2O_4$ nanopowders were examined by Raman spectroscopy and powder X-ray diffraction (PXRD). Raman spectra were collected using a Horiba LabRAM HR spectrometer covering the range from 150 to 1100 cm^{-1} for all the samples.

Room temperature powder X-ray diffraction (RT-PXRD) patterns were collected on a PANalytical X'Pert Pro X-ray diffractometer using a Cu anode (λ = 1.54 Å for Kα1 radiation). Powder samples were placed on a zero-background Si holder, and diffraction patterns were collected in the angular range from 10 to 80 degrees in $2\theta°$ with a $0.02°/min$ step size. Powder X-ray diffraction patterns were processed by using the PANalytical X'PertHighScore Plus software (version 3.0.5) and structure refinement was performed by the Rietveld method using the FullProf program. A six-coefficient polynomial function was used to model the background, whereas the peak shape was described by a pseudo-Voigt function.

The dielectric properties of the samples and capacitors were measured using a CONCEPT 40 Broadband Dielectric Spectrometer (Novocontrol Technologies GmbH & Co. KG, Montabaur, Germany) equipped with an Alpha-A high performance frequency analyzer. The measurements were conducted in the frequency range from 1 Hz to 10 MHz, respectively, in a closed temperature cell in a nitrogen atmosphere in order to avoid moisture effects. Thin film capacitors were fabricated on glass substrates, and the top and bottom electrodes were patterned using sputtered metal coatings through the Quorum Sputter Coater, Model: Q150TES. Top electrodes were deposited using a mask with a 1.5 mm in diameter. Both Ag electrodes were deposited by sputtering, with a thickness of 20 nm. The thickness of both electrodes and dielectric film were measured by scanning electron microscopy, in cross section, after cryogenic fracturing. The geometrical parameters were used to evaluate the dielectric permittivity through BDS analyses.

The dielectric properties of $ZnCr_2O_4$ nanoparticle-based pellets were also investigated. In order to increase the pellets' density and to improve their mechanical properties, polyvinyl alcohol (PVA) was used as binder. Thus, $ZnCr_2O_4$ nanoparticles were mixed with a few drops of 10 wt% polyvinyl alcohol (PVA) aqua solution and subsequently pressed into pellets with a thickness of about 0.4 mm and 5 mm in diameter. Silver electrodes with a thickness of 20 nm were deposited on both sides of the pellets, ensuring good electrical contact.

3. Results and Discussion

3.1. Raman Spectroscopy

The Raman spectra exhibit well defined peaks at 180, 450, 510, 605 and 685 cm^{-1} corresponding to the $^1F_{2g}$, E_g, $^2F_{2g}$, $^3F_{2g}$ and A_{1g}, respectively active modes of the $ZnCr_2O_4$ spinel (Figure 1). These results are in good agreement with the previously published data, revealing the high crystallinity of the samples and the absence of secondary phases [26–29].

However, it is evident that the nanoparticles treated at 500 °C show only an initial formation of spinel structure because the main peak (A_{1g}) located at 684 cm^{-1} is broad and small. The $^1F_{2g}$ mode at 182 cm^{-1} is also barely visible, thereby indicating the formation of the $ZnCr_2O_4$ spinel structure. In contrast, the bands located around 348 and 550 cm^{-1} are most likely attributed to the E_g and A_{1g} modes of Cr_2O_3 [30,31]. This suggests that annealing at 500 °C for 5 h is insufficient for the formation of crystalline $ZnCr_2O_4$ nanoparticles, the chemical reaction being incomplete. Consequently, the samples underwent a second annealing treatment at different temperatures, namely 700, 800 and 900 °C for 7, 9 and 11 h, respectively, to allow the formation of the single phase, crystalline zinc chromite nanopowders. The Raman spectra of the nanopowders also reveal that the peak around 450 cm^{-1}, corresponding to the E_g mode, is slightly visible and its absorption value increases with temperature. This is to be expected because as the annealing temperature increases, the size of nanoparticles becomes comparable to those in the bulk $ZnCr_2O_4$ crystal. The intensity of the peak corresponding to the E_g mode is small in the bulk crystal [26], making it even difficult to identify in nanocrystalline $ZnCr_2O_4$.

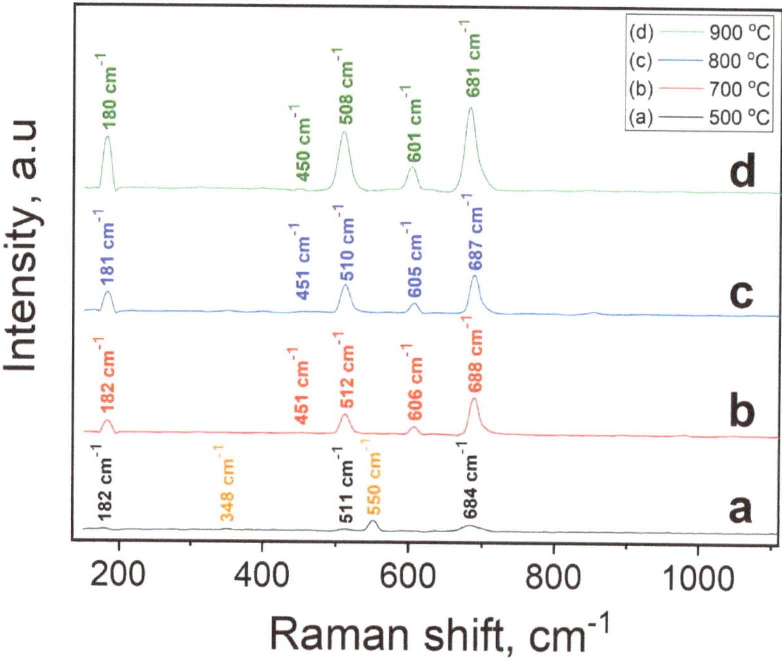

Figure 1. Raman spectra of $ZnCr_2O_4$ samples after annealing under 500 °C (**a**), 700 °C (**b**), 800 °C (**c**) and 900 °C (**d**) for 5, 7, 9 and 11 h, respectively.

3.2. Structure Analysis

The phase purity and crystallinity of the nanopowdered $ZnCr_2O_4$ samples were analyzed by powder X-ray diffraction. The corresponding diffractograms confirm the Raman analysis, showing a similar trend in the formation of spinel type nanoparticles. Figure 2a represents the reference XRD spectrum (JCPDS No. 22-1107) [2], whereas Figure 2b–e correspond to $ZnCr_2O_4$ samples annealed at 500, 700, 800 and 900 °C in air, respectively.

The sample treated at 500 °C exhibits the presence of a secondary phase corresponding to the Cr_2O_3 structure (marked with ★, with the reference pattern of Cr_2O_3 [32]). In contrast, all other samples annealed at 700, 800 and 900 °C confirm the formation of the pure $ZnCr_2O_4$ spinel phase. To further investigate the crystal structure and atomic site distribution of the ions in the zinc chromite nanopowders, the powder X-ray diffraction data were fitted using the Rietveld analysis in the cubic system (space group $F\bar{d}3m$). Figure 3 shows the Rietveld refinement of the $ZnCr_2O_4$ nanopowders annealed in air at 700 °C for 7 h.

It can be easily observed that the fitted curve matches very well with the experimental one and the positions of the Bragg reflections are very similar to those corresponding to the indexed peaks in the spinel structure. The calculated atomic coordinates, lattice parameter and reliability factors are listed in Table 1. The refined cubic lattice parameter a = 8.327(7) Å matches well with the value of the lattice parameter of the bulk standard material (a = 8.327(5) Å; JCPDS 22-1107). Similar values for the crystallographic parameters have been obtained for the samples annealed at 800 °C and 900 °C, as can be seen in the ESI (See Tables S1 and S2 from ESI).

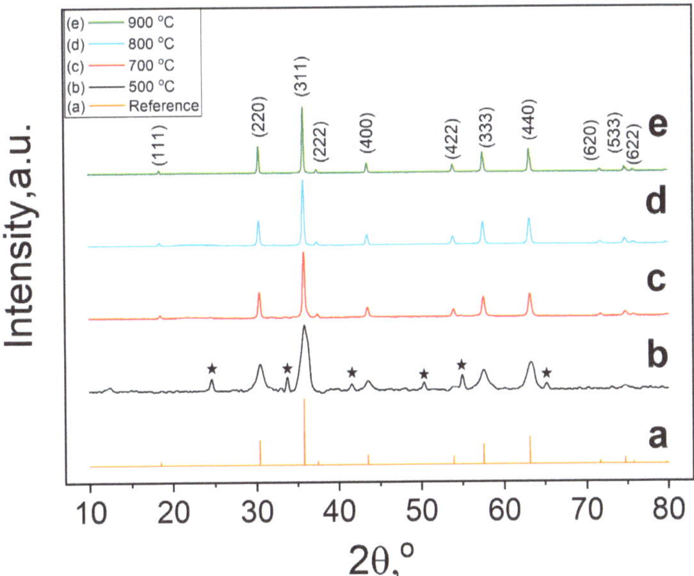

Figure 2. XRD diffractograms of ZnCr₂O₄ samples: reference pattern (**a**), and powder XRD patterns of nanoparticles annealed at 500, 700, 800 and 900 °C (**b**–**e**), respectively. The reference pattern of Cr₂O₃ are marked with ★.

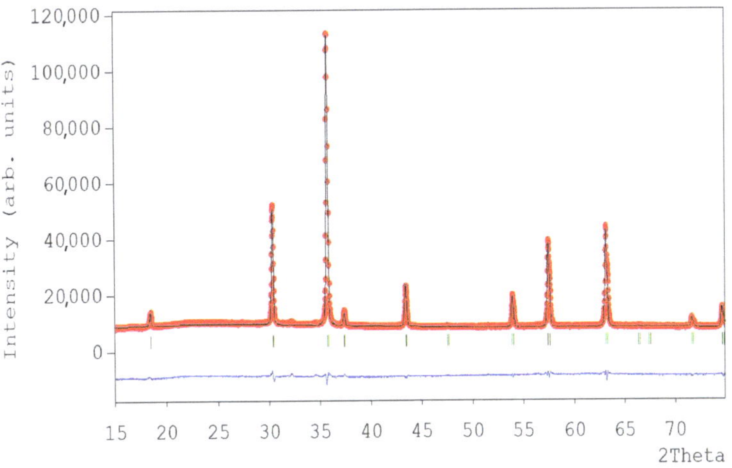

Figure 3. Structure refinement by Rietveld method of nanostructured ZnCr₂O₄ samples annealed in air at 700 °C for 7 h (**Red curve**). The blue curve represents the difference $I_{obs}-I_{calc}$ pattern, whereas the green vertical bars represent the positions of the Bragg reflections.

Table 1. O_4 nanopowder annealed at 700 °C in air for 7 h.

Phase	Lattice Parameters (Å)	Atomic Coordinates			Occupancy	B_{iso}	R Factors
$ZnCr_2O_4$	$a = b = c$	Ion	$x = y = z$	Wyckoff			$R_p = 9.57\%$; $R_{wp} = 7.15\%$, $R_{exp} = 2.42$, $\chi^2 = 2.76$
	8.327(7)	Zn^{2+}	0.125	8a	1.0	0.72	
		Cr^{3+}	0.500	16d	1.0	0.45	
		O^{2-}	0.259(8)	32e	1.0	0.79	

3.3. Morphology and Elemental Analysis

The examination of the field-emission scanning electron microscopy (FE-SEM) micrograph presented in Figure 4 suggests a strong dependence of the morphology of as-synthesized $ZnCr_2O_4$ nanoparticles on the annealing temperature, whereby the increase of the treatment temperature promotes the growth of the nanoparticles along with the development of well-defined faces, eventually resulting in the formation of octahedral-like nanoparticles.

Figure 4. FE–SEM micrographs recorded on $ZnCr_2O_4$ after thermal treatment at 500 °C (**a**), 700 °C (**b**), 800 °C (**c**) and 900 °C (**d**), respectively.

Specifically, the detailed analysis of the SEM images revealed that whereas the nanopowders obtained after a post-synthesis annealing at 500 °C for 5 h in air contain shapeless nanoparticles with an average size of 10 ± 2 nm (Figure 4a), subsequent heat treatments of the same sample at 700 °C for 7 h, 800 °C for 9 h and 900 °C for 11 h lead to an increase of the size of the nanoparticles to 30 ± 10 nm, 90 ± 20 nm and 350 ± 32 nm, respectively (Figure 4b–d). EDX analysis was performed on all the samples to determine the elemental distribution and weight percent in the nanoparticles. To this end, the samples were deposited on silver-coated glass, and during analysis the sample that annealed at 500 °C for 5 h in air showed the presence of four chemical elements, namely C, Zn, Cr and O. Other elements such as Si, Ca, Mg and Ag or Pt identified by EDX analysis were not taken into consideration as they originated from the substrate. Further details on the EDX analysis can be found in Figure S3.

In the second step of the thermal treatment at 700 °C in air for 7 h, not only did the particles grow, stabilizing the spinel structure, but also the purity of the sample increased as the amount of carbon residuals in the sample (estimated empirically from the EDX analysis) decreased from 5.4 to 2.2 wt. %.

Figure S4 provides information about the EDX elemental map, the EDS spectrum and weight percent elemental distribution in the sample. The sample that annealed at 800 °C for 9 h is free of secondary phases, as no other elements were detected (see Figure S5). Nanoparticles annealed at 900 °C for 11 h were also analyzed by EDX. Similar to the previous annealing step, the main elements contained in the nanoparticles are Zn, Cr and O, respectively (see Figure S6). Moreover, a detailed study was conducted on the morphology and elemental distribution of a single nanoparticle by using the nanopowdered sample annealed at 900 °C in air. A high magnification image of the $ZnCr_2O_4$ nanoparticle revealed the existence of an octahedral shape with a slightly trunked edge (Figure 5a). EDX mapping confirmed that nanoparticle contains three elements, namely Zn, Cr and O (Figure 5b–d), and the quantitative analysis of the spectrum of the detected elements in the $ZnCr_2O_4$ particle matched the expected molar Zn:Cr, ratio (Figure 5e).

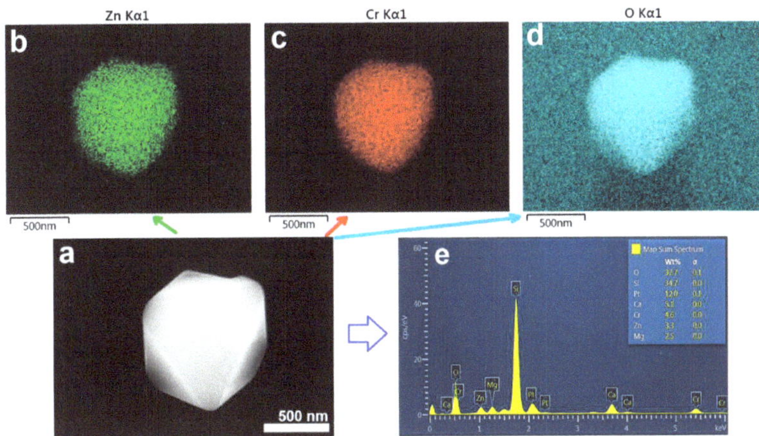

Figure 5. FE–SEM image of an octahedral single particle from a sample annealed at 900 °C (**a**), elemental mapping (**b–d**) and EDX spectrum with weight distribution of elements (**e**).

The weight percent ratio of Zn:Cr extracted from the EDX analysis are shown in Table 2. The closest Zn:Cr ratio to the expected value (0.628), has been obtained for the samples treated at 900 °C (0.648).

Table 2. 700, 800 and 900 °C for 5, 7, 9 and 11 h, respectively.

Element	(500 °C) wt%	(700 °C) wt%	(800 °C) wt%	(900 °C) wt%
Zn	3.1	10.3	2.0	3.5
Cr	4.0	14.7	2.8	5.4

3.4. Dielectric Properties of $ZnCr_2O_4$ Nanopowders

The electrical behavior of the $ZnCr_2O_4$ nanopowders was performed by using dielectric spectroscopy. Prior to the analysis, the samples were pressed into pellets by mixing the nanoparticles with polyvinyl alcohol (PVA) aqua solution, used with a binding agent. As expected, the real part of the permittivity increased with the size of nanoparticles, except for the sample annealed at 500 °C in air, which exhibits poor dielectric characteristics, presumably due to the presence of Cr_2O_3 as a secondary phase and does not follow the experimentally observed trend of variation of the permittivity. Thus, the lowest value of the dielectric permittivity was observed for single-phase spinel-type nanopowders obtained after heat treatment at 700 °C for 7 h, whereby the dielectric constant increased from $\varepsilon = 20$ at 10 MHz to $\varepsilon = 600$ at 1 Hz for nanoparticles with an average size of about 30 nm (Figure 6a). The sample containing $ZnCr_2O_4$ nanoparticles with an average size of 90 nm that was annealed at 800 °C exhibited dielectric permittivity values ranging from $\varepsilon = 36$ to $\varepsilon = 300$, depending on the measuring frequency. At $\nu = 10$ Hz, it was observed that the dielectric permittivity values for the samples annealed at 700 and 800 °C crossover, which can be explained by the polarization of electrodes as indicated by the higher dielectric loss for the sample annealed at 700 °C, 800 °C and 900 °C (Figure 6b). The low frequency values of the dielectric loss (tanδ) for the samples that annealed at 500, then at 700 and 800 °C, were 3.3; 3.3 and 1.74, respectively, with a dramatic decrease in value for the sample subjected to an additional heat treatment at 900 °C for 11 h. In this latter case the value of the dielectric loss was found to be tan$\delta = 5 \times 10^{-4}$ at $\nu = 1$ Hz, which increases to tan$\delta = 2.5 \times 10^{-3}$ for frequencies as high as 10 Mhz.

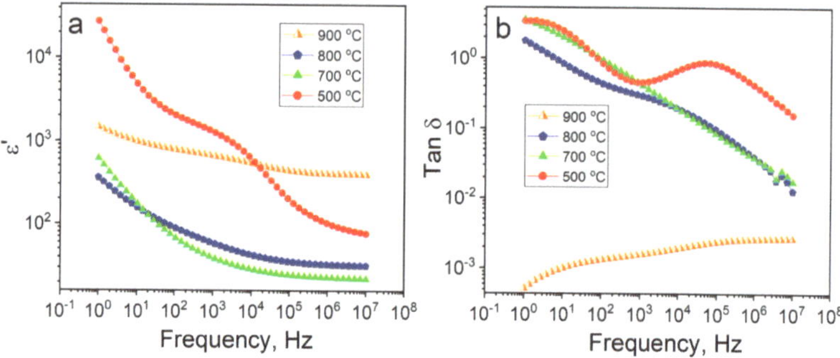

Figure 6. Frequency dependence of the real part of dielectric permittivity (**a**) and Tan δ (**b**), recorded on nanoparticles annealed at various temperatures mixed with PVA binding agent measured at room temperature.

For intermediate post-synthesis annealing of the as-prepared samples, the loss tangent values were found to decrease with increasing the frequency of the external applied electric field. These results strongly suggest that the shape nanoparticles and the development of well-defined facets leads to a dramatic improvement of their dielectric properties as a result of the successive post-synthesis heat treatments.

All in all, the sample annealed at 900 °C for 11 h appears to be the most promising for device fabrication, as it is a single phase and exhibits a stable dielectric permittivity value across the entire frequency range, with a dielectric constant of $\varepsilon = 400$ at high frequency and $\varepsilon = 1500$ at low frequency, respectively. The highest dielectric permittivity value at low frequency was found for the sample annealed at 500 °C for 5 h in air, which can be attributed to the presence of secondary phases.

3.5. Fabrication of ZnCr₂O₄ Nanoparticle Based Thin Films

As discussed in the previous section, the sample annealed at 900 °C for 11 h in air was found to be the most promising for the fabrication of nanoparticle-based thin films. Thin film structures were fabricated by dispersing zinc chromite nanoparticles into an SBS solution in toluene with the concentration of 0.1 mg mL^{-1} and the resulting mixture was sonicated for 10 min to ensure the thorough dispersion of the nanoparticles.

Once the $ZnCr_2O_4$ nanoparticles were dispersed in the SBS solution, the solution was left undisturbed for 15 min to allow the precipitation of the largest nanoparticles. The supernatant was then collected after decantation, and thin films were prepared by using the drop-casting method on silver-coated glass substrates. The thickness of the films varied with the volume of the solution applied to the substrate surface, which ranged from 25 to 150 µL, respectively. This procedure allowed the fabrication of nanoparticle-based $ZnCr_2O_4$ films with a thickness of varying from 0.5 to 2 µm (Figure 7).

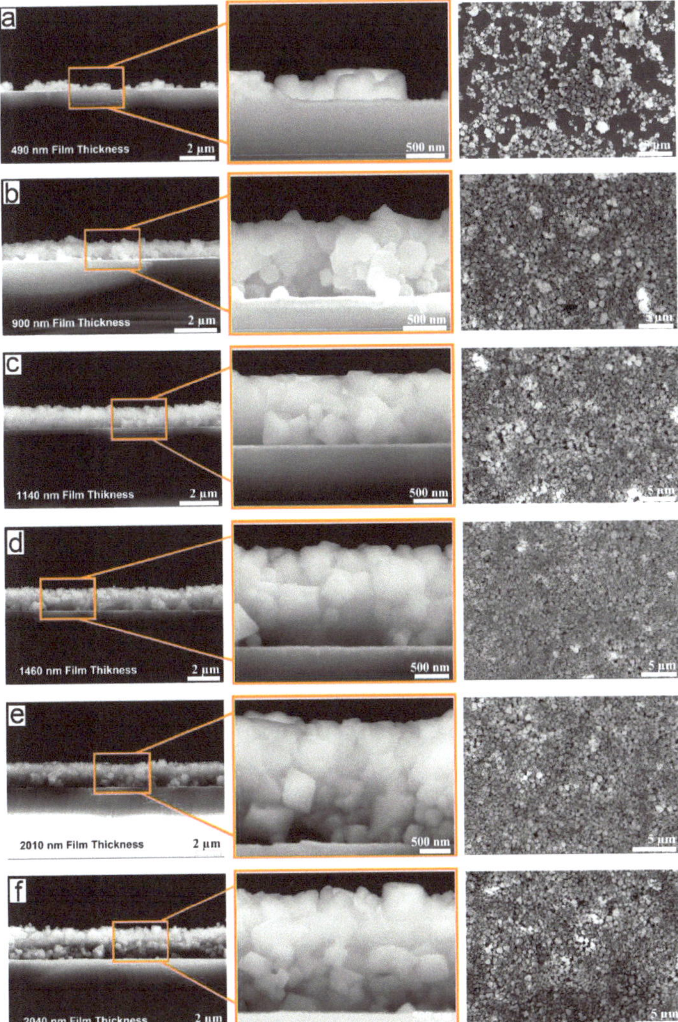

Figure 7. Cross section and top view FE-SEM images of nanoparticulate $ZnCr_2O_4$ thin films with different thicknesses: (**a**) 490 nm; (**b**) 900nm; (**c**) 1140 nm; (**d**) 1460 nm; (**e**) 2010 nm and (**f**) 2040 nm, respectively.

Both top-view and cross-section SEM images of the film with a thickness of 490 nm obtained from 25 µL volume of the solution revealed that the film was nonuniform and presented voids (Figure 7a). This indicates that a volume of 25 µL of solution is insufficient for forming a continuous film. As seen in Figure 7b–f, upon increasing the volume of the solution to 50, 75, 100, 125 and 150 µL, respectively, the quality of the films improved considerably, although some slight porosity was still observed in all the films fabricated by this method. This suggested that a higher concentration of SBS in the solution was needed to evenly fill the pores with the polymer. Therefore, a suspension of zinc chromite nanoparticles with an SBS concentration of about 0.45 mg mL^{-1} was prepared to obtain nanoparticle-based $ZnCr_2O_4$ thin films for device fabrication.

3.6. Device Fabrication

Nanoparticulate $ZnCr_2O_4$ thin film structures were deposited on silver-coated glass substrates and subsequently incorporated as the dielectric layer into capacitors. The $ZnCr_2O_4$-SBS solution was prepared using spinel nanopowders annealed at 900 °C in air for 11 h. The solution was drop cast on the substrate followed by the evaporation of the solvent and the formation of the thin film structures. The top view SEM image of $ZnCr_2O_4$-SBS film suggests that the nanoparticles have been uniformly mixed with the SBS polymer (Figure 8a). Cross-section analysis (Figure 8b) revealed that the film has a thickness of approximately 2 μm, which was subsequently used to calculate the permittivity of the capacitor.

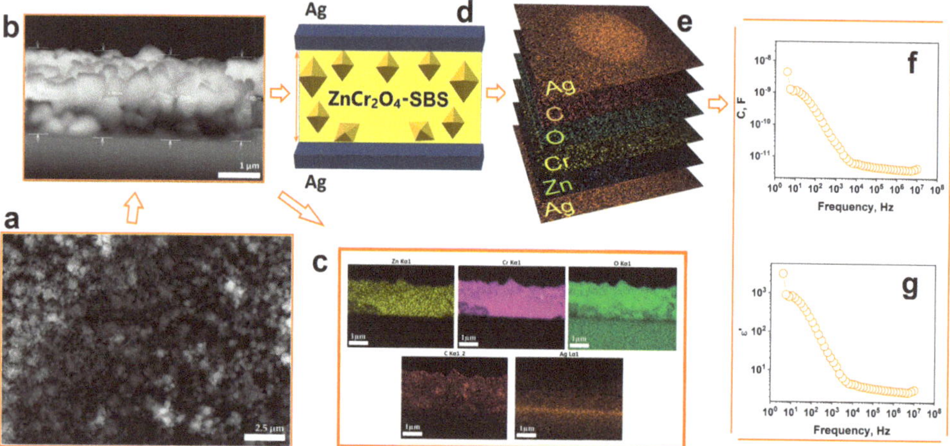

Figure 8. FE–SEM images of thin film of $ZnCr_2O_4$–SBS nanoparticles top view (**a**); cross–section SEM image (**b**), EDX elemental mapping of cross–section (**c**), illustration of produced high–k capacitor (**d**), EDX elemental maps of capacitor (**e**), graph of capacitor capacity (**f**) and dielectric constant (**g**).

Furthermore, the EDX mapping analysis confirmed the presence of elemental Zn, Cr and O in a molar ratio corresponding to $ZnCr_2O_4$ and C from SBS, elements which were found to be uniformly distributed throughout the film. As can been seen in Figure 8c, the bottom electrode was also clearly visible. After the formation of the nanoparticle-based films, top electrodes were sputtered on these films in a planar capacitor geometry (Figure 8d). EDX analysis was therefore conducted to investigate the chemical composition of the capacitor. Layered EDX maps of the elemental composition of the planar capacitors are presented in Figure 8e illustrating the elemental distribution across different layers of the device. Capacitance and dielectric permittivity measurements of the capacitors were performed within the frequency range of ν = 1 Hz to ν = 10 MHz. The values of the capacitance of the device were found to vary from 10^{-11} F at high frequencies to 10^{-9} F at low frequencies (Figure 8f), whereas its permittivity decreased from about ε = 1000 at low frequencies to ε = 5 ν = 10^4 Hz, respectively, reaching a plateau in the high frequency range (Figure 8g).

At high frequencies, the dielectric constant values of the thin films are influenced by polarization processes [33,34]. Specifically, at lower frequencies the dipoles tend to align themselves with the oscillating electric field, whereas when the frequency increases, it becomes more challenging for the dipoles to align rapidly with the polarity of the electric field. At lower frequencies, the dielectric response of the sample is dominated by space charge polarization due to the accumulation of charges at the boundaries between the spinel zinc chromite nanoparticles and between the nanoparticles and the SBS elastomer, thereby leading to higher values of the dielectric constant. The onset of observed Debye dielectric dispersion of the dielectric constant for the $ZnCr_2O_4$-SBS-based capacitor could be tentatively explained by Koop's theory, which is based on the Maxwell–Wagner model [35,36]. The dielectric performance of the nanoparticle-based $ZnCr_2O_4$-SBS planar capacitors is comparable and even superior to that of spinel ferrite-based devices reported in the literature. For example, $CoFe_2O_4$ thin film materials were reported to have a value of dielectric permittivity that varies from $\varepsilon = 117$ to about $\varepsilon = 100$ at frequencies between 20 Hz and 1 MHz [37]. Bellino and colleagues reported values of the dielectric constant for $MgFe_2O_4$ thin films, which decreased from $\varepsilon = 140$ to $\varepsilon = 1$ when the measuring frequency increased from $\nu = 20$ Hz to $\nu = 2$ MHz, respectively [33], whereas others reported values below $\varepsilon = 14$ for $MgAl_2O_4$ [38], $NiFe_{2-x}Cr_xO_4$ [39], $Mg_{0.6}Cu_{0.2}Ni_{0.2}Cr_2O_4$ [39], and $CoFe_2O_4$ thin films [40].

4. Conclusions

To conclude, $ZnCr_2O_4$ nanopowders with a high dielectric constant and well-defined morphology of the constituting nanoparticles have been successfully synthesized by a simple, yet highly reliable solution-based method. The post-synthesis annealing temperature was found to play a crucial role on the morphology of the nanoparticles and, implicitly, the dielectric characteristics of the $ZnCr_2O_4$ nanopowders. Specifically, the annealing temperatures, which resulted in the formation of single-phase $ZnCr_2O_4$ nanopowders, range between 700 °C and 900 °C and the annealing time of 7 h, whereas lower annealing temperatures resulted in the formation of a secondary phase, identified as Cr_2O_3. However, a well-defined octahedral morphology of the nanoparticles is obtained at an annealing temperature of 900 °C for at least 11 h. Overall, this work provides valuable insights into the synthesis, characterization, and possibility for the use of nanostructured spinel chromites in advanced capacitor technologies.

Supplementary Materials: The following supporting information can be downloaded at: https://www.mdpi.com/article/10.3390/mi14091759/s1.

Author Contributions: V.M. and M.M. synthesized and characterized the nanoparticles; G.C., A.G., P.F., I.F., G.-M.R. and A.R. conceived the experiments; V.M. fabricated and characterized the devices and performed scanning probe microscopy; V.M. and A.R. performed Raman and dielectric spectroscopy experiments; G.C., A.G., P.F., I.F., G.-M.R. and A.R. analyzed the results. All authors have read and agreed to the published version of the manuscript.

Funding: This work was supported by the Executive Unit for the Financing of Higher Education, Research, Development and Innovation (Romanian Research Grant (PN-III-P4-ID-PCCF-2016-0175: HighKDevice). The work of Vasyl Mykhailovych was supported by the project "PROINVENT", Contract no. 62487/03.06.2022—POCU/993/6/13—Code 153299, financed by The Human Capital Operational Program 2014–2020 (POCU), Romania. P.F. acknowledges the financial support of Simons Foundation (Award Number: 1030286).

Data Availability Statement: Data are available from the authors on reasonable request.

References

1. Zhao, Q.; Yan, Z.; Chen, C.; Chen, J. Spinels: Controlled Preparation, Oxygen Reduction/Evolution Reaction Application, and Beyond. *Chem. Rev.* **2017**, *117*, 10121–10211. [CrossRef]
2. O'Neill, H.S.C.; Dollase, W.A. Crystal structures and cation distributions in simple spinels from powder XRD structural refinements: $MgCr_2O_4$, $ZnCr_2O_4$, Fe_3O_4 and the temperature dependence of the cation distribution in $ZnAl_2O_4$. *Phys. Chem. Miner.* **1994**, *20*, 541–555. [CrossRef]

3. Wang, Y.L.; An, T.; Yan, N.; Yan, Z.F.; Zhao, B.D.; Zhao, F.Q. Nanochromates MCr_2O_4 (M = Co, Ni, Cu, Zn): Preparation, Characterization, and Catalytic Activity on the Thermal Decomposition of Fine AP and CL-20. *ACS Omega* **2020**, *5*, 327–333. [CrossRef] [PubMed]

4. Arslan, M.T.; Ali, B.; Gilani, S.Z.A.; Hou, Y.L.; Wang, Q.; Cai, D.L.; Wang, Y.; Wei, F. Selective Conversion of Syngas into Tetramethylbenzene via an Aldol-Aromatic Mechanism. *ACS Catal.* **2020**, *10*, 2477–2488. [CrossRef]

5. Dong, S.; Niu, A.; Wang, K.; Hu, P.; Guo, H.; Sun, S.; Luo, Y.; Liu, Q.; Sun, X.; Li, T. Modulation of oxygen vacancy and zero-valent zinc in $ZnCr_2O_4$ nanofibers by enriching zinc for efficient nitrate reduction. *Appl. Catal. B Environ.* **2023**, *33*, 122772. [CrossRef]

6. Song, H.Q.; Laudenschleger, D.; Carey, J.J.; Ruland, H.; Nolan, M.; Muhler, M. Spinel-Structured $ZnCr_2O_4$ with Excess Zn Is the Active ZnO/Cr_2O_3 Catalyst for High-Temperature Methanol Synthesis. *ACS Catal.* **2017**, *7*, 7610–7622. [CrossRef]

7. Sonkusare, V.N.; Chaudhary, R.G.; Bhusari, G.S.; Mondal, A.; Potbhare, A.K.; Mishra, R.K.; Juneja, H.D.; Abdala, A.A. Mesoporous Octahedron-Shaped Tricobalt Tetroxide Nanoparticles for Photocatalytic Degradation of Toxic Dyes. *ACS Omega* **2020**, *5*, 7823–7835. [CrossRef]

8. Xiong, R.; Zhou, X.; Chen, K.; Xiao, Y.; Cheng, B.; Lei, S. Oxygen-Defect-Mediated $ZnCr_2O_4/ZnIn_2S_4$ Z-Scheme Heterojunction as Photocatalyst for Hydrogen Production and Wastewater Remediation. *Inorg. Chem.* **2023**, *62*, 3646–3659. [CrossRef]

9. Kavasoglu, N.; Kavasoglu, A.S.; Bayhan, M. Comparative study of $ZnCr_2O_4$-K_2CrO_4 ceramic humidity sensor using computer controlled humidity measurement set-up. *Sens. Actuators A-Phys.* **2006**, *126*, 355–361. [CrossRef]

10. Liang, Y.C.; Cheng, Y.R.; Hsia, H.Y.; Chung, C.C. Fabrication and reducing gas detection characterization of highly-crystalline p-type zinc chromite oxide thin film. *Appl. Surf. Sci.* **2016**, *364*, 837–842. [CrossRef]

11. Liang, Y.C.; Hsia, H.Y.; Cheng, Y.R.; Lee, C.M.; Liu, S.L.; Lin, T.Y.; Chung, C.C. Crystalline quality-dependent gas detection behaviors of zinc oxide-zinc chromite p-n heterostructures. *CrystEngComm* **2015**, *17*, 4190–4199. [CrossRef]

12. Guo, R.; Shang, X.; Shao, C.; Wang, X.; Yan, X.; Yang, Q.; Lai, X. Ordered large-pore mesoporous $ZnCr_2O_4$ with ultrathin crystalline frameworks for highly sensitive and selective detection of ppb-level p-xylene. *Sens. Actuators B Chem.* **2022**, *365*, 131964. [CrossRef]

13. Yu, G.L.; Wang, W.R.; Jiang, C. Linear tunable NIR emission via selective doping of Ni^{2+}ion into ZnX_2O_4 (X = Al, Ga, Cr) spinel matrix. *Ceram. Int.* **2021**, *47*, 17678–17683. [CrossRef]

14. Dixit, T.; Agrawal, J.; Muralidhar, M.; Murakami, M.; Ganapathi, K.L.; Singh, V.; Rao, M.S.R. Exciton Lasing in ZnO-$ZnCr_2O_4$ Nanowalls. *IEEE Photonics J.* **2019**, *11*, 4501307. [CrossRef]

15. Dixit, T.; Agrawal, J.; Ganapathi, K.L.; Singh, V.; Rao, M.S.R. High-Performance Broadband Photo-Detection in Solution-Processed ZnO-$ZnCr_2O_4$ Nanowalls. *IEEE Electron. Device Lett.* **2019**, *40*, 1143–1146. [CrossRef]

16. Peng, C.; Gao, L. Optical and Photocatalytic Properties of Spinel $ZnCr_2O_4$Nanoparticles Synthesized by a Hydrothermal Route. *J. Am. Ceram. Soc.* **2008**, *91*, 2388–2390. [CrossRef]

17. Gao, H.J.; Wang, S.F.; Fang, L.M.; Sun, G.A.; Chen, X.P.; Tang, S.N.; Yang, H.; Sun, G.Z.; Li, D.F. Nanostructured spinel-type $M(M = Mg, Co, Zn)Cr_2O_4$ oxides: Novel adsorbents for aqueous Congo red removal. *Mater. Today Chem.* **2021**, *22*, 100593. [CrossRef]

18. Kagomiya, I.; Kohn, K.; Toki, M.; Hata, Y.; Kita, E. Dielectric anomaly of $ZnCr_2O_4$ at antiferromagnetic transition. *J. Phys. Soc. Jpn.* **2002**, *71*, 916–921. [CrossRef]

19. Javed, M.; Khan, A.A.; Khan, M.N.; Kazmi, J.; Mohamed, M.A. Investigation on Non-Debye type relaxation and polaronic conduction mechanism in $ZnCr_2O_4$ ternary spinel oxide. *Mater. Sci. Eng. B-Adv. Funct. Solid-State Mater.* **2021**, *269*, 115168. [CrossRef]

20. Naz, S.; Durrani, S.K.; Mehmood, M.; Nadeem, M. Hydrothermal synthesis, structural and impedance studies of nanocrystalline zinc chromite spinel oxide material. *J. Saudi Chem. Soc.* **2016**, *20*, 585–593. [CrossRef]

21. Dey, J.K.; Majumdar, S.; Giri, S. Coexisting exchange bias effect and ferroelectricity in geometrically frustrated $ZnCr_2O_4$. *J. Phys.-Condens. Matter* **2018**, *30*, 235801. [CrossRef]

22. Shafqat, M.B.; Ali, M.; Atiq, S.; Ramay, S.M.; Shaikh, H.M.; Naseem, S. Structural, morphological and dielectric investigation of spinel chromite (XCr_2O_4, X = Zn, Mn, Cu & Fe) nanoparticles. *J. Mater. Sci.-Mater. Electron.* **2019**, *30*, 17623–17629. [CrossRef]

23. Binks, D.J.; Grimes, R.W.; Rohl, A.L.; Gay, D.H. Morphology and structure of $ZnCr_2O_4$ spinel crystallites. *J. Mater. Sci.* **1996**, *31*, 1151–1156. [CrossRef]

24. Xiao, X.; Liu, X.; Zhao, H.; Chen, D.; Liu, F.; Xiang, J.; Li, Y. Facile shape control of Co_3O_4 and the effect of the crystal plane on electrochemical performance. *Adv. Mater.* **2012**, *24*, 5762–5766. [CrossRef] [PubMed]

25. Mancic, L.; Marinkovic, Z.; Vulic, P.; Moral, C.; Milosevic, O. Morphology, structure and nonstoichiometry of $ZnCr_2O_4$ nanophased powder. *Sensors* **2003**, *3*, 415–423. [CrossRef]

26. D'Ippolito, V.; Andreozzi, G.B.; Bersani, D.; Lottici, P.P. Raman fingerprint of chromate, aluminate and ferrite spinels. *J. Raman Spectrosc.* **2015**, *46*, 1255–1264. [CrossRef]

27. Wang, Z.W.; Lazor, P.; Saxena, S.K.; Artioli, G. High-pressure Raman spectroscopic study of spinel ($ZnCr_2O_4$). *J. Solid State Chem.* **2002**, *165*, 165–170. [CrossRef]

28. Marinković Stanojević, Z.V.; Romčević, N.; Stojanović, B. Spectroscopic study of spinel $ZnCr_2O_4$ obtained from mechanically activated ZnO–Cr_2O_3 mixtures. *J. Eur. Ceram. Soc.* **2007**, *27*, 903–907. [CrossRef]

29. Kant, C.; Deisenhofer, J.; Rudolf, T.; Mayr, F.; Schrettle, F.; Loidl, A.; Gnezdilov, V.; Wulferding, D.; Lemmens, P.; Tsurkan, V. Optical phonons, spin correlations, and spin-phonon coupling in the frustrated pyrochlore magnets $CdCr_2O_4$ and $ZnCr_2O_4$. *Phys. Rev. B* **2009**, *80*, 214417. [CrossRef]

30. Xie, D.J.; Luo, Q.Y.; Zhou, S.; Zu, M.; Cheng, H.F. One-step preparation of Cr_2O_3-based inks with long-term dispersion stability for inkjet applications. *Nanoscale Adv.* **2021**, *3*, 6048–6055. [CrossRef]

31. Zuo, J.; Xu, C.; Hou, B.; Wang, C.; Xie, Y.; Qian, Y. Raman Spectra of Nanophase Cr_2O_3. *Raman Spectrosc.* **1996**, *27*, 921–923. [CrossRef]

32. Yang, J. Structural analysis of perovskite LaCr1—xNixO3 by Rietveld refinement of X-ray powder diffraction data. *Acta Crystallogr. Sect. B* **2008**, *64*, 281–286. [CrossRef]

33. Nazli, H.; Ijaz, W.; Kayani, Z.N.; Razi, A.; Riaz, S.; Naseem, S. In-Situ oxidation time dependent structural, magnetic and dielectric properties of electrodeposited magnesium-iron-oxide thin films. *Mater. Today Commun.* **2023**, *35*, 106045. [CrossRef]

34. Behera, B.; Nayak, P.; Choudhary, R.N.P. Structural and impedance properties of $KBa_2V_5O_{15}$ ceramics. *Mater. Res. Bull.* **2008**, *43*, 401–410. [CrossRef]

35. Azam, M.; Riaz, S.; Akbar, A.; Naseem, S. Structural, magnetic and dielectric properties of spinel $MgFe_2O_4$ by sol–gel route. *J. Sol-Gel Sci. Technol.* **2015**, *74*, 340–351. [CrossRef]

36. Batoo, K.M.; Mir, F.A.; Abd El-sadek, M.S.; Shahabuddin, M.; Ahmed, N. Extraordinary high dielectric constant, electrical and magnetic properties of ferrite nanoparticles at room temperature. *J. Nanoparticle Res.* **2013**, *15*, 2067. [CrossRef]

37. Bagade, A.A.; Rajpure, K.Y. Development of $CoFe_2O_4$ thin films for nitrogen dioxide sensing at moderate operating temperature. *J. Alloys Compd.* **2016**, *657*, 414–421. [CrossRef]

38. Nazli, H.; Anjum, R.; Iqbal, F.; Awan, A.; Riaz, S.; Kayani, Z.N.; Naseem, S. Magneto-dielectric properties of in-situ oxidized magnesium-aluminium spinel thin films using electrodeposition. *Ceram. Int.* **2020**, *46*, 8588–8600. [CrossRef]

39. Chavan, A.R.; Somvanshi, S.B.; Khirade, P.P.; Jadhav, K.M. Influence of trivalent Cr ion substitution on the physicochemical, optical, electrical, and dielectric properties of sprayed $NiFe_2O_4$ spinel-magnetic thin films. *RSC Adv.* **2020**, *10*, 25143–25154. [CrossRef]

40. Gutiérrez, D.; Foerster, M.; Fina, I.; Fontcuberta, J.; Fritsch, D.; Ederer, C. Dielectric response of epitaxially strained $CoFe_2O_4$ spinel thin films. *Phys. Rev. B* **2012**, *86*, 125309. [CrossRef]

 *micromachines*

Article

A Highly Efficient Electromagnetic Wave Absorption System with Graphene Embedded in Hybrid Perovskite

Haitao Yu [1], Hui Liu [2], Yao Yao [3], Ziming Xiong [3], Lei Gao [4], Zhiqian Yang [3], Wenke Zhou [3,5,*] and Zhi Zhang [4,*]

[1] Field Engineering College, Army Engineering University of PLA, Nanjing 210007, China
[2] Unit of 32399 of PLA, Nanjing 211131, China
[3] State Key Laboratory for Disaster Prevention & Mitigation of Explosion & Impact, Army Engineering University of PLA, Nanjing 210007, China
[4] Position Engineering Research Office, Army Engineering University of PLA, Nanjing 210007, China
[5] Electromagnetic Environmental Effects Laboratory, Army Engineering University of PLA, Nanjing 210007, China
* Correspondence: zhou.w.k@163.com (W.Z.); zhangnjjn@163.com (Z.Z.)

Abstract: To cope with the explosive increase in electromagnetic radiation intensity caused by the widespread use of electronic information equipment, high-performance electromagnetic wave (EMW)-absorbing materials that can adapt to various frequency bands of EMW are also facing great demand. In this paper, $CH_3NH_3PbI_3$/graphene (MG) high-performance EMW-absorbing materials were innovatively synthesized by taking organic–inorganic hybrid perovskite (OIHP) with high equilibrium holes, electron mobility, and accessible synthesis as the main body, graphene as the intergranular component, and adjusting the component ratio. When the component ratio was 16:1, the thickness of the absorber was 1.87 mm, and MG's effective EMW absorption width reached 6.04 GHz (11.96–18.00 GHz), achieving complete coverage of the Ku frequency band. As the main body of the composite, $CH_3NH_3PbI_3$ played the role of the polarization density center, and the defects and vacancies in the crystal significantly increased the polarization loss intensity; graphene, as a typical two-dimensional material distributed in the crystal gap, built an efficient electron transfer channel, which significantly improved the electrical conductivity loss strength. This work effectively broadened the EMW absorption frequency band of OIHP and promoted the research process of new EMW-absorbing materials based on OIPH.

Keywords: organic–inorganic hybrid perovskite; graphene; electromagnetic absorption; conductive network

Citation: Yu, H.; Liu, H.; Yao, Y.; Xiong, Z.; Gao, L.; Yang, Z.; Zhou, W.; Zhang, Z. A Highly Efficient Electromagnetic Wave Absorption System with Graphene Embedded in Hybrid Perovskite. *Micromachines* **2023**, *14*, 1611. https://doi.org/10.3390/mi14081611

Academic Editor: Amir Hussain Idrisi

Received: 29 July 2023
Revised: 11 August 2023
Accepted: 14 August 2023
Published: 16 August 2023

1. Introduction

With the extensive use of electronic equipment in human life, the electronic equipment used in various frequency bands is increasingly abundant [1–3]. In order to deal with the electromagnetic interference and radiation caused by electronic equipment with different electromagnetic frequency bands, it is more urgent to study new EMW-absorbing materials with wide influential electromagnetic frequency bands [4–6]. In previous work, it has been proved that organic–inorganic hybrid perovskite (OIHP) is a potential new EMW absorbing material [7–10]. Still, it has some problems, such as a high doping ratio, thick absorbing body thickness, narrow adequate EMW absorption frequency bandwidth, and low EMW absorption intensity [11]. These problems limit the further application of it in the field of EMW-absorbing materials.

As a typical two-dimensional carbon material, graphene has received much attention and applications in the fields of new energy, photoelectric devices, and high-resistance materials due to its extremely low resistivity, high light transmittance, and excellent mechanical properties [12–15]. Graphene also attracted wide attention in EMW absorption due to its outstanding dielectric properties [16]. Xiaowei Yin et al. fabricated graphene/ZnO composites [17], and Man He et al. designed $TiO_2/Ti_3C_2T_x/RGO$ composites [14], achieving good EMW absorption effects. Meanwhile, due to its unique two-dimensional structure, graphene has a high specific surface area, which helps to improve the interface contact between graphene and other components, thus enhancing the interface polarization effect. Jianping He et al. significantly improved the polarization strength and impedance matching performance of Fe_3O_4/graphene composites by introducing the laminated graphene structure between Fe_3O_4 nanoparticles. The improved polarization loss strength improves the absorption strength of Fe_3O_4/graphene, and the improved impedance matching performance broadens the effective EMW absorption frequency bandwidth, which made the composites achieve better EMW absorption performance [18].

These studies proved that, by adjusting the type and proportion of components, the impedance matching characteristics of composites can be effectively adjusted while maintaining a high level of attenuation loss characteristics, and the effective EMW absorption bandwidth of absorbing material can be significantly widened [19–22]. In addition, by introducing carbon materials, an efficient conductive network can be formed inside the composites, which also promotes the reduction of the amount of absorbing materials [23–26]. Therefore, by introducing graphene into OIHP materials, a new type of OIHP/graphene-composite-absorbing material with OIHP as the primary material and graphene as the supplement is expected to reduce the addition amount and realize broadband absorption of EMW.

In this paper, $CH_3NH_3PbI_3$/graphene composites (MPI/graphene) were prepared using an anti-solvent method. Firstly, the physical phase properties and microstructure of MPI/graphene composites were characterized, and the effect of graphene on the crystallinity and crystalline phase of MPI was investigated. Then, the optimal EMW absorption properties of MPI/graphene composites were determined by adjusting the component ratios of MPI and graphene. Finally, the mechanism of EMW absorption performance of MPI/graphene composites was analyzed to accumulate relevant experience for the composite of MPI and two-dimensional high dielectric materials.

2. Materials and Methods

2.1. Materials

Graphene was obtained from Nanjing Xianfeng Nano Co., Ltd. (Nanjing, China), while the particle size of graphene was 1–5 μm, the thickness was 1–5 nm, the specific surface area was more significant than 100 m^2/g, and the electrical conductivity was 1000–1500 S/cm. Lead iodide (PbI_2, 99%), methylamine iodine (MAI, 99%), Gamma-butyrolactone (γ-GBL, 99%), and anisole (C_7H_8O, 99.9%) were supplied by Advanced Election Technology Co., Ltd. (Dalian, China).

2.2. Synthesis of MPI/Graphene Composites

The anti-solvent method prepared MPI/graphene composites (MG) [8]. Firstly, the MAI powder and PbI_2 powder were dissolved in γ-GBL solution and stirred continuously at 80 °C for 1 h to prepare 0.8 mol/mL of MPI precursor solution. Graphene powder was dispersed in γ-GBL solution at 10 mg/mL concentration for 3 h at 25 °C with continuous stirring. Then, the MPI precursor solution was mixed with the graphene suspension in a specific ratio and stirred continuously at 25 °C for 1 h. The mixing ratio of MPI precursor solution and graphene suspension is shown in Table 1. After the reaction, the mixture of MPI/graphene solution was quickly added to the excessive anisole solution dropwise, and a large number of black suspensions could be observed during the reaction process. After the drip process, the anisole solution was placed in a nitrogen atmosphere for 12 h. Finally, the underlying precipitate in the anisole solution was washed several times by centrifugation with isopropanol and hexane, and the product was dried in a vacuum at 80 °C for 24 h. The resulting black sample is the MPI/graphene composite.

Table 1. The mixing ratio of MPI precursor solution and graphene suspension.

MPI Precursor Solution (mL)	Graphene Suspension (mL)	MPI/Graphene	Sample Number
0.806	2.4	24:1	MG-1
0.806	3.75	16:1	MG-2
0.806	4.95	12:1	MG-3
0.806	7.5	8:1	MG-4
0.806	9.9	6:1	MG-5

2.3. Characterization

The X-ray diffraction (XRD) was carried out with Cu Kα radiation (λ = 1.5406 Å) at 40 kV, and SEM images were obtained in an S-4800 (Hitachi, Tokyo, Japan) machine. Raman spectroscopy (Raman) and steady-state photoluminescence (PL) spectrums were recorded with a Renishaw InVia Basis Raman Spectrometer (Renishaw, London, UK) and an FLS980 Series of Fluorescence Spectrometers, respectively. A precision LCR meter (TH2829C) and current density tester (SDM-200) were used to research the electrical conductivity (EC) of samples. The crystal features were tested by X-ray photoelectron spectroscopy (250XI, Thermo Scientific, Berlin, German). The electromagnetic properties of the pieces were measured by the Agilent PNA N5244A vector network analyzer with the coaxial probe method [27,28], and the electromagnetic parameters of the sample in the range of 2–18 GHz were obtained, including the real (ε' and μ') and imaginary (ε'' and μ'') parts of permittivity and permeability. To measure the electromagnetic properties, the samples were uniformly mixed with paraffin in a mass ratio of 2:3 and pressed into coaxial rings, which were toroidal shapes with an outer diameter of 7 mm and an inner diameter of 3.04 mm.

3. Results and Discussion

X-ray diffraction analysis technology analyzed MG composites' phase and crystallinity. For comparison purposes, the XRD patterns of MG composites, MPI crystals, and graphene are all given in Figure 1a. The XRD pattern of MPI crystals shows that the characteristic diffraction peaks of MPI appeared at 2θ = 14.1°, 23.4°, 24.4°, 28.4°, and 31.9°, which corresponds to the lattice plane of (002), (211), (202), (220), and (222), respectively. These diffraction peaks mean that MPI crystals exist in a tetragonal structure [29,30]. For the XRD pattern of the MG-1 sample, five typical characteristic peaks corresponding to the tetragonal phase structure of MPI crystals can be clearly seen, which indicates that MPI in the MG-1 sample still exists in a specific tetragonal phase structure. At this ratio, the addition of graphene did not have a significant impact on the growth crystallization process of MPI. For the XRD pattern of MG-3 samples, with the increase in the proportion of graphene material in MG composites, the characteristic diffraction peaks of corresponding tetragonal-phase MPI crystals disappeared, and the intensity of diffraction peaks at 2θ = 24.4° significantly increased, which indicates that the grain integrity of MPI crystals in MG-3 samples was seriously damaged. There may be many defects and vacancies on the crystal surface. It also can be seen that the characteristic diffraction peaks of MPI crystals cannot be found at the ratio of MG-5 samples. According to the above results, it can be shown that the addition of graphene has an inhibitory effect on the growth and crystallization process of MPI. With the increase in graphene content, the crystallinity of MPI crystals decreased, and the crystals began to appear with vacancies and defects. When graphene was excessive, the crystallization process of MPI was utterly destroyed, and MG composites could not be synthesized. In the XRD patterns of MG samples, the characteristic diffraction peak corresponding to the graphene (2θ = 24.1°) could not be seen. This is because the crystallinity of the graphene material is much lower than that of MPI crystals. Therefore, to analyze the existence form and state of the graphene samples in MG materials, Raman spectra were introduced to further study MG composites.

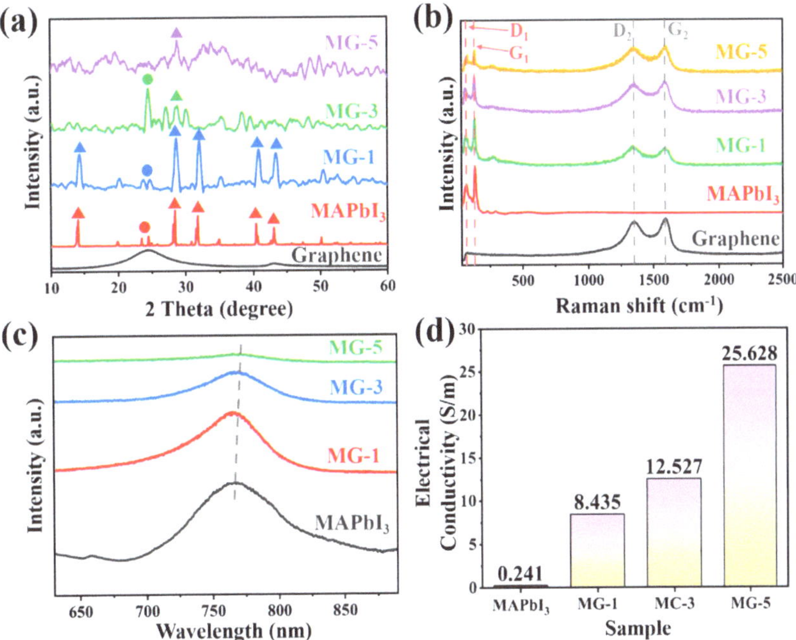

Figure 1. The XRD patterns (**a**), Raman spectrum (**b**), photoluminescence (PL) spectrum (**c**), and electrical conductivity diagram (**d**) of MG composites, MPI crystals, and graphene.

Figure 1b shows the Raman spectra of MG composites, MPI crystals, and graphene. Two distinct sets of absorption peaks can be seen in the Raman spectrum of the MG composites. The D_1 and G_1 peaks at 73.2 cm^{-1} and 135.5 cm^{-1} correspond to the characteristic absorption peaks of MPI crystals, while the D_2 and G_2 peaks at 1350.3 cm^{-1} and 1589.4 cm^{-1} correspond to the characteristic absorption peaks of graphene [31–33]. The presence of two sets of characteristic peaks demonstrates the existence of MPI and graphene in the MG composites. Meanwhile, the intensity ratio between the D and G peaks also reflects the lattice integrity of the test samples, with an increase in the I_D/I_G ratio, indicating more vacancies and defects in the lattice of that group of test samples. With the increase in graphene content in MG composites, it can be seen that the ratio of I_{D1}/I_{G1} increased from 0.61 of MG-1 to 0.85 of MG-3 and then to 0.95 of MG-5. This indicates that with the increase in graphene content, the lattice integrity of MPI in MG composites significantly decreased, which is consistent with the XRD analysis. Compared with the obvious change of I_{D1}/I_{G1} values, I_{D2}/I_{G2} values remained stable in all proportions (MG-1 = 0.96, MG-3 = 0.94, MG-5 = 0.96). Therefore, it can be inferred that graphene not only existed statically in MG composites but also did not react with MPI crystals without any change in lattice state.

The steady-state photoluminescence spectra of MG composites are shown in Figure 1c. As a kind of fluorescent material, hybrid perovskite has a very typical characteristic fluorescence peak. As an essential method to analyze the crystallinity of perovskite materials, the PL spectrum was also used to analyze the lattice structure of MPI crystals in MG composites. According to Figure 1c, the intensity of the characteristic fluorescence peak of the MG composites decreased, and the half-peak width narrowed as the graphene content in the MG increased, indicating an increase in the number of vacancies and defects within the MPI crystals. At the same time, a slight rightward shift of the characteristic fluorescence peak can also be observed, suggesting that the disruption of the MPI crystal structure leads to an increase in its band gap, which may result in a shift in the electromagnetic absorption band [34]. Good electrical conductivity is an essential condition for dielectric loss of EMW. From Figure 1d, the electrical conductivity of MPI crystals was low, while introducing graphene material will significantly improve the electrical conductivity of MG composites. This is due to the excellent electrical conductivity of the graphene and the fact that the two-dimensional layer structure of the graphene allows for a larger surface area of contact between the graphene and the MPI, which facilitates the rapid movement of free electrons between the MPI crystals and the MG composites.

The XPS survey spectra of MG composites, MPI crystals, and graphene are shown in Figure 2a. Graphene material is mainly composed of C elements. The C peak at 284.8 eV corresponds to the C–C bond in the aromatic ring structure of graphene. In comparison, the O peak at 533.0 eV indicates a small number of oxygen-containing functional groups on the surface or at the edges of the graphene [35]. As for the MPI crystals, two Pb 4f7 and Pb 4f5 peaks were located at 138.0 eV and 143.0 eV, respectively. The peak at 285.0 eV corresponded to C 1s, and the height at 402.0 eV corresponded to N 1s. The 533.0 eV and 619.0 eV peaks corresponded to O 1s and I 3d, respectively [36,37]. At the same time, the characteristic peaks of the above elements can also be observed in the XPS spectra of MG composites, indicating the stable existence of MPI crystals and graphene in MG composites. However, it should be noted that with the increase in graphene content, except for the characteristic peaks of C and O elements, the intensity of other elemental peaks decreased significantly. On the one hand, due to the influence of graphene, the crystal structure of MPI crystals was destroyed and decomposed. On the other hand, XPS analysis had limited sample penetration, typically to a depth of 5–10 nm. The reduced intensity of the characteristic elemental peaks for Pb and I reflected the increased graphene content wrapped around the outer layers of the MPI crystals, which the electron beam from the XPS analyzer could not accurately detect.

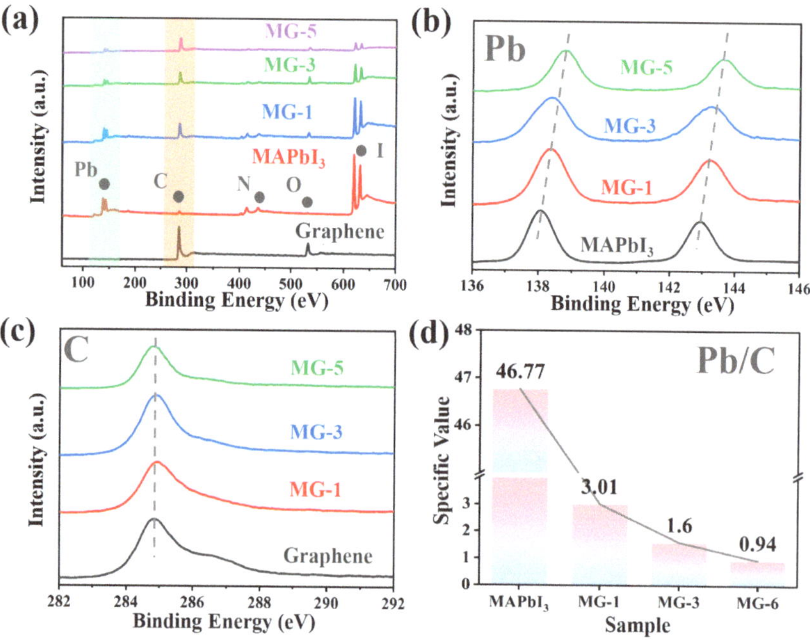

Figure 2. The XPS spectra of MG composites, MPI crystals, and graphene (**a**); the XPS spectra of Pb (**b**) and C (**c**); the Pb/C content ratio of MG composites (**d**).

Figure 2b shows the XPS spectra of the Pb element. It can be seen that with the increase in graphene content, compared with MPI crystals, the characteristic peaks of Pb 4f₇ and

Pb 4f$_5$ in MG composites shifted significantly to the right, and the degree of rightward shift gradually increased. According to Figure 2c, the characteristic peak of element C was more consistently located near 285.0 eV, which is further evidence of the high stability of graphene in MG composites, where the lattice morphology and surface structure were not significantly affected by the MPI crystals. The XPS energy spectrum and date of C1s sub-peak of MG-3 have are provided in Figure S1 and Table S1. As shown in Figure 2d, the Pb/C content ratio of MPI crystal was 46.771, and that of MG composite decreased from 3.012 (MG-1) to 0.936 (MG-5). This proved that the content of graphene in MG composites was steadily increasing. The MG composites with different components were successfully prepared by the anti-solvent method.

Figure 3 shows the SEM images of the MPI crystals, graphene, MG-1, MG-3, and MG-5 composites. In Figure 3d, the individual MPI crystals can be seen as three-dimensional blocks with a size of about 2–3 μm. As shown in Figure 3e, the graphene exhibited a distinctive two-dimensional layered structure with a large specific surface area, a monolayer thickness of less than 5 nm, and layer-to-layer folds [38]. According to Figure 3a–c, in MG-1, it can be seen that the MPI crystals still maintained a three-dimensional massive structure and had an increased grain size compared to the MPI crystals. The graphene material was distributed around the MPI crystals and had less contact area with the MPI crystals. A change in the morphology of the MPI crystals can be seen in MG-2, with a shift from a three-dimensional bulk to a one-dimensional rod structure. At the same time, the MPI micron rod was wrapped by the graphene, which indicates that the contact area of the graphene and MPI crystals significantly increased at this component ratio. No noticeable MPI crystals were observed in MG-3, while the graphene material was abundantly distributed. This confirmed the XPS analysis: the MPI crystals in MG-5 were partially destroyed, and graphene was wholly wrapped around some of the defective MPI crystals. The elemental mapping images of MG-3 are shown in Figure 3f–i. The C, Pb, and I elements were uniformly distributed in MG-3, which means that the MPI crystals and graphene were successfully combined to form a stable composite material. This is consistent with the XRD, Raman, XPS analysis results, and verifies the feasibility of the anti-solvent method for the preparation of MG composites.

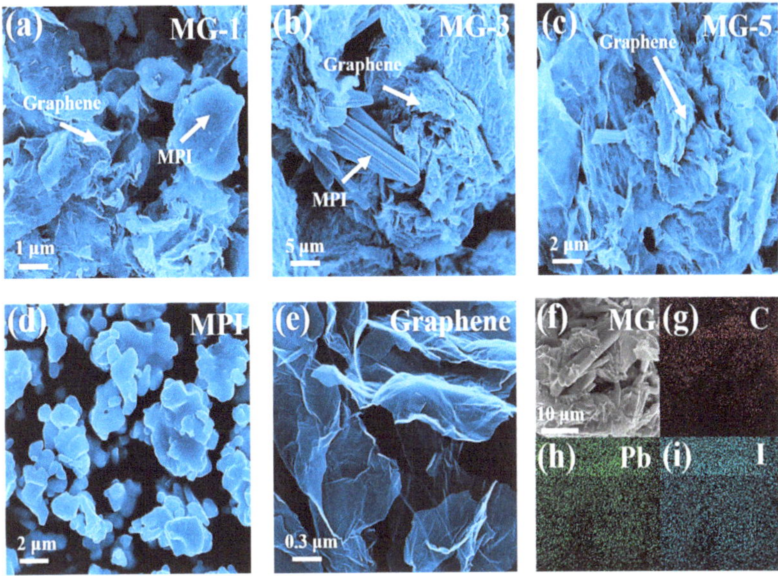

Figure 3. The SEM images of MG-1 (**a**), MG-3 (**b**), MG-5 (**c**), the MPI crystals (**d**), and graphene (**e**); the elemental mapping images of MG-3 (**f**), C (**g**), Pb (**h**), and I (**i**).

To further explore the dissipative ability of composite materials on EMW, the electromagnetic parameters of MPI crystals, graphene, and MG composites within 2–18 GHz were studied under the coaxial-ring doping amount of 40 wt%. The relevant test results are shown in Figure 4 and Figure S2, respectively. The electromagnetic parameters of MG composites with different component proportions are given in Figure 4. On the one hand, with the increase in graphene content in MG composites, the ε' value and ε'' The value of MG composites increased regularly. According to the equation ($\varepsilon'' \approx 1/\pi\varepsilon_0\rho f$) [39], there was a negative correlation between the resistivity (ρ) and ε'', the increase in the ε'' value indicates a decrease in resistivity and an increase in electrical conductivity of the MG samples, suggesting that the addition of graphene enhanced the electrical conductivity loss of the MG composites. On the other hand, the ε' value and ε'' value of MG composites affected by the graphene also showed a trend of decreasing values with the increase in EMW frequency in the range of 2–18 GHz. In comparison, the μ' value of MG composites at each component ratio was approximately 1.08 and the μ'' value was approximately 0, suggesting that MG composites have strong dissipation ability to the electric energy carried by the incident EMW but weak dissipation to the magnetic energy.

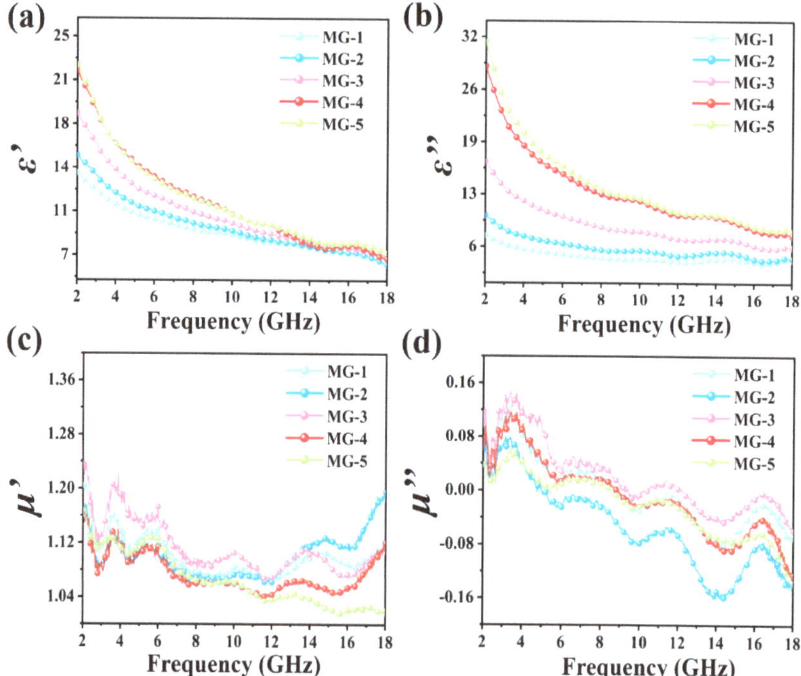

Figure 4. Complex permittivity and permeability of the MG composites: ε' (**a**), ε'' (**b**), μ' (**c**), and μ'' (**d**).

The dielectric loss tangent ($tan\varepsilon_r$) and magnetic loss tangent ($tan\mu_r$) are shown in Figure 5, which can be calculated as follows [40]:

$$tan\varepsilon_r = \frac{\varepsilon''}{\varepsilon'} \tag{1}$$

$$tan\mu_r = \frac{\mu''}{\mu'} \tag{2}$$

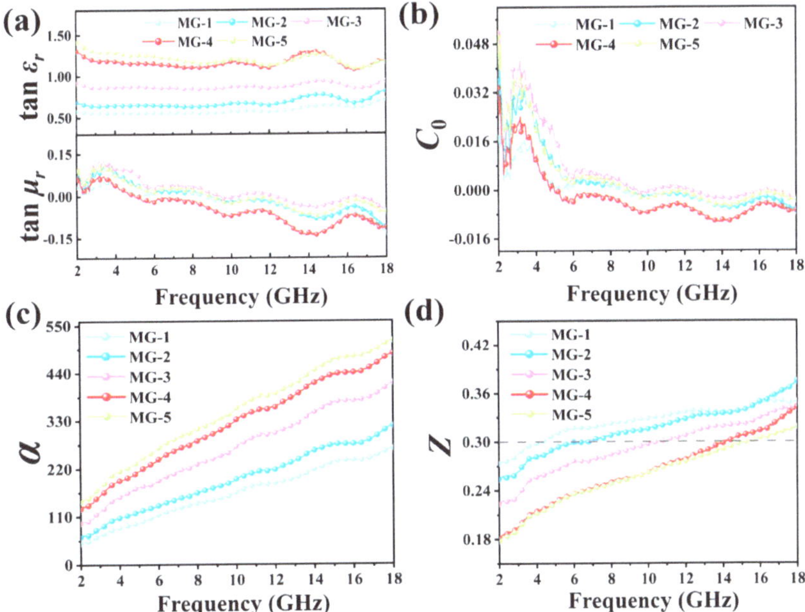

Figure 5. Dielectric loss tangent and magnetic loss tangent (**a**), C_0 (**b**), the attenuation constant (**c**), and intrinsic impedance ratio (**d**) of the MG composites.

The $tan\varepsilon_r$ and $tan\mu_r$ represent the dissipation capacity of a material to electric energy and magnetic energy carried by incident EMW, respectively. In Figure 5a, the addition of graphene improved the dissipation ability of MG composites to electric energy while the dissipation ability of magnetic energy was not significantly affected. An analysis of the magnetic loss properties of MG can be obtained in supporting information. Figure 5b is the eddy current loss coefficient of MG composites. The study of magnetic loss characteristics of MG composites can be obtained from Supporting Information.

The attenuation loss constant (α) reflects the ability of a material to attenuate the EMW entering its interior, and previous studies demonstrated that when the attenuation loss constant is higher than 100, the material can achieve effective attenuation of the incident EMW. The attenuation loss constant can be calculated as follows [41]:

$$\alpha = \frac{\sqrt{2}\pi f}{c} \times \sqrt{(\mu''\varepsilon'' - \mu'\varepsilon') + \sqrt{(\mu''\varepsilon'' - \mu'\varepsilon')^2 + (\mu'\varepsilon'' + \mu''\varepsilon')^2}} \tag{3}$$

The intrinsic impedance ratio Z reflects the difficulty of EMW entering the material's interior. It was found that when the intrinsic impedance ratio of the material is more significant than 0.3, EMW can enter the material's interior in large quantities. The inherent impedance ratio Z can be calculated as follows [42]:

$$Z = Z_r / Z_0 \tag{4}$$

$$Z_r = Z_0 \sqrt{|\mu_r| / |\varepsilon_r|} \tag{5}$$

As shown in Figure 5c, the attenuation loss coefficient of MG composites at each component ratio was relatively high. While the attenuation loss coefficient of some composites at the low-frequency band (2–5 GHz) was lower than 100, the remaining attenuation loss coefficients were all over 100, which indicates that MG composites at each component ratio can achieve good attenuation loss characteristics for incident EMW. In Figure 5d, the impedance matching ratio of MG-1 and MG-2 are relatively high and higher than 0.3 in the 6–18 GHz range. This indicates that only MG-1 and MG-2 composites allow many EMW to enter the interior of the absorbing materials. Graphene content (MG-3, MG-4, MG-5) that is too high will lead to many EMW reflected or refracted on the surface of the absorbing materials, which can not be effectively attenuated. It can be inferred that the absorption performance of MG materials is regulated and affected by the intrinsic impedance ratio, and it is expected that MG-1 and MG-2 composites will show better EMW absorption performance. Combined with the analysis of the attenuation loss coefficient and intrinsic impedance ratio of MG samples, the ability of MG materials to attenuate and lose EMW gradually increased as the content of graphene increased, and the difficulty of EMW being able to enter inside the absorbing material gradually increased. This shows that adding graphene improved the polarization loss strength and conductance loss strength inside MG materials. And the MPI crystal, as the center of electric polarization composite with graphene, can also effectively adjust the impedance matching of the composite material, which is conducive to improving the absorption performance.

In order to analyze the polarization relaxation phenomenon of the dielectric loss-absorbing materials, the Debye relaxation process of MG composites with different component ratios is described by the Cole–Cole diagram. Cole–Cole diagrams are calculated and plotted according to the following Debye relaxation formula [41]:

$$\left(\varepsilon' - \frac{\varepsilon_s + \varepsilon_\infty}{2}\right)^2 + (\varepsilon'')^2 = \left(\frac{\varepsilon_s - \varepsilon_\infty}{2}\right)^2 \tag{6}$$

where ε_s is the static dielectric constant and ε_∞ is the relative dielectric constant of the high-frequency limit. A semicircle equation with $(\varepsilon_s - \varepsilon_\infty)/2$ as the radius and $(0, (\varepsilon_s + \varepsilon_\infty)/2)$ as the circle's center can be established from the Debye relaxation equation, and the resulting circle is known as a Cole–Cole semicircle. Usually, each semicircle represents a relaxation process within the material, and the number of semicircles represents the intensity of the relaxation polarization. Figure 6 shows Cole–Cole semicircular graphs of MG composites with different component ratios, and these Cole–Cole graphs show similar trends. In the high-frequency band, the five samples all have two semi-circles, but with the increase in graphene content, the circle's radius gradually decreases. On the one hand, this phenomenon shows that MG samples in the high-frequency band will produce pronounced relaxation polarization under the electromagnetic field, which is conducive to improving the dielectric loss strength of materials. On the other hand, it also shows that excessive graphene will cause the relaxation polarization intensity to decrease. When the radius of the Cole–Cole circle decreases and is close to a straight line, the electrical conductivity loss dominates. This conclusion is also applicable to the positive correlation between the real and imaginary parts of the dielectric constant of MG samples in the low-frequency band, indicating that the electrical conductivity loss of MG samples is the dominant form of low-frequency EMW loss, with the Debye relaxation polarization making a more minor contribution to the dielectric loss intensity. The irregularity of the Cole–Cole semicircle suggests that high-frequency EMW will lead to Debye relaxation polarization and multiple polarization loss processes, including interface polarization, electron polarization, and dipole polarization.

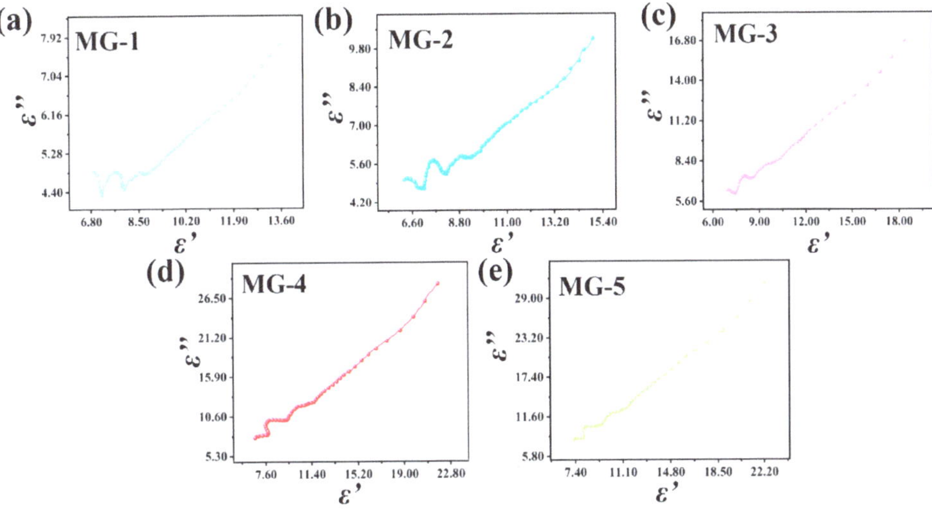

Figure 6. Cole–Cole plots of MG composites: MG-1 (**a**), MG-2 (**b**), MG-3 (**c**), MG-4 (**d**), MG-5 (**e**).

Based on the transmission line theory, the value of reflection loss (RL) can be calculated by the following equations [43,44]:

$$Z_{in} = Z_0 \sqrt{\frac{\mu_r}{\varepsilon_r}} \tanh \left(j \frac{2\pi f d}{c} \sqrt{\mu_r \varepsilon_r} \right) \tag{7}$$

$$RL(dB) = 20 \lg \left| \frac{Z_{in} - Z_0}{Z_{in} + Z_0} \right| \tag{8}$$

where Z_{in} is the normalized input impedance corresponding to the thickness of the absorber, d is the thickness of the absorber, and h is the Planck constant. Figure 7 shows the RL of MG samples in the 2–18 GHz range when the doping amount is 40 wt%, and the thickness is 1 mm to 5 mm. In the figure, the area in the red line indicates that the RL is less than −10 dB, and the yellow dotted line indicates that the RL is less than −20 dB. The area where the RL is less than −10 dB is usually called the effective electromagnetic absorption band (EAB) of the absorbing material. According to Figure 7, the overall EMW absorption performance of MG materials shows a decreasing trend as the graphene content increases. Figure 7f shows the EAB of MG-1, MG-2, and MG-3 composites with corresponding absorption thicknesses. The EAB of MG-2 sample reaches 6.04 GHz (11.96–18.00 GHz) at a 1.87 mm absorber thickness, which is wider than MG-1 and MG-3. This is consistent with the conclusion obtained from the analysis of the attenuation loss coefficient and intrinsic impedance ratio. MG-3 sample shows better absorbing performance in the high-frequency range with the EAB of 4.08 GHz at 1.62 mm thickness. For MG-4 and MG-5, there is no significant EAB in the 2–18 GHz range due to the large amount of graphene resulting in samples with intrinsic impedance ratios well below 0.3 and a large amount of EM waves being reflected at the absorber surface. Combined with the characterization results of MG materials, it can be seen that when the crystallinity of MPI crystals in the material is very low, the incident EMW cannot be effectively absorbed by polarization. MG's impedance matching performance (IMP) can be obtained in supporting information.

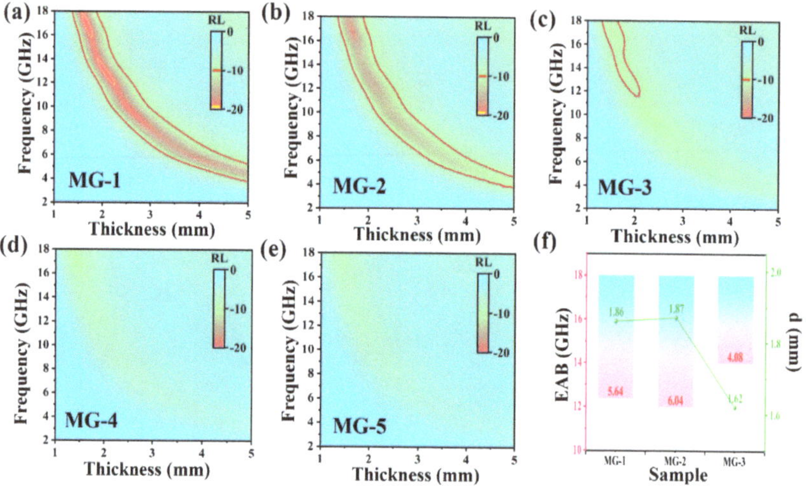

Figure 7. Reflection loss curves of MG-1 (**a**), MG-2 (**b**), MG-3 (**c**), MG-4 (**d**), and MG-5 (**e**); the EAB of MG composites with corresponding absorption thicknesses (**f**).

EAB and EMW absorption intensity are two key indexes to evaluate the properties of absorbing materials. Based on the outstanding EAB properties of MG-1, MG-2, and MG-3 samples, the absorption intensity of the three samples is further studied in Figure 8. The correlation between maximum reflection loss (RL), normalized input impedance (Z_d), and the absorption thickness (T_m) is shown in Figure 8. The absorption thickness (T_m) can be calculated by [39]:

$$T_m = \frac{n\lambda}{4} = \frac{nc}{\left(4f_m\sqrt{|\mu_r||\varepsilon_r|}\right)}, (n = 1,3,5,\ldots)$$

(9)

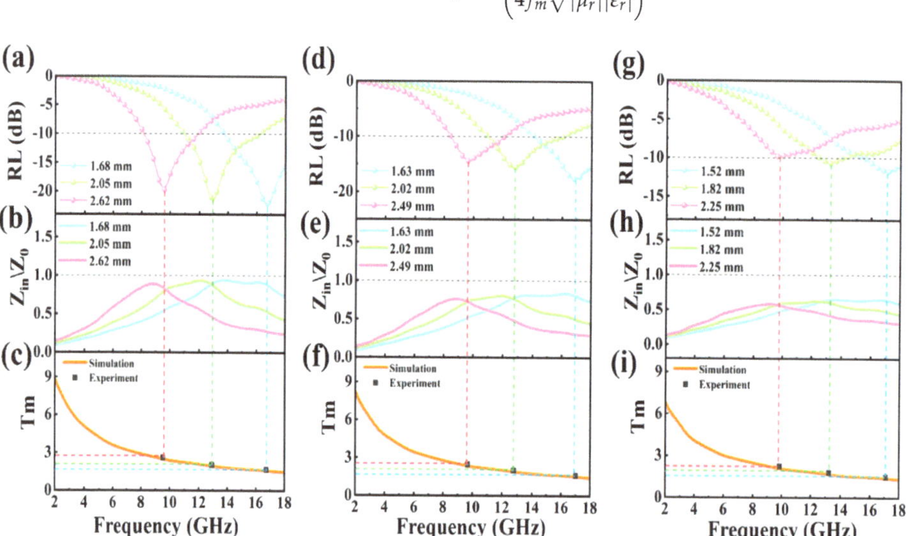

Figure 8. MG-1: the correlation between the maximum RL (**a**), normalized input impedance (**b**), and the absorption thickness (**c**); MG-2: the correlation between the maximum RL (**d**), normalized input impedance (**e**), and the absorption thickness (**f**); MG-3: the correlation between the maximum RL (**g**), normalized input impedance (**h**), and the absorption thickness (**i**).

MG-1's maximum RL value of -22.689 dB at 16.72 GHz is found at a thickness of 1.68 mm. The second largest RL value of -21.878 dB appears at 12.92 GHz, corresponding to a thickness of 2.05 mm. When the absorption thickness is 2.62 mm, the RL value of MG-1 at 19.52 GHz is -19.929 dB. Combined with Figure 8b, it can be found that when the normalized input impedance value at the peak reflection loss frequency is closer to 1.0, the RL intensity of the absorbing material is higher. This shows that good impedance matching is beneficial to enhance the EMW absorption performance of the material. In addition, as the thickness of the material increases, the maximum RL peak of the MG-1 material moves towards lower frequencies, which the quarter-wavelength theory can explain. The orange line in Figure 8c represents the t_m simulation curve of MG-1 material, and the gray box represents the thickness of the absorber corresponding to the maximum reflection loss. The grey squares are evenly distributed around the simulation curve, indicating that the absorption mechanism of MG-1 material conforms to the quarter-wavelength theory. The conclusions in the MG-1 material can also explain the relevant phenomena in the MG-2 and MG-3 materials. The EMW absorption properties of some MG samples are given in Table 2. Collectively, the MG-1 and MG-2 samples exhibited superior EMW absorption properties.

Table 2. EMW absorption properties of MG samples.

Sample	Point	$|RL|$ (dB)	d (mm)	f (GHz)
	m_1	22.69	1.68	16.72
MG-1	m_2	21.88	2.05	12.92
	m_3	19.93	2.62	9.52
	m_1	17.68	1.63	17.00
MG-2	m_2	15.74	2.02	12.76
	m_3	14.53	2.49	9.68
	m_1	11.75	1.52	17.08
MG-3	m_2	10.78	1.82	13.20
	m_3	9.97	2.25	9.84

The EMW absorption properties of MG materials result from the combined effect of MPI crystals and graphene, which improves the impedance matching of MG samples by adjusting the ratio of the two components. As a typical dielectric loss absorbing material, graphene wraps and connects several MPI crystals in series, which creates conditions for free electrons to move freely between the MPI crystals, forming a large number of field-induced microcurrents inside the MG materials under an alternating electric field, which effectively converts electromagnetic energy into heat by electrical conductivity losses. In addition, the polarization loss effect is sufficiently enhanced to improve the EMW absorption performance of MG composites substantially: Firstly, the addition of graphene disrupts the crystal structural integrity of the MPI, generating a large number of surface defects and vacancies and enhancing the space charge polarization effect of the MG materials significantly under the action of external electric field. Secondly, because the MPI crystals and graphene are different in dielectric constant and electrical conductivity, this will lead to charge accumulation at the interface of the two dielectric materials, resulting in an interface polarization effect.

4. Conclusions

In this paper, MPI crystals and graphene were successfully compounded by an anti-solvent method to fabricate MG composites with different component proportions. The phase composition, lattice morphology, and micro-morphology of MG materials with different proportions were analyzed by XRD, Raman, PL, and SEM. The EMW absorption performance of MG materials was fully investigated in terms of the type of EMW-absorbing materials, absorption mechanism, effective EMW absorption bandwidth, and reflection loss strength. It was found that for MG composites, MG materials at MG-1 and MG-2 ratios exhibited better EMW absorption performance than other ratios. This work demonstrates that introducing two-dimensional carbon materials plays an important role in broadening the effective EMW absorption band of MPI crystals and reducing the amount of EMW-absorbing materials added within the absorber.

Supplementary Materials: The following supporting information can be downloaded at: https://www.mdpi.com/article/10.3390/mi14081611/s1, Figure S1. XPS energy spectrum of C1s sub peak of MG-3; Figure S2. Complex permittivity and permeability of the MPI crystals and graphene: ε' (a), ε'' (b), μ' (c), and μ'' (d); Figure S3. The impedance matching performance of the MG samples: MG-1 (a), MG-2 (b), MG-3 (c), MG-4 (d), and MG-5 (e). Table S1. The date of C1s sub peak of MG-3. References [4,45] are cited in the supplementary materials.

Author Contributions: Conceptualization, Z.Z. and Y.Y.; methodology, H.Y. and H.L.; software, L.G.; validation, Z.X., H.L. and Z.Y.; formal analysis, H.Y.; investigation, Z.Z.; resources, W.Z.; writing—original draft preparation, H.Y.; writing—review and editing, Z.Z.; visualization, W.Z. All authors have read and agreed to the published version of the manuscript.

Funding: This research was funded by the Nature Science Foundation of Jiangsu Province (No. BK20221055), the Chinese Postdoctoral Science Fund (No. 2020M683756), and by the State Key Laboratory of Disaster Prevention & Mitigation of Explosion & Impact (No. LGD-SKL-202203).

Data Availability Statement: Data are contained within the article or supplementary material. The data presented in this study are available in article and supplementary material.

Conflicts of Interest: The authors declare no conflict of interest.

References

1. Zeng, X.; Zhao, C.; Jiang, X.; Yu, R.; Che, R. Functional Tailoring of Multi-Dimensional Pure MXene Nanostructures for Significantly Accelerated Electromagnetic Wave Absorption. *Small* **2023**, e2303393. [CrossRef] [PubMed]
2. Wang, Y.; Yao, L.; Zheng, Q.; Cao, M.-S. Graphene-wrapped multiloculated nickel ferrite: A highly efficient electromagnetic attenuation material for microwave absorbing and green shielding. *Nano Res.* **2022**, *15*, 6751–6760. [CrossRef]
3. Cao, M.; Han, C.; Wang, X.; Zhang, M.; Zhang, Y.; Shu, J.; Yang, H.; Fang, X.; Yuan, J. Graphene nanohybrids: Excellent electromagnetic properties for the absorbing and shielding of electromagnetic waves. *J. Mater. Chem. C* **2018**, *6*, 4586–4602. [CrossRef]
4. Wang, F.; Gu, W.; Chen, J.; Wu, Y.; Zhou, M.; Tang, S.; Cao, X.; Zhang, P.; Ji, G. The point defect and electronic structure of K doped LaCo$_{0.9}$Fe$_{0.1}$O$_3$ perovskite with enhanced microwave absorbing ability. *Nano Res.* **2021**, *15*, 3720–3728. [CrossRef]
5. Miao, P.; Yang, J.; Liu, Y.; Xie, H.; Chen, K.-J.; Kong, J. Emerging Perovskite Electromagnetic Wave Absorbers from Bi-Metal–Organic Frameworks. *Cryst. Growth Des.* **2020**, *20*, 4818–4826. [CrossRef]
6. Miao, P.; Qu, N.; Chen, W.; Wang, T.; Zhao, W.; Kong, J. A two-dimensional semiconductive Cu-S metal-organic framework for broadband microwave absorption. *Chem. Eng. J.* **2023**, *454*, 140445. [CrossRef]
7. Zhang, Z.; Yao, Y.; Zhang, J.; Ma, Y.; Xu, P.; Zhang, P.; Yang, Z.; Zhou, W. Two-dimensional (PEA)$_2$PbBr$_4$ perovskite modified with conductive network for high-performance electromagnetic wave absorber. *Mater. Lett.* **2022**, *326*, 132926. [CrossRef]
8. Zhang, Z.; Xiong, Z.; Yao, Y.; Wang, D.; Yang, Z.; Zhang, P.; Zhao, Q.; Zhou, W. Constructing Conductive Network in Hybrid Perovskite for a Highly Efficient Microwave Absorption System. *Adv. Funct. Mater.* **2022**, *32*, 2206053. [CrossRef]
9. Zhang, Z.; Xiong, Z.; Yao, Y.; Shi, X.; Zhang, P.; Yang, Z.; Zhao, Q.; Zhou, W. Inorganic Halide Perovskite Electromagnetic Wave Absorption System with Ultra-Wide Absorption Bandwidth and High Thermal-Stability. *Adv. Electron. Mater.* **2022**, *9*, 2201179. [CrossRef]
10. Qin, F.; Chen, J.; Liu, J.; Liu, L.; Tang, C.; Tang, B.; Li, G.; Zeng, L.; Li, H.; Yi, Z. Design of high efficiency perovskite solar cells based on inorganic and organic undoped double hole layer. *Sol. Energy* **2023**, *262*, 111796. [CrossRef]
11. Guo, H.; Yang, J.; Pu, B.; Chen, H.; Li, Y.; Wang, Z.; Niu, X. Excellent microwave absorption of lead halide perovskites with high stability. *J. Mater. Chem. C* **2018**, *6*, 4201–4207. [CrossRef]
12. Lai, R.; Shi, P.; Yi, Z.; Li, H.; Yi, Y. Triple-Band Surface Plasmon Resonance Metamaterial Absorber Based on Open-Ended Prohibited Sign Type Monolayer Graphene. *Micromachines* **2023**, *14*, 953. [CrossRef]
13. Chen, Z.; Cai, P.; Wen, Q.; Chen, H.; Tang, Y.; Yi, Z.; Wei, K.; Li, G.; Tang, B.; Yi, Y. Graphene Multi-Frequency Broadband and Ultra-Broadband Terahertz Absorber Based on Surface Plasmon Resonance. *Electronics* **2023**, *12*, 2655. [CrossRef]
14. Tong, Y.; He, M.; Zhou, Y.; Nie, S.; Zhong, X.; Fan, L.; Huang, T.; Liao, Q.; Wang, Y. Three-Dimensional Hierarchical Architecture of the TiO$_2$/Ti$_3$C$_2$T$_x$/RGO Ternary Composite Aerogel for Enhanced Electromagnetic Wave Absorption. *ACS Sustain. Chem. Eng.* **2018**, *6*, 8212–8222. [CrossRef]
15. Feng, J.; Hou, Y.; Wang, Y.; Li, L. Synthesis of Hierarchical ZnFe$_2$O$_4$@SiO$_2$@RGO Core-Shell Microspheres for Enhanced Electromagnetic Wave Absorption. *ACS Appl. Mater. Interfaces* **2017**, *9*, 14103–14111. [CrossRef] [PubMed]

16. Wang, M.; Lin, M.; Li, J.; Huang, L.; Zhuang, Z.; Lin, C.; Zhou, L.; Mai, L. Metal-organic framework derived carbon-confined Ni$_2$P nanocrystals supported on graphene for an efficient oxygen evolution reaction. *Chem. Commun.* **2017**, *53*, 8372–8375. [CrossRef] [PubMed]
17. Han, M.; Yin, X.; Kong, L.; Li, M.; Duan, W.; Zhang, L.; Cheng, L. Graphene-wrapped ZnO hollow spheres with enhanced electromagnetic wave absorption properties. *J. Mater. Chemisry A* **2014**, *2*, 16403–16409. [CrossRef]
18. Sun, X.; He, J.; Li, G.; Tang, J.; Wang, T.; Guo, Y.; Xue, H. Laminated magnetic graphene with enhanced electromagnetic wave absorption properties. *J. Mater. Chemisry C* **2013**, *1*, 765–777. [CrossRef]
19. Zhang, C.; Chen, G.; Zhang, R.; Wu, Z.; Xu, C.; Man, H.; Che, R. Charge modulation of CNTs-based conductive network for oxygen reduction reaction and microwave absorption. *Carbon* **2021**, *178*, 310–319. [CrossRef]
20. Hu, Q.; Yang, R.; Mo, Z.; Lu, D.; Yang, L.; He, Z.; Zhu, H.; Tang, Z.; Gui, X. Nitrogen-doped and Fe-filled CNTs/NiCo$_2$O$_4$ porous sponge with tunable microwave absorption performance. *Carbon* **2019**, *153*, 737–744. [CrossRef]
21. Gao, Z.; Zhao, Z.; Lan, D.; Kou, K.; Zhang, J.; Wu, H. Accessory ligand strategies for hexacyanometallate networks deriving perovskite polycrystalline electromagnetic absorbents. *J. Mater. Sci. Technol.* **2021**, *82*, 69–79. [CrossRef]
22. Wang, K.; Zhang, S.; Chu, W.; Li, H.; Chen, Y.; Chen, B.; Chen, B.; Liu, H. Tailoring conductive network nanostructures of ZIF-derived cobalt-decorated N-doped graphene/carbon nanotubes for microwave absorption applications. *J. Colloid Interface Sci.* **2021**, *591*, 463–473. [CrossRef] [PubMed]
23. Wang, R.; Yang, E.; Qi, X.; Xie, R.; Qin, S.; Deng, C.; Zhong, W. Constructing and optimizing core@shell structure CNTs@MoS$_2$ nanocomposites as outstanding microwave absorbers. *Appl. Surf. Sci.* **2020**, *516*, 146159. [CrossRef]
24. Wu, F.; Shi, P.; Yi, Z.; Li, H.; Yi, Y. Ultra-Broadband Solar Absorber and High-Efficiency Thermal Emitter from UV to Mid-Infrared Spectrum. *Micromachines* **2023**, *14*, 985. [CrossRef]
25. Zheng, Y.; Yi, Z.; Liu, L.; Wu, X.; Liu, H.; Li, G.; Zeng, L.; Li, H.; Wu, P. Numerical simulation of efficient solar absorbers and thermal emitters based on multilayer nanodisk arrays. *Appl. Therm. Eng.* **2023**, *230*, 120841. [CrossRef]
26. Liang, S.; Xu, F.; Li, W.; Yang, W.; Cheng, S.; Yang, H.; Chen, J.; Yi, Z.; Jiang, P. Tunable smart mid infrared thermal control emitter based on phase change material VO$_2$ thin film. *Appl. Therm. Eng.* **2023**, *232*, 121074. [CrossRef]
27. Weir, W.B. Automatic Measurement of Complex Dielectric Constant and Permeability at Microwave Frequencie. *Proc. IEEE* **1974**, *62*, 33–36. [CrossRef]
28. Nicolson, A.M.; Ross, G.F. Measurement of the Intrinsic Properties of materials by Time-Domain Techniques. *IEEE Transsactions Instrum. Meas.* **1970**, *19*, 377–382.
29. Liu, Y.; Yang, Z.; Cui, D.; Ren, X.; Sun, J.; Liu, X.; Zhang, J.; Wei, Q.; Fan, H.; Yu, F.; et al. Two-Inch-Sized Perovskite CH$_3$NH$_3$PbX$_3$ (X = Cl, Br, I) Crystals: Growth and Characterization. *Adv. Mater.* **2015**, *27*, 5176–5183. [CrossRef]
30. Zhang, C.; Mu, C.; Xiang, J.; Wang, B.; Wen, F.; Song, J.; Wang, C.; Liu, Z. Microwave absorption characteristics of CH$_3$NH$_3$PbI$_3$ perovskite/carbon nanotube composites. *J. Mater. Sci.* **2017**, *52*, 13023–13032. [CrossRef]
31. Chang, S.; Chen, C.; Chen, L.; Tien, C.; Cheng, H.; Huang, W.; Lin, H.; Chen, S.; Wu, G. Unraveling the multifunctional capabilities of PCBM thin films in inverted-type CH$_3$NH$_3$PbI$_3$ based photovoltaics. *Sol. Energy Mater. Sol. Cells* **2017**, *169*, 40–46. [CrossRef]
32. Chen, X.; Xu, Y.; Wang, Z.; Wu, R.J.; Cheng, H.L.; Chui, H.C. Characterization of a CH$_3$NH$_3$PbI$_3$ perovskite microwire by Raman spectroscopy. *J. Raman Spectrosc.* **2022**, *53*, 288–296. [CrossRef]
33. Pérez-Osorio, M.A.; Milot, R.L.; Filip, M.R.; Patel, J.B.; Herz, L.M.; Johnston, M.B.; Giustino, F. Vibrational Properties of the Organic–Inorganic Halide Perovskite CH$_3$NH$_3$PbI$_3$ from Theory and Experiment: Factor Group Analysis, First-Principles Calculations, and Low-Temperature Infrared Spectra. *J. Phys. Chem. C* **2015**, *119*, 25703–25718. [CrossRef]
34. Jia, X.; Hu, Z.; Zhu, Y.; Weng, T.; Wang, J.; Zhang, J.; Zhu, Y. Facile synthesis of organic–inorganic hybrid perovskite CH$_3$NH$_3$PbI$_3$ microcrystals. *J. Alloys Compd.* **2017**, *725*, 270–274. [CrossRef]
35. Zhang, Z.; Lv, X.; Chen, Y.; Zhang, P.; Sui, M.; Liu, H.; Sun, X. NiS$_2$@MoS$_2$ Nanospheres Anchored on Reduced Graphene Oxide: A Novel Ternary Heterostructure with Enhanced Electromagnetic Absorption Property. *Nanomaterials* **2019**, *9*, 292. [CrossRef]
36. Tseng, W.S.; Jao, M.H.; Hsu, C.C.; Huang, J.S.; Wu, C.I.; Yeh, N.C. Stabilization of hybrid perovskite CH$_3$NH$_3$PbI$_3$ thin films by graphene passivation. *Nanoscale* **2017**, *9*, 19227–19235. [CrossRef] [PubMed]
37. Li, Y.; Xu, X.; Wang, C.; Ecker, B.; Yang, J.; Huang, J.; Gao, Y. Light-Induced Degradation of CH$_3$NH$_3$PbI$_3$ Hybrid Perovskite Thin Film. *J. Phys. Chem. C* **2017**, *121*, 3904–3910. [CrossRef]
38. Yang, W.; Ni, M.; Ren, X.; Tian, Y.; Li, N.; Su, Y.; Zhang, X. Graphene in Supercapacitor Applications. *Curr. Opin. Colloid Interface Sci.* **2015**, *20*, 416–428. [CrossRef]
39. Quan, B.; Liang, X.; Xu, G.; Cheng, Y.; Zhang, Y.; Liu, W.; Ji, G.; Du, Y. A permittivity regulating strategy to achieve high-performance electromagnetic wave absorbers with compatibility of impedance matching and energy conservation. *New J. Chem.* **2017**, *41*, 1259–1266. [CrossRef]
40. Zhao, B.; Shao, G.; Fan, B.; Zhao, W.; Zhang, R. Investigation of the electromagnetic absorption properties of Ni@TiO$_2$ and Ni@SiO$_2$ composite microspheres with core-shell structure. *Phys. Chem. Chem. Phys.* **2015**, *17*, 2531–2539. [CrossRef]
41. Xie, A.; Sun, M.; Zhang, K.; Jiang, W.; Wu, F.; He, M. In situ growth of MoS$_2$ nanosheets on reduced graphene oxide (RGO) surfaces: Interfacial enhancement of absorbing performance against electromagnetic pollution. *Phys. Chem. Chem. Phys.* **2016**, *18*, 24931–24936. [CrossRef] [PubMed]
42. Sun, X.; Yuan, X.; Li, X.; Li, L.; Song, Q.; Lv, X.; Gu, G.; Sui, M. Hollow cube-like CuS derived from Cu$_2$O crystals for the highly efficient elimination of electromagnetic pollution. *New J. Chem.* **2018**, *42*, 6735–6741. [CrossRef]

43. Wang, Y.; Han, X.; Xu, P.; Liu, D.; Cui, L.; Zhao, H.; Du, Y. Synthesis of pomegranate-like $Mo_2C@C$ nanospheres for highly efficient microwave absorption. *Chem. Eng. J.* **2019**, *372*, 312–320. [CrossRef]

44. Cheng, Y.; Li, Z.; Li, Y.; Dai, S.; Ji, G.; Zhao, H.; Cao, J.; Du, Y. Rationally regulating complex dielectric parameters of mesoporous carbon hollow spheres to carry out efficient microwave absorption. *Carbon* **2018**, *127*, 643–652. [CrossRef]

45. Zhao, B.; Shao, G.; Fan, B.; Zhao, W.; Xie, Y.; Zhang, R. Facile preparation and enhanced microwave absorption properties of core-shell composite spheres composited of Ni cores and TiO_2 shells. *Phys. Chem. Chem. Phys. PCCP* **2015**, *17*, 8802–8810. [CrossRef]

micromachines

MDPI

Communication

A Flexible Supercapacitor Based on Niobium Carbide MXene and Sodium Anthraquinone-2-Sulfonate Composite Electrode

Guixia Wang [1,*], Zhuo Yang [1], Xinyue Nie [1], Min Wang [2,*] and Xianming Liu [1]

[1] Henan Key Laboratory of Function-Oriented Porous Materials, College of Chemistry and Chemical Engineering, Luoyang Normal University, Luoyang 471934, China; niexinyue0609@163.com (X.N.)
[2] School of Pharmaceutical Sciences, Chongqing University, Chongqing 401331, China
* Correspondence: author: guixia0116@163.com (G.W.); wang_min@cqu.edu.cn (M.W.)

Abstract: MXene-based composites have been widely used in electric energy storage device. As a member of MXene, niobium carbide (Nb_2C) is a good electrode candidate for energy storage because of its high specific surface area and electronic conductivity. However, a pure Nb_2C MXene electrode exhibits limited supercapacitive performance due to its easy stacking. Herein, sodium anthraquinone-2-sulfonate (AQS) with high redox reactivity was employed as a tailor to enhance the accessibility of ions and electrolyte and enhance the capacitance performance of Nb_2C MXene. The resulting Nb_2C–AQS composite had three-dimensional porous layered structures. The supercapacitors (SCs) based on the Nb_2C–AQS composite exhibited a considerably higher electrochemical capacitance (36.3 mF cm^{-2}) than the pure Nb_2C electrode (16.8 mF cm^{-2}) at a scan rate of 20 mV s^{-1}. The SCs also exhibited excellent flexibility as deduced from the almost unchanged capacitance values after being subjected to bending. A capacitance retention of 99.5% after 600 cycles was observed for the resulting SCs, indicating their good cycling stability. This work proposes a surface modification method for Nb_2C MXene and facilitates the development of high-performance SCs.

Keywords: niobium carbide; MXene; sodium anthraquinone-2-sulfonate; supercapacitors; energy storage; composite electrode

Citation: Wang, G.; Yang, Z.; Nie, X.; Wang, M.; Liu, X. A Flexible Supercapacitor Based on Niobium Carbide MXene and Sodium Anthraquinone-2-Sulfonate Composite Electrode. *Micromachines* **2023**, *14*, 1515. https://doi.org/10.3390/mi14081515

Academic Editors: Hong Li, Azam Iraji zad and Amir Hussain Idrisi

Received: 23 June 2023
Revised: 25 July 2023
Accepted: 26 July 2023
Published: 28 July 2023

1. Introduction

With the increasing demand for electric energy, electric energy storage has attracted increasing attention from researchers [1–3]. There are several electric energy storage sources, including aqueous Zn–ion batteries [4,5], lithium–selenium batteries [6], Li–ion batteries [7,8], Zn–air batteries [9], ammonium-ion batteries [10], and supercapacitors (SCs) [11–13]. Among them, SCs possess useful characteristics such as ultrahigh power density, long lifetime, and environmental sustainability, and thus, they have become promising electrical energy storage materials for portable electronics and other electric devices [14]. SCs can be divided into electrical double-layer capacitors (EDLCs) and pseudocapacitors in accordance with the energy storage mechanism. Compared with EDLCs, pseudocapacitors usually exhibit much higher capacitances and energy densities through Faradaic redox reactions [1,15]. Pseudocapacitive materials include conducting polymers, electrochemically active organic molecules, and transition metal compounds. As the representative electrochemically active organic molecules, quinones and their derivatives exhibit high redox reactivity, high capacitance, and good adjustable electrochemical reversibility, and they have been used as electrode materials of SCs [16–21]. However, the capacitive performance and long-term cycle stability of SCs based on organic quinones or their derivatives are relatively low due to their low electronic conductivity. To enhance the electronic conductivity of quinones and their derivatives, the Shi group [22] and Zhu group [23] proposed a strategy of using reduced graphene oxide (rGO) as a conductive support of sodium anthraquinone-2-sulfonate (AQS). The resulting SCs based on AQS–rGO composites had enhanced capacitive performance and high capacity retention.

MXenes are a new family of multifunctional two-dimensional (2D) materials comprising transition metal carbides, nitrides, and carbonitrides [24,25]. Titanium carbide MXene-based nanocomposites exhibit good structural stability, electrical properties, and electrocatalytic activity, which makes them very suitable for good matrices of SCs. Niobium carbide (Nb_2C) MXene has a similarly high electronic conductivity to that of titanium carbide, excellent dispersibility in various solvents, and good compatibility with various components. Recently, Nb_2C MXene has been revealed to be a good electrode candidate for energy storage due to its high specific surface area and high conductivity [26–28]. Both single Nb_2C MXene (denoted as Nb_2CT_x) [29] as well as composites consisting of Nb_2C MXene delaminated with isopropylamine with carbon nanotubes [30] can be used as electrode materials with excellent cycle stability and Li capacities.

Herein, the organic active molecule AQS was used as a tailor to modify 2D Nb_2C MXene and the resulting Nb_2C–AQS composite was employed to construct a novel SC with high capacitance. The SO_3^- functional group in AQS can render AQS soluble in aqueous solutions as well as facilitate the combination of AQS and Nb_2C MXene on the molecular level. Then, the electrochemical capacitance and cycle stability of the SC based on the as-prepared Nb_2C–AQS composite were evaluated and also compared with bare Nb_2C MXene.

2. Materials and Methods

2.1. Materials and Reagents

Nb_2AlC powder was purchased from Forsman Scientific (Beijing, China) Co., Ltd. Hydrofluoric acid (HF, 40%) was from Tianjin Deen Chemical Reagents Co., Ltd., Tianjin, China. Tetrapropylammonium hydroxide (TPAOH, 25 wt%) was from J&K Scientific Co., Ltd., Beijing, China. Sodium anthraquinone-2-sulfonate (AQS, 99%) was purchased from Sigma–Aldrich (St. Louis, MO, USA). Sodium sulfate and other chemicals were all reagent grade quality or better and used without further purification. Aqueous solutions were prepared from deionized water (>18 M·cm, Milli-Q purification system).

2.2. Methods

2.2.1. Synthesis of Nb_2C MXene

Two-dimensional Nb_2C MXene was prepared by a modified chemical exfoliation method [28]. Nb_2AlC powder (ca. 5 g) was slowly added to 30 mL of 40% aqueous HF (Note: HF is highly corrosive) and stirred for 20 h at room temperature. After collection by centrifugation (3500 rpm) and washing with water and anhydrous ethanol, the precipitate was dispersed in 30 mL of aqueous tetrapropylammonium hydroxide under stirring for 10 h at room temperature. The raw Nb_2C materials were collected by centrifugation and washed $3\times$ with ethanol and water to remove the residual TPAOH. Then, the precipitate was dispersed in 100 mL of deionized water and sonicated at 200 W for 30 min. The supernatant was collected and dried in a vacuum oven at 80 °C for 6 h. Thus, the resulting Nb_2C MXene was obtained.

2.2.2. Preparation of Nb_2C–AQS Composite

The Nb_2C–AQS composite was prepared by a simple one-step hydrothermal method. Aqueous Nb_2C and AQS (30 mL) with a mass ratio of 1:3 was stirred for 6 h at room temperature. Then, the mixture was transferred into a 50-mL Teflon-lined stainless-steel autoclave and heated at 180 °C for 12 h. Nb_2C–AQS composite was obtained after washing the aforementioned Nb_2C–AQS solution with deionized water $2\times$ and freeze-drying under vacuum.

2.2.3. Characterization

Scanning electron microscopy (SEM) images were acquired on a field-emission Sigma 500 microscope (Carl Zeiss, Jena, Germany). X-ray diffraction (XRD) was measured with a D8 Advance (Bruker, Mannheim, Germany) system. Fourier-transform infrared spectra (FTIR) were obtained with a Nicolet 6700 spectrometer (Thermo Nicolet, Madison, WI, USA).

The electrochemical performance of as-prepared Nb$_2$C–AQS was carried out with a three-electrode system in 0.1 mol L^{-1} Na$_2$SO$_4$, with platinum foil as the counter electrode and Ag/AgCl as the reference electrode. The working electrode was fabricated as follows: the as-prepared active material (40 mg), acetylene black, and polytetrafluoroethylene were first mixed in a mass ratio of 75:15:10 in a small quantity of absolute ethanol in a manner that formed a homogeneous slurry. Then, the slurry was coated on a piece of nickel foam (1.0 cm × 1.0 cm), which was dried in a vacuum oven at 80 °C for 12 h and pressed before measurement. Cyclic voltammetry (CV) curves and galvanostatic charge/discharge (GCD) were all performed with a CHI 660E electrochemical workstation (CH Instruments).

3. Results and Discussion

3.1. Synthesis and Characterization of Nb$_2$C–AQS Composite

Figure 1 illustrate the synthesis process of the Nb$_2$C–AQS composite. The morphology of Nb$_2$C and the corresponding Nb$_2$C–AQS composite was characterized by SEM. Nb$_2$C has an accordion-like lamellar structure similar to that of Ti$_3$C$_2$ MXene [24,31], and its layered structure has a large lateral size with almost no defects (Figure 2a). Compared with Nb$_2$C, the Nb$_2$C–AQS composite maintained the layered structure of Nb$_2$C, but it has 3D porous structures (Figure 2b), attributable to the decorated AQS molecules as spacers that suppress aggregation of Nb$_2$C MXene. The porous structures of the Nb$_2$C–AQS composite can act as ion buffers that facilitate ion transport between the Nb$_2$C layers. The exiting of the S element also indicates the successfully assembly of AQS molecules on Nb$_2$C MXene surface (Figure 2d–h).

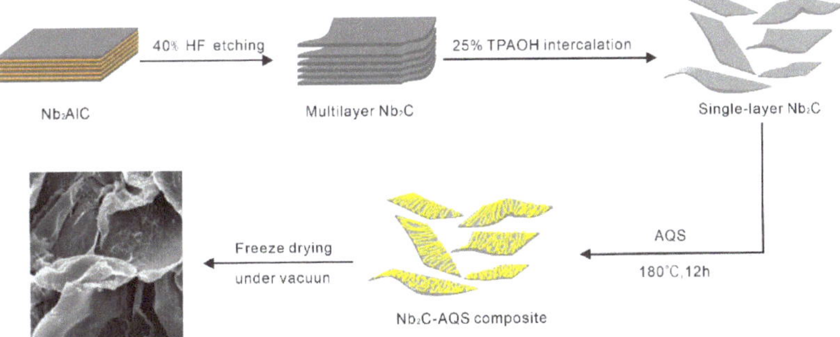

Figure 1. Schematic diagram for the fabrication of Nb$_2$C–AQS composite.

3.2. Chemical Structure

The XRD patterns indicate formation of Nb$_2$C materials (Figure S1, Supporting Information). Only one intense peak of Nb$_2$C at ca. 25° was observed, and most of the non-basal plane peaks of Nb$_2$AlC bulk (red line in Figure S1, Supporting Information), including the most intense diffraction peak at ca. 39°, were essentially no longer evident after HF treatment and TPAOH intercalation (black line in Figure S1, Supporting Information). Compared with the diffraction patterns of AQS (blue line in Figure 3a), the Nb$_2$C–AQS composite had similar XRD patterns, but the peak intensities all decreased, due to interaction of Nb$_2$C MXene and AQS molecules (black line in Figure 3a). The FTIR spectra of the Nb$_2$C–AQS composite exhibited characteristic peaks of AQS at 1215 and 1660 cm^{-1} (Figure 3b), attributable to the asymmetric stretching vibration of –SO$_3^-$ and the stretching vibration of –C=O from AQS, respectively [32,33]. The zeta potential of the Nb$_2$C–AQS composite was −31.8 mV, a much more negative value than that of the Nb$_2$C films (−0.19 mV). The results strongly demonstrate the anchoring of AQS onto Nb$_2$C MXene.

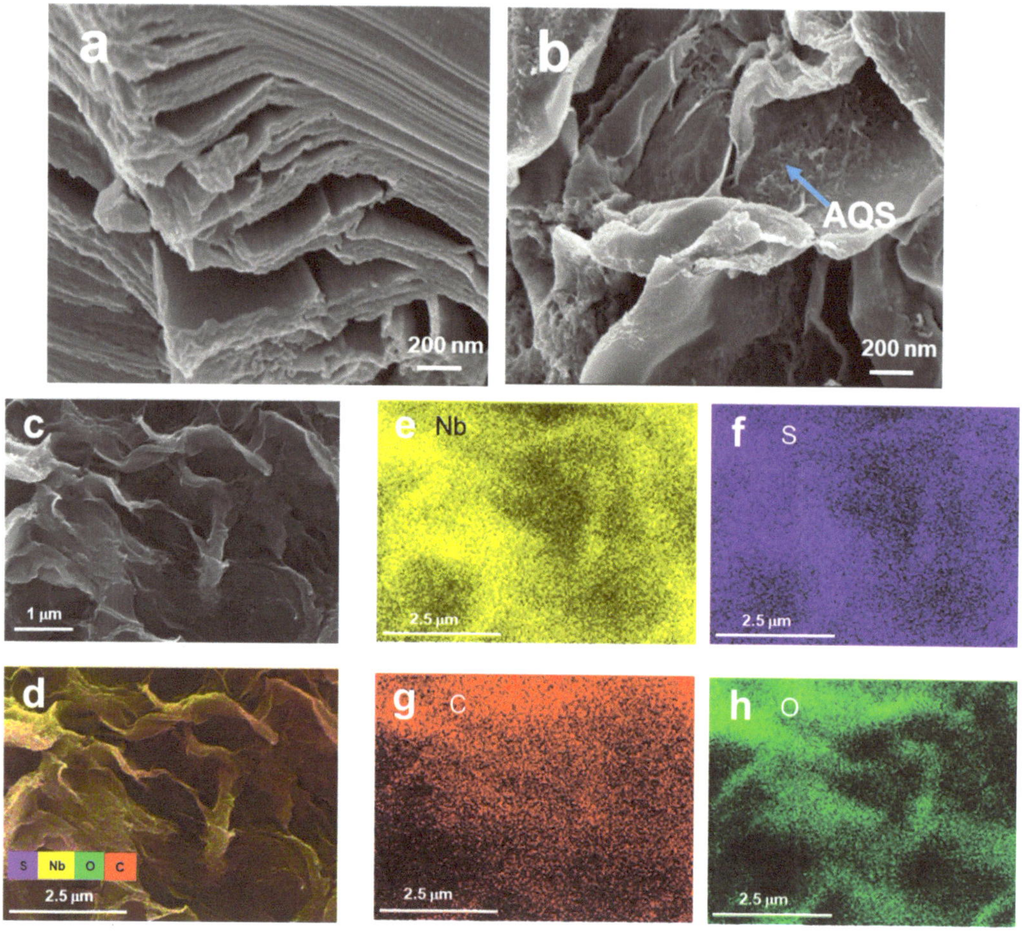

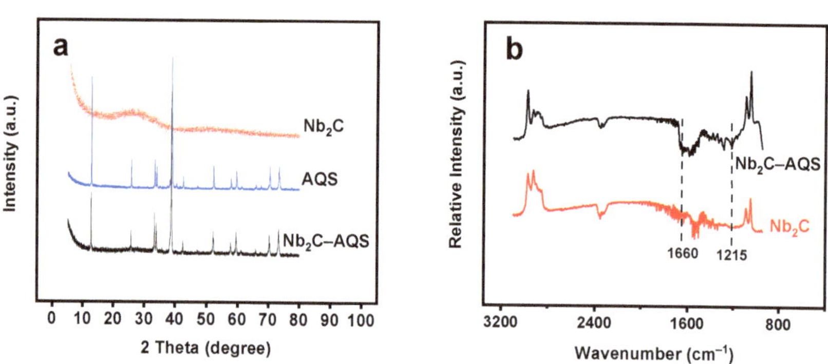

Figure 2. SEM images: (**a**) Nb$_2$C layers; (**b**) and (**c**): Nb$_2$C–AQS composite; (**d–h**) element mapping of (**c**).

Figure 3. (**a**) XRD patterns of Nb$_2$C–AQS, AQS, and Nb$_2$C MXene. (**b**) FTIR spectra of Nb$_2$C–AQS and Nb$_2$C MXene.

3.3. Electrochemical Performance

The obtained Nb_2C–AQS composite had good conductivity with an equivalent-series resistance (ESR) value of ca. 0.5 Ω derived from its Nyquist plots (Figure S2, Supporting Information). The CV profiles of the Nb_2C electrode exhibited a nearly rectangular shape at low scan rates in the range of 10–100 mV s^{-1} (Figure 4a), which is characteristic of EDLC behavior. Compared with bare Nb_2C, the resultant device based on the Nb_2C–AQS composite (denoted as-prepared) exhibited both Faradaic and EDLC capacitance with a pronounced single pair of peaks at ca. 0.22/0.45 V versus Ag/AgCl (at 20 mV s^{-1}), originating from the redox reactions of the AQS molecules [22] (Figure 4b). Moreover, the CV curves of Nb_2C–AQS remained well-defined and symmetric even at a high scan rate of 100 mV s^{-1}, suggesting excellent reversibility and stability of the Nb_2C–AQS electrode. Notably, the peak potentials of the redox peaks slightly shifted with increasing scan rate, which might be due to partial proton-coupled charge transfer between AQS and Nb_2C. Circa 70% of the capacitance was retained when the scan rate was increased to 100 mV s^{-1} (25.4 mF cm^{-2}; refer to the Supporting Information for the calculation methods of the specific capacitance), indicating its good rate capability. The Nb_2C–AQS composite electrode exhibited a considerably higher electrochemical capacitance (36.3 mF cm^{-2} at 20 mV s^{-1}) than the bare Nb_2C electrode (16.8 mF cm^{-2} at 20 mV s^{-1}) (Figure 4c, Table 1); this is mainly due to the high pseudocapacitance of AQS as well as the enhanced accessibility of ions and electrolyte, facilitated by the unique porous structure of the Nb_2C–AQS composite. The 42.7 mF cm^{-2} of specific capacitance for the resulting Nb_2C–AQS capacitor is higher than that of graphene-based micro-supercapacitors (micro-SCs) (Table 2). The as-prepared SCs display similar capacitance performance as those in some other published works [34–41].

Table 1. Electrochemical parameters for as-prepared MSCs in different conditions.

Electrode Materials	Areal Capacitance (mF cm^{-2})	Scan Rate
Nb_2C–AQS	36.3	20 mV s^{-1}
Nb_2C	16.8	20 mV s^{-1}
Nb_2C–AQS	42.7	10 mV s^{-1}
Concavely bent	48.1	10 mV s^{-1}
Convexly bent	40.0	10 mV s^{-1}

Table 2. Supercapacitance comparison of several composite electrode materials.

Electrode Materials	Specific Capacitance	Test Condition	Ref.
Nb_2C–AQS	36.3 mF cm^{-2}	20 mV s^{-1}	This work
PET-CGO-LGO	0.756 mF cm^{-2}	20 mV s^{-1}	[42]
MPG	0.0807 mF cm^{-2}	-	[43]
Ultrathin rGO	0.462 mF cm^{-2}	0.1 µA g^{-1}	[44]

The CVs of the SCs exhibited similar shapes to that of as-prepared SCs when subjected to bending at a scan rate of 10 mV s^{-1} (Figure 4d). The specific capacitances were 48.1 and 40.0 mF cm^{-2} for concavely bent and convexly bent SCs (Table 1, bending modes in Figure S3, Supporting Information), respectively, which are almost the same as that of as-prepared SCs (42.7 mF cm^{-2}). Thus, the as-prepared SCs have excellent flexibility.

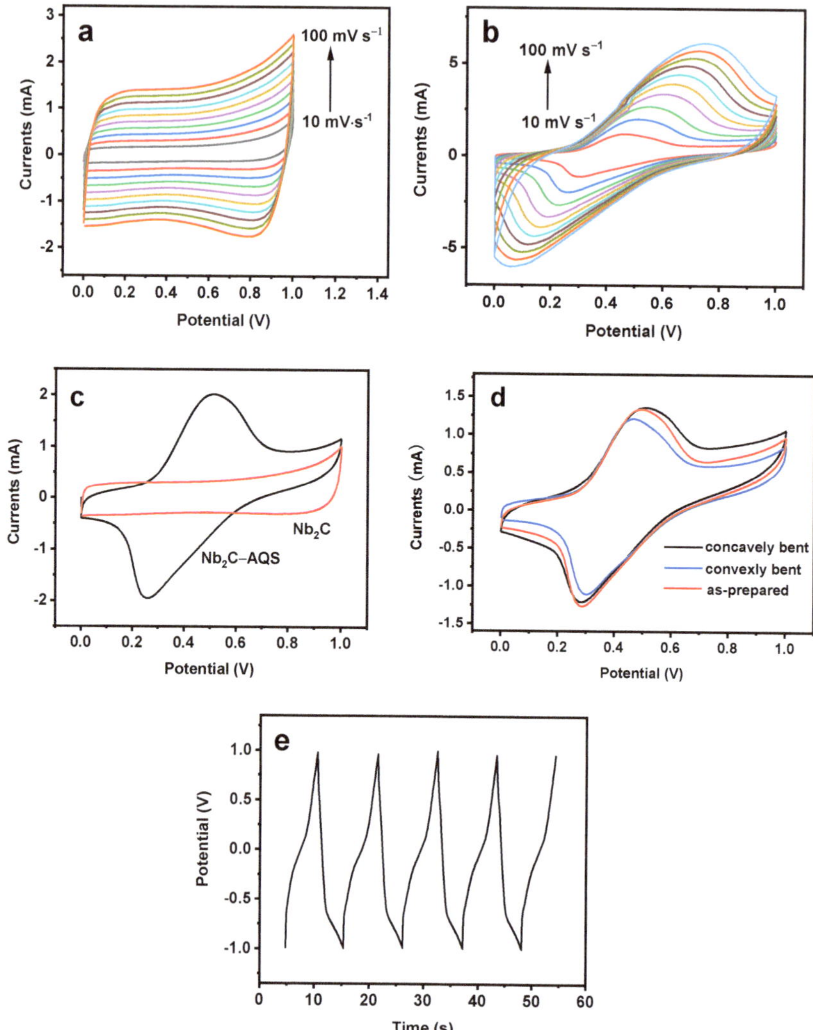

Figure 4. CV and GCD curves: (**a**) CV curves of Nb$_2$C MXene and (**b**) Nb$_2$C–AQS composite-based SCs at various scan rates. (**c**) CV curves of Nb$_2$C MXene (red line) and Nb$_2$C–AQS composite (black line) at a scan rate of 20 mV s^{-1}. (**d**) CV curves of Nb$_2$C–AQS-based SCs (as-prepared, red line), concavely bent (black line), and convexly bent (blue line) at a scan rate of 10 mV s^{-1}. (**e**) GCD curves of the as-prepared SCs at a current density of 15 mA cm^{-2}. All the electrochemical measurements were performed in 0.1 mol L^{-1} Na$_2$SO$_4$.

The GCD curves of the Nb$_2$C–AQS composite included two parts (Figure 4e): a pair of charge–discharge plateaus attributable to the anchored AQS, and an oblique line stage due to the EDLC behavior of Nb$_2$C, which coincides well with the CV curves in Figure 3b. The specific capacitance of the resulting SCs was 36.0 mF cm^{-2} at a current density of 15 mA cm^{-2} (for the calculation of the specific capacitance, refer to the Supporting Information). In addition, the device exhibited a capacitance retention of 99.5% of its initial capacitance after 600 cycles (Figure S4, Supporting Information), indicating good cycling stability.

4. Conclusions

In summary, AQS was anchored successfully onto the surface of Nb_2C MXene, and the obtained composite had good conductivity as well as a porous structure. SCs based on the Nb_2C–AQS composite exhibited a high specific capacitance and good cycling stability. The SCs maintained their initial capacitance when subjected to bending, indicating good flexibility. Compared with bare Nb_2C MXene, the Nb_2C–AQS composite electrode provides higher capacitance, which was mainly ascribed to Faradaic redox reactions of the anchored AQS molecules as well as the enhanced accessibility of ions and electrolyte. This work proposes a surface functionalization method for Nb_2C MXene and also provides a strategy for enhancing the capacitance performance of pure MXene, which will facilitate the development of high-performance SCs.

5. Patents

The results of the patent (a preparation method of a supercapacitor based on 2D niobium carbide nanocomposites, No. ZL2019109099814, China) are from the work reported in this manuscript.

Supplementary Materials: The following supporting information can be downloaded at: https://www.mdpi.com/article/10.3390/mi14081515/s1, Figure S1: XRD patterns of Nb_2C nanosheets and Nb_2AlC powder; Figure S2: Nyquist plots of the Nb_2C–AQS nanocomposite-modified electrode in 10 mmol L^{-1} $K_4Fe(CN)_6$ and $K_4Fe(CN)_6$ (molar ratio 1:1) from 1–10 mHz at 10 mV sinusoidal signal; Figure S3: Two bending modes of the Nb_2C–AQS based SC: (a) concavely bent, and (b) convexly bent; Figure S4: Capacitance retention at a current density of 15 mA cm^{-2} for the Nb_2C–AQS-based micro-SC in 0.1 mol L^{-1} Na_2SO_4.

Author Contributions: Conceptualization: G.W. and X.L.; methodology, G.W. and M.W.; investigation, Z.Y.; data curation, G.W. and X.N.; writing—original draft preparation, G.W.; writing—review and editing, X.L. and M.W.; supervision and funding acquisition, G.W. All authors have read and agreed to the published version of the manuscript.

Funding: This work was financially supported by the National Natural Science Foundation of China (21305060) and the Training Project for Youth Backbone Teachers in Colleges and Universities of Henan Province (2018GGJS127).

Data Availability Statement: All data that support the findings of this study are available within the article and its Supplementary Materials.

Conflicts of Interest: The authors declare no conflict of interest.

References

1. Wu, Z.; Li, L.; Yan, J.M.; Zhang, X.B. Materials Design and System Construction for Conventional and New-Concept Supercapacitors. *Adv. Sci.* **2017**, *4*, 1600382. [CrossRef] [PubMed]

2. Li, L.; Hu, C.; Liu, W.; Shen, G. Progress and Perspectives in Designing Flexible Microsupercapacitors. *Micromachines* **2021**, *12*, 1305. [CrossRef]

3. Sung, J.; Shin, C. Recent Studies on Supercapacitors with Next-Generation Structures. *Micromachines* **2020**, *11*, 1125. [CrossRef]

4. Lobinsky, A.A.; Kaneva, M.V.; Tenevich, M.I.; Popkov, V.I. Direct Synthesis of $Mn_3[Fe(CN)_6]_2 \cdot nH_2O$ Nanosheets as Novel 2D Analog of Prussian Blue and Material for High-Performance Metal-Ion Batteries. *Micromachines* **2023**, *14*, 1083. [CrossRef] [PubMed]

5. Deng, W.N.; Xu, Y.X.; Zhang, X.C.; Li, C.Y.; Liu, Y.X.; Xiang, K.X.; Chen, H. $(NH_4)_2Co_2V_{10}O_{28} \cdot 16H_2O/(NH_4)_2V_{10}O_{25} \cdot 8H_2O$ heterostructure as cathode for high-performance aqueous Zn-ion batteries. *J. Alloys Compd.* **2022**, *903*, 163824. [CrossRef]

6. Deng, W.N.; Li, Y.H.; Xu, D.F.; Zhou, W.; Xiang, K.X.; Chen, H. Three-dimensional hierarchically porous nitrogen-doped carbon from water hyacinth as selenium host for highperformance lithium–selenium batteries. *Rare Met.* **2022**, *41*, 3432–3445. [CrossRef]

7. Li, D.J.; Guo, H.T.; Jiang, S.H.; Zeng, G.L.; Zhou, W.; Li, Z. Microstructures and electrochemical performances of TiO_2-coated Mg-Zr co-doped NCM as a cathode material for lithium-ion batteries with high power and long circular life. *New J. Chem.* **2021**, *45*, 19446–19455. [CrossRef]

8. Ren, L.; Wang, L.; Qin, Y.; Li, Q. High Cycle Stability of Hybridized $Co(OH)_2$ Nanomaterial Structures Synthesized by the Water Bath Method as Anodes for Lithium-Ion Batteries. *Micromachines* **2022**, *13*, 149. [CrossRef] [PubMed]

9. Zhou, W.; Zeng, G.L.; Jin, H.T.; Jiang, S.H.; Huang, M.J.; Zhang, C.M.; Chen, H. Bio-Template Synthesis of V_2O_3@Carbonized Dictyophora Composites for Advanced Aqueous Zinc-Ion Batteries. *Molecules* **2023**, *28*, 2147. [CrossRef] [PubMed]

10. Wen, X.Y.; Luo, J.H.; Xiang, K.X.; Zhou, W.; Zhang, C.F. Han Chen High-performance monoclinic WO_3 nanospheres with the novel NH_4^+ diffusion behaviors for aqueous ammonium-ion batteries. *Chem. Eng. J.* **2023**, *458*, 141381. [CrossRef]
11. Xiao, J.L.; Li, H.L.; Zhang, H.; He, S.J.; Zhang, Q.; Liu, K.M.; Jiang, S.H.; Duan, G.G.; Zhang, K. Nanocellulose and its derived composite electrodes toward supercapacitors: Fabrication, properties, and challenges. *J. Bioresour. Bioprod.* **2022**, *7*, 245–269. [CrossRef]
12. Yang, C.; Li, P.; Wei, Y.; Wang, Y.; Jiang, B.; Wu, W. Preparation of nitrogen and phosphorus doped porous carbon from watermelon peel as supercapacitor electrode material. *Micromachines* **2023**, *14*, 1003. [CrossRef] [PubMed]
13. Qian, Y.; Lyu, Z.; Zhang, Q.; Lee, T.H.; Kang, T.K.; Sohn, M.; Shen, L.; Kim, D.H.; Kang, D.J. High-Performance Flexible Energy Storage Devices Based on Graphene Decorated with Flower-Shaped MoS_2 Heterostructures. *Micromachines* **2023**, *14*, 297. [CrossRef] [PubMed]
14. Xue, Q.; Gan, H.; Huang, Y.; Zhu, M.; Pei, Z.; Li, H.; Deng, S.; Liu, F.; Zhi, C. Boron element nanowires electrode for supercapacitors. *Adv. Energy Mater.* **2018**, *8*, 1703117. [CrossRef]
15. Han, Y.; Ge, Y.; Chao, Y.; Wang, C.; Wallace, G.G. Recent progress in 2D materials for flexible supercapacitors. *J. Energy Chem.* **2018**, *27*, 57–72. [CrossRef]
16. Xie, J.; Shi, H.Z.; Shen, C.; Huan, L.; He, M.X.; Chen, M. Heteroatom-bridged pillar[4]quinone: Evolutionary active cathode material for lithium-ion battery using density functional theory. *J. Chem. Sci.* **2021**, *133*, 1–11. [CrossRef]
17. Wu, Q.; Sun, Y.; Bai, H.; Shi, G. High-performance supercapacitor electrodes based on graphene hydrogels modified with 2-aminoanthraquinone moieties. *Phys. Chem. Chem. Phys.* **2011**, *13*, 11193–11198. [CrossRef] [PubMed]
18. Chen, X.; Wang, H.W.; Yi, H.; Wang, X.F.; Yan, X.R.; Guo, Z.H. Anthraquinone on porous carbon nanotubes with improved supercapacitor performance. *J. Phys. Chem. C* **2014**, *118*, 8262–8270. [CrossRef]
19. Brousse, K.; Martin, C.; Brisse, A.L.; Lethien, C.; Simon, P.; Taberna, P.L.; Brousse, T. Anthraquinone modification of microporous carbide derived carbon films for on-chip micro-supercapacitors applications. *Electrochim. Acta* **2017**, *246*, 391–398. [CrossRef]
20. Xu, L.; Shi, R.; Li, H.; Han, C.; Wu, M.; Wong, C.P.; Kang, F.; Li, B. Pseudocapacitive anthraquinone modified with reduced graphene oxide for flexible symmetric all-solid-state supercapacitors. *Carbon* **2017**, *127*, 459–468. [CrossRef]
21. Boota, M.; Chen, C.; Bécuwe, M.; Miao, L.; Gogotsi, Y. Pseudocapacitance and excellent cyclability of 2,5-dimethoxy-1,4-benzoquinone on graphene. *Energy Environ. Sci.* **2016**, *9*, 2586–2594. [CrossRef]
22. Shi, R.Y.; Han, C.P.; Duan, H.; Xu, L.; Zhou, D.; Li, H.F.; Li, J.Q.; Kang, F.Y.; Li, B.H.; Wang, G.X. Redox-active organic sodium anthraquinone-2-sulfonate (AQS) anchored on reduced graphene oxide for high-performance supercapacitors. *Adv. Energy Mater.* **2018**, *8*, 1802088. [CrossRef]
23. Zhu, C.Y.; Zhang, W.J.; Li, G.; Li, C.L.; Qin, X.H. Ultra-simple and green two-step synthesis of sodium anthraquinone-2-sulfonate composite graphene (AQS/rGO) hydrogels for supercapacitor electrode materials. *J. Alloys Compd.* **2021**, *862*, 158472. [CrossRef]
24. Naguib, M.; Kurtoglu, M.; Presser, V.; Lu, J.; Niu, J.; Heon, M.; Hultman, L.; Gogotsi, Y.; Barsoum, M.W. Two-dimensional nanocrystals produced by exfoliation of Ti_3AlC_2. *Adv. Mater.* **2011**, *23*, 4248–4253. [CrossRef]
25. Kosnan, M.A.; Azam, M.A.; Safie, N.E.; Munawar, R.F.; Takasaki, A. Recent Progress of Electrode Architecture for MXene/MoS2 Supercapacitor: Preparation Methods and Characterizations. *Micromachines* **2022**, *13*, 1837. [CrossRef] [PubMed]
26. Hu, J.P.; Xu, B.; Ouyang, C.Y.; Zhang, Y.; Yang, S.Y. Investigations on Nb_2C monolayer as promising anode material for Li or non-Li ion batteries from first-principles calculations. *RSC Adv.* **2016**, *6*, 27467–27474. [CrossRef]
27. Xin, Y.; Yu, Y.X. Possibility of bare and functionalized niobium carbide MXenes for electrode materials of supercapacitors and field emitters. *Mater. Des.* **2017**, *130*, 512–520. [CrossRef]
28. Lin, H.; Gao, S.; Dai, C.; Chen, Y.; Shi, J.L. A two-dimensional biodegradable niobium carbide (MXene) for photothermal tumor eradication in NIR-I and NIR-II biowindows. *J. Am. Chem. Soc.* **2017**, *139*, 16235–16247. [CrossRef]
29. Naguib, M.; Halim, J.; Lu, J.; Cook, K.M.; Hultman, L.; Gogotsi, Y.; Barsoum, M.W. New two-dimensional niobium and vanadium carbides as promising materials for Li-ion batteries. *J. Am. Chem. Soc.* **2013**, *135*, 15966–15969. [CrossRef] [PubMed]
30. Mashtalir, O.; Lukatskaya, M.R.; Zhao, M.Q.; Barsoum, M.W.; Gogotsi, Y. Amine-assisted delamination of Nb_2C MXene for Li-ion energy storage devices. *Adv. Mater.* **2015**, *27*, 3501–3506. [CrossRef] [PubMed]
31. Wang, G.X.; Yang, Z.; Wu, L.N.; Wang, J.M.; Liu, X.M. Studies on improved stability and electrochemical activity of titanium carbide MXene-polymer nanocomposites. *J. Electroanal. Chem.* **2021**, *900*, 115708. [CrossRef]
32. Han, Y.Q.; Wang, T.Q.; Li, T.X.; Gao, X.X.; Li, W.; Zhang, Z.; Wang, Y.; Zhang, X.Q. Preparation and electrochemical performances of graphene/polypyrrole nanocomposite with anthraquinone-graphene oxide as active oxidant. *Carbon* **2017**, *119*, 111–118. [CrossRef]
33. Xu, C.; Cao, Y.; Kumar, R.; Wu, X.; Wang, X.; Scott, K. A polybenzimidazole/sulfonated graphite oxide composite membrane for high temperature polymer electrolyte membrane fuel cells. *J. Mater. Chem.* **2011**, *21*, 11359–11364. [CrossRef]
34. Duan, G.G.; Xiao, J.L.; Chen, L.; Zhang, C.M.; Jian, S.J.; He, S.J.; Wang, F. Zinc gluconate derived porous carbon electrode assisted high rate and long cycle performance supercapacitor. *J. Energy Storage* **2023**, *67*, 107559. [CrossRef]
35. Cao, L.H.; Li, H.L.; Liu, X.L.; Liu, S.W.; Zhang, L.; Xu, W.H.; Yang, H.Q.; Hou, H.Q.; He, S.J.; Zhao, Y. Nitrogen, sulfur co-doped hierarchical carbon encapsulated in graphene with "sphere-in-layer" interconnection for high-performance supercapacitor. *J. Colloid Interface Sci.* **2021**, *599*, 443–452. [CrossRef] [PubMed]

36. Li, H.L.; Cao, L.H.; Zhang, H.J.; Tian, Z.W.; Zhang, Q.; Yang, F.; Yang, H.Q.; He, S.J.; Jiang, S.H. Intertwined carbon networks derived from Polyimide/Cellulose composite as porous electrode for symmetrical supercapacitor. *J. Colloid Interface. Sci.* **2022**, *609*, 179–187. [CrossRef] [PubMed]

37. Guo, W.C.; Guo, X.T.; Yang, L.; Wang, T.Y.; Zhang, M.H.; Duan, G.G.; Liu, X.H.; Li, Y.W. Synthetic melanin facilitates MnO supercapacitors with high specific capacitance and wide operation potential window. *Polymer* **2021**, *235*, 124276. [CrossRef]

38. Wang, F.; Chen, L.; Li, H.L.; Duan, G.G.; He, S.J.; Zhang, L.; Zhang, G.Y.; Zhou, Z.P.; Jiang, S.H. N-doped honeycomb-like porous carbon towards high-performance supercapacitor. *Chin. Chem. Lett.* **2020**, *31*, 1986–1990. [CrossRef]

39. Duan, G.G.; Zhao, L.Y.; Chen, L.; Wang, F.; He, S.J.; Jiang, S.H.; Zhang, Q. $ZnCl_2$ regulated flax-based porous carbon fibers for supercapacitors with good cycling stability. *New J. Chem.* **2021**, *45*, 22602–22609. [CrossRef]

40. Yang, J.J.; Li, H.L.; He, S.J.; Du, H.J.; Liu, K.M.; Zhang, C.M.; Jiang, S.H. Facile Electrodeposition of $NiCo_2O_4$ Nanosheets on Porous Carbonized Wood for Wood-Derived Asymmetric Supercapacitors. *Polymers* **2022**, *14*, 2521. [CrossRef]

41. Duan, G.G.; Zhang, H.; Zhang, C.M.; Jiang, S.H.; Hou, H.Q. High mass-loading α-Fe_2O_3 nanoparticles anchored on nitrogen-doped wood carbon for high-energy-density supercapacitor. *Chin. Chem. Lett.* **2023**. [CrossRef]

42. Wang, G.X.; Babaahmadi, V.; He, N.; Liu, Y.; Pan, Q.; Montazer, M.; Gao, W. Wearable supercapacitors on polyethylene terephthalate fabrics with good wash fastness and high flexibility. *J. Power Sources* **2017**, *367*, 34–41. [CrossRef]

43. Wu, Z.; Parvez, K.; Feng, X.; Müllen, K. Graphene-based in-plane micro-supercapacitors with high power and energy densities. *Nat. Commun.* **2013**, *4*, 2487. [CrossRef] [PubMed]

44. Niu, Z.; Zhang, L.; Liu, L.; Zhu, B.; Dong, H.; Chen, X. All-solid-state flexible ultrathin micro-supercapacitors based on graphene. *Adv. Mater.* **2013**, *25*, 4035–4042. [CrossRef]

MDPI AG
Grosspeteranlage 5
4052 Basel
Switzerland
Tel.: +41 61 683 77 34

Micromachines Editorial Office
E-mail: micromachines@mdpi.com
www.mdpi.com/journal/micromachines